Pyrrhic Progress

Critical Issues in Health and Medicine

Edited by Janet Golden, Rutgers University–Camden, and Rima D. Apple, University of Wisconsin–Madison

Growing criticism of the US healthcare system is coming from consumers, politicians, the media, activists, and healthcare professionals. Critical Issues in Health and Medicine is a collection of books that explores these contemporary dilemmas from a variety of perspectives, among them political, legal, historical, sociological, and comparative, and with attention to crucial dimensions such as race, gender, ethnicity, sexuality, and culture.

For a list of titles in the series, see the last page of the book.

Pyrrhic Progress

The History of Antibiotics in Anglo-American Food Production

CLAAS KIRCHHELLE

Rutgers University Press

New Brunswick, Camden, and Newark, New Jersey, and London

Library of Congress Cataloging-in-Publication Data
Names: Kirchhelle, Claas, 1987– author.
Title: Pyrrhic progress: the history of antibiotics in Anglo-American food production /
    Claas Kirchhelle.
Description: New Brunswick: Rutgers University Press, [2020] | Series: Critical issues in
    health and medicine | Includes bibliographical references and index.
Identifiers: LCCN 2019009912 | ISBN 9780813591476 (pbk.: alk. paper) | ISBN
    9780813591483 (cloth: alk. paper)
Subjects: | MESH: Anti-Bacterial Agents—history | Drug Resistance, Microbial | Food
    Safety | Legislation, Drug—history | Agriculture—history | History, 20th Century |
    History, 21st Century | United Kingdom | United States
Classification: LCC RM267 | NLM QV 11 FA1 | DDC 615.7/922—dc23
LC record available at https://lccn.loc.gov/2019009912

A British Cataloging-in-Publication record for this book is available from the British Library.

www.rutgersuniversitypress.org

Manufactured in the United States of America

Für Charlotte

Contents

List of Abbreviations

AAA	Agricultural Adjustment Act
AAFC	Antibiotics in Animal Feeds Subcommittee (NAFDC)
ABPI	Association of the British Pharmaceutical Industry
ACMSF	Advisory Committee on the Microbiological Safety of Food
AGP	Antibiotic Growth Promoter/Promotion
AHI	Animal Health Institute
AMR	antimicrobial resistance
APF	animal protein factor
AR	*Agricultural Research*
ARC	Agricultural Research Council
BF	*British Farmer*
BFS	*British Farmer and Stock Breeder*
BMJ	*British Medical Journal*
BSE	bovine spongiform encephalopathy
BVA	British Veterinary Association
BVJ	*British Veterinary Journal*
BVM	Bureau of Veterinary Medicine
CAP	Common Agricultural Policy
CAST	Council for Agricultural Science and Technology
CCC	Commodity Credit Cooperation
CEJ	*Coronado Eagle and Journal*
CIA	critically important antibiotic
CL	*Country Life*
CLM	Countway Library of Medicine
CSM	Committee on Safety of Medicines
CVM	Center for Veterinary Medicine
DaMa	*Daily Mail*

DARC	DEFRA Antimicrobial Resistance Coordination
DEFRA	Department for Environment, Food and Rural Affairs
DM	*Daily Mirror*
DRB	Drug Research Board (NRC-NAS)
DS	*Desert Sun*
EARSS	European Antimicrobial Resistance Surveillance System
EC	European Committee
EEC	European Economic Community
EPA	Environmental Protection Agency
EPHLS	Emergency Public Health Laboratory Service
ESBL	extended-spectrum beta-lactamases
EU	European Union
FBNews	*Farm Bureau News*
FDA	Food and Drug Administration
FDC	Federal Food, Drug, and Cosmetic Act
FEDESA	European Federation of Animal Health
FedReg	*Federal Register*
FJ	*Farm Journal*
FQPA	Food Quality Protection Act
FSA	Food Standards Agency
FT	*Financial Times*
FW	*Farmers Weekly*
FWR	*Farmers Weekly Review*
GFI	Guidance for Industry (FDA)
GRAS	generally recognized as safe
GRE	glycopeptide-resistant enterococcus (GRE)
HEW	US Department of Health, Education, and Welfare
HHS	US Department of Health and Human Services
HT	*Healdsburg Tribune*
JSC	Joint Sub-Committee on Antimicrobial Substances
LF	*Lancaster Farming*
LIN	*London Illustrated News*
MAFF	Ministry of Agriculture, Fisheries and Forestry
ME	*Mother Earth*
MERL	Museum of English Rural Life
MH	Ministry of Health
MMB	Milk Marketing Board
MoF	Ministry of Food
MRC	Medical Research Council
MRSA	methicillin-resistant *Staphylococcus aureus*
MT	*Madera Tribune*

NAFDC	National Advisory Food and Drug Committee
NARA	National Archives and Records Administration
NARMS	National Antimicrobial Resistance Monitoring System
NAS	National Academy of Sciences
NDA	New Drug Application
NEJM	*New England Journal of Medicine*
NFU	National Farmers Union (UK)
NOAH	National Office for Animal Health
NR	*National Review*
NRC	National Research Council (NAS)
NRDC	National Resources Defense Council
NYT	*New York Times*
OF	*Organic Farmer*
OGF	*Organic Gardening and Farming*
OMB	Office of Management and Budget
OTA	Office of Technology Assessment
PAMTA	Preservation of Antibiotics for Medical Treatment Act
PF	*Progressive Farmer*
PHLS	Public Health Laboratory Service
PML	Pharmacy and Merchants' List
PPM	parts per million
PrFa	*Prairie Farmer*
RCVS	Royal College of Veterinary Surgeons
RSPCA	Royal Society for the Prevention of Cruelty to Animals
RUMA	Responsible Use of Medicines in Agriculture Alliance
SAP	Scientific Advisory Panel (MAFF)
SBS	*San Bernardino Sun*
SciAm	*Scientific American*
SF	*Successful Farming*
SGFS	Steering Group on Food Surveillance
STS	science and technology studies
TNA	The National Archives (UK)
TSA	Therapeutic Substances Act
vCJD	variant Creutzfeldt-Jakob disease
VMD	Veterinary Medicines Directorate
VPC	Veterinary Products Committee
VR	*Veterinary Record*
VRC	Veterinary Residues Committee
VRE	vancomycin-resistant enterococci
WC	Wellcome Collections
WF	*Wallaces Farmer*

WHO	World Health Organization
WP	*Washington Post*
WSJ	*Wall Street Journal*
YBL	Yale Beinecke Library

Pyrrhic Progress

1

The Sound of
Coughing Pigs

In winter 2015, a door unexpectedly opened on a group of pigs. Where a moment before the room had been filled with the sound of grunts, squeals, and squeaks, the author and his wife were greeted by an expectant silence as sixty healthy-looking pigs turned toward the door, sat down on their behinds—and coughed. The surprisingly human sound of sixty coughing pigs has stayed with us to this day. For the purposes of this book, the sound is doubly significant because these coughing pigs had not received any antibiotics. Instead, they were part of an experiment by the pig farmer, who was giving us an impromptu two-hour tour of the farm's facilities.

Having heard of my research, the farmer insisted that I get an inside look at how conventional agriculture *really* works. The family farm specializes in fattening piglets and selling them for slaughter. While its capacity of 4,000 pigs is small compared to US concentrated animal feeding operations, the farm we were touring is a fairly typical example of pig production in northern Europe. One of only two surviving farms in the village, it had formerly also produced crops, raised cattle, and bred horses. However, following the 1960s, it had heeded the maxim of "get big or get out" and specialized in pig production. While it still grows cereals to feed its pigs, the farm's cattle have long since gone and horses are now mostly kept as lodgers paid for by wealthy urbanites. As part of a fine-tuned just-in-time production system, the farmer starts the day by checking international commodity prices on the Chicago Board of Trade's website. If the price is right, pigs are bought and sold with a click. Despite ingrained agricultural pessimism about market outlooks, business is going well. The farmer is a well-respected member of the local community and has just built another high-tech pig sty. And yet, the farm's future seems uncertain.

One of our farmer's main concerns is the increasing political focus on husbandry practices in conventional agriculture. Getting planning permission for the new sty was challenging enough because of local complaints about the smell. However, in an accurate premonition, the farmer is most concerned about mounting pressure on European Union (EU) regulators to restrict routine antibiotic use on farms. And this is where we come back to our coughing pigs: caught between competing expert narratives, the coughing pigs are the farmer's own DIY experiment to gauge whether antibiotics are really necessary for the farm's mode of production. It seems that they are. Although no costly disease diagnosis was made, the other piglets reaching the farm on the same lorry received antibiotics prophylactically and did not fall ill. In a business where the difference between profit and loss is decided by the length of time and the amount of feed it takes to produce an animal, coughing pigs are a problem. So what is the farmer to do? Sacrifice a business model that has worked for more than forty years or support lobbying efforts to delay the implementation of stricter antibiotic regulations? This was the dilemma the pig farmer posed to us. What follows is the history of how this antibiotic dilemma came about.

The Antibiotic Dilemma

Antibiotics are part of a wider family of antimicrobial drugs with activity against a variety of microorganisms, including bacteria, viruses, fungi, and other eukaryotic parasites. Revolutionizing the medical marketplace from the early twentieth century onwards, modern antimicrobials comprise synthetic antimicrobials (e.g., sulfonamides), biological antibiotics—substances produced by microorganisms to act against other microorganisms (e.g., penicillin), and semisynthetic or modified biological antimicrobials (e.g., methicillin).[1] In public discourse, the terms antibiotic and antimicrobial are frequently conflated. For the sake of simplicity, this book uses the most well-known term: *antibiotic*.

The modern prominence of antibiotics is hard to exaggerate. In schools, children learn the story of Alexander Fleming's 1928 discovery of the antibacterial qualities of the *Penicillium notatum* mold, museums feature exhibitions on "yellow magic," and patients and doctors routinely take and prescribe antibiotics for various ailments. So common and important have antibiotics become that recent books even talk of an "antibiotic era"[2] in human medicine from the 1930s onward. What is, however, often forgotten is that antibiotics have also come to play a significant role in food production. In fact, more than 50 percent of global antibiotic production is not destined for human use.[3]

The origins of non-human antibiotic use lie in the interwar period. Starting during this time, dramatic changes began to transform livestock production. Over the following decades, sizes and animal concentrations grew rapidly while new breeds and production systems changed the biological rhythms of livestock production. Although there were significant variations between different sectors and regions, growing numbers of animals disappeared into confined high-input housing systems or mixed indoor-outdoor systems. Breeding programs and fierce competition resulted in the dominance of a small number of animal breeds that were particularly efficient at converting feed into meat. Meanwhile, concentration processes increased both the productivity of animal husbandry and the investments necessary to survive oversaturated markets. In a process known as vertical integration, many producers now contract or work directly for larger corporations, which often control not only animal production but also feed production, slaughtering, and processing.[4] Changes were particularly impressive in the poultry industry. Whereas a meat-producing broiler chicken took 112 days to reach an average market weight of 2.8 pounds in 1935, it only needed 47 days to reach a live market weight of 4.7 pounds in 1995.[5] Chickens' bodies changed accordingly. Starting in the 1950s, heavier broiler chickens began to replace older varieties like the Rhode Island Red.[6] Since then, poultry meat has become a cheap and popular food with a small number of companies dominating international production.[7] Pig and cattle production have also become more concentrated—although the process was often more fragmented and confinement slower to develop.[8] In the twenty-first century, the intensive (confined and concentrated) and industrial (integrated) production of animals is fast becoming the global norm.[9]

One of the most formidable obstacles faced by expanding animal production is infectious disease. Prior to the interwar period, farmers had already attempted to increase herd densities. However, despite the use of antibacterial compounds like organoarsenics, infectious disease remained a serious threat. This situation changed during the 1940s. Within a decade, cheap antibiotics became routine components of the agricultural fight against bacterial infection.[10] Farmers soon found that antibiotics could also be used prophylactically to prevent infections from spreading in the first place. A significant third factor contributing to agriculture's antibiotic adoption was that even small doses of some antibiotics—if fed regularly—allegedly enabled animals to metabolize feedstuffs more efficiently. The mechanics behind the so-called antibiotic growth effect remain unclear. While postwar researchers believed that antibiotics optimized the microbial flora in animals' digestive systems,[11] contemporary theories posit (1) that by inhibiting bacterial digestion, antibiotics maximize the amount of available sugar, (2) that feeding antibiotics favors vitamin-producing bacteria and combats toxin-producing bacteria, and (3) that antibiotics favorably change the acidity of animals' stomachs.[12]

Although the past seven decades have seen consistent controversies over the mechanisms, extent, and very existence of the antibiotic growth effect,[13] antibiotics' tripartite function of combating and preventing infection and saving feedstuffs was a winning combination. Alongside other similarly important interventions like new vaccines, improved housing, nutrition, and breeds,[14] antibiotics aided a significant reduction of animal mortality. Whereas mortality in US broiler production was 10 percent in 1945, it sank to 8 percent in 1950, 6 percent in 1960, and 4.8 percent in 2015. Meanwhile, feed efficacy improved from 4 pounds of feed consumed to produce one pound of meat in 1945 to 2.5 pounds in 1960 and 1.82 pounds in 2018.[15] Then as now, the boundaries between therapeutic, prophylactic, and growth promotional antibiotic use frequently blurred. Antibiotics also entered other areas of food production. In addition to a brief career as food preservatives, they are still used to combat bacterial infections of crops and fruit and to protect bees.[16] While it is important to note that agricultural industrialization would have occurred with or without their discovery,[17] cheap antibiotics thus greatly facilitated the monoculture-like concentration of organisms and the substitution of human labor in modern livestock production. Over time, antibiotics' extensive use created significant infrastructure-like physical and cultural dependencies in global food production.[18]

While antibiotic infrastructures continue to influence contemporary husbandry and disease management systems, the resulting chemical cornucopia has also come at a price. Agricultural antibiotics face three different strands of criticism. First, according to some critics, antibiotics have enabled a neglect of animal welfare and allow inhumane factory farms to profit from animals' suffering. Second, many consumers and health authorities are also concerned about drug residues in food, water, and the environment. Some antibiotics are allergenic and can either sensitize individuals or trigger existing antibiotic allergies.[19] Allergic reactions can range from stomach irritations to the eruption of painful hives on the skin. In the worst case, allergies can trigger a life-threatening anaphylactic shock when antibiotics are administered in higher doses. A third and increasingly vocal group of critics focuses on antibiotics' selection for antimicrobial resistance (AMR).[20]

Although agriculture's exact contribution to the global AMR burden remains contested, bacteria's increasing ability to "resist" antibiotics is now widely held to be one of the most pressing global health challenges of the twenty-first century. The causes of AMR are complex. Bacteria can be either intrinsically resistant to an antibiotics or acquire resistance to it. In the case of intrinsic resistance, the natural characteristics of a bacterium's biology (e.g., cell wall or metabolism) can make it "immune" to certain antibiotics. For example, the iconic penicillin G is ineffective against most gram-negative bacteria, which possess double cell walls. Other bacteria may also already "naturally" possess the mechanisms with which to "resist" antibiotics, such as enzymes that deactivate

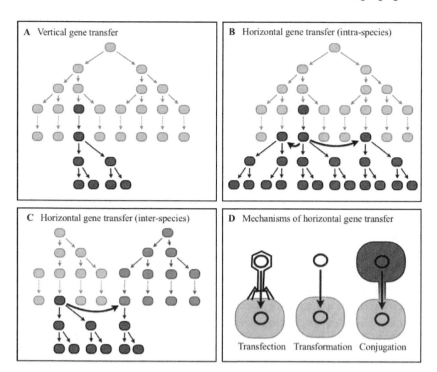

FIGURE A.1 Modes and mechanisms of AMR transfer.

antibiotics, the ability to modify antibiotic target sites, or efflux pumps with which to pump antibiotics out of the bacterial cell. By contrast, acquired resistance arises from spontaneous bacterial mutations and the acquisition of resistance-conferring mobile genes. Increased tolerance to an antibiotic will give a bacterium an evolutionary advantage over its peers the next time it is exposed to the respective antibiotic or to substances with similar effects (co-selection).[21]

Intrinsic and acquired resistance can pass from one bacteria generation to the next (vertical gene transfer). However, bacteria also possess the remarkable ability to "communicate" information on how to resist antibiotics among each other in a process called horizontal gene transfer. Whereas mutations bring new AMR genes into the world, horizontal gene transfer is the major force spreading these genes across the globe. Horizontal gene transfer can either occur as a result of *transduction,* during which resistance genes (R-factors) are inserted into the bacterial genome by viral bacteriophages; *transformation,* during which bacteria absorb free-floating resistance-encoding DNA sequences; and as a result of *conjugation* ("bacterial sex"), during which bacteria exchange small circular DNA strands called plasmids, which encode R-factors.[22]

It is difficult to exaggerate the implications of horizontal gene transfer. Because genetic information can be exchanged between bacteria of different species, an antibiotic-resistant yet innocuous *Escherichia coli* (*E. coli*) bacterium can pass on its resistance to a pathogenic *Salmonella* bacterium. Significantly, it is often not just resistance against one but against several antibiotics that is transferred *en bloque*. This *en bloque* transfer occurs because resistance genes tend to be stockpiled in mobile regions of the bacterial genome that are more easily transferred to other bacteria—the mobilome—such as plasmids and integrons. Any bacterium receiving such a genomic island via horizontal gene transfer can immediately resist multiple antibiotics. Exposing this bacterium to one of these antibiotics or related substances will automatically co-select for all of the other resistance information encoded on the genomic island.[23] Even sublethal concentrations of antibiotics, such as diluted antibiotics in rivers or in animal feeds and medicines, can select for clinically relevant resistance genes.[24]

Once established, resistant organisms and genes routinely cross geographic and species borders. With approximately 60 percent of all infectious diseases (and 75 percent of emerging infectious diseases) affecting humans shared by other vertebrate animals, the selection of resistant pathogens in one population can have serious consequences for the other.[25] This is not only true for established zoonotic pathogens but also for seemingly harmless bacteria. In the case of livestock-associated methicillin-resistant *Staphylococcus aureus* (CC398 MRSA), researchers found that the strain had originated as an antibiotic susceptible human *Staphylococcus aureus* strain that had jumped to livestock, where it had acquired resistance to methicillin and tetracycline (probably as a result of agricultural antibiotic use) before reinfecting humans in contact with farm animals.[26] Horizontal gene transfer means that a similar ecological blurring of AMR risks is also occurring because of the spread of AMR genes selected in one species and geographic environment to another.

Decades of continuous antibiotic use have led to a significant enrichment and selection for AMR in the global microbiome. It is uncertain whether there is a way back from this Anthropocene of the cell.[27] Although the higher energy cost of maintaining AMR means that some bacteria may regain sensitivity once selection pressure is decreased, others may not. The case of the mobilized colistin resistance-1 (mcr-1) allele serves as a warning. In November 2015, *Lancet Infectious Diseases* reported that resistance against colistin, a member of the old antibiotic class of the polymyxins, had been detected in *E. coli* samples from Chinese pigs. Polymyxins were commonly used in Chinese agriculture but also serve as last-resort antibiotics in Western hospitals. Chromosomal polymyxin resistance via point mutations was well known and had relatively high fitness costs. What was alarming about the 2015 resistance detection was that the *mcr-1* allele was encoded on a plasmid (i.e., prone to horizontal gene transfer) and associated with significantly lower fitness costs. Because polymyxins were rarely used in Chinese hospitals, AMR in local human populations had probably arisen in agriculture; *mcr-1* spread rapidly.[28] Within a year, researchers had detected it in human and animal samples from five continents. Although plasmids containing mcr-1 have, at the time of writing, failed to accumulate resistance to carbapenems, the other group of antibiotics used to treat resistant *Enterobacteriaceae*, this is likely only a matter of time.[29] The case of *mcr-1* highlights the potential of untreatable pandemic strains emerging as a result of local antibiotic exposure in humans, animals, or the environment.

Farmyard Realities

Faced with surging AMR and no new antibiotic classes since the 1980s, a growing number of politicians, experts, and consumers agree that the combined price of AMR, residues, and inadequate welfare outweighs the productivity benefits conferred by agricultural antibiotic use. Others disagree. Back on the farm, our farmer's answer is mixed. In the case of residue and welfare allegations, the farmer denies that there are serious problems. Although individual black sheep may ignore guidelines, strict European testing and new legislation have solved previous problems. The farmer is more defensive when it comes to AMR: Yes, AMR is a significant problem. But who is to say that agricultural antibiotics are responsible for it? Aren't patients and doctors more to blame for antibiotic overuse? Don't consumers drive agricultural antibiotic use by demanding cheap meat? And why should the farmer stop using antibiotics if foreign competitors continue to use them?

The farmers' opinions are legitimate and should be taken seriously. They also reflect more than seventy-five years of disputes over agricultural antibiotic regulation and increasingly fragmented notions of antibiotic risk. Understanding the factors driving this fragmentation of perceptions and resulting impacts on antibiotic regulation is the central aim of this book. Breaking with previous historiography's emphasis on human antibiotic use, it provides a unifying lens through which to analyze the often contradictory actions and perceptions of multiple actors and groups. What emerges is a familiar story of spreading dependencies and narrowly conceived regulatory solutions. Perhaps more importantly, it is also a story of how one group of substances acquired multiple meanings in different communities. This second story not only complicates moralistic accounts of "Big Ag" or "Big Pharma," it also serves as a cautionary tale about short-term thinking when it comes to current drug licensing and antibiotic stewardship.[30]

Two observations on the nature of risk and group formation help the book tackle its topic. The first observation is that differing risk cultures (epistemes) have played a crucial role in allowing global antibiotic use and AMR to rise to current levels. Because risk epistemes are difficult to reconstruct by looking at one nation in isolation, the book compares the development of antibiotic risk epistemes in the United States and the United Kingdom. Both nations were key to structuring the antibiotic infrastructure underpinning modern food production. However, the comparative analysis shows that distinct cultural priorities, agricultural infrastructures, and regulatory path dependencies produced very different risk epistemes. The book's second central observation is that each national risk episteme was strongly influenced by how antibiotic risk was staged in three overlapping social spheres: the general public, the agricultural community, and the regulatory community. Communities in each sphere developed their own understanding of antibiotics' risks and benefits. Over time, diverging understandings fostered the formation of rival groups and triggered increasingly fierce clashes over how agricultural antibiotics should be viewed, used, and regulated. Clashes occurred both within and between spheres. The wider national risk episteme was a continuously evolving reflection of these domestic clashes.

The book's twin focus on risk epistemes and communicative spheres is inspired by sociologists Ulrich Beck and Niklas Luhmann. In his seminal *Risikogesellschaft* (1986), Ulrich Beck highlighted the importance of risk for driving societal action. According to Beck, risk is the probability-based anticipation of catastrophe but not the catastrophe itself. Societies and individuals constantly strive to avoid predicted risks. Because everything can theoretically be interpreted as a risk, there is a constant struggle over whether a risk is real, negligible, or urgent.[31] Risk's infinite nature also means that no definitive expertise is possible. In the virtual world of risk, "no one is an expert or everyone is an expert."[32] As a consequence, the "objectivity of a risk is the product of its [cultural] perception and [social/objective] staging."[33] Within every society, existing inequalities mean that certain groups will be more successful at staging their version of risk and distributing resulting burdens than others. Often, the poorest and weakest end up bearing the greatest risk. Understanding how risk-based group identities emerge and interact is crucial if one wants to unravel the patchwork of national and international regulations that has emerged around antibiotic use. It is here that the communicative sociology of Niklas Luhmann can serve as a useful tool. According to Luhmann, every society comprises multitudes of smaller subsystems with relatively autonomous modes of communication: the smaller the group, the more distinct its communication and attitudes. Each subsystem or sphere attempts to frame or stage its version of reality for other spheres. A society's values, identities, and risk definitions are the contingent result of this potpourri of competing worldviews.[34]

As the anecdotal example of our pig farmer and recent social sciences research show, perceptions of antibiotic risk have certainly begun to differ between farmers and the wider public as well as between different countries—even experts' risk heuristics have been found to be affective and culturally biased.[35] These findings should not surprise us. Over the past three decades, science and technology studies (STS) research has revealed the rootedness of risk in distinct—often national—contexts. Sheila Jasanoff has described how cultural values, legal systems, and differing modes of knowledge production have created unique "civic epistemologies."[36] Historians have also explored who gets to decide whether risks are relevant, can be regulated, and should be monitored. In this context, Alexander Schwerin has highlighted the role of national "risk epistemes"—the complex nexus of political interests, expert epistemologies, institutional path dependencies, and technological heuristics—in deciding over the perception and management of environmental and technological hazards.[37]

Emphasizing the role of risk evaluation and management is important. In contrast to an unavoidable calamity, risk is both a hazard and an opportunity. Historically, formal risk definitions have played a significant role in transforming hazardous technologies from unpredictable dangers into calculable and thus manageable entities.[38] Phenomena that can be predicted can be accounted for and integrated into capitalist modes of redistribution and commodification. In the context of chemical and pharmaceutical regulation, numerous historians and STS scholars have pointed to the role of threshold models and boundary values in allowing hazardous substances to permeate workspaces, environments, and human bodies. In contrast to a zero-tolerance attitude toward risk, exposure to chemicals, radiation, and other hazards is deemed acceptable if it stays below a defined level. Large parts of industrial production now depend on the normalization of risk by a specialized regulatory branch of science.[39] For those who are unwilling to accept this normalization of risk, a lucrative market offers ways to further reduce exposure via insurance, "pure" food, or "safe" housing. However, not everybody can afford to opt out of normal risk. In most countries, risk exposure varies significantly along familiar socioeconomic divides.[40]

What is generally accepted as normal risk can also shift. Tolerance of technologies is often linked to historically fluid cultural notions of purity and naturalness.[41] According to Sheila Jasanoff: "The demarcation between the natural and the unnatural in any society is not given in advance but is crafted through situated, culturally specific forms of boundary work."[42] For regulators, the shifting results of public, industrial, and scientific boundary work pose a constant challenge: What if some experts say that the theoretical risk of a new technology is too great for it to be licensed but others disagree? Should a potentially beneficial technology be licensed even if prevalent cultural sentiment is against it? Should one wait until concrete proof of harm arises or should one listen to societal "gut instincts" and enforce precautionary bans? What if a technology is already on the market and new evidence challenges previous safety models? Over time, two contested philosophies have developed to address this epistemic challenge. One philosophy holds that concrete evaluations of proven costs and benefits should decide over technological regulation. The other philosophy holds that precautionary bans are justifiable if there is strong reason to suspect harm—even without conclusive evidence of it having occurred.[43]

The long conflict about agricultural antibiotics takes us right to the heart of the competitive staging of risk, its distribution, and the intricacies of precautionary and cost-benefit regulation. From the 1940s onward, societies had to decide which of antibiotics' benefits to enjoy at the cost of which risk. Meanwhile, changing conceptions of food safety and evolving AMR science repeatedly led to redefinitions of antibiotics' "true risk." Caught between agro-industrial interests and vocal medical, consumer, and animal welfare activists, regulators struggled to define boundary values for safe agricultural antibiotic use. One person's meat was another person's poison. Over time, different risk epistemes led to divergent regulation: reacting to industry pressure and public concerns, US regulators focused on monitoring food and milk for antibiotic residues. Meanwhile, new surveillance technologies and popular concerns made British regulators prioritize AMR over residue hazards. In 1969, the British Swann report recommended precautionary bans of low-dosed growth promoter feeds containing medically relevant antibiotics. Swann-inspired growth promoter bans were subsequently adopted in Britain and the European Economic Community (EEC) but not in the United States, where regulators continued to focus on preventing residues and favored cost-benefit evaluations. In recent decades, the regulatory gap has widened with the EU expanding antibiotic growth promoter bans between 1997 and 2006 and announcing plans to ban prophylactic and forms of metaphylactic (control treatment of animal groups) antibiotic use by 2022 in late 2018.[44] Although the effects of bans on AMR remain contested, EU agricultural antibiotic sales roughly stagnated between 2011 and 2014 with approximately 9,909 tons consumed in 2014 before falling by about 20 percent to 7,860 tons in 2016.[45] By contrast, the United States did not enact full statutory bans. While sales increased by 22 percent to 15,576 tons between 2009 and 2015, voluntary feed restrictions, industry reforms, regulatory threats, and changing consumer preferences have, however, led to a significant drop of consumption by about 30 percent to 10,933 tons between 2015 and 2017.[46]

This book's four-part structure reflects its transnational scope and the importance of the three domestic spheres—public, agricultural, and regulatory—in shaping national risk epistemes. Each part studies a distinct phase of British or US antibiotic history and is divided into three chapters. Of these, one focuses on the public, one on the agricultural, and one on the regulatory perception and management of agricultural antibiotics. Inspired by Michelle Mart's analysis of US attitudes toward pesticides, each chapter explores dominant themes and mutually reinforcing discourses on antibiotics.[47]

Analyzed sources include national and regional newspapers, consumer and activist publications, fashion and lifestyle magazines, cookery books, organic and conventional farming magazines, agricultural and veterinary manuals, industry publications, and archival material from regulatory agencies. However, there remain limits to what these sources can tell us. Many of them are elite and were written *about* rather than *by* actors themselves. Meanwhile, media reporting on antibiotics was often monolithic and conservative in its framing of mainstream opinion. There is also no guarantee that information in handbooks, newspapers, or official publications was taken seriously by producers or consumers. Finally, the sheer wealth of material means that it is impossible to cover all of the nuances of antibiotic opinion. This means that depictions of consumers, regulators, and farmers will often remain ideal types—a fact exacerbated by the book's transnational comparison, which to a certain degree depends on generalizable categories. However, these limitations should not deter curiosity or research. Although their coverage remains eclectic, keyword-searchable databases now give historians access to an unprecedented wealth of material on actors' lives and opinions. The digital repositories informing this book have been chosen to reflect as many political and regional perspectives as possible. While no single source group can fully reconstruct actors' opinions and actions, grouping as many as possible together provides a more detailed— if still grainy—picture.

The book's four parts are ordered chronologically. Following a brief introduction to antibiotics and AMR, part 1 studies the emergence of agricultural antibiotic use in the United States. Chapter 2 describes the initial public optimism about agricultural antibiotics before analyzing the effects of 1950s concerns about chemical additives, carcinogens, and food purity on substances' image. The chapter also highlights the divide between risk perceptions of AMR in medical and agricultural settings. Chapter 3 shows that antibiotics quickly acquired infrastructural relevance within US agriculture and were seamlessly incorporated into existing feed industry marketing channels. Focusing on risk perceptions among conventional and organic farmers, it highlights that US farmers knew little about the new miracle drugs' effects or potential hazards. Chapter 4 then describes the challenges US officials faced when regulating antibiotics. It shows that agricultural antibiotics' risks were initially deemed negligible by a regulatory matrix that focused on "classic" toxic and carcinogenic hazards. Lacking resources and a coherent policy framework, regulators later struggled to enforce drug compliance and respond to AMR warnings.

Part 2 focuses on Britain. Chapter 5 reconstructs how British public perceptions of antibiotic risk differed from those in the United States. During the 1960s, traditional concerns about animal welfare and new warnings about infectious resistance on factory farms created a powerful alliance for AMR-focused reform. Chapter 6 studies antibiotics' hesitant adoption by Britain's diverse agricultural community. It discusses the importance of close corporatist ties between the National Farmer Union (NFU) and government officials in overcoming initial skepticism regarding antibiotics and ensuring that later reforms did not harm farmers. Chapter 7 reconstructs how British officials first licensed antibiotics and then reviewed their use three times during the 1960s. Although their demands were repeatedly watered down, public health experts' use of a technology called bacteriophage typing to track the spread of resistant organisms played a decisive role in forcing officials to implement precautionary yet narrow restrictions of antibiotic feeds as a result of the 1969 Swann report.

Part 3 returns to the United States. Chapter 8 asks why US perceptions of agricultural antibiotics remained relatively unaffected by European AMR concerns. Between the 1970s and the mid-1990s, public opinion on antibiotic stewardship remained divided. This division was caused by ongoing residue fears, concerns about "stagflation," and the increasingly partisan nature of environmentalist politics. Chapter 9 highlights how US farmers struggled to come to terms with non-agricultural criticism of antibiotic use during a time of increasing economic pressure. It shows how hostility toward government intervention and a seeming lack of alternative production methods made most conventional farmers support lobbyists' campaigns against antibiotic restrictions. Although the mid-1980s saw organic intensification and financial pressure facilitate a market-driven conciliation with conventional agriculture, it did not trigger a wider reform of production methods. Chapter 10 analyzes US officials' reaction to new risk scenarios involving horizontal gene transfer. It shows that officials were divided in their assessment of precautionary European bans and underestimated industry opposition to antibiotic reform. After failing to restrict agricultural antibiotic use four times during the 1970s, government officials reverted to earlier policies of nonstatutory reform.

Part 4 covers the era following the implementation of the Swann report in Britain. Chapter 11 reconstructs the 1970s fragmentation of public pressure for antibiotic reform, the re-emergence of concerns about agricultural AMR selection around 1980, and the 1996 bovine spongiform encephalopathy (BSE) crisis's role in reigniting national reform campaigns. Chapter 12 studies the limited impact of initial antibiotic restrictions on British farming. After a brief reduction during the early 1970s, agricultural antibiotic use soon recovered and continued to increase until the late 1990s. Similar to the United States, most British farmers continued to see antibiotics as an important component of modern production and risk management. It was only when corporatist decision-making began to fracture during the 1980s that AMR and residue hazards were discussed more extensively. Following the 1996 BSE crisis, agricultural opposition to further growth promoter bans was limited. Chapter 13 asks why the

Swann report failed to curb either antibiotic use or AMR. It emphasizes the importance of interministerial rivalry and corporatist ties in watering down many proposals of the original report. With British efforts stagnating, European integration played a decisive role in reinvigorating British antibiotic reform from the 1980s onward.

Returning to our farmer and the coughing pigs, chapter 14 recapitulates the tumultuous history of agricultural antibiotics' perception, use, and regulation. It asks whether this history can provide insights for current regulators. The chapter tells four cautionary stories about narrow reforms, infrastructural entanglements, epistemic fragmentation, and short-termism before arguing for a new policy emphasis on microbial resilience.

Part I

USA

From Industrialized Agriculture
to Manufactured Hazards,
1949–1967

Despite antibiotics' European roots, it was in America that the "antibiotic era" truly began. Profiting from early and reliable access to cheap antibiotics, US consumers, farmers, and regulators played a crucial role in deciding which forms of nonhuman antibiotic use were commercially viable and publicly acceptable. The three chapters of Part One map the early highs and lows of antibiotics' career in US food production. They show that agricultural antibiotics' rapid adoption was enabled by their seamless integration into agro-pharmaceutical supply chains, their nontoxic properties, and the neo-Malthusian framework of Cold War politics. The emergent antibiotic infrastructure soon began to cause problems for consumers and farmers alike. However, reforms proved difficult. Facing a growing cost-price squeeze, farmers felt unable to reduce antibiotic use. Attempting to reconcile consumer and producer interests, officials struggled to adapt regulatory frameworks. Between the 1950s and mid-1960s, a lack of cohesive antibiotic policies and the prioritization of antibiotic residues over AMR selection led to a "narrow" official focus on guaranteeing pure milk and establishing "safe" residue tolerances for meat. Although external pressure and new scenarios of horizontal gene transfer triggered a review of agricultural antibiotic use in 1965, the residue-focused US risk episteme remained intact.

2

Picking One's Poisons

Antibiotics and the Public

This chapter traces the turbulent history of agricultural antibiotics' public image in the United States. Analyzing newspaper reports, fashion magazines, and cookbooks, it shows that agricultural antibiotics' initial status as guarantors of plenty and signifiers of progress became double-edged during the late 1950s. This crisis of trust began with reports of antibiotic residues in food and milk and was exacerbated by a series of scandals surrounding the US Food and Drug Administration (FDA). Significantly, public awareness of antibiotic hazards other than residues remained limited. In the case of AMR, concerns centered on medical overuse and seldom touched on agricultural antibiotic use.

Euphoria

Although the use of naturally occurring antimicrobial remedies is probably as old as humanity itself, the first half of the twentieth century saw the revolutionary introduction of effective, industrially produced antimicrobial drugs. Launched in 1910, Paul Ehrlich and Sahachiro Hata's antisyphilitic Salvarsan (arsphenamine) triggered a large-scale hunt for further "magic bullets" against disease. Scientists searched for substances that only targeted *prokaryotic* bacteria cells while leaving *eukaryotic* animal cells unharmed. In the 1930s, the discovery of Prontosil (Sulfamidochrysoïdine) by Gerhard Domagk and his colleagues at Bayer and the subsequent synthesis of chemically related drugs seemingly answered this challenge. The synthetic sulfonamides cured previously fatal bacterial infections and achieved prominence by saving the life of President Roosevelt's son in 1936 and of Winston Churchill in 1943. Sulfonamides' arrival also marked a high point for a new form of industrial pharmaceutical research. Housed in pharmaceutical laboratories, microbiologists screened compounds for antibacterial effects while chemists purified, mass-produced, and modified promising compounds.[1]

By the 1940s, many sulfonamides' toxic properties and growing bacterial resistance resulted in a second round of antibacterial research. A growing number of researchers now grew interested in the antimicrobial substances produced by certain microorganisms, which Selman Waksman termed "anti-biotic" ("against-life") in 1941.[2] Of this second generation of antibiotics, penicillin is undoubtedly the most iconic. Isolated and refined from the fungus *Penicillium notatum* in Oxford and subsequently modified and mass-produced in the United States, penicillin cured many sulfonamide-resistant infections. The vast resources deployed by the Allies to upscale penicillin production also made it exemplary of a new kind of planned "Big Science."[3]

With production of both sulfonamides and unpatented penicillin expanding rapidly after 1945, prices collapsed and pharmaceutical companies began to search for new, patentable antibiotics. They did not have to look for long: employing mass-screening techniques, companies experienced an antibiotic gold rush. In 1943, a team surrounding Selman Waksman had already discovered streptomycin. In 1946, Parke-Davis isolated chloramphenicol. One year later, the Lederle Laboratories branch of American Cyanamid launched aureomycin (chlortetracycline). In 1949, Pfizer isolated terramycin (oxytetracycline), and Selman Waksman discovered neomycin. Meanwhile, Dorothy Crowfoot Hodgkin's decryption of penicillin's molecular structure indicated that a bountiful age of tailored antibiotics might be forthcoming.[4]

The resulting 1950s antibiotic euphoria is best captured in a series of paintings commissioned from American artist Robert Thom by drug manufacturer Parke-Davis in 1957. Reminiscent of Norman Rockwell's parallel portrayals of American plenty, Painting 39 of Thom's series is titled "The Era of Antibiotics" and features a woman screening inhibitory substances beneath a portrait of Alexander Fleming. On the other side of a glass window, a technician tends to gleaming fermentation tanks. The painting's caption reads: "Intensive research continues to find antibiotics that will conquer more of man's microbial enemies."[5] Thom's martial confidence in antibiotics' efficacy was indicative of wider enthusiasm about the alleged defeat of microbial scourges like pneumonia, tuberculosis, gonorrhea, syphilis, and typhoid. With reproductions of Thom's paintings hanging in their offices and waiting rooms, physicians confidently prescribed a rapidly expanding array of new antibiotics to patients. Despite reports of overprescription, allergic reactions, and the fatal side effects of chloromycetin (chloramphenicol), public and medical trust in the new drugs

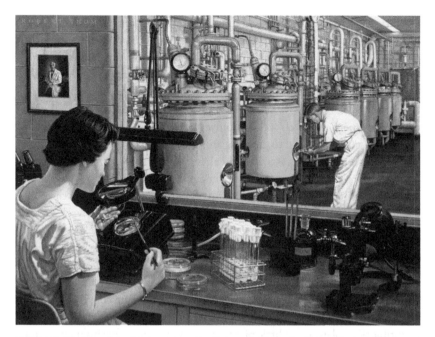

FIGURE 2.1 *The Era of Antibiotics*, painted by Robert A. Thom for Parke, Davis & Co. APhA Foundation.

remained unshaken.[6] Reacting to and further stoking public demand, pharmacists and drug companies advertised both medical antibiotic use and supposedly prescription-free antibiotic lozenges, nose sprays, mouthwash, toothpaste, and shaving balms in newspapers and magazines like *Good Housekeeping*.[7] Bastions of upper-middle-class taste like *Vogue* and *Life* soon featured adverts for Hi and Dri neomycin deodorants: "Finally . . . a deodorant that *really* works because it does the job *medically*!"[8]

The rapid growth of public and medical demand was welcome news for major antibiotic manufacturers like Lederle Laboratories and Pfizer, the majority of whose profits soon came from antibiotic sales.[9] However, the flood of new antibiotics also increased competition. Keen to maximize profits, manufacturers looked for additional antibiotic outlets. The veterinary market seemed particularly promising. Discussed in more detail in chapter 3, sulfonamides and antibiotics like gramicidin had been used against bacterial udder infections in cows (mastitis) and gastro-enteric poultry infections since the late 1930s. While many of the new biological antibiotics were initially deemed too expensive for use on animals, the postwar price decline meant that both pet owners and farmers could increasingly afford antibiotic treatments.[10] Nontherapeutic antibiotic applications proved equally popular. Since the 1910s, medicated feeds and metabolism-enhancing supplements had become increasingly popular on

US farms.[11] This market increased dramatically during the 1940s. In 1948, the identification of vitamin B$_{12}$ as the hitherto mysterious animal protein factor (APF) and the discovery that antibiotic fermentation wastes were B$_{12}$-rich strengthened lucrative links between antibiotic producers and the US feed industry.[12] This link was further strengthened in 1949 when Lederle researchers Thomas Jukes and Evan Ludvik (E. L.) Robert Stokstad claimed that low— or subtherapeutic—levels of aureomycin in fermentation wastes caused additional gains when fed to growing animals.[13]

Announced in April 1950, the so-called antibiotic growth effect not only opened a large new outlet for the oversupplied antibiotic market but also received glowing coverage in national and regional newspapers across the political spectrum.[14] In 1951, the *Washington Post* gushed: "Each week 80,000 chicks are produced and moved to [Armour and Co.'s] Ches-Peake farm . . . . By scientific feeding, controlled temperatures, germ-killing rays, water treated with such drugs as terramycin, aureomycin, and antibiotics [*sic*], the birds are ready for slaughter in three-fourths the time by ordinary methods."[15] Meanwhile, *Time* magazine described how Pfizer was shaking up conservative farmers with the help of "synthetic sow's milk spiked with terramycin."[16] Thanks to antibiotics, "platoons of little pigs were enjoying a peril-free infancy . . . none are trampled or eaten; no luckless runts are left teatless."[17] On Pfizer farms, piglet mortality had declined from between 21 and 33 percent to 5 percent.[18] Sows could be "put back to work" immediately instead of "performing no other service than can be performed by the milking machine at the nearest dairy."[19] Impressed by rapidly rising sales in 1954, Dow Jones Company–owned *Barron's National Business and Financial Weekly* sagely predicted "a steadily rising market for animal medicines in the foreseeable future."[20] While American children entered home-made growth promotion experiments on rats at local science fairs,[21] penicillin co-discoverer Alexander Fleming predicted that penicillin's use as a growth promoter might someday exceed therapeutic uses.[22]

Trust in the new "miracle drugs" was such that antibiotics' use as food preservatives and plant sprays was greeted with similar enthusiasm.[23] In 1953, the *Washington Post* rejoiced: "antibiotics are becoming wonder drugs to save food crops . . . give us more and tastier meats, even aid in making beer and whisky."[24] *Better Homes and Gardens* speculated that streptomycin could well "prove to be the long-heralded 'injection' to cure plant diseases."[25] According to the *Desert Sun* from Palm Springs, antibiotics not only "promise[d] housewives food that will keep for weeks"[26] but also "toothsome whaleburgers" and "whale steaks," which would make "'mighty fine eating' if properly preserved and prepared."[27] American Cyanamid and Pfizer began to market antibiotic preservatives from the mid-1950s onward. Described by Maryn McKenna, Cyanamid's Acronization process (dipping poultry in chlortetracycline) was celebrated in the national and regional media, was explicitly advertised on packaging, and was awarded *Good Housekeeping*'s vaunted Seal of Approval. By 1958, more than half of US slaughterhouses had purchased Acronize licenses. Fish wholesalers soon also began using antibiotic preservatives.[28]

Nowhere was antibiotic enthusiasm greater than among the authors of *Scientific American*: proclaiming an "antibiotic age"[29] in 1952, the magazine noted that the wholesale market value of antibiotic and vitamin B_{12} feed supplements was already worth from $40 million to $50 million. With production costs of antibiotics like penicillin falling from $20 to 4 cents per 100,000 units between 1943 and 1951, it was obvious that agricultural antibiotic use would continue to expand.[30] Another article claimed that chemical technologies like antibiotics would advance "agricultural efficiency at least as much as machines have in the past 150 years."[31] By taming capricious nature, a chemical revolution was finally allowing humans "to free [them]selves from the dismal philosophy of Robert Malthus."[32]

Scientific American's attack on "dismal" Malthusianism supported a central tenet of postwar US politics. According to eighteenth-century English cleric Robert Malthus, the human population's exponential growth would always exceed the linear growth of agricultural productivity. As a consequence, population growth would eventually be halted either by preventive checks on fertility or positive checks—such as rising mortality through famine.[33] After reaching dismal heights during the interwar period, the 1950s saw Malthusian rhetoric challenged by new agricultural technologies and population control programs. According to the *New York Times*, the "chemical revolution on the farm" had "all but [wiped] out the Malthusian fear."[34] Significantly, providing plenty was also seen as an effective way to contain communism. With communist parties performing well in European elections and making gains in Asia, American planners began to equate prosperity and development with social stability and pro-US values. In his inaugural 1949 "Four Point Speech," President Truman promised to "embark on a bold new program for making the benefits of our scientific advances and industrial progress available for the improvement and growth of underdeveloped areas."[35] For patriotic US researchers, politicians, and journalists, promulgating agricultural plenty and efficiency-boosting technologies like antibiotics became a moral duty.[36] Speaking at the 1953 meeting of the National Farm Chemurgic Council, biochemist H. J. Prebluda claimed that applying antibiotics in farm soil "may do for crops what penicillin and other antibiotics do for animals... this may solve most of the world's hunger problems, thus eliminating one of the causes of unrest upon which communism has tried to capitalize."[37] In a similar vein, major newspapers like the *New York Times* and *Washington Post* equated rising US meat and antibiotic consumption with American leadership:[38] "Nowadays the doctor arrives with a station wagon full of hypos, stimulators, pills and penicillin and Buttercup gets the benefit of modern medicine."[39] According to the

FIGURE 2.2 *Life* also celebrated AGPs. Art Shay, "Pigs of Different Diets," *Life*, October 1953. ©Art Shay/Art Shay Archive.

San Bernardino Sun, even Soviet delegates were impressed by the extensive use of antibiotic growth promoter feeds (AGPs) on American farms.[40]

With antibiotic factories and supplies already serving as a popular form of US foreign aid, AGPs were also mobilized as Cold War assets for human nutrition. Speaking at the 1950 meeting of the American Chemical Society, Washington State College researchers reported that young rats whose mothers had been fed antibiotics grew faster because of residues in mother's milk. It might also "be desirable to feed antibiotic concentrates to physically retarded children in order to increase their growth rate."[41] Following promising domestic tests on navy recruits, prison inmates, and children in mental institutions and public schools, antibiotic supplements were trialed as foreign aid. During the 1950s, US researchers trialed antibiotic diets for children in Kenya, Jamaica, Haiti, and Central America.[42] The trials received positive regional and national news coverage. In 1951, Nevin Scrimshaw, founding director of the International Institute of Nutrition of Central America and Panama, enthused that

FIGURE 2.3 Cyanamid advertisement, "2,000,000 More of You to Feed," 1953. AGPs were part of US campaigns against hunger and communism. Lederle, *Life*, February 1953.

malnourished "children may some day be eating aureomycin candy to improve their diets."[43] Foreign feed trials continued until 1958 when the difficulty of assessing whether antibiotics were solving malnutrition or simply curing low-level infection led to an end of research.[44]

Overall, it is hard to exaggerate 1950s US optimism about antibiotics in medicine and food production. Whether one looks at Robert Thom's paintings or national and regional media from across the political spectrum, antibiotics were presented as a sign of Western progress. To most observers it seemed as though the constant stream of new antibiotic blockbusters would not only eliminate "man's microbial enemies" but also break the Malthusian trap and defeat poverty-fueled communism. That this antibiotic feast came at a price was rarely discussed.

Residues and Allergies

As discussed in chapter 1, agricultural antibiotic hazards can be roughly divided into residues, AMR selection, and the facilitation of welfare abuse. However, not all hazards received equal attention. By far the most potent American concerns about agricultural antibiotics centered on residues in food and milk. Residue concerns mirrored and reinforced popular fears about the invisible contamination of food, the environment, and citizens' bodies. The success of muckraking bestsellers like *The Jungle* (1906) or *100,000,000 Guinea Pigs* (1933) shows that contamination concerns were already deep-seated in Americans' collective consciousness. However, postwar research and debates significantly increased public wariness of long-term exposure to invisible hazards.[45]

Chemical residues in food seemed particularly problematic. In 1949, Republican representative Frank B. Keefe successfully lobbied for a congressional Select Committee to Investigate the Use of Chemicals in Food and Cosmetics. Following Keefe's death, Democratic representative James J. Delaney took over the committee's chair.[46] In 1951, the select committee's report attacked the use of inadequately tested synthetic substances and demanded legislation to protect the public from carcinogens and latent poisoning.[47] Published during a time of rising concern about cancer rates, the report received widespread media coverage and focused Americans' attention on potential long-term hazards of the many new substances being used in houses, gardens, and food production (chapter 4).[48]

Although Delaney's committee did not address mostly nontoxic and non-carcinogenic agricultural antibiotics, this did not mean that antibiotics' reputation was safe. During the 1940s, it emerged that some people were allergic to certain antibiotics. Allergies to β-lactam antibiotics like penicillin were particularly frequent, and allergic reactions could range from mild skin irritations to painful hives or a lethal anaphylactic shock. Between 1953 and early 1957, 1,070 "life-threatening" allergic reactions to antibiotics and 1,925 "severe reactions" were reported to the FDA. The cases included 72 penicillin-related deaths.[49] Meanwhile, studies on nurses revealed that constant antibiotic exposure could lead to hypersensitivity.[50]

Concerns about exposure to allergenic antibiotics in hospitals were, however, slow to spread to the agricultural sphere. In what was to become one of the main features of US antibiotics reporting, commentators frequently treated risks posed by antibiotic use in medical settings as distinct from those posed by use of or exposure to the same substances in nonmedical settings. This was initially also true for antibiotic contaminated food and milk.[51] During the early 1950s, publications like *True Republican* from Illinois warned rural readers that "indiscriminate use of drugs [was] dangerous"[52] in medicine but did not mention nonmedical forms of antibiotic use on farms. In 1951, *New York Times* journalist Jane Nickerson regarded antibiotic residues in milk as an "interesting, if not too serious"[53] annoyance, which merely complicated cheese production by inhibiting bacteria cultures.

Wider attitudes towards antibiotic residues only began to change following the 1956 publication of an FDA survey, which found that up to 10 percent of US milk might be contaminated with penicillin. Although officials claimed that detected levels were insufficient to create new allergies, residues could trigger existing allergies.[54] The fact that residues had been found in milk made the FDA survey particularly poignant. Described by Kendra Smith-Howard, milk held a special place in the minds of US consumers. A symbol of health, milk was associated with feeding the young, infirm, and vulnerable.[55] Prior to the Second World War, measures to secure milk purity had centered on the eradication of pathogens like tuberculosis.[56] However, after 1945, definitions of milk safety and purity increasingly encompassed the absence of chemical and radioactive adulterants. This made the FDA's penicillin residue detections particularly damaging and reduced public commentators' previous tolerance of US farmers' direct access to high-dosed antibiotics.[57]

In the wake of the 1956 milk scandal, US media reports on food hazards and cancer risks increasingly lumped agricultural antibiotics together with other dubious "chemicals" like DDT and chemical dyes. Antibiotic residues also emerged on the problem radar of the surging US consumer movement.[58] In 1956, the *New York Times* reported on the International Union Against Cancer's symposium in Rome. Speaking at the symposium, William Hueper, head of the US National Cancer Institute's Environmental Cancer Section, listed chemical hazards in food production. The list included dyes, thickeners, synthetic sweeteners, preservatives, bleaches, fat substitutes, pesticide residues, chemical sterilizers, wrapping materials, estrogens, *and* antibiotics.[59] In the same year, journalist James Rorty and physician N. Philip Norman published the second edition of their popular *Tomorrow's Food*. The book's preface warned about new hazards posed by the "tremendous postwar increase in the use of highly toxic pesticides by farmers, the employment of hormone fatteners . . . and the adulteration of processed foods with a multitude of new and inadequately tested chemical additives."[60] Whereas Delaney's committee had mostly ignored agricultural antibiotics in 1950, the authors now discussed the residue hazards of antibiotic feeds and sprays and criticized new residue tolerances for legalizing "the 'calculated risks' of mass poisoning."[61]

Agricultural antibiotics' cultural association with carcinogenic chemicals became especially damaging when food safety concerns reached fever pitch around 1960. In 1959, the so-called Delaney Clause of the 1958 Food Additives Amendment (chapter 4) forced the FDA to take action against millions of pounds of cranberries produced with the herbicide and suspected carcinogen aminotriazole ahead of Thanksgiving, the most important date in cranberry growers' year.[62] Hitting the US cranberry industry hard, the scandal heightened popular fears that farmers were abusing dangerous chemicals—including antibiotics. Concerns were reinforced by the FDA's nearly simultaneous disclosure that 3 percent of analyzed US milk samples contained "substantial residues"[63] of pesticides and that 3.7 percent contained penicillin residues.[64] With residue fears damaging the US cranberry and dairy industries, wary consumers also started abandoning new antibiotic preservatives amidst complaints about the dubious quality of preserved produce. In reaction, some companies began to advertise "non-acronized" chickens.[65]

A series of bestselling exposés fanned the flames of public distrust of agricultural antibiotic use. In 1960, journalist and future Pulitzer Prize winner William Longgood published *The Poisons in Your Food*. Longgood's introduction invited consumers to inspect their shopping baskets: "Then there's the milk you give the children . . . . But did you know the odds are . . . one to ten it contains antibiotics? . . . Sunday's chicken may have traces of antibiotics, arsenic and artificial sex hormones . . . . The roasts or steaks probably have traces of hormones, antibiotics and the inevitable poisons that went into the cattle's diet."[66] Similar to *Tomorrow's Food*, Longgood accused congressionally mandated residue tolerances of legalizing "mass poisoning . . . by granting FDA the right to determine how much poison residue may remain on marketed food."[67] For Longgood, agricultural antibiotics were no longer miracle substances but sinister contaminants. Contrary to official assurances, cooking did not destroy antibiotic preservatives, and the 1956 milk scandal had shown "how precarious the public's margin of safety is when a dangerous drug is placed in the hands of laymen . . . who are expected to exercise their sense of responsibility at the risk of losing money."[68] Linking them to rising allergic reactions, Longgood also claimed that antibiotic residues acted as "vitamin antagonists" and masked disease in slaughtered animals.[69]

Longgood's book provoked angry reactions. In *Science*, William J. Darby, an influential nutritionist from Vanderbilt University, attacked *The Poisons in Your Food* as "an all-time high in 'bloodthirsty pen-pushing'" from the "bias of the non-scientific, natural food-organic cult."[70] Longgood's "authorities" were "the cult leaders . . . or a few true scientists whose work or expressions have been taken either out of context or out of time."[71] Dispensing with a bibliography, some of Longgood's claims were indeed sketchy. However, the fact that his book managed to elicit a review in *Science* showed that the days of wholesale chemical optimism were over. In popular culture, the promised chemical cornucopia of the 1950s was acquiring the bitter aftertaste of invisible residues.

Longgood was not the only one to fan concerns about chemicals and antibiotics in food and the environment. Authors associated with the US organic movement were equally active. Based in Pennsylvania, publicist Jerome Irving (J. I.) Rodale was a long-standing critic of conventional agriculture (chapter 3). For much of the 1950s, Rodale's advocacy of organic food production, healthy nutrition, and self-improvement had, however, only attracted a fringe readership.[72] This changed during the 1960s. Between 1961 and the end of the decade, readership of *Organic Gardening and Farming* nearly doubled from 270,000 to over 500,000 while readership of the consumer-oriented *Prevention* reached about 725,000.[73] During this time, antibiotics played an increasingly prominent role in Rodale's criticism of conventional nonorganic agriculture. In 1961, Rodale's *The Complete Book of Food and Nutrition* summarized several years of *Prevention*'s dietary advice. The book echoed Longgood by warning about antibiotic-contaminated milk but also referred to agricultural AMR selection, spreading allergies, the dangers of antibiotic food preservation, and alleged links between penicillin residues and blood clots. According to Rodale's increasingly influential publishing empire, healthy animal produce only came from "antibiotic free" organic farms.[74]

A further reputational blow for agricultural chemicals' image fell in 1962 when *Silent Spring*, the iconic bestseller by marine biologist Rachel Carson, launched a frontal attack on chemical polluters and on DDT in particular. Similar to antibiotics, the insecticide DDT was initially regarded as a success story of wartime science and postwar commercialization.[75] DDT's similarity to antibiotics did not end here: it quickly became clear that DDT use could select for resistance in insect populations and result in residues, which accumulated in animal tissues. DDT concentrations were especially high toward the top of the food chain. In the case of America's heraldic animal, the bald eagle, DDT resulted in thinner eggshells, which were unable to support the weight of brooding parents. Significantly, Carson also accused DDT and other chemicals of causing cancer. Of *Silent Spring*'s seventeen chapters, five were devoted to pesticides' and herbicides' carcinogenicity.[76] Profiting from earlier bestsellers like *The Poisons in Your Food* and anarchist Murray Bookchin's contemporaneous *Our Synthetic Environment*,[77] *Silent Spring*'s fusion of environmental and health concerns led to heated American debates about regulation of private, official, and agricultural chemical use.[78] In December 1962, the *Washington Post* claimed that *Silent Spring* had turned chemical use into "the most controversial non-political subject in American agriculture."[79] Resulting debates about DDT and persistent pesticides also affected the cultural framing of chemically unrelated substances. Although Carson did not discuss them in her book, American public debates about agricultural antibiotics would for a long time be framed according to the residue-focused language of *Silent Spring*.

Growing consumer concerns about invisible chemical and antibiotic residues in US food were paralleled by a series of pharmaceutical corruption scandals and safety scares, which resulted in a loss of public confidence in the integrity of American drug manufacturers and their regulators. The 1950s had been a golden era for the US pharmaceutical sector. Venerated as "merchants of life,"[80] companies' value had more than quadrupled to $2,200,000,000 between 1945 and 1958.[81] However, companies' behavior had on occasion also been questionable. Between 1959 and 1962, investigations by the Senate's Antitrust and Monopoly Subcommittee shed a harsh light on dubious mark-up prices, marketing practices, and attempts to drive generic drug producers out of business. The most damaging findings came to light in May 1960, when Democratic senator Carey Estes Kefauver's subcommittee announced that it was investigating extra income received by the head of the FDA's Antibiotics Division, Henry Welch. Between 1953 and 1960, Welch had received $287,142 for his role as editor-in-chief of the journals *Antibiotics and Chemotherapy* and *Antibiotic Medicine and Clinical Therapy*. Financed by industry, the widely distributed journals contained articles designed to advertise antibiotic products—sometimes prior to their licensing by Welch's division.[82] Industry representatives had even edited some of Welch's official speeches—in one case, a Pfizer slogan had been written into a speech to "jazz it up."[83] A defiant Welch was forced to resign from the FDA in mid-May 1960.[84]

Although Secretary of Health Arthur Flemming ordered a review of all of Welch's licensing decisions,[85] the 1961 thalidomide scandal added to public concerns about FDA and industry standards. In 1957, the West German company Chemie Grünenthal had launched a new substance called thalidomide as a sedative and soporific suitable for pregnant women. Despite evidence linking thalidomide to neural damage and fetal malformation, Grünenthal continued to market thalidomide until November 1961. By then, fetal exposure to the teratogen was believed to have caused an estimated 10,000 malformations and several hundred deaths.[86] Fortunately, thalidomide had not been licensed for US markets. Despite repeated requests, FDA reviewer Frances Oldham Kelsey had deemed industry data insufficient. Kelsey's heroic story, however, also highlighted how lucky Americans had been. With no requirements for manufacturers to submit clinical trials or report adverse effects, Kelsey's doubts had been the only thing standing between thalidomide and the US market.[87] Reacting to the thalidomide scandal and Kefauver hearings in 1962, President Kennedy awarded Kelsey the President's Award for Distinguished Federal Civilian Service[88] and signed the so-called Kefauver-Harris Amendment. The 1962 amendment mandated prelicensing efficacy tests of new drugs via controlled clinical trials and required manufacturers to report adverse effects from 1963 onward.[89]

However, the reputational damage had been done. Not only had contemporary food, chemical, and pharmaceutical scandals revealed gaps in consumer protection, they had also tainted post-war narratives of agricultural and chemical progress. During the 1960s, this reputational shift led to a curious divide in US reporting on agricultural antibiotics. On the one hand, fears of global overpopulation and communism continued to engender cross-party and cross-regional support for antibiotic-fueled plenty. On the other hand, programs of defending democracy with agricultural chemicals now had to be justified against a growing list of environmental and health warnings. Even optimistic reports on chemicals' ability to tame nature now demanded that farmers contain "synthetic" substances so as not to adulterate "natural" food and bodies. In 1963, *Scientific American* claimed that underdeveloped countries still depended on Western food imports produced with the help of "finely calculated diets and rations, synthetic hormones, ... drugs and vaccines to control disease."[90] However, the magazine was now also forced to assuage consumer concerns by using the classic argument that harmless substances did not exist: "there are only harmless ways of using them."[91]

A defensive line of reasoning also began to characterize industry rhetoric. In 1961, the US Manufacturing Chemists' Association published a booklet entitled *Food Additives: What They Are, How They Are Used*. The booklet was designed to help physicians and nutritionists inform consumers about the benefits of the ongoing "food revolution" and the role of chemical additives as "an important 'tool' of food science."[92] However, this revolution was in danger. Although Americans were "better fed and in better health than at any time in history," "unprecedented population growth"[93] threatened food security. Further chemical use was key if Americans wanted to preserve a "plentiful, varied, nutritious, safe, economical and good-tasting"[94] food supply. Attacking irrational faddists and organic gardeners and employing Cold War rhetoric,[95] the booklet praised "natural" antibiotic preservatives without which it would be impossible to feed the crews of "our atomic submarines" which "can cruise for months in a hostile environment without surfacing even once."[96]

It is hard to judge just how seriously the average US consumer took industry advice or residue warnings. On the one hand, a keyword analysis of popular cookery books does not reveal a marked rise of chemical concerns.[97] Specialist publications like the *Principles of Nutrition* (1959) included positive discussions of antibiotic growth promotion, and 1960s regional reporters continued to exhort readers to "give thanks to science for all that lush meat" produced by "pampered"[98] animals. On the other hand, the flood of new US publications either attacking or defending agricultural chemicals is indicative of a wider shift in public discourse, which also affected antibiotics. Once enjoying near universal support, agricultural antibiotics' had become culturally tarnished via association with other suspicious chemicals. This association had significant implications for the public staging of antibiotic risk. With both the media and whistle-blowing bestsellers focusing on antibiotic residues in food and milk, other aspects of antibiotic risk like agricultural AMR selection were consistently overshadowed.

Resistance

Throughout the 1950s and 1960s, there was a surprising dearth of American public debates about AMR selection on farms. This was not because of lacking knowledge. Although many AMR mechanisms were still unknown, experts were well aware of the general phenomenon. During the 1930s, physicians had already noted that certain bacteria developed resistance against sulfonamides.[99] In 1940—one year ahead of penicillin's first clinical trial—Oxford researchers described penicillin-resistant staphylococci.[100] Echoing earlier warnings by gramicidin discoverer René Dubos,[101] penicillin discoverer Alexander Fleming used his 1945 Nobel Prize lecture to warn about antibiotic overuse and AMR.[102] However, prior to the discovery of bacteria's ability to exchange genes conferring AMR (R-factors) via horizontal gene transfer (chapter 1), most experts remained confident that AMR could be contained by avoiding "irrational" antibiotic use, employing drug combinations, and improving infection control. If all of these measures failed, the development of new antibiotics would eventually solve problems.[103]

During the 1950s, American experts and media outlets thus responded to rising AMR with campaigns for "rational" antibiotic use. Although most commentators saw "no reason to think antibiotics are on the way out,"[104] papers like the *Madera Daily News* and *San Bernardino Sun* warned that one could "develop a sort of tolerance"[105] to penicillin and that "Your Neighbors' Wonder Drugs Can Make You Sick; Careless Use Builds Up Resistant Germs."[106] Mid-1950s hospital outbreaks involving resistant pathogens like *Staphylococcus* 80/81 added urgency to "rational" antibiotic use campaigns.[107] In 1958, Surgeon General Leroy Burney categorized antibiotic resistant staphylococci as a "problem of national significance."[108] Described by historian Scott Podolsky, US infectious disease experts like Maxwell Finland from Boston City Hospital used the opportunity to attack sales of popular but inefficacious fixed dose combinations of multiple antibiotics as well as American physicians' accommodativeness to pharmaceutical marketing.[109] In 1959, Finland warned: "physicians who are overconfident of germ-killing wonder drugs are living in a fool's paradise where their patients may die."[110] While American medical practitioners ultimately escaped calls for greater statutory oversight, AMR concerns were one of the reasons behind the US government's decision to move against inefficacious products with the 1962 Kefauver-Harris Amendment (chapter 4).[111]

Curiously, widespread concern about AMR selection in medical settings did not translate into alarm about similar processes on the farm or in the environment. Throughout the 1950s and 1960s, most American commentators treated AMR in the hospital as distinct from AMR on the farm. With the exception of a 1952 reader's letter to the *Washington Post*,[112] none of the analyzed US media sources addressed the fact that resistant organisms could emerge in animals and spread to humans or spread from humans to animals.[113] The closest an article came to addressing agricultural AMR selection was in the *Madera Daily Tribune* in 1962. Titled "Mass Murders Have Changed Bacteria,"[114] the article cited research by bacteriologist George Eastman according to which large-scale antibiotic use had allowed resistant salmonella to "overgrow"[115] sensitive species but failed to mention instances of antibiotic use beyond human medicine. The epistemic divide between human and nonhuman antibiotic use and microbiota also characterized whistle-blowing bestsellers. Although William Longgood and Lewis Herber warned about AMR selection in humans as a result of residues in food and milk, they did not connect their residue-oriented criticism of agricultural antibiotics with AMR selection in the environment or with contemporary campaigns for "rational" antibiotic use.[116]

It was only in 1966 that the curious epistemic divide of reporting on medical and agricultural AMR began to be challenged. Following reports by *Scientific American*,[117] the prestigious *New England Journal of Medicine* (*NEJM*) warned about the dangers of horizontal AMR proliferation on August 4, 1966. Instead of merely passing on resistance to subsequent generations, bacteria could exchange AMR blueprints—R factors—horizontally across species borders (chapter 1). Horizontal gene transfer meant that locally emerging resistance could spread rapidly throughout the regional and global environment. Ecological proliferation scenarios also implied that AMR selection on farms could be just as dangerous as AMR selection in hospitals. Referring to R-factor transfer as "infectious drug resistance," *NEJM*'s editorial blamed the "precipitous rise in frequency of R factors" on the "increasing use of antibiotics not only in clinical practice but also in the care and feeding of livestock."[118] Relying strongly on British research (chapter 5) and invoking what would become a standard apocalyptic genre of AMR warnings, *NEJM* specifically accused low-dosed antibiotic growth promoters (AGPs) of "providing a constant selection pressure on R factors that can readily be transferred to man"—"unless drastic measures are taken very soon, physicians may find themselves back in the preantibiotic Middle Ages."[119]

Early American media reactions to the *NEJM* editorial were surprisingly muted. Eight days after the editorial was published, the *New York Times* intuitively compared the potential implications of horizontal AMR warnings to the impact of Carson's *Silent Spring*: "the available evidence suggests that the development of such hardy microbes is greatly facilitated by the widespread feeding of antibiotics . . . . Put bluntly, people may be paying for cheaper and better meat by suffering more and graver infectious diseases."[120] However, this harsh attack on antibiotic feeds remained exceptional. Most other US newspapers initially greeted *NEJM*'s warnings with silence. The silence was in part due to the pending release of an FDA report on veterinary medical and nonmedical uses of antibiotics. Published three weeks after the *NEJM* editorial, the report acknowledged that antibiotics were being misused but limited AMR warnings to the presence of residues in food. Calling for an end to antibiotic food preservation, stricter punishment of residue offenders, and more research on antibiotics' ecological effects, the FDA report effectively refocused public attention on residue rather than AMR hazards (chapter 4).[121]

Most media outlets followed the FDA by either downplaying or linking concerns about agricultural AMR selection to already familiar residue-focused risk scenarios. *Time* magazine described the "contagious cuddling" between bacteria but relativized *NEJM*'s warnings of pre-antibiotic Middle Ages: some experts were "calmly argu[ing] that laboratories are producing new antibiotics too fast for germs to catch up."[122] Drawing an analogy to Upton Sinclair's bestseller, the *Washington Post* criticized the FDA for having allowed an "antibiotic jungle"[123] to spread with regards to food preservation but failed to focus on AMR. Further articles in the *New York Times* also rehashed analogies between antibiotics, *Silent Spring*, and residue scandals.[124] Equating the absence of residues with a reduction of AMR, a reader of the *Washington Post* encouraged farmers to capitalize on consumer insecurities: "There are quite a few of us who go out of our way to buy such pure foods . . . —at a price."[125]

The 1960s were thus a confusing time for the American public: their sense of risk heightened by the Kefauver hearings, *Silent Spring*, and residue scandals, US citizens remained exposed to an unattenuated stream of optimistic reports about agricultural plenty and the necessity of curbing threats like overpopulation-induced hunger and communism. Newspaper subscribers could read about chemical dangers in one issue only to encounter praise for "push-button farming"[126] and "coddled swine" getting "plenty of food, shots, pills [and] antibiotics"[127] in the next. In terms of agricultural antibiotics, pervasive fears of invisible poisoning made consumer activists and the media prioritize risk scenarios focusing on residues and contaminated food. Although reports on allergic reactions gradually linked debates about penicillin exposure in medical and agricultural settings, more abstract fears of AMR remained focused on medicine. This epistemic fragmentation between AMR selection in human and nonhuman settings would have a lasting influence on the US antibiotic risk episteme. Commentators only gradually connected the two spheres of agricultural and medical AMR selection following *NEJM*'s popularization of British research on horizontal resistance proliferation. However, as reactions to the 1966 FDA report show, the dominance of residue-focused risk scenarios subordinated concerns about agricultural AMR selection to fears of invisible antibiotic residues. As a consequence, public opposition to agricultural antibiotic use remained fragmented and uncoordinated. Frustrated by US consumers' seeming complacency about food-related dangers in 1964, *Washington Post* journalist Sue Cronk noted: "the biggest worry the American housewife has when she shops for meat is likely to be how much it will cost—not whether it will be safe for her family to eat."[128]

3

Chemical Cornucopia

Antibiotics on the Farm

This chapter reconstructs antibiotics' adoption and ensuing conflicts in the US agricultural sphere. After 1945, antibiotics were rapidly integrated into all areas of US food production. This rapid introduction was facilitated by growing pressure to expand and intensify agricultural production and by a sophisticated sales network connecting pharmaceutical producers with farms. By the early 1960s, antibiotics had acquired infrastructural importance in many production sectors. Although US farmers shared public antibiotic optimism, they soon expressed concern about the negative effects of chemical-driven intensification on smaller producers. However, all-out rejection of agricultural chemicals remained limited to organic farmers. While conventional farmers attempted to curb antibiotic residues in sensitive products like milk, growing public concerns about chemicals instead provoked angry outbursts against "irrational" faddists. Similar to the US public sphere, AMR hazards were rarely discussed. Despite warnings by veterinary bacteriologists, most agricultural commentators did not discuss potential public health implications of AMR selection on farms.

The Origins of Agricultural Antibiotic Use

US agriculture's rapid adoption of antibiotics was no coincidence. Beginning in the interwar period, the Taylorian logic of Henry Ford's factories gradually spread throughout the countryside. A new generation of agricultural experts, officials, and producers wanted to apply the principles of quantification and mechanization to US farms. Already farming larger acreages and producing more animals than Europeans, US farmers further expanded and also began to rationalize production. Interwar farms employed accounting techniques alongside new technologies like tractors, hybrid seeds, and pesticides.[1] Farmers also invested in more intensive ways of producing livestock. Conceptualizing animals as machine-like feed converters, farmers began to purchase premixed fortified rations from specialist producers like the Commercial Solvents Corporation or Pfizer. While farmers' motivation was to produce more meat with less feed, manufacturers saw farms as lucrative outlets for industrial surplus and by-products. Connecting farmers with pharmaceutical manufacturers was an increasingly sophisticated network of veterinarians, local feed-mixers, and government extension officials.[2]

Rising output soon exceeded demand. Attempting to maintain incomes despite sinking commodity prices, US farmers increased their production by 13 percent between 1917 and 1929. Unsurprisingly, prices continued to sink. By the end of the 1920s, it cost more to produce many commodities than farmers earned from selling them. Unable to service their debts, many farmers suffered bankruptcy, and the US farm population declined from 32.5 million in prewar years to 30 million in 1930. Only very efficient or large farming operations remained profitable due to lower production costs. When commodity prices declined by another 37 percent during the Great Depression, even the most efficient producers struggled to survive.[3]

Reacting to farmers' plight, the Roosevelt administration launched a comprehensive program of agricultural aid. Passed in May 1933, the Agricultural Adjustment Act (AAA) was designed to reduce surpluses, stabilize prices, and enhance farmers' purchasing power. The AAA allowed the United States Department of Agriculture (USDA) to administer adjustment payments to farmers, who in turn agreed to reduce production of surplus commodities. Together with compensated slaughter programs and other initiatives, the AAA was supposed to restore the relative purchasing power—parity—of agricultural goods to prewar levels. However, in attempting to alleviate the Great Depression's impact, New Deal measures increased agricultural intensification pressure and subsidy dependence: by 1941, one-third of US gross farm income was derived from direct or indirect federal payments.[4] Paying producers to slaughter animals and take land out of production also incentivized them to produce more with remaining assets—thereby putting larger farmers at an advantage. As a consequence, the farms that survived the Great Depression were culturally and economically geared to strive for factory-like efficiency, scale, and technological sophistication.[5]

When commodity prices recovered, production boomed. Reacting to America's entry into the Second World War, Congress passed the 1942 Emergency

Price Control Act and the Steagall Amendment. Legislators guaranteed commodity prices at around full parity for the duration of hostilities and for two years afterward to incentivize farmers to maximize production and invest in productivity increases. Ensuing transformations were particularly dramatic in the livestock sector: whereas New Dealers had ordered the compensated slaughter of about 6 million hogs in 1934, the new price guarantees encouraged farm investment and a significant rise in production.[6]

Parallel war-induced grain, protein, and labor shortages, however, soon threatened rising outputs. US researchers were hastily commissioned to find solutions. Described by historian Mark Finlay, one of these researchers was Damon Catron. At Purdue University's Work Simplification Laboratory, Catron launched a systematic attempt to overcome shortages in the pig sector with efficiency increases. Farrowed in spring, animals were fattened on pastures during summer and autumn, and mass slaughtered ahead of winter. The resulting pork glut often overwhelmed processing facilities and depressed prices. By contrast, Catron's vision for production resembled an integrated car assembly plant that divided a pig's life into distinct stages: breeding, farrowing, weaning, rebreeding, and finishing. Removed from pastures, animals were to be "assembled" all-year-round in optimized indoor environments and fed tailored rations prior to their final disassembly in an abattoir.[7]

This Fordist vision of animal production faced significant challenges. While intensification was already underway in poultry farming, the overwhelming majority of US pigs were still held in outdoor or mixed indoor-outdoor systems. Most US cattle were kept on pastures.[8] Convincing these producers to abandon their low-cost systems and invest in confined intensive production proved challenging. Disease posed another obstacle. On both sides of the Atlantic, previous attempts to increase animal densities in more confined systems had been stunted by a corresponding growth of disease pressure: infections had wiped out herds or diminished productivity.[9] It was here that the advent of antibiotics had a significant impact.

During the late 1930s, producers had already used organoarsenics, inorganic sulfur compounds, sulfonamides, and biological antibiotics like gramicidin to treat individual animals (e.g., mastitis in cows) or larger groups of animals. However, drug prices were high, and incorrect dosages could poison animals. This situation began to change during the 1940s. The American poultry industry was a leader of this change. In 1939, Cornell veterinarian P. Philip Levine reported that the only recently marketed sulfanilamide had shown efficacy against coccidiosis—a protozoal infection of the intestine. Researchers soon reported the successful use of other sulfonamides like sulfamethazine, sulfadiazine, and sulfaguanidine to both treat and prevent coccidiosis in poultry and foulbrood in bees. Developed with the support of pharmaceutical companies, safe sulfonamide ratios in water and feed soon enabled the mass medication of entire flocks and were also used against other diseases like fowl typhoid (*Salmonella gallinarum*) and Pullorum disease (*Salmonella pullorum*).[10] What had once been devastating herd and flock diseases seemed increasingly controllable.

With the end of the war easing military demand, veterinary antibiotic treatments became cheaper and more widely available. Described by Susan Jones, ads for Lederle's sulfathiazole began to appear in poultry magazines by 1946. In the same year, Merck patented sulfaquinoxaline. Originally developed as an antimalarial but licensed against coccidiosis in 1948, sulfaquinoxaline became the first antibiotic product officially approved for inclusion in animal feeds and proved extremely lucrative. Poultry farmers could now medicate entire flocks with only minimal (and often without) veterinary supervision.[11]

While sulfonamides paved the way for routine therapeutic antibiotic use, it was the nonspecific application of antibiotics that would transform agropharmaceutical relations. There was already a lucrative market for animal feed supplements in the United States (chapter 2).[12] However, researchers continued to puzzle over why certain feeds supported animal growth and others did not. The animal protein factor (APF—also known as anti–pernicious anemia factor) was at the center of this puzzle. Nutritionists had long known that feeds enriched with fishmeal, cow and chicken manure, or fermentation wastes were more effective at promoting growth than feeds containing cheap vegetable protein alone. Meanwhile, expensive liver extracts and cod oil had been found to be both effective against pernicious anemia in humans and to promote animal growth.[13] With animal protein imports from Japan and Norway interrupted during the Second World War, researchers redoubled efforts to isolate and find alternative APF sources.[14]

The APF hunt led to unexpected results: in 1946, a University of Wisconsin team including Peter Moore and the colorful Thomas Donnell (later Sir Samurai) Luckey reported that a combination of sulfasuxidine and streptomycin increased the growth of chicks when fed alongside folic acid.[15] The Wisconsin researchers had been studying the role of B-vitamins for animal growth by using antibiotics to "knock out" parts of the digestive system. They had expected streptomycin to sterilize the gut and create a growth-retarding vitamin deficiency. To their surprise, the opposite had happened. However, the Wisconsin team and other academic groups researching the feeding of waste products like mycelium from penicillin production failed to realize the practical and commercial implications of their observations.[16]

This changed three years later. In 1948, research by the University of Maryland and Merck as well as by Glaxo in the United Kingdom led to the isolation of vitamin B_{12} and its identification as the mysterious APF. In a lucrative spinoff, Merck not only discovered that *Streptomyces griseus* (the organism producing streptomycin) produced B_{12} but that industrial fermentation wastes accruing after streptomycin extraction still contained large amounts of the vitamin. Far cheaper than other animal protein sources, fermentation wastes of biological antibiotics could be fed directly to animals as APF growth promoters. There was a particularly large market for B_{12}/APF supplements in the Midwest where animal husbandry and the production of cheap vegetable protein in the form of hybrid corn and soybeans were expanding rapidly.[17]

APF feeds also excited other companies. Working at Lederle Laboratories' Pearl River station, Thomas Jukes and Robert Stokstad investigated whether vitamin B_{12} produced by Lederle's *Streptomyces aureofaciens* (the organism producing chlortetracycline) could be commercialized too. Both Jukes and Stokstad were experienced B-vitamin researchers: trained as a biochemist, the group's leader, Jukes, had previously studied relationships between B-complex vitamins and their effects on deficiency diseases and animal growth. During the 1930s, he had identified pantothenic acid and choline as growth factors in chickens and turkeys. Stokstad was a trained animal nutritionist whose research would lead to the isolation of vitamin K and who had previously isolated and purified folic acid.[18] Over Christmas 1948, Jukes and Stokstad fed chicks a deficiency diet consisting of 75 percent soybean meal and supplemented it with sterilized *S. aureofaciens*. To their surprise, they found that chicks eating *S. aureofaciens* feeds grew quicker than those eating feeds only supplemented with purified vitamin B_{12}. Antibiotics seemed to be "promoting" animals' growth.

In contrast to the Wisconsin team, their previous work and commercial setting made Jukes and Stokstad keenly aware of their findings' implications. To their dismay, they were, however, denied access to further *S. aureofaciens*, which was needed for Lederle's aureomycin production. For their feed experiments, they turned to alternative aureomycin sources like acetone cake—dried acetone solvent left over from purifying aureomycin fermentation liquid—and at one point dug out fermentation wastes from the Lederle dump. Befriended researchers who received unmarked feed samples soon began to confirm "growth promotion" in other species and also reported combating bloody diarrhea (scours) in pigs.[19] The reliability of these early studies is debatable. Although this was not uncommon during the period, AGP trials were conducted on a small number of animals for short periods of time and did not produce sufficient data to determine statistical significance. Results were also not reproducible in other countries and later studies (chapter 6).[20] Most contemporaries, however, trusted Lederle's claims.

Triggering a commercial storm, Jukes and Stokstad publicly announced the "antibiotic growth effect" in 1950. Despite the earlier Wisconsin report, it was the serendipitous combination of their observation, commercial focus, and access to fermentation wastes that gave birth to a new era of antibiotic mass consumption. The effects of this new era reverberated around the globe. While it is important to note that US animal production would have continued to intensify without them, antibiotics' alleged ability to boost metabolisms, control disease, and reduce labor previously devoted to caring for individual animals made many agricultural commentators soon consider them indispensable.[21]

Antibiotic Infrastructure

Economically, antibiotics' mass introduction to US agriculture could not have come at a better time. Between 1940 and 1945, American farmers' average per capita net income had increased from $706 to $2,063.[22] Feeding the United States and large parts of postwar Europe and encouraged by the Korean War's promise of stable prices, farmers paid off debts and invested in new technologies.[23] Expensive animal protein supplements made them particularly well disposed toward new APF sources that would help turn cheap vegetable protein into lucrative animal products.[24]

Manufacturers were happy to oblige and launched advertising campaigns for "APF growth factors," "AGP miracle additives," and therapeutic applications in farming magazines.[25] Using what Rima Apple has called "reason why" and "negative appeal"[26] strategies to lure or scare farmers, early AGP commercials were nearly identical to previous APF and sulfonamide commercials. However, it was often unclear what farmers were actually buying. Discussed in more detail in chapter 4, Lederle's decision to avoid FDA licensing by marketing AGPs as APF with an additional vaguely specified growth factor meant that drug concentrations varied. Without conducting assays for either B_{12} or aureomycin, Lederle sold "tankcars of brine containing residues from [aureomycin] fermentation"[27] to feed merchants, who then repackaged the wastes. This procedure caused significant confusion. The history of the first AGP advertisement in the popular Iowan magazine *Wallaces Farmer* is telling. Featuring a proud farmer holding a feisty piglet, a June 1950 Gooch Feeds advertisement reported "amazing results" achieved with the new "Aureomycin APF" "wonder-worker."[28] However, it soon emerged that Gooch Feeds' "Genuine Lederle Aureomycin APF"[29] was not always effective. Two weeks after printing the ad, *Wallaces Farmer* warned readers: "crystalline aureomycin is not available at the present time to either the feed industry or the farmer."[30] According to a competing merchant: "no statement should be made ... concerning the presence of the antibiotic since it is naturally inherent in the ingredient."[31]

The story of Gooch's advertisement is indicative of the gold rush atmosphere surrounding AGPs. In 1985, co-discoverer Thomas Jukes remembered: "The demand was such that the available supply was prorated among customers. On one occasion, we had to deal with a complaint from Senator Wherry of Nebraska that the supplies of APF were all going to Iowa rather than Nebraska.... In Austin, Minnesota, a local pharmacist purchased Lederle APF in bulk, repackaged it, and sold it at an inflated price. Allegedly, he made so much money

GOOCH'S BEST
PIG AND SOW MEAL
fortified with
AUREOMYCIN APF

"GIVES MY PIGS A HEAD START"

Always A Leader

GOOCH FEEDS were first to bring Iowa and Nebraska hog raisers advantages of Aureomycin APF. This "Wonder-Worker" with pigs is now considered the most important advancement in Swine Nutrition in the last 25 years.

Leading farm papers have already brought you the amazing results reported by State Colleges and other experimental institutions.

Act now to keep your pigs doing their best right through the critical weaning period. Be sure you get GOOCH'S BEST Pig & Sow Meal properly fortified with

Genuine LEDERLE AUREOMYCIN APF.

These advantages are also yours in GOOCH'S BEST
SARDINE FISH SOLUBLES . . . NATURAL APF
RIBOFLAVIN-NIACIN-PANTOTHENIC ACID

The "B"-Complex Vitamins

See Your Dependable Gooch Dealer For

GOOCH'S BEST
PIG AND SOW MEAL

FIGURE 3.1 Gooch's APF/AGP Feed, *Wallaces Farmer*, 1950.

that he retired and went to live in Florida."[32] Other companies were "right on [Lederle's] heels,"[33] and aureomycin was soon joined by a bewildering array of competing AGPs. For a while, it seemed as though farmers would buy any feed as long as it contained preferably large doses of antibiotics. While companies like Ful-O-Pep or Kraft advertised their own antibiotic supplements,[34] Allied Mills promised that its AGP would turn a "scrawny runt" into a "husky hog" in "just 81 days."[35] For farmers unwilling to trust only one antibiotic, a company called Occident advertised multimycin, an unspecified "combination of miracle antibiotics" offering "up to 18% greater gains than with *single* antibiotic feeds."[36] Trying to ward off competition, Lederle soon claimed that aureomycin was "the *only* antibiotic that has been proved highly effective for swine, poultry, calves and several kinds of small animals".[37]

US farmers trusted these claims. Although later surveys indicated that producers' reasons for purchasing them were varied (chapter 9), antibiotic supplements had become standard ingredients of broiler and turkey mashes by mid-1951.[38] Calves and pigs also received large amounts of low-dosed antibiotics. One year after the antibiotic growth effect was announced, 110,000 kilograms of antibiotics—about 16 percent of total US antibiotic sales—were already being used for unspecified nontherapeutic purposes.[39] However, chaos over what constituted effective growth promotion and what might be therapeutically relevant persisted even after the FDA introduced AGP dosage requirements in 1951. Initially, officials recommended antibiotic dosages of up to 50 grams per ton of finished feed. However, industry-sponsored trials soon made some farmers disregard guidelines in favor of higher feed dosages, which could also be used to prevent and treat disease. Manufacturers also experimented with higher-dosed penicillin, bacitracin, and chlortetracycline implants. On farms, the boundaries between growth promotion and treatment were fast blurring. By the early 1960s, experts resignedly noted that "legally precise [dosage] boundaries are easier to establish but not always easier to maintain than biologically precise boundaries."[40]

Reacting to exaggerated marketing claims and widespread confusion, agricultural advisors rushed to provide farmers with expertise. In popular farming magazines, articles promoted "rational" and cost-effective antibiotic use. Most experts agreed that antibiotics would reveal their true potential only on hygienic and modern farms. Soon familiar messages included: farmers should "follow-through"[41] with treatments; antibiotics were no substitute for good management;[42] using drugs to maintain outmoded husbandry systems would not pay off—"drugs can't whip old lots."[43]

A similar promotion of "rational" antibiotic use took place in contemporary agricultural manuals for both therapeutic and nontherapeutic purposes. As late as 1944, US pig and cattle manuals had stressed preventive health care but recommended little in the way of effective DIY therapeutics against microbial infection.[44] However, by 1947, newer publications like the *Eastern States Farmers Handbook* featured therapeutic sulfonamide use.[45] Following the announcement of the antibiotic growth effect, manuals soon reclassified antibiotics as a routine part of animal nutrition. In 1951, Interstate's *Livestock Feeding Manual* contained guidance and exercises for pig farmers to calculate the right amount of "vitamin B_{12} or some antibiotic"[46] for feeds. In the same year, the *Midwest Farm Handbook* contained an entire section devoted to B_{12} and AGPs. Penicillin, streptomycin, bacitracin, aureomycin, and terramycin were praised for boosting growth and controlling diseases like scours. Reflecting the drugs' popularity, readers were explicitly cautioned that "a deficiency symptom does not develop by omitting antibiotics."[47] The next few years saw a further proliferation of handbooks giving advice on when, how much, and how long to administer antibiotics.[48] In 1952, *Hog Profits for Farmers* listed "antibiotic and B_{12} supplements"[49] among its seven essentials of a complete pig ration. The drugs would also act as an "insurance"[50] against runts and many other problems. Government advisors joined the chorus with USDA and Farm Credit Administration publications praising AGPs and feeding antibiotics alongside cheap cottonseed rations.[51]

Although the boundaries between antibiotic growth promotion and treatment were always fluid, falling prices also triggered a boom of explicitly therapeutic applications. In pig husbandry, the postwar period saw antibiotics being used against atrophic rhinitis, suspected paratyphoid, edema, and respiratory diseases.[52] In feedlots, beef cattle were given penicillin against bacterial footrot. Antibiotics were also used to aid artificial insemination and against infections of cows' udders (bovine mastitis).[53] Caused by several bacteria species and spread via hands and inadequately sterilized milking equipment, bovine mastitis is painful, occasionally fatal, reduces productivity, and can taint the flavor of milk. Drinking tainted milk can also cause septic sore throat and food poisoning in humans.[54] In 1955, *Farmers Weekly Review* from Illinois estimated that preventable US mastitis losses amounted to $225 million.[55] Given its bacterial causes, antibiotic mastitis treatments proved popular. In the 1940s, articles advised US farmers to "lick mastitis"[56] with sulfonamide, penicillin-sulfonamide, or streptomycin infusions. In 1949, American Cyanamid began marketing collapsible broadspectrum antibiotic tubes for intermammary mastitis control. "Ready-to-use-one-treatment tube[s]" could be purchased without a veterinary prescription and were soon regularly advertised in US farming magazines.[57] By 1953, a USDA survey reported that American farmers had access to a wide variety of antibiotic treatments containing sulfonamides, nitrofurazone, tyrothricin, penicillin, streptomycin, aureomycin, terramycin, neomycin, bacitracin, polymyxin, and chloromycetin: "combinations of the various antibiotics and sulfonamides are being widely used."[58]

THEY'RE "FENCING OUT"
YOUR INVISIBLE ENEMIES

Outside the fence in the picture above—blown up 1,000 times so you can see them—are some of the most vicious killers of pigs and calves. They are one of the many kinds of bacteria which cause scours.

These germs, and ones that cause a host of other common animal and poultry diseases, are so small that millions of them can be carried into the feedlot on a puff of air or bit of litter stuck to a shoe sole. Since you can't very well fence them off your farm, veterinary scientists have sought ways to build barriers against disease *within the animals themselves.*

Why not, they asked, put disease-fighting amounts of antibiotic right in the feed so as to provide a fast, low-cost way to protect the whole herd or flock?

The feed industry saw a new chance to be of service. Their veterinarians and poultry pathologists, working with others at college experiment stations and the Pfizer Research Center, tested the most effective antibiotics . . . the most effective levels of use . . . the best times to introduce these special feeds into the feeding program.

The result: new disease-fighting feeds fortified with high levels of Terramycin are helping veterinarians and farmers to feed away scours and shipping fever—keep blue comb out of their flocks—maintain weight gains and high egg production in the presence of chronic respiratory disease and other mixed infections.

Other special medicated feeds are effective against coccidiosis and parasites that once robbed farmers of a big part of their profits. And these feeds, like your regular feeds, also have the scientifically balanced nutrients your birds and animals need *at all times* for the highest level of health and resistance.

What next from Pfizer and your feed man? They're working on still more ways to *bring science to the farm in a feed bag*—new ways to take still more of the risk out of livestock and poultry production—and put more profit in.

Science Comes to the Farm in a Feed Bag

Pfizer Agricultural Research and Development Department

FIGURE 3.2 Boundaries between AGPs and therapeutics soon blurred. Pfizer advertisement, *Farm Journal and Country Gentleman,* 1956.

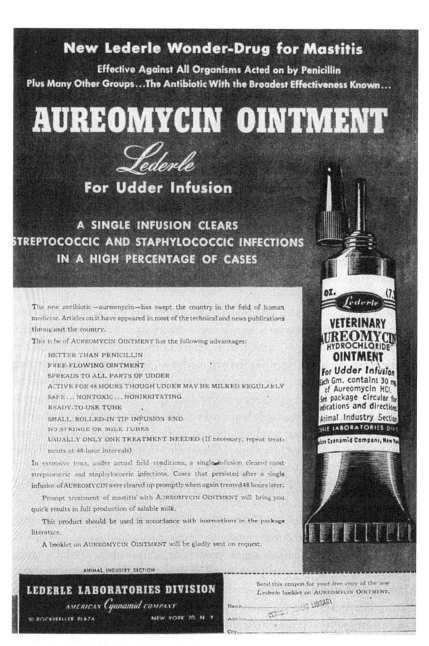

FIGURE 3.3 Cyanamid advertisement, *Wallaces Farmer*, 1949.

Antibiotic mastitis treatments proved so popular that dairies and creameries were complaining about drug residues in raw milk as early as 1948.[59] According to *Wallaces Farmer*, "antibiotics not only kill mastitis germs, but also kill bacteria which ferment milk."[60] In April 1951, the "cheese state" Wisconsin ruled that mastitis ointments carry labels on drug withdrawal times. In addition to public concerns about allergies and invisible poisoning (chapter 2), agricultural commentators warned that excessive antibiotic use would select for resistant pathogens in udders and that residues might select for AMR when ingested by humans.[61] However, such warnings could not dampen antibiotic enthusiasm. By 1956, US farmers annually used 75 tons of antibiotics against mastitis.[62] Faced with widespread residue problems, Ohio State University researchers began to experiment with antibiotic-resistant lactic acid starter cultures to produce cheese from residue-laden milk.[63] Meanwhile, Canadian studies revealed that *Staphylococci* and *Streptococci* isolated from cheese were becoming increasingly resistant against penicillin and dihydro-streptomycin—both popular mastitis treatments.[64]

Interestingly, US veterinarians had little control over the rapid proliferation of agricultural antibiotic use. Veterinary researchers' success in developing effective antibiotic treatments "unwittingly created difficulties for their brethren practicing in the field."[65] Unlike human medicine, where antibiotics were prescription only drugs from 1951 onward, labeled drugs like penicillin could be sold to farmers without a prescription (chapter 4). As a consequence, pharmaceutical companies increasingly bypassed veterinarians by selling and advertising easy-to-use antibiotic products directly to farmers. This strategy was not uncontroversial. Thomas Jukes later remembered that the decision to sell pure aureomycin to farmers was "strongly opposed by the veterinarians at Lederle"[66] but was supported by Lederle president Wilbur G. Malcolm. Subsequent attempts by the American Veterinary Medical Association (AVMA) to convince officials to restrict antibiotic access via prescription requirements similarly failed.[67] As a consequence, postwar veterinarians soon found themselves competing for farmers' custom against pharmaceutical salesmen, feed merchants, and a new group of experts specializing in mass animal health management. Many veterinarians reacted by leaving the livestock sector. Others expanded their traditional purview to include preventive services against subclinical diseases and for animal productivity.[68]

While US veterinarians had reason to feel ambivalent, agricultural scientists, farming magazine commentators, and American officials all endorsed broadening antibiotic access and use. In 1951, Damon Catron conceded: "we don't know why antibiotics do what the experiments indicate. But we do know that they prevent scours, increase rate of gains and reduce feed requirements."[69] There was also pride in antibiotics as part of a new American-led push for agricultural efficiency and nutritional plenty. According to *Wallaces Farmer*, AGPs and new production systems enabled poultry farmers to devote only "ten seconds per bird per day" and raise "flock profits by 110 per cent."[70] In Illinois, *Farmers Weekly Review* reported on awestruck British visitors' reactions to Pfizer's "Miracle Drug Pigs,"[71] new breeds of "antibiotic-age chicks,"[72] and trials of antibiotic-doused earth to boost crop production.[73] In Pennsylvania, *Lancaster Farming* informed readers on antibiotic tree sprays and using antibiotic-doused bees to combat fire blight.[74] Major companies like Merck, Cyanamid, and Pfizer did their best to promote this enthusiasm by regularly publishing glossy brochures, free manuals, and annotated bibliographies on antibiotics' many benefits.[75]

Antibiotics' popularity is best expressed in numbers: around 1955, the USDA estimated that 50 percent of American formula feeds for poultry, hogs, and cattle contained antibiotics and vitamin B$_{12}$. Ninety-three percent of poultry feed manufacturers, 60 percent of hog feed manufacturers, 22 percent of dairy feed (mostly calf feed) manufacturers, and 4 percent of beef cattle feed manufacturers added antibiotics to their products. Drug concentrations ranged from an average of 16.5 grams and 27 grams (mostly broadspectrum antibiotics) per ton in pig and calf feeds to 2.7 to 3.5 grams (mostly penicillin) per ton in poultry feeds.[76] Having focused on large businesses, surveyors believed that antibiotic use was even more common at the local small business level. Between 1951 and 1960, US sales of nontherapeutic antibiotic applications grew seven-fold from about 110,000 to 770,000 tons. Although total US antibiotic production grew significantly, the proportion of total production sold for nontherapeutic purposes expanded from 16 percent to 36 percent.[77] Resulting profits encouraged manufacturers to professionalize with many larger feed producers investing in new research facilities.[78] By 1962, it was estimated that an astounding 99 percent of US poultry, 90 percent of pigs, and 30 percent of beef cattle were receiving antibiotic-supplemented feeds.[79]

Entering nearly all areas of animal production and also entering food preservation and plant protection, antibiotics had rapidly achieved infrastructural relevance in US food production. The new antibiotic age was nearly universally welcomed. In the eyes of farmers and large parts of the US agricultural establishment, antibiotics had seemingly overcome the bacterial limitations previously imposed on the size of production facilities and the productivity of their inhabitants. Hardly anybody worried that the new antibiotic era might come at a price.

The Costs of Plenty

In 1951, US meat packer Swift sponsored large ads calling on farmers to throw aside fears of overproduction and produce as much meat as possible. The new ABCs of animal nutrition—*A* standing for antibiotics—would guarantee rising production and profits.[80] According to Swift: "The problem's never surplus meat—you can't raise more than we can eat."[81] This trust in American appetites proved misguided.

Following the Korean War, agricultural commodity prices began to sink and the Eisenhower administration became concerned about overproduction and expensive subsidies. Between 1953 and 1954 alone, the US government's Commodity Credit Cooperation (CCC) was forced to purchase $1.5 billion of agricultural surpluses. However, CCC purchases were not enough to shield farmers from a return of the interwar cost-price squeeze. Forced to maintain federal subsidies, the Eisenhower Administration attempted to dispose of surpluses abroad with the 1954 Food for Peace program—an opportune side effect of Cold War diplomacy.[82] The 1956 Agricultural Act recycled the New Deal idea of paying farmers to reduce production. However, agricultural production continued to grow by an annual average of 2.1 percent throughout the 1950s.[83] Concerned about annual CCC expenditure of $4 billion and daily storage costs of about $1 million,[84] the new Kennedy administration established the US food stamp program and expanded the Food for Peace and lunch and milk programs in schools. Kennedy also reduced US acreage *and* the quantity of marketed produce.[85]

The re-emerging cost-price squeeze hit the rural community hard. As in the Great Depression, small American farmers were worst affected. With polls showing that farmers themselves were upsizing definitions of a "family farm,"[86] the number of US farms decreased from 3,710,503 in 1959 to 2,730,250 in 1969 while the average farm size increased from 302.8 to 389.5 acres.[87] In the livestock sector, developments in the poultry production were indicative of things to come: since the early 1960s, large national vertically integrated firms controlled as much as 90 percent of now mostly confined US broiler production.[88] US pig production also began to change. Although confinement remained far from ubiquitous, a growing number of hog producers experimented with new housing systems to save labor and reduce animal movement.[89] In the beef sector, the 1960s saw an increasing number of cattle concentrated and fattened on large feedlots prior to slaughtering. This development had already gathered steam during the 1950s. As a result of cheap cereals and the fencing off of range land, a growing number of cattle were fed grain diets and new additives like diethylstilbestrol (DES) to fatten them quicker. By February 1955, 6 million US cattle were "on feed." During the 1960s, capital injections and conflicts over ranching on public lands facilitated the spread of commercial feedlots from California to Colorado, the Texas panhandle, Kansas, and Nebraska. By 1970, nearly two-thirds of the US calf crop was placed in feedlots prior to slaughter.[90]

Interestingly, most agricultural commentators did not blame new production systems for the ensuing cost-price squeeze. Despite bemoaning family farms' decline,[91] the "factory farm" remained a utopia rather than a dystopia.

In an age of superpower rivalry and overpopulation, technology-driven inten-sification was presented as essential for farmers' and the nation's long-term sur-vival.[92] This trajectory included further increases of antibiotic use. In 1956, *Farm Journal and Country Gentleman* wrote that farmer Hugh Fussell was getting everything right: "Detroit's automobile factories have nothing on Hugh Fussell. This Georgia farmer raises hogs on a truly assembly-line basis. Every two weeks Fussell is on the market with 50 to 60 head of No. 1 hogs."[93] Significantly, Fussell was also a "fanatic on disease control": every day, each of his finishing barn pens was disinfected; Fussell's pigs were vaccinated, their "feeds [were] well laced with vitamins and antibiotics."[94] With further "mighty new germ killer[s]"[95] on their way, who could blame farmers if—for their peace of mind—they invested in continuous antibiotic use to reduce feed costs and stay ahead of their competitors and infection?[96]

Only rarely was antibiotic use described not as a solution but as a problem. While *Lancaster Farming* erroneously thought that new antibiotic food pre-servatives would slow the advance of refrigeration,[97] *Farm Journal* viewed them as a further step down the road to universal low-cost competition: "Acronize is doing it. The cheaper broiler areas can now sell anywhere. . . . It's now one big national market with broiler prices, like water, seeking one level."[98] The arti-cle's ambivalence is telling. Caught in a cost-price squeeze, the majority of US farmers, however, felt that they could no longer afford to stop and reconsider the antibiotic infrastructure growing around them. By the end of the 1950s, a path to dependency was emerging: falling prices led to greater herd densities associated with higher productivity and greater antibiotic use, which in turn led to a further fall of commodity prices. The price of this chemical cornuco-pia was agricultural insecurity. When public concerns about chemical expo-sure became more pronounced during the late 1950s, many farmers found themselves torn between shared health concerns and the perceived necessitude of further production increases and antibiotic consumption.

The 1956 scandal surrounding penicillin residues in milk was the first to shake public trust in agricultural antibiotic use (chapter 2). Initially, it did the same in agricultural circles. Attempting to restore trust, experts exhorted farm-ers to adhere to withdrawal times and identify bacterial strains prior to using antibiotics. Not only would cows recover sooner, farmers would also stop pay-ing for ineffective antibiotics and prevent stricter regulations.[99] *Farm Journal* warned that the FDA was merely asking "farmers to cooperate": "If that doesn't work, . . . they may either order that drug companies put dyes in mastitis treat-ments . . . or put a ban on penicillin."[100] Despite mentioning allergenic hazards, most commentators, however, described antibiotic residues as an isolated prob-lem. Events seemingly proved them right. Bolstered by sinking residue find-ings and blaming black sheep, dairy farmers averted statutory antibiotic restrictions.[101]

Public trust proved more difficult to win back. Despite agricultural campaigns to curb residue levels, US chemical criticism reached fever pitch in the wake of the 1959 cranberry scare and bestsellers like *The Poisons in Your Food*. Whereas agricultural commentators had formerly presented episodes like the 1956 penicillin scandal as a credible hazard to public health, the continuous expansion of public chemical fears was increasingly interpreted as a threat to modern agriculture. Commentators were especially apprehensive about public fears leading to substance bans. In 1959, *Wallaces Farmer* warned that a bigger "clamp down on all farm chemicals" was only a question of time: "a small army of FDA inspectors ... have orders from Washington to go from farm to farm, if necessary, to find [antibiotic] violators."[102] *Lancaster Farmer* cautioned producers to exercise chemical self-control: "In light of the tremendous publicity accorded to the recent cranberry situation, dairy industry leaders are very much concerned about the great damage which could be done to milk if FDA officials are forced to file lawsuits against dairies or producers in order to enforce rules."[103]

Over time, ongoing public clashes about the safety of agricultural chemicals led to a radicalization of agricultural rhetoric. In the US farming media, experts and editors increasingly resorted to painting black and white pictures of efficient, hard-pressed family farmers falling victim to an "irrational" anti-chemistry campaign. In 1959, *Wallaces Farmer* accused consumers and officials of stirring a "Big Ruckus" about the Cranberry Scare and publicizing the "incident entirely out of proportion to the dangers involved."[104] According to *Progressive Farmer* from Alabama, FDA officials were guilty of spreading "fear and disfavor for the entire production of an industry."[105] The magazine warned: "the nation is being harassed by a number of food cranks who insist that a food is good only if no chemicals were used in growing it" and "No nation in the world has a more abundant food supply, one that is cleaner, safer, or more nutritious than ours. . . . unless farmers look out, the 'food cranks' and other misinformed people may pressure Congress into passing unreasonable restrictions—restrictions that may do serious damage to our food supply and to national welfare."[106] Commenting on *The Poisons in Your Food*, *Lancaster Farming* compared agricultural chemical use to the plowing of "virgin prairie sod": "The use of hormonized and fortified feeds with antibiotics" was just as necessary if the "producer is to stay in meat production."[107] While other articles complained about official tests detecting "what is almost 'less than zero' amounts of residue,"[108] *Wallaces Farmer* proclaimed a "battle for farmers"[109] in 1962. A "worrisome new movement" no longer just included cranks but also ordinary people, "well-meaning for the most part, who have become overly alarmed at our growing use of chemicals in food production."[110]

Attacks on "cranks" and overzealous inspectors did not mean that the American agricultural community was naïve about potential side-effects of the chemical technologies it was employing. In farming magazines, there was usually a sharp contrast between the fierce rhetoric directed against "external" critics and the concerned tone adopted for "internal" safety advice. Reacting to scientific warnings, commentators promoted cost-benefit strategies to reduce personal health risks without foregoing chemicals' economic benefits. "Rational" farmers were expected to follow labeling instructions and limit direct exposure. If personal exposure remained below critical thresholds, agricultural chemical use was deemed safe.[111] However, even agricultural expert advice could be contradictory. In June 1960, an issue of *Progressive Farmer* contained two very different articles: whereas one commentator advocated using various chemicals to fight yard pests on page 76,[112] page 78 contained an article warning about "harmful residues"[113] of similar chemicals on homegrown fruits and vegetables. The effects of such mixed messaging on farmers' personal attitudes are difficult to judge. In 1964, *Wallaces Farmer* conducted a poll to see whether chemical warnings had changed farmers' habits. According to the poll, half of farmers regularly using pesticides and insecticides reported having taken more precautions because of hazards to crop, livestock, and personal health. One interviewee confessed, "These chemicals are beginning to scare me to death and I wouldn't be surprised if only experts will be allowed to apply them in the near future.'"[114] However, with overall pesticide, herbicide, and antibiotic use continuing to increase,[115] most producers seem to have viewed personal safety measures as sufficient to contain potential hazards and justify ongoing use.

The only agricultural community to wholeheartedly endorse growing public chemical criticism were organic producers. US organic producers were a small and heterogeneous community throughout the 1950s and 1960s. Within the community, publications by Jerome Irving (J. I.) Rodale (*Organic Farmer, Organic Farming and Gardening, Prevention*) enjoyed high visibility. Rodale, a former accountant and electrical equipment manufacturer, had purchased a farm in Emmaus, Pennsylvania, in 1940 and turned it into an organic experiment station. He also began publishing books and magazines on organic farming, personal health, and self-improvement in what would become a veritable publishing and direct-advertising empire. Inspired by English activists like Alfred Howard and Eve Balfour and relying on a mix of experimental evidence, anecdotal accounts, and pseudo-science (chapter 6),[116] Rodale was convinced that healthy food could only grow on "living" organic soil.[117] He thus promoted "natural" farming methods like composting, mulching, and crop rotation in opposition to "artificial" production methods involving chemical fertilizers, pesticides, or hormones, which he blamed for problems ranging from polio to cavities.[118] Although the wider ideals driving organic agriculture were more complex, *Organic Farmer*'s subtitle "farming without chemicals" was one of the budding movement's most important maxims.

Because of their "natural" roots, biological antibiotics presented a classificatory problem for post-war organic producers. Rodale and his readers initially interpreted antibiotics not as chemicals but as proof of the health-giving properties of "living" soil: "One variety of actinomycete is known to produce vitamin B-12.... But of far greater importance is their production of antibiotics.... there is a very strong antibiotic action in a soil loaded with organic matter, and thus many plant diseases are licked before they can even get near the plant."[119] "Natural" soil's antibiotic benefits could allegedly also be passed on to humans and animals. In 1952, an *Organic Farmer* article noted that organic crops enabled a Michigan family to ingest "the benefits of... health-giving antibiotics from stable manures put into the soil."[120] In 1954, retired chemist Leonard Wickenden reiterated some of these claims in his popular *Gardening with Nature*.[121] However, with antibiotic use growing in conventional agriculture, organic producers were soon forced to differentiate between beneficial and bad antibiotic exposure. For a while, this led to confusion. In 1952, *Organic Farmer* warned that "chicks fed antibiotics can poison consumers" by developing "resistant bacteria that will give food poisoning, enteritis and typhoidal infections to people eating their meat."[122] However, one year later, the magazine printed a feature on antibiotic soil's benefits for poultry: "Nature has placed, and is constantly manufacturing antibiotics—germ killers—in the soil to promote the health of poultry. All one has to do to reap the benefits of such germ killers is to make sure that the range is not contaminated."[123] According to the article, "natural" antibiotics had to be distinguished from "artificially" manufactured antibiotics: "It is wrong to jump off the deep end in believing only in the merits of synthetically produced antibiotics, for Nature has given us antibiotics for lo, these many years."[124]

Within the loosely organized organic community, a firmer front against agricultural antibiotics emerged only after the 1956 residue scandal. In addition to increasing attacks on "poisonous" and allergenic antibiotic residues,[125] organic criticism also began to regularly encompass concerns about agricultural AMR selection. According to a 1959 article, AGPs caused senility and sterility, upset breeding patterns, and destroyed natural disease resistance in animals. For humans, risks resulting from agricultural antibiotic use included the destruction of sensitive organisms by "highly resistant strains of dangerous bacteria, notably a very lethal staphylococcus."[126] While organic farmers were producing a growing supply of "safe" chemical free meat, the conventional farmer "will find himself facing economic ruin—if his own mal-nourished [sic] body hasn't given up the fight first."[127] Citing British research, *Organic Farmer* also began to warn about the on-farm selection of tetracycline-resistant *E. coli* from pigs and resistant mastitis in cows.[128] This early joint discussion of residue and AMR hazards was exceptional both within the US agricultural and public spheres.

In conventional agriculture, only a small group of mostly veterinary bacteriologists warned about AMR as a public health hazard. In November 1951, University of California researchers Mortimer P. Starr and Donald M. Reynolds published a remarkable study on streptomycin AGPs' effect on the microflora of turkeys. Feeding an oil meal ration containing 50 milligrams of streptomycin per kilogram of feed, the authors conducted sensitivity tests on *E. coli* isolated from bird's feces. Although resistant organisms were also isolated from control birds, *E. coli* from streptomycin-fed birds quickly proved "generally highly resistant to the drug."[129] The authors noted that AMR was probably of little relevance to poultry producers since it would be cheaper to cull than to treat flocks in the case of disease. However, they worried that antibiotics could foster the spread of resistant pathogens like salmonella. Thinking about public health more widely, the authors wondered "just how much indiscrimination should be permitted in the use of new chemotherapeutic agents."[130] AMR against penicillin and sulfonamides had proven "inevitable" even with the "best planned scheme of drug administration":

> It would be unfortunate if a large reservoir of drug-fast enteric pathogens potentially harmful to man accumulated unchecked in the poultry population. We hope that those charged with the protection of the public health will objectively evaluate the situation. . . . We grant that the poultry industry cannot readily forego [AGPs'] great economic advantages . . . . But . . . a few years of research are likely to elucidate the fundamental mechanism underlying this growth promoting effect and . . . will permit agriculturalists to secure more rapid animal growth without inflicting potential hazards on the public health.[131]

Corroborated by a 1953 Minnesota study on AGP-fed rats and Canadian research on AMR in oxytetracycline-fed pigs,[132] Starr and Reynolds' warnings, however, failed to inspire regulatory or agricultural action.

Instead, most commentators in conventional agriculture described AMR not as a general threat to public health but as an economic problem of inefficient overuse to be overcome not by bans but via "rational therapeutics"—a movement that was also gaining ground in human medicine.[133] In US veterinary circles, the 1950s saw a growing number of textbooks and magazines complain about "irrational" antibiotic use on farms caused by ignorance, lacking sanitation, and hasty disease diagnosis.[134] The 1957 textbook *Veterinary Pharmacology and Therapeutics* also complained about manufacturer's "dangerous tendency to add excessive levels of antibiotics"[135] to animal feeds. However, most veterinary observers did not believe that resulting agricultural AMR selection was a serious health threat. Despite containing a separate section on

AMR in 1961, the sixth edition of *Veterinary Bacteriology and Virology* maintained that AMR was "natural" and not an existential threat.[136]

Most non-veterinary agricultural observers also remained complacent about AMR.[137] Of thirty analyzed US farming manuals and reviews published between 1955 and 1966, all stressed antibiotics' benefits, many cautioned about inefficient drug use, but only two warned about potential health hazards resulting from AMR selection.[138] And even these two warnings stressed that ill effects would be contained by rationalizing antibiotic use. According to veterinary pharmacologist L. Meyer Jones, "no patient should be deprived of the benefit of antibiotic therapy solely because of fear of inducing resistance in the disease germ."[139] Warning about inadequately dosed mastitis treatments and storing unsterilized needles in antibiotic bottles,[140] a 1962 manual on *Milking Machines and Mastitis* similarly stressed that "rational" veterinary supervision would solve problems: "The only people "benefiting from [inefficient antibiotic use] are the sellers and advertisers of antibiotics."[141]

The focus on AMR as a problem not of health but of inefficiency was mirrored in contemporary debates about fluctuating AGP performance. Despite widespread endorsement in farming magazines and manuals, varying feed trials were leading some observers to speculate whether AMR was diminishing AGPs' efficacy. Industry reacted fiercely. In 1956, a Cyanamid booklet claimed that "reports that antibiotics are 'losing' their effect cannot be taken seriously."[142] According to Cyanamid, reductions in growth promotion were due to improved sanitation on farms—antibiotics simply had less bacteria to control. Three years later, the company repeated that "there is no evidence of a diminishing trend in production with time. On the contrary, the conversion of feedstuffs to pork has improved considerably throughout a decade of swine feeding."[143] Taking stock of the situation in 1962, *Farmers Weekly Review* acknowledged that "certain antibiotics used in a swine herd for several months may become less effective."[144] Ignoring the potential health implications of AMR selection, the article, however, noted that new antibiotics or antibiotic combinations still produced "a significant boost in rate of gain when fed to growing-finishing pigs."[145] In 1965, Iowa State University animal nutritionist Virgil Hays emphasized: "antibiotics are definitely of value in 98 percent of our farm situations."[146] The increasing use of higher-dosed AGPs was simply due to sinking antibiotic prices.

Sales figures show that routine antibiotic use remained popular on US farms. Between 1960 and 1970, sales of antibiotics added to animal feed and other applications grew by 330 percent from about 770,000 to 3,310,000 kilograms (about 43 percent of total US antibiotic sales).[147] With the exception of organic farmers, an overwhelming majority of agricultural experts, commentators, and producers remained committed to the ideal of industrialized plenty—and to the chemical helpers enabling it. The antibiotic-facilitated drive for efficiency and growth remained stable despite the re-emerging cost-price squeeze and increasingly visible side effects like the decline of family farms and safety concerns about new agricultural chemicals. In the minds of many agricultural commentators, unfettered access to antibiotics had acquired almost infrastructural importance for productivity. The cultural and physical importance of expanding antibiotic infrastructures also meant that producers and commentators reacted first with alarm and then with indignation to "irrational" external antibiotic criticism. Mirroring public critics' focus on residue hazards, US agricultural commentators rarely touched on AMR throughout the 1950s and early 1960s. Although the farming media reported on resistant pathogens in human medicine,[148] AMR selection on farms was mostly portrayed as an efficiency problem to be overcome with "rational" drug use. As late as the mid-1960s, commentators reassured farmers that antibiotic use and medicated feeds would produce "little or no resistance, even when used over long periods of time."[149] This sanguine assessment seemed justified following the publication of the FDA's 1966 report on veterinary and non-veterinary antibiotics (chapter 4). Industry journal *Feedstuffs* was happy to report: "scientific data now available do not show reason for alarm."[150] Overall, the US agricultural community remained confident in its ability to manage antibiotic risk.

4

Toxic Priorities

Antibiotics and the FDA

For the American FDA, antibiotic regulation posed unique challenges: not only were regulators forced to reconcile demands for agricultural plenty with consumer protection, they also had to account for antibiotics' hybrid identity as medical therapeutics and nontherapeutic production enhancers, preservatives, and plant sprays. This chapter explores how a traditional focus on toxic hazards and a "gatekeeper" mode of drug regulation facilitated the licensing of many antibiotic applications. Under pressure from Congress, zero-tolerance policies for residues were abandoned in favor of thresholds below which antibiotics were allowed on meat, poultry, fish, and plants—but not in milk. Ensuring that FDA antibiotic guidelines were followed proved more difficult. By the late 1950s, lack of resources, corruption scandals, and the absence of a cohesive antibiotic policy resulted in an overburdening of FDA regulators and a temporary breakdown of antibiotic licensing and enforcement. Public pressure to reduce residues and FDA regulators' focus on toxic and carcinogenic hazards also made officials repeatedly downplay evidence pointing to rising AMR on American farms.

Opening the Gates

In many ways, the FDA's preoccupation with antibiotic residues was a logical continuation of long-standing regulatory traditions. Reacting to the rapid development of novel chemicals, the late 1800s saw US officials attempt to protect the public from dangerous levels of toxicity. From the beginning, officials struggled to establish a model of substance regulation that protected the general public without stifling industrial growth. Following the enactment of the 1906 Pure Food and Drug Act, the USDA's Division of Chemistry became responsible for enforcing new bans on the interstate sale and transport of "adulterated" or "misbranded" food and drugs. However, the division failed to win congressional support for regulations designed to force industry to prove that substances were safe and would not cause acute or long-term poisoning.[1] The rejection of this precautionary licensing approach meant that US drug regulation became based on a more industry-friendly philosophy of thresholds. Influential industrial hygienists asserted that humans' inevitable exposure to chemicals only became dangerous once it toppled the body's "natural homeostasis." If it remained below this threshold, chemical exposure was acceptable. During the interwar period, the renamed Food, Drug, and Insecticide Administration—from 1930 onward Food and Drug Administration (FDA)—thus focused on using toxicology to establish the point at which "natural" turned into "unnatural" exposure.[2]

In 1938, FDA competencies were significantly strengthened by the Federal Food, Drug, and Cosmetic Act (FDC). Passed in the wake of the 1937 sulphanilamide tragedy during which diethylene glycol contaminated cough syrup killed over 100 people, the FDC allowed regulators to ban dangerous substances and required manufacturers to file New Drug Applications (NDAs). NDAs contained information on drugs' composition, manufacturing process, intended use, and evidence of safety. Transferred to the Federal Security Agency in 1940, the FDA had at least sixty days to evaluate evidence and approve or deny NDAs.[3]

The FDA's unification of responsibilities for consumer protection, food security, and drug regulation was unique. In contrast to Europe where powers were often divided between different ministries, FDA regulators could (a) act as gatekeepers for the licensing of veterinary and human drugs, (b) define when food safety was threatened by microbial or chemical adulteration, and (c) evaluate the safety of agricultural chemicals and food additives. However, this significant accumulation of powers stood on clay feet. Constrained by budget cuts and lacking a large science department, the FDA remained heavily reliant on external research to assess NDAs. Resources were also insufficient to guarantee enforcement at the retail or farm level. Regulators responded to constraints by emphasizing their role as premarket gatekeepers, issuing advisory opinions, establishing precedent via targeted prosecutions, and using legal gray areas to exert authority—as in the case of FDA rulings on pesticides and food additives, which were not covered by NDA requirements. According to historian Dan Carpenter, a large part of early FDA power was thus exercised via reputation.[4]

There were several downsides to this mode of reputational gatekeeping: regulators' reliance on legal ambiguities and individual rulings could favor ad hoc rather than coherent risk policies. Resulting incoherence not only fostered contradictory rulings by different FDA departments but also led to a lack of oversight, which made regulators focus on curbing well-known rather than unfamiliar or emerging hazards. Meanwhile, inadequate resources and close industry ties complicated reforms of already licensed drug use. Between the 1940s and 1960s, this mix of limited gatekeeping, incoherent policies, and a "narrow" focus on toxic and carcinogenic hazards would not only open the gates for mass antibiotic use but also constrain FDA responses to resulting hazards.[5]

With the 1937 sulphanilamide tragedy still a recent memory, the FDA first assessed agricultural sulfonamides and antibiotics in light of familiar toxicity hazards. Initial concerns seem to have been minimal and trust in industry safety and efficacy claims high. In 1985, parasitologist Ashton Cukler recounted how he and a colleague had traveled to Washington by train in 1948 to apply for a license to include and sell Merck's sulfaquinoxaline in feeds: an informal go-ahead was granted on the same day and formal approval followed soon afterward.[6] However, the ensuing boom of sulfonamide use soon led to a partial reversal of *laissez-faire* regulation. Reacting to reports in 1949 that farmers were using agricultural sulfonamides for self-medication and accidently poisoning themselves and their flocks, the FDA restricted access to larger quantities of pure sulfonamides and established compulsory maximum concentrations in premixed feeds. In a significant move, the agency also transferred existing NDA requirements for human medicines to agricultural settings by mandating that safety labels should be attached to sulfonamide feeds and concentrates.[7] Sulfonamide labeling served the dual purpose of exerting control without disrupting the pharmaceutical market: the mandatory attachment of guidelines on safe use to products containing veterinary drugs legally entitled the FDA to punish drug misuse. Meanwhile, the official assumption that "rational" consumers would understand and follow safety labels allowed most veterinary drugs to stay on the market and enabled expensive farm-level enforcement to stay minimal. With manufacturers involved in establishing official guidelines, only very few veterinary drugs for which no reasonable guidelines of safe lay use could be developed were restricted by the FDA and assigned prescription-only status, which was formally established in 1951.[8]

The FDA's regulatory matrix for synthetic sulfonamides was also applied to biological antibiotics, which had to be accompanied by safety labels when sold to farmers for therapeutic purposes. However, the post-1949 boom of AGP sales soon forced the FDA to create new regulatory categories for "nontherapeutic" antibiotic products—especially since the 1938 FDC only mandated safety tests for drugs but not for nontherapeutic chemicals.[9] As described in chapter 3,

Lederle had initially marketed its AGPs containing aureomycin fermentation wastes as B$_{12}$/APF feeds to avoid "registration problems"[10] regarding feeds' antibiotic content. Lederle subsequently used the rapid uncontrolled rise of APF/AGP sales to present the FDA with a *fait accompli* licensing application when officials started to become concerned about the varying dosages of feed additives being sold in interstate commerce. Recounting Lederle's strategy, AGP co-discoverer Thomas Jukes claimed that "approval was readily granted . . . a most important step included a conversation with Dr. Elmer Nelson of the FDA during which he asked what level of aureomycin should be authorized for use as an animal feed supplement; we suggested 50g/ton, to which he agreed."[11] Aware of toxicity-based FDA sulfonamide regulations, Lederle representatives pointed out that the daily administration of 0.5 grams of aureomycin for periods of up to two years had not triggered toxic reactions in human subjects. Residues resulting from AGPs were therefore unlikely to cause problems. Although Jukes later referred to discussions with FDA officials about AMR, there is no archival evidence indicating that deliberations of potential allergenic or AMR hazards took place.[12]

In April 1951, the FDA consented to retrospectively legalize the already booming AGP market. Probably applying toxicity oriented threshold thinking and trusting industry safety assurances, the agency does not seem to have conducted further AMR, allergy, or toxicity tests of low-dosed AGPs. It also decided to license a wide range of AGPs in bulk. In addition to Lederle's aureomycin (chlortetracycline), the 1951 *Federal Register* announcement licensed penicillin, bacitracin, streptomycin, di-hydro-streptomycin, and chloramphenicol (soon to be linked to aplastic anemia) for use in feeds. If marketed solely as subtherapeutic feed supplements and not as therapeutics,[13] the listed antibiotics were, moreover, exempted from both NDA and federal batch certification requirements (see below).[14] According to the 1951 guidelines of the Association of American Feed Control Officials, supplements were to be rendered unsuitable for human use with denaturants and labeled "for feeding use only" and to include no more than 50 grams of antibiotics per ton—a weight limitation that could, however, be waived from 1953 onward.[15] With the FDA focusing on toxicity thresholds, AGPs flew right under officials' hazard radar. Thomas Jukes later noted that had the antibiotic growth effect been discovered in 1985, AGPs would not have been licensed.[16]

Strengthened by the 1951 Delaney report, which did not mention antibiotics despite having heard evidence on AMR and allergenic residues (chapter 2),[17] the FDA also launched a policy of actively facilitating antibiotic access. The agency's first move was to oppose perceived attempts by veterinarians to monopolize the antibiotic market. After the AVMA asked the Committee on Public Health to review antibiotic feed additives in August 1952 (see below), FDA officials launched a broadside against the "movement on the part of veterinarians" to restrict antibiotics and noted: "this Administration has always insisted that drugs for veterinary use, to the extent practicable, be not restricted to professional use."[18] Deputy commissioner George P. Larrick reassured mill owners: "We have consistently followed the course of placing no obstacles in the way of self-medication when the medicines employed can be safely and intelligently used by lay persons. The same principles apply to our regulation of livestock remedies."[19] Tasked with protecting "rational" consumers, regulators would not intervene if drugs could be used safely—what happened if consumers chose to ignore guidelines and act "irrationally" was not frequently discussed.

The FDA also attempted to remove regulatory barriers for manufacturers. From 1945 onward, biological antibiotics had been regulated through a certification procedure under Section 507 of the FDC. Following the joint definition of standards, identity, strength, quality, and purity for an antibiotic, each batch of an antibiotic product had to be certified by the FDA—producers could choose to have their products certified or to have them treated like other new drugs.[20] While the 1951 regulations had eliminated batch certification requirements for certifiable listed feed antibiotics, feed mills producing antibiotic feeds for purposes other than growth promotion theoretically still required federal batch certification. Initially, each batch of an animal feed "containing therapeutic levels of [antibiotics] for therapeutic purposes as a new drug"[21] had to be certified individually by the Division of Antibiotics. Because such a procedure "would be impracticable,"[22] the FDA exempted low-dosed antibiotic feeds, which were marketed for therapeutic purposes, from NDA and batch certification requirements in 1953.[23] Other exemptions followed rapidly. By 1955, the list of approved combinations of certifiable antibiotics and other drugs listed in the *Federal Register* was nearly two pages long. Two years later, the FDA amended its 1951 certification exemptions to include all certifiable antibiotics for use in properly labeled animal feeds additives.[24] The boundaries between therapeutic and nontherapeutic antibiotic use were blurring.

Normalizing Risk

The resulting expansion of US antibiotic sales took place against a backdrop of minimal FDA controls. Throughout the 1950s, officials not only struggled to make manufacturers adhere to NDA requirements but also failed to convince hesitant US attorneys to prosecute offenders.[25] In 1952, a frustrated regulator confessed: "Personally, I am of the opinion that many of the therapeutic representations made for antibiotics in feed bases are exaggerated and not based on adequately controlled experiments."[26] Lacking resources meant that there was even less control over drug use on farms. Resulting problems triggered mixed responses. While officials initially favored case-by-case rulings against antibiotic residue offenders, 1950s FDC amendments led to a policy of "normalizing" risk by delineating boundaries within which residues were "tolerated."[27] Because there was no objective way to define the exact boundary between tolerable and intolerable residues, the FDA had to strike a delicate balance between industry safety claims, toxicity concerns, and the cultural values attached to different foodstuffs—AMR was mostly ignored. As evidenced by residue bans in milk as opposed to tolerances in meat, fish, poultry, and plants, this balancing act produced paradoxical results.

In 1948, a policy announcement in the *Federal Register* had mandated that drugs effecting food animals' physiological functions were to be approved only if they did not leave deleterious or poisonous residues in food.[28] This ruling gave FDA regulators considerable scope to dictate dosages and withdrawal times for antibiotic products. However, in practice, officials' reliance on industry data and inability to assay many drugs meant that safeguards were limited. Although they warned that residue offenders might be prosecuted for adulteration, officials could do little more than hope that producers would use drugs responsibly and that potential residues would be too low-dosed to harm consumers. More coherent residue policies emerged in response to antibiotic food preservatives. Faced with the prospect of intentional residues, regulators drew a red line. In February 1953, the *Federal Register* announced that "careful consideration" had made the FDA decide that using antibiotic drugs as food preservatives "constitutes a public-health hazard."[29] Residues could sensitize consumers and "result in the emergence of strains of pathogenic microorganisms resistant to these drugs."[30] Neither intentional nor unintentional residues in food could be deemed safe.

FDA regulators' stand was short lived. Within a year, the 1954 Miller Pesticides Chemical Amendment overthrew FDA polices. Passed in reaction to the recent Delaney hearings (chapter 2), the Miller amendment tasked the FDA with streamlining policies and establishing tolerances for many pesticide residues on raw food. As Nancy Langston and Sarah Vogel have shown, this focus on tolerances had an ambivalent impact. Instead of preventing exposure to hazardous chemicals per se, the Miller amendment enabled official risk management based on—often industry-friendly—calculations of "acceptable" exposure.[31] In the case of antibiotics, the amendment opened the door for antibiotic use in food preservation, plant protection, and whaling. Starting in 1954, FDA officials began distinguishing between residues resulting from the use of antibiotics as pesticides (i.e., to control bacteria on *raw* agricultural commodities) and the use of antibiotics in or on *processed* food. In the case of processed foods, residue tolerances remained difficult to obtain because section 406 of the FDC required proof that an added chemical was necessary for production. However, in the case of raw food, antibiotics fell under the Miller amendment, and manufacturers were allowed to apply for official residue tolerances.[32]

Industry applications arrived quickly. Following a series of studies, the FDA overturned its 1953 residue ban and legalized the preservation of poultry meat with chlortetracycline (Cyanamid's Acronize) in November 1955 and with oxytetracycline (Pfizer's Biostat) in October 1956. In order to extend their shelf life by as much as 50 percent,[33] batches of about 200 eviscerated poultry carcasses were dipped for fifteen minutes into tanks of slush-ice containing 10 parts per million of chlortetracycline or oxytetracycline. Tolerances of 7 parts per million were established on raw poultry.[34] In 1959, the FDA established similar tolerances for the preservation of fish via antibiotic ice or dipping solutions. Scallops and shrimp could also be preserved via antibiotics.[35] Preservation trials for milk, beef, and eggs were ultimately abandoned.[36] Retrospectively, perhaps the most bizarre antibiotic application was the use of antibiotics to preserve whale meat. Tested by Pfizer in Norway and Iceland, injections, harpoons, and the pipes used to inflate carcasses were loaded with Biostat. Whalers were supposed to release an antibiotic equivalent of about 1 part per million of a whale's weight into its circulatory system to delay the usually rapid spoilage of slow-cooling carcasses. Additional antibiotic preservation was suggested in the form of dipping solutions and antibiotic ice. Company representatives hoped that preservatives would increase whale meat yields for extracts and animal feeds by 30 to 10 percent respectively[37] and allow whale meat to "become plentiful in [American] grocery stores."[38]

Although whale meat failed to conquer US dinner tables, antibiotic preservatives for fish and poultry became common (chapter 2). Justifying their licensing decisions, FDA officials claimed that cooking would degrade antibiotics and that labels on raw food would inform consumers about antibiotics' presence.[39] In effect, regulators no longer guaranteed "pure" meat but made consumers responsible for preparing poultry and fish in a way that would destroy legal, yet undesirable residues. On the part of manufacturers, safety concerns were limited to the fact that inadequate storage could lead to premature spoilage.[40] Contemporary reports of strong variations in the antibiotic dosages used to preserve food and studies on AMR selection on carcasses and in abattoirs did not lead to corrective action.[41]

In plant production, the Miller amendment enabled the licensing of antibiotic sprays and paints for use against bacterial infections. In 1946, initial tests of penicillin against bacterial fire blight had proven unsuccessful. However, six years later, researchers using streptomycin achieved spectacular results against fire blight in orchards in Missouri, Mississippi, Delaware, and California.[42] The trials triggered widespread industry interest. Within a year of the Miller amendment, the FDA was analyzing applications for mostly streptomycin-based sprays and paints to combat a wide range of bacterial diseases in apples, pears, walnuts, peaches, beans, tobacco, tomatoes, peppers, cherries, spinach, lettuce,

FIGURE 4.1 USDA antibiotic spraying, Ben Howard, May 1959, USDA Forest Service Region 6, Forest Health Protection. Collection: Region 6, Forest Health Protection slide collection, Portland, Oregon.

and potatoes.[43] Sprays were deemed harmless as long as antibiotic residues did not reach consumers.

Despite the Miller amendment's "normalization" of specified antibiotics in US meat, poultry, fish, and plants, not all antibiotic residues were legalized. During the 1950s, FDA regulators were faced with increasing reports of illegal antibiotic residues in US food. In 1955, they warned about possible "exposure of large segments of the population to a multiplicity of antibiotics." "It was discovered that tens of thousands of chickens were being injected in the neck tissues with a preparation which left an insoluble residue of active drug. . . . It was further found that when such antibiotic containing tissue was baked or fried the concentration of drug was not appreciably diminished."[44] Still relying on gatekeeper regulations and running on 1940s budgetary and manpower levels,[45] officials, however, struggled to expand residue controls and prioritized the protection of certain foodstuffs over others.

While expert assurances that low-dosed residues in meat would not cause harm facilitated relative official complacency about antibiotics in meat, poultry, fish, and plants,[46] similar residue detections in culturally sensitive milk triggered a robust regulatory response (chapter 2). The FDA had been aware of residue problems in milk since 1948 and had reformed labeling requirements for mastitis tubes in 1951.[47] Beginning in 1954, officials reacted to growing concerns by testing a variety of milk products for antibiotic residues. Initially, 3.2 percent of samples tested positive for penicillin.[48] One year later, 11.6 percent of samples tested positive. Tasked by the FDA to undertake an impromptu risk assessment,[49] medical experts warned that "the ingestion of the amounts of penicillin found in milk might conceivably cause a reaction in an extremely sensitive individual."[50] In 1955, an FDA official openly attacked antibiotic overuse in the dairy sector: "Larger and larger quantities of veterinary preparations were being used, dosages were increasing, and it appeared that some individuals were marketing milk from treated cows too soon after treatment. Furthermore, in some instances it was believed that unscrupulous individuals were deliberately adding penicillin to milk to upgrade it."[51] The FDA also launched educational campaigns to curb residues. In 1956, detections fell to 6.9 percent of analyzed milk samples—but overall residue concentrations were much higher. In 1957, the FDA responded by mandating labels on withdrawal times on drug containers and by limiting prepackaged mastitis medications to 100,000 units per dose instead of the 1,500,000 units per dose in some older products.[52] Although only 3.7 percent of milk samples tested positive in 1958, the FDA remained dissatisfied. Under particular public pressure to guarantee the purity of milk, the agency introduced a pioneering sanction-based interstate monitoring program for penicillin in milk.[53] Starting in 1960, monitoring was based on a two-and-a-half-hour test using *Bacillus subtilis*, whose growth would be inhibited if penicillin was present.[54] Other antibiotics were not tested for. Within one year, penicillin detections fell to 0.5 percent of tested milk samples.[55]

The example of milk shows how cultural notions of purity strongly affected official antibiotic regulation. Resulting in the installation of residue monitoring well ahead of other foodstuffs, US scientists, consumers and farmers all agreed that antibiotic residues in milk were taboo. By contrast, similar antibiotic residues became tolerable in meat, fish, poultry, and plants. Although tolerances can partially be explained by the belief that antibiotics would degrade as a result of cooking, these foodstuffs were also not hedged by cultural taboos similar to that surrounding milk. Regulatory calculations of antibiotics' risks and benefits thus mirrored and reinforced cultural perceptions of risk.

A similar bias also affected early FDA responses to AMR. Because AMR did not match established cultural and regulatory priorities, 1950s officials repeatedly downplayed the risks posed by AMR selection on farms and instead focused limited resources on curbing residues. In August 1952, a Special Committee on Veterinary Public Health debated potential hazards resulting from medicated feeds. Authored by James H. Steele, head of the CDC's Veterinary Public Health Section, the committee's report discussed possible AMR selection through therapeutic and nontherapeutic antibiotic use. However, it limited warnings to traditional scenarios involving the presence of antibiotic residues and called for more research on wider AMR selection. Five years later, USDA and CDC officials expressed concerns about FDA plans to limit the dosage of commercial mastitis tubes. Lower dosages would prolong treatment and select for AMR. Clearly prioritizing residue over AMR hazards, the FDA's Division of Antibiotics curtly notified the other agencies that it had not requested their advice.[56]

Contemporary expertise did little to challenge officials' relative complacency about AMR. At the 1955 International Conference on Agricultural Antibiotic Use, Boston-based antibiotics expert Maxwell Finland upheld an epistemological divide between AMR in humans and animals: "In contrast to the human experience, disease-producing strains have not been found to emerge among the types of animals that are raised primarily for market on antibiotic-supplemented feeds."[57] Bolstering contemporary threshold models, Finland and other experts claimed that AGPs in particular were simply too low-dosed to select for harmful resistance.[58]

At first glance, it seems strange that a leading campaigner for "rational" antibiotic use in human medicine like Finland (chapter 2) would endorse unrestricted antibiotic use in agriculture. Two factors can, however, explain Finland's position. The first factor was the contemporary understanding of AMR proliferation solely in "vertical" terms (chapter 1): natural or mutational AMR would only be passed on from one bacterial generation to the next. According to this "organismal" scenario of AMR proliferation, containing individual strains and reducing selection pressure would curb AMR. Perceived biological differences between animal and human strains and spatial distances between farms and hospitals further reinforced an epistemic divide between risk assessments of AMR in humans and animals. A second much more prosaic factor behind Finland's endorsement of agricultural antibiotics were his close contacts with the US pharmaceutical industry. After being asked to present a critical review on the "Emergence of Resistant Strains in Chronic Intake of Antibiotics"[59] at the upcoming antibiotics conference, Finland had contacted AGP co-discoverer Thomas Jukes at American Cyanamid. "Tom" was only too happy to supply "Max" with published and unpublished data, slides, and a copy of his forthcoming book *Antibiotics in Nutrition*,[60] which subsequently provided the "main source"[61] for Finland's bibliography. Driven from New York to Pearl River via company limousine,[62] Finland also talked to other Cyanamid researchers and was allowed to borrow company figures and slides, which he failed to return.[63] Speaking at the international antibiotics conference a few weeks later, Finland in effect presented a Cyanamid-review of antibiotic hazards.

Finland's case was not unique. Assembling the international elite of antibiotic expertise, the entire 1955 NAS conference had been lavishly financed by the pharmaceutical industry: companies sponsored cocktail receptions, hotel expenses, and a seven-day post-conference tour of the United States for speakers, with diverse recreational activities.[64] The first—and for a long time only—conference of its kind, the 1955 Conference on Agricultural Antibiotic Use had lasting effects on perceptions of antibiotic risk and forged an international community of mostly industry-friendly antibiotic experts.

Flying by the Seat of Your Pants

With the exception of penicillin residues in milk, the 1950s thus saw few challenges to the FDA's gatekeeper mode of antibiotic regulation. This situation changed between 1958 and 1965 when new legislation, corruption scandals, and residue scandals stretched FDA regulators to their limits and strained formerly close relations with industry. Lacking financial resources, statutory powers, and a clear regulatory philosophy, FDA morale plummeted and regulators struggled to enforce guidelines and reform drug licensing.

In 1958, the passage of the Food Additives Amendment and its so-called Delaney Clause placed a great strain on FDA resources. The Delaney Clause was a direct result of lobbying by Congressman James Delaney, who remained dissatisfied with regulations resulting from his 1951 report.[65] Similar to the 1954 Miller amendment, proposed food additives regulations were, however, watered down by industry. Instead of enacting a mandatory testing program for new substances, Congress passed a weaker bill, which required unspecified proof of additives' safety and tasked regulators with establishing new tolerances for additives in food.[66] Any veterinary drug leaving residues in food would also be treated as an additive.[67] The only exception to this dose-response dominated approach was the Delaney Clause, which established a precautionary zero-tolerance policy for carcinogens.[68]

Congress's attempt to create a comprehensive food additives framework pushed the FDA to its organizational limits. It also cast a harsh light on the confused nature of previous ad hoc licensing. Since 1938, NDAs had generally been approved on a case-by-case basis. This meant that manufacturers had to file supplemental applications if they changed any component of accepted NDAs. In 1958, officials were already struggling to keep up with the rapidly increasing number of NDAs and supplemental NDAs. The 1958 amendment's demands for additional prelicensing data on the occurrence and harmfulness of drug residues, which would lead feeds to be assessed both as drugs and as additives, and the parallel introduction of efficacy reviews of drugs added to food and water threatened to overwhelm FDA capacities.[69] The situation was even more complicated regarding already licensed drugs. Fearing economic damage, legislators tasked the FDA with compiling a list of substances which were generally recognized as safe (GRAS) via scientific consensus or long experience. GRAS substances would not require new NDA certification.[70] A "grandfather clause" also exempted NDAs licensed prior to 1958 from further review—a rule that caused severe problems in the case of carcinogenic diethylstilbestrol (DES), which had been licensed as a growth promoter before 1958.[71]

Byzantine complexity also characterized the regulation of agricultural antibiotics. Although existing 7 parts per million tolerances for chlortetracycline and oxytetracycline were grandfathered,[72] new NDA requirements, GRAS exceptions, and the absence of a medicated feed rule compendium soon resulted in universal confusion when it came to licensing and labeling procedures.[73] As described above, batch certification requirements had been waived for many popular certifiable antibiotics like penicillin, chloramphenicol, bacitracin, chlortetracycline, and streptomycin at doses below 50 grams per ton between 1951 and 1957. Where requirements had not been waived, manufacturers had to file a Form 10, which was identical to an NDA but required additional proof of efficacy. However, a different set of regulations applied to preparations containing noncertifiable antibiotics, which had to be licensed via an NDA. This continued to be the case for tylosin, hygromycin, novobiocin, oleandomycin, and nystatin. However, in the case of oxytetracycline, neomycin, and several sulfonamides, long-standing experience turned them into GRAS drugs. Producers could use these drugs according to GRAS guidelines without filing extra Form 10 or NDA applications.[74] AGP labeling rules proved even more arcane: to stop producers from advertising excessive amounts of antibiotics, the FDA had banned quantitative AGP labels in October 1953.[75] However, many manufacturers did not know how new AGP labels should look[76] and some used the absence of quantitative labeling to sell deficient feeds.[77]

Frustration soon ran high on all sides. After a meeting of the Pharmaceutical Manufacturers Association in April 1959, a representative noted: "one feels that, perhaps, the problem of medicated feeds has been regarded as a stepchild. . . . [and the] administration approach . . . has been one of flying by the seat of your pants."[78] Manufacturers also complained about fragmented responsibilities: "we now have the following groups of the administration . . . concerned with the medicated feeds: Front Office . . . Veterinary Medical Branch, New Drug Branch, Division of Antibiotics, Division of Pharmacology, State Relations Division, and, now where a tolerance in a meat product might be concerned, the Food Additives Division."[79] Fragmented responsibilities, licensing backlogs, and new testing requirements also meant that speeds of FDA drug licensing began to vary considerably. Because the Division of Veterinary Medicine's average review for a Form 1800 NDA exceeded one year, some manufacturers added certifiable antibiotics to products so that they could file a Form 10 with the quicker Division of Antibiotics. With hardly anybody able to navigate existing drug rules, a 1959 FDA memo dreaded new regulations because there was already "so much confusion and misinformation."[80]

Facing mounting pressure to streamline licensing procedures and combat residues, there was a strong temptation for the FDA to relax rather than to rethink regulations. By the late 1950s, officials were increasingly forced to concede that an overreliance on labels and under-resourced enforcement had allowed drug and feedstuff abuse to become rampant. Initial responses to noncompliance nonetheless remained lenient. In July 1959, the FDA's Division of Pharmacology warned George Larrick, who had been promoted to commissioner in 1954, that reducing the amount of drugs entering the agricultural market would lead to misuse: it did not "take a great deal of foresight to predict . . . that a veterinarian can (and no doubt will) prescribe new drugs currently marketed for human use."[81] Mail advertisements and reports on "uncontrolled studies" would result in veterinary misuse of potent pharmaceuticals—"misuse over which we (FDA) have very little (or no) control."[82] Significantly, the division did not call for stricter enforcement but for the ability to grant exemptions from the 1958 amendment's treatment of veterinary residues as additives to facilitate licensing if products were deemed reasonably safe under normal conditions of use.[83] On farms, the agency followed a similar strategy of relaxing rather than enforcing rules. Although penalties for feed violations remained relatively low,[84] Larrick publicly stressed that the FDA would not intensify prosecutions in 1959.[85]

The question of how to deal with increasing numbers of industry applications divided American regulators. In 1959, Massachusetts's official chemist, John Kuzmeski, called for a general reform of FDA drug licensing: "It is a well known fact . . . that [withdrawal warnings are] largely ignored. . . . If a farmer feeds a medicated feed to his chickens up to the day of slaughter, and undesirable residues remain in the flesh as a consequence, you, I and a host of other people will be eating those residues."[86] Because it was impossible for officials to "stand over every farmer," it was necessary to stop trusting in "rational" consumers and "to assume that many farmers will not heed"[87] guidelines. Before licensing drugs, the FDA should therefore always consider "what danger to public health exists when widespread disregard of necessary warning statements has been established."[88] Officials should also insist on the availability of reliable assay methods to discern drug levels. In 1957, an official review of thirty veterinary drugs had shown that "reliable methods for analysis in the finished feeds [were] only available for less than half of them."[89] Because industry opposition had prevented the release of information contained in reformed NDAs to state feed control officials in 1958, the FDA had to ask for a voluntary industry release of assay methods.[90]

The results of lenient licensing and patchy surveillance were thrown into stark relief by a limited program of farm and feed mill inspections, which began in 1960.[91] Inspection reports revealed that antibiotic noncompliance was rampant. During a 1962 meeting with industry representatives, FDA officials blamed problems on ignorance and willful violations by the growing number of small feed mill operators: "Their time is limited and often their capabilities to interpret the regulations . . . are also. . . . [Operators] have looked upon inspection programs with fear and distaste.[92] In September of the same year, an FDA inspector described the breakdown of guidelines on an integrated turkey farm: "Field men, most of whom are not trained veterinarians, are employed to check the flocks [of about 300,000 birds] daily and diagnose disease conditions . . . the field men prescribe drugs and/or antibiotics for control or prevention. As the medicated feeds are not resold, the firm does not apparently feel it comes under the scope of the new drug or antibiotic regulations. The usual amounts of medication are not adhered to."[93] Drugs were "frequently purchased in [as] large amounts as 25 kgs. penicillin."[94] It was not uncommon for "a coccidiostat, blackhead preventive, and antibiotic . . . all [to] be fed at the same time."[95] Drug residues in meat were likely. However, when proactive officials tried to take action against violative shipments of drug premixes, the FDA general counsel warned that the agency "w[as] not on sound legal ground to take equal enforcement action against the majority of the violative shipments of new drug premixes and . . . advised . . . not [to] approve any more actions in this area until the problem could be resolved."[96] Responding to rampant noncompliance, an internal FDA memo suggested that unless industry improved observance, legalizing illegal residues via new antibiotic tolerances might be the only course of regulatory action.[97]

In October 1962, the FDA's situation was further complicated by the passage of the Kefauver-Harris Amendment. The 1962 amendment tasked the already overburdened FDA with establishing a distinct licensing process for veterinary drugs and mandated the registration—and regular control—of veterinary drug manufacturers and feed mills.[98] Controversially, the amendment also contained a new feed additive law (the so-called DES-proviso) which allowed regulators to circumvent the Delaney Clause's zero tolerance provision for carcinogens by stating that a drug was safe if there was reasonable certainty that labels would be followed and it would not leave residues in food. Passed just as *Silent Spring* was beginning to make headlines, the proviso was fiercely criticized by consumer activists but supported by commissioner Larrick, who also launched parallel campaigns against alleged "food faddists" and the Rodale Press.[99]

The combined burden of the 1958 and 1962 amendments overstretched FDA veterinary drug regulation. Between 1958 and 1962, commissioner Larrick's regulators had experienced a rapid expansion of their workload, demoralizing reports about noncompliance and residues, sustained public criticism, and corruption charges against senior officials (chapter 2). Scandals and problems not only tarnished the FDA's public image but also lowered internal morale. While senior officials remained loyal to commissioner Larrick, others warned about the degree to which industry and regulatory science had become intertwined.[100] For an organization whose power was built on reputation, the increasing chaos surrounding veterinary drug regulation was becoming harmful.[101] It also led to a neglect of emerging risks like AMR.

Throughout the 1950s and 1960s, FDA regulators' preoccupation with the increasing chaos surrounding drug compliance, licensing, and residues meant that the risks posed by agricultural AMR selection remained low on their agenda. In contrast to Britain, where public health warnings triggered the first AGP review in 1960 (chapter 7), US officials downplayed studies indicating potential health hazards and an increase of AMR in *Staphylococci*, *Salmonellae*, and *E. coli* on US farms.[102] Complacency about nonhuman AMR risks was also shared by many US scientists. In 1959, John T. Logue from the University of Missouri insisted that agricultural AMR hazards were limited to immediate exposure to antibiotics via residues or plant sprays. Although field workers using streptomycin plant sprays had experienced strong allergic reactions and transient AMR selection, general agricultural antibiotic use was too low-dosed to pose an AMR hazard.[103] In 1960, a book coauthored by the soon-to-be-ousted head of the FDA Antibiotics Division, Henry Welch, similarly stated that "use of antibiotic drugs as preservatives, as crop sprays, in the nutrition of animals, and in the treatment or prophylaxis of disease (except mastitis) . . . are in the best interests of the people's health."[104]

Still expressing confidence in "rational" agricultural antibiotic use, FDA officials also downplayed international AMR warnings. In 1961, Dr. Antonio Santos Ocampo Jr. from Arenata University in the Philippines wrote to the FDA asking for advice: "For instance, people here are just plain crazy about the use of antibiotics to stimulate egg production and to prevent CRD. We are literally flooded with literatures (of course by Pfizer people) regarding the efficacy. The Terramycin egg formula and the anti-germ 77 sells like hot cake here."[105] However, Ocampo had "always entertained doubts as to the wisdom of the indiscriminate use of antibiotics." "I fear that the microbial flora in animals might become resistant to antibiotics and when the time comes this antibiotic will no longer have any value."[106] In response, the FDA assured Ocampo: "So far no one has produced any conclusive evidence that this is the case in poultry."[107]

The specter of agricultural AMR selection was raised again in response to British and WHO expert evaluations of AGPs in 1962 (chapter 7).[108] Debating an extension of AGPs to mature animals, the FDA's Division of Veterinary Medicine referred to Britain's Netherthorpe Committee and cautioned that "there may be a build-up of resistant organisms when adult animals are fed low levels of antibiotics continuously, Great Britain does not allow their use for this reason."[109] However, no wider reform of agricultural antibiotic use resulted from these deliberations. In September 1962, the FDA elaborated on its static assessment of agricultural AMR risks in response to a constituent enquiry submitted by Democratic senator Hubert H. Humphrey.[110] The future vice president and former pharmacist was one of commissioner Larrick's main critics and had coauthored the 1951 Durham-Humphrey Amendment, which introduced the legal distinction between over-the-counter drugs and prescription-only medicines.[111] Humphrey's constituent enquiry had been authored by James S. Collins, a PhD in animal breeding and former employee of the feed company Nutrena (Cargill Inc.). While he was working for Nutrena, Collins had actively campaigned against AGPs and discovered that "our animals are now carrying a heavy infection of antibiotic resistant pathogens."[112] In a remarkable statement, Collins criticized the residue-centered FDA view that products were safe "if no antibiotic turns up in [animals'] tissue"[113] and warned: "It would seem to me that we are not only laying our animal population wide open for disaster as well as providing reservoirs of pathogens to invade man."[114]

Responding to Senator Humphrey, FDA deputy commissioner Harvey defended his agency's policies: "experts regard use of drugs and chemicals . . . as necessary in order to . . . assure adequate food as the Nation's population increases while the acreage of productive farmland decreases."[115] Referring to the "indirect hazard" of AMR, Harvey claimed that FDA "scientists are keeping abreast of developments in this field."[116] The FDA had contacted the

British Netherthorpe Committee, which had "concluded that although there are problems resulting from the use of . . . antibiotics in animal feeds, such use should be allowed to continue."[117] Regarding over-the-counter sales of higher-dosed mastitis tubes, Harvey waspishly reminded Senator Humphrey of the wording of his own 1951 FDC Amendment: "there is nothing in the Act itself dealing specifically with the question of whether a veterinary drug may be restricted to veterinary prescription dispensing."[118]

Interestingly, FDA AMR risk assessments were different when it came to companion animals. In 1962, the Division of Pharmacology warned about antibiotic feeds for pets and "the possible danger to man from development of transmissible antibiotic-resistant strains of bacteria (particularly staphylococci)."[119] In contrast to their relatively benign assessment of AGPs in food production, regulators agreed to move against "rations containing certifiable antibiotics which are marketed for continuous feeding to household pets."[120]

Antibiotic Reform

The FDA's 1960s deadlock regarding veterinary drug regulation only began to loosen ahead of the retirement of George Larrick and other closely allied senior FDA officials in 1965.[121] Larrick's successor, James L. Goddard, fulfilled congressional demands for a medically qualified FDA commissioner and had previously headed the CDC. Starting as commissioner in January 1966 and soon known as "Go-Go Goddard," Goddard embarked on a fundamental restructuring of the FDA. Under Goddard, FDA drug recalls grew by about 75 percent, and the NAS was contracted for an efficacy review of drugs licensed prior to 1962—the Drug Efficacy Study Implementation (DESI).[122]

FDA leadership changes also affected the regulation of agricultural antibiotics. In February 1965, the FDA had decided to install an ad hoc Committee on the Veterinary Medical and Nonmedical Uses of Antibiotics.[123] Despite an archival gap and an elusive final report, circumstantial evidence makes it possible to reconstruct the ad hoc committee's proceedings. The committee had been formed because of the above-mentioned FDA surveys on feedstuff compliance and parallel detections of penicillin in US red meat.[124] Responsible for meat inspections, the USDA had begun to call attention to antibiotic residues in 1964. According to the USDA's inspection manual, assays (bacterial inhibition tests) were to be conducted at the USDA's central laboratory in Beltsville if carcasses showed injection lesions: "the antibiotic may have been administered to alleviate or disguise acute symptoms of disease or as a preventive measure, but in any event, the animal is often marketed prior to complete absorption of the oil base antibiotic."[125] FDA officials were also becoming concerned about allegations that antibiotic preservatives in fish and poultry were selecting for resistant spoilage organisms and food-borne pathogens.[126]

Established as a result of these concerns, the ad hoc committee was primarily residue-oriented, and members initially referred to it as the "Committee to Consider the Public Health Implications of the Presence of Antibiotic Residues in Food and the Use of Antibiotics as Food Preservatives."[127] The committee was headed by Mark Lepper, professor of preventive medicine at the University of Illinois, and also counted Maxwell Finland among its members. Members met for the first time in early May 1965 and submitted their final report one year later.[128] Major changes occurred during this period: in addition to commissioner Larrick's resignation, British warnings about "infectious AMR" fundamentally challenged existing antibiotic policies (chapter 7).[129] No longer limited to the vertical proliferation of resistant organisms, the horizontal exchange of AMR genes undermined spatial and biological differentiations between medical and agricultural settings.

However, because of its focus on residues, the ad hoc committee barely addressed British AMR research. Although the first paper on horizontal AMR appeared in the *Lancet* three months ahead of their first meeting,[130] committee members did not include it in their preparatory reading list[131] but discussed the matter later on.[132] Correspondence with British AMR researchers like Ephraim Saul Anderson also failed to sway antibiotic risk assessments.[133] In August 1966, the ad hoc committee's report expressed "concern" about the "possibility of microorganisms in animals developing resistance or of strains being selected that are resistant."[134] However, despite calling for more research, the final report reinforced traditional risk scenarios by limiting concrete AMR warnings to antibiotics' immediate presence as residues in meat. According to the 1966 report, preservatives should no longer consist of or give rise to cross-resistance against therapeutically relevant antibiotics. Ideally, antibiotic food preservation should be banned completely. The FDA should also increase efforts to prevent illegal residues in edible tissues, reevaluate dosages and withdrawal times, and make "'warning' statement[s] used in veterinary chloramphenicol [chloromycetin] labeling . . . more emphatic"[135] to prevent its illegal use in food production. Reinforcing the US risk episteme's traditional residue focus and presenting risks as containable, the report did not call for a wider reevaluation of agricultural antibiotic use.

The FDA quickly committed to implement the report and used it to assuage the limited public concerns about AMR that had emerged in the wake of the *New England Journal of Medicine*'s 1966 attack on agricultural antibiotics (chapter 2). Within five days of *NEJM*'s editorial, the *Federal Register* announced the following policy measures: producers of licensed antibiotic products were to submit new data on "whether or not such antibiotics and their metabolites are present as residues in edible tissues, milk, and eggs from treated animals."[136] Should they fail to submit data within 180 days, producers could lose their product licenses.[137] Citing AMR and hygiene concerns, in

September 1966 the FDA also banned the antibiotic preservation of poultry, fish, and shellfish and commissioned the NAS to organize a conference on agricultural antibiotics.[138] Despite barely addressing the new AMR concerns, it seemed as though FDA commissioner Goddard was in full control of the situation.

The resulting 1967 NAS symposium on the Use of Drugs in Animal Feeds reinforced this view by gathering many well-known antibiotic supporters from the 1955 NAS antibiotics conference. In his presentation, AGP co-discoverer Thomas Jukes attacked the "emotional phraseology used in [the *NEJM* editorial] . . . that led the *New York Times* to . . . threaten us with the propaganda device of a new *Silent Spring*."[139] Meanwhile, Maxwell Finland maintained that there was "little evidence to implicate food as a source of infections caused by organisms resistant to antimicrobial agents."[140] Finland found it equally "difficult to implicate"[141] AGPs in British reports on enteric "infective resistance." Downplaying warnings of horizontal resistance transfer, most American participants called for more research. Former FDA ad hoc committee head Mark Lepper acknowledged that current drug licensing was too crude. While no "major catastrophe" seemed to be "around the corner," "the use of drugs in feeds could be influencing . . . the background level of organism resistance, without any of us being aware of the fact."[142] Although he used the opportunity to call for an ambitious AMR surveillance program, James Steele, head of the CDC's Veterinary Public Health Section, also confirmed that "no one is seriously recommending that use or production of antibiotics be discontinued."[143]

Americans' wait-and-see attitude was criticized by European attendees. British veterinary researcher Herbert Williams Smith noted: "There is no essential difference between the emergence of resistant strains of bacteria as a result of the use of drugs in the treatment of clinical disease and as a result of the use of drugs as feed additives."[144] Prolonged exposure to low-dosed AGPs was especially conducive to AMR proliferation: "A strong case, therefore, exists for limiting the number of different kinds of drugs that can be used for 'nutritional' purposes."[145] Dutch researcher E. H. Kampelmacher similarly warned: "one cannot and should not deny that antibiotic addition to feeds, as applied on a large scale in husbandry today, has increased the proportion of resistant pathogenic and non-pathogenic bacteria in animals . . . . Infectious drug resistance has complicated the problem."[146]

However, European critics failed to find official US support. Most conference attendees remained convinced that global malnutrition posed a far graver threat than nonhuman AMR selection. In his presentation, commissioner Goddard repeated that the FDA was taking the ad hoc committee's report seriously. Goddard did not dwell long on AMR and instead stressed FDA progress against antibiotic residues. Although past inaction and a lack of reliable data was hampering progress, Goddard was proud of recent FDA attempts to recall residue-prone antibiotic products and of "the denial of certification for oil-based injectable penicillin products, which required an unrealistic withholding time" (chapter 10).[147] Goddard was also optimistic about antibiotics' general future in US agriculture. Because it was "vital to keep the industry moving ahead . . . and to protect the supply of food,"[148] the FDA would carefully consider risks but also "eliminate, wherever possible, purely administrative delays in the introduction of new drugs for animal use."[149]

With the ad hoc report justifying the FDA's existing risk priorities, the agency also announced the establishment of a new national surveillance program for antibiotic residues in meat in cooperation with the USDA in 1967. A mixture of targeted and random meat sampling would allow officials to gain an overview of contamination levels: the USDA would annually test a total of 5,200 samples (3,900 red meat and 1,300 poultry).[150] In a further step, FDA inspectors would randomly sample meat at the retail level. Regulators also launched educational campaigns warning farmers to "use medicated feeds carefully and wisely," "protect the public health . . . avoid economic loss."[151]

The first program of its kind, FDA-USDA residue monitoring marked a decisive break with earlier gatekeeping policies. However, it remained unclear whether testing would enable the effective prosecution of offenders and whether combating antibiotic residues would be sufficient to curb AMR. Interviewed in September 1967, the head of the FDA's recently restructured Bureau of Veterinary Medicine (BVM), Cornelius Donald Van Houweling stressed the importance of conducting more AMR research: "We seem to be at a stage where reasons can be advanced that there will be, or will not be, a public health hazard with continued use of medicated feeds."[152] FDA decisions would be made "on the basis of the best scientific evidence available."[153] Should FDA officials err, they would naturally do so "on the side of public protection."[154]

After eighteen years of expanding agricultural antibiotic use, it remained to be seen whether Goddard's reformist officials would be able to reorder the fragmented path dependencies of 1950s drug regulation. Focusing on toxic hazards, relying on industry compliance, and lacking a coherent policy framework, the under-resourced FDA had failed to maintain control over the booming antibiotic market. Although officials tried to protect consumers from residues, the 1954 Miller and 1958 Food Additives amendments replaced nascent zero-tolerance policies with a system of thresholds below which antibiotics were deemed safe. Once established, official residue tolerances resulted in a temporary "normalization" of antibiotics in poultry, fish, and plants—but not in culturally sensitive milk. The era of normalization did not last long. By the late 1950s, compliance problems, industry complaints, and public criticism led to a gradual abandonment of early antibiotic gatekeeping policies. Officials tried to streamline drug licensing and established residue monitoring programs for milk and meat. Their narrow focus on residue hazards mirrored and reinforced the wider US risk episteme. Agricultural AMR selection was not considered a primary hazard. In 1952 and 1966, official reviews called for more research but limited AMR risks to the presence of residues. Despite new scenarios of horizontal gene transfer and research highlighting AMR selection on farms, the FDA remained committed to a narrow residue-focused rather than a broader AMR-oriented view of antibiotic risk.

Part II

Britain

From Rationing to Gluttony, 1945-1969

The three chapters of this part reconstruct the development of antibiotic perceptions and regulations in Britain. With the exception of the dairy and poultry industries, British agriculture was comparatively slow to adopt antibiotics. AGPs were only licensed in 1953. Close corporatist ties between officials and the National Farmers Union (NFU) subsequently played an important role in overcoming initial agricultural hesitancy regarding antibiotics, which were intended to boost national productivity and reduce imports. Similar to the United States, the resulting rise of agricultural antibiotic use was paralleled by increasing public ambivalence regarding the new substances. Significantly, the emerging British risk episteme centered on AMR and animal welfare concerns rather than residues. Once again, the specific constellation of the domestic risk episteme had a powerful effect on regulatory trajectories. During the 1960s, public concerns triggered an important review of British farm animal welfare and three separate AMR-focused reviews of agricultural antibiotic use. In 1969, prolonged public pressure and new data on horizontal resistance transfer led to the publication of the so-called Swann report. The report called for pioneering precautionary restrictions of medically relevant antibiotics in AGPs. However, its corporatist origins also meant that the Swann report tried to fix rather than reduce existing antibiotic use. By narrowly focusing on farmers' direct access to antibiotics, the corporatist report ignored contemporary calls for a more systematic reform of Britain's antibiotic infrastructure.

5

Fusing Concerns

Antibiotics and
the British Public

This chapter traces how public views of agricultural antibiotics evolved in Britain. Following a wave of enthusiasm, perceptions of antibiotic hazards developed in a nuanced fashion. Whereas US concerns focused on residues (chapter 2), long-standing British concerns about animal welfare and expert warnings about agricultural AMR selection led to a more holistic staging of antibiotic risk. During the 1960s, fused concerns about horizontal AMR spreading from "factory farms" led to three government reviews of agricultural antibiotic use. Although commentators remained divided in their assessment of resulting hazards, all agreed on the necessity of some kind of AMR-focused reform of agricultural antibiotic use.

Great British Antibiotics

Following the deprivations of the Second World War, Britain struggled to stem the costs of decommissioning large parts of her military while rebuilding the national industry. Trying to prevent a rise in expensive food and feed imports following bad harvests, the government embarked on a program of subsidized agricultural expansion (chapter 6). Meanwhile, public consumption was held in check by maintaining wartime rationing. Ultimately, the prolonged disruption of international trade, costly colonial campaigns, and the Korean War made postwar rationing last longer than the entire Second World War. Food availability actually decreased between 1946 and 1948. It was only in 1954 that the British Ministry of Food (MoF) was dismantled along with its rationing system.[1] By this time, consumers were craving meat: between 1950 and 1970 UK meat consumption increased by 33.1 percent.[2]

Similar to the United States, officials and farmers did their best to satisfy rising demand. However, trade deficits and currency devaluations meant that the government had to balance the need for expanded animal production with the need to reduce expensive imports like feedstuffs (chapter 7). It is thus no coincidence that the end of British meat rationing coincided with the legalization of antibiotic feed supplements. In 1947, the Penicillin Act had attempted to curb AMR and protect limited supplies by making biological antibiotics only available via medical and veterinary prescriptions.[3] However, in 1953, the Therapeutic Substances (Prevention of Misuse) Act (TSA) exempted the low-dosed nontherapeutic use of antibiotics for feed purposes from prescription requirements.

Initial public optimism about antibiotics' introduction to agriculture matched that in the United States. Already accustomed to celebrating "British" miracle drugs,[4] commentators had keenly followed US research and welcomed antibiotic use as a progressive way of increasing animal productivity and welfare. Following Jukes and Stokstad's report on the antibiotic growth effect in early 1950, the left-leaning tabloid *Daily Mirror* published an article titled "A New Drug Speeds Pork Chops to Dining Table."[5] In Parliament, Conservative MP Rupert De la Bère immediately asked Labour's minister of agriculture whether AGPs would also be tested in Britain.[6] Although early feed trials proved inconclusive (chapter 6), British optimism about the new AGPs continued to grow. In 1951, the politically conservative *Times* described them as a "strange nutrition"[7] with the potential to speed up animal growth and save feed. By the time of Parliament's reading of the TSA in 1953, the clause licensing AGPs was popularly known as the "penicillin for pigs clause."[8] Enthusiastic reports subsequently appeared in the left-leaning *Observer* and the conservative-liberal *Financial Times*.[9] British commentators also greeted antibiotics' use in plant and egg production as well as in fodder and food preservation.[10] In 1956, the *Times* claimed that antibiotic preservatives marked "the greatest advance in the field of processing perishable foods since the advent of refrigeration."[11] The *Financial Times* covered the landing of the first antibiotic-preserved fish in Britain. US pharmaceutical manufacturers had worked with the British government to trial ice containing 3 to 5 parts per million of tetracycline to retard spoilage and keep fleets on the water for longer. Trials took place aboard the government's research vessel the *Sir William Hardy*, which would later be sold to Greenpeace as the first *Rainbow Warrior*. Antibiotic-preserved cod, haddock, and flatfish landed by the vessel were proudly displayed in the ports of Aberdeen, Grimsby, and Hull.[12]

While most media reports mirrored US enthusiasm (chapter 2), there were also subtle differences. For one, British observers tended to emphasize antibiotics' benefits for nutritional independence rather than their use against famine-fueled communism. Anxious about an alleged brain drain,[13] others downplayed US contributions and viewed biological antibiotics as a quintessentially British technology that was now lining American pockets.[14] Stagnating penicillin sales exacerbated this view. Following a dramatic increase of British production between 1947 and 1950,[15] the US government's decision to increase penicillin production in the face of the Korean War saturated 1950s markets. With revenues declining, British manufacturers had to find new antibiotic outlets. Major producers like Glaxo lobbied the British government to imitate the United States and loosen restrictions on antibiotic feeds for humans and animals.[16] However, international competition remained stiff even after Britain legalized AGPs with US broad-spectrum drugs proving more popular than British-manufactured penicillin, streptomycin, and chloramphenicol (chapter 6).

Americans' success was in part due to skillful marketing. Responding to antibiotic patriotism, US producers subcontracted British companies like Boots to produce tetracyclines, which could now be branded as British.[17] When Pfizer opened a terramycin (oxytetracycline)-plant in Sandwich in 1955, Pfizer's vice president stressed the plant's *Britishness*: "although the installation was financed by the United States it was partly designed and wholly built and operated by the British."[18] Courting clients inside and outside agricultural circles, US companies also placed expensive advertisements in national newspapers. In 1953, Lederle purchased a large advertisement section in the *Times* ahead of the launch of its chlortetracycline-based feed AUROFAC 2A.[19] Sales personnel were also in high demand: in 1956, Pfizer announced that the "world's largest producer of antibiotics" was "expanding its Agricultural Sales Force" and looking for male British personnel with an agricultural background and experience in "modern sales techniques."[20] Three days later, Lederle announced that it too was looking for "top-class Sales Representatives who will sell Animal Feed additives such as Aurofac."[21] Celebrating its new Gosport plant in 1958, Cyanamid claimed that AUROFAC and other products were "bringing untold benefits to almost every sphere of life," "Cyanamid contrives to make a new discovery almost every day, transmuting the hopes of yesterday into the realities of today."[22]

Increasing antibiotic consumption and improving economic outlooks gradually dispersed anti-American sentiments in the British press. By the early 1960s, positive media reports presented both US and UK antibiotics as part of a safe and efficient movement to industrialize food production and to secure influence in a rapidly decolonizing world.[23] In 1962, the *Financial Times*

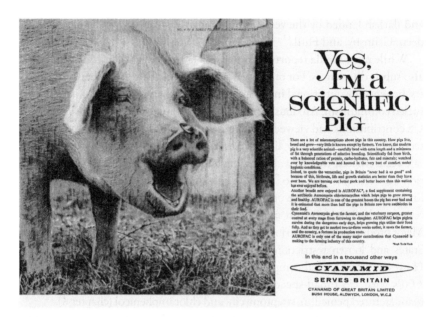

FIGURE 5.1 Cyanamid advertisement, "Yes, I'm a Scientific Pig," *The Times* (London), April 14, 1961.

reported on a Pig Industry Development Authority survey of 20,000 litters comprising about 200,000 pigs: whereas pigs fed no antibiotics weighed 38.1 pounds after eight weeks, those fed antibiotics weighed 38.7 pounds. According to industry figures, antibiotic supplements created 1s 6d additional worth per piglet as a result of saved feedstuffs.[24] Similar reports on an antibiotic-optimized era of livestock production appeared in the *Times*.[25] Drug manufacturers also continued to hammer home the message of antibiotics and modern agriculture as preconditions to ethical, safe, and plenty food. In 1961, Cyanamid started an aggressive advertisement campaign for aureomycin. One ad showed a laughing pig exclaiming, "Yes, I'm a Scientific Pig" and presented agricultural antibiotics as a progressive way of improving animals' well-being and farmers' profits: "Indeed, to quote the vernacular, pigs in Britain 'never had it so good' . . ."[26] Further commercials featured grateful cows cured of mastitis and praised aureomycin's role in preventing any "disastrous rise of mortality"[27] in modern poultry production.

A Plethora of Concerns

Not everybody was happy about the rapid expansion of agricultural antibiotic use. During the 1950s, three interrelated yet distinct strands of antibiotic criticism emerged in Britain's public sphere: one group of critics attacked antibiotics' adulterating presence in foodstuffs; a second group of critics began to warn about the spread of AMR on farms; a third group condemned antibiotics as accomplices to the deplorable conditions of animals in intensive housing units. Depending on one's position within the various opposition camps, antibiotics' image could vary from dangerous adulterator to endangered miracle substance or partner in cruelty. While disparate risk perceptions fragmented early protest, their fusion during the 1960s would pose a far more systemic challenge to agricultural antibiotic use than "narrow" residue criticism in the United States.

A groundswell of elite British antibiotic criticism had existed even before the 1953 TSA. In 1951, former Labour Parliamentary Secretary Lord Douglas of Barloch warned his peers in the House of Lords against "poisonous chemicals in the growing and preparation of foodstuffs."[28] Focusing on antibiotics, DDT, and hormones, Barloch called "for strict control over all processes which might affect the natural quality of food."[29] Two years later, Barloch's Conservative colleague Lord John Justyn Llewellin expressed concern about allergic reactions caused by penicillin residues in meat.[30] Similar fears were voiced in the House of Commons. In February 1953, Conservative MP Anthony Hurd asked whether there was sufficient evidence that AGP residues would not endanger the public.[31] One week later, Conservative MP Dodds asked Conservative Minister of Agriculture Thomas Dugdale how consumers could be protected when "famous experts . . . have declared that more harm than good"[32] would result from the TSA. Seconding his colleague, Conservative MP Colonel Gomme-Duncan asked "whether we have all gone mad to want to give penicillin to pigs to fatten them?"[33] Concerns were not limited to Parliament. Following a 1952 report on the planned TSA,[34] readers of the social-liberal *Observer* worried that AGPs might destroy the intestinal flora's capacity for producing vitamins and lead to infertility and degeneration.[35]

However, in contrast to the United States, fears of antibiotic residues failed to dominate public perceptions of antibiotic risk. British consumers had a long history of remaining less concerned about invisible microbial or chemical hazards than other countries.[36] This does not mean that residue fears were absent—they were simply less pronounced. During the 1950s, deaths resulting from pesticide-contaminated Welsh flour and the radioactive contamination of milk following the 1957 Windscale fire did make national headlines.[37] British supporters of "pure" and "natural" food had also published numerous books with colorful titles like *Constipation and Our Civilisation*[38] on the dangers of artificial additives, chemicals, and drugs since the interwar period. However, as in the case of 1950s detections of penicillin in British milk (chapter 6), residues did not provoke widespread anxiety.

It was only around 1960 that agricultural antibiotics' "toxic association" with other hazardous chemicals became more pronounced. Often enough, British commentators imitated contemporary US coverage. Referring to American residue scandals, a series of articles on "What's in our food" in the conservative tabloid the *Daily Mail* warned readers that there were two kinds of modern food-related dangers: malnutrition and the overconsumption of food sprayed with "poisonous insecticides and weedkillers" and produced by animals, which had been reared in darkness and fed "tranquilisers, antibiotics, hormones."[39] In 1961, a *Times* interview exhorted farmers and veterinarians to protect consumers from drinking "diluted pus with noxious additions such as penicillin."[40] Media concerns about residues peaked in 1963 when the Milk and Milk Products Technical Advisory Committee reported that 14 percent of English and 11.6 percent of Scottish milk tested positive for antibiotics.[41] Although officials introduced US-style monitoring (chapter 7), the 1963 scandal triggered a longer-term increase of public criticism of conventional agriculture and antibiotic adulterants.[42] According to the *Guardian*'s Michael Winstanley, the days of "pure" food were gone.[43] Meanwhile, purveyors of antibiotic-free organic food like the Soil Association and Cranks restaurant in London experienced an unprecedented surge of media coverage—often also in conservative publications like the *Spectator* or *Daily Mail*.[44]

Residue warnings in the media were accompanied by a series of critical bestsellers. In 1960, physician Franklin Bicknell published *Chemicals in Food and in Farm Produce*.[45] Rehashing warnings from his *The English Complaint* (1952)[46] and attacking a wide range of chemicals, Bicknell devoted a whole chapter to agricultural antibiotics and the long-term hazards of antibiotic exposure via milk, meat, and eggs.[47] Remarkably, he also linked discussions of residues with warnings about wider agricultural AMR selection: "by 1956 penicillin-resistant staphylococci were present in 47 per cent of samples of milk in Dorset, and resistant [*E. coli*] strains have been found in the faeces of 36 per cent of pigs going to a bacon factory. . . . Urban England could cynically ignore this farming problem were not our farms still the basis of our food and therefore unavoidably a possible source of human infection."[48] Warning that regulators were only focusing on hospitals as places of AMR selection, Bicknell was especially concerned about antibiotic preservatives' creation of "reservoirs of resistant bacteria": "all the fish sheds and butchers' shops and poultry packing stations will harbour resistant bacteria; . . . however carefully the use of antibiotics is in theory controlled, in practice they will just be sloshed over any food."[49] Similar concerns were voiced by natural foods advocate Doris Grant. Quoting Bicknell, Grant's *Housewives Beware* (1958) and *Your Bread and Your Health* (1961) warned about AGPs and antibiotic preservatives. In chapters titled "Beware of the Dragon"[50] and "The Poisons on Your Plate,"[51] Grant discussed a possible correlation between AGP use and animal illness as well as the use of antibiotics to mask disease in abattoirs. According to Grant, "the only way to escape this particular aspect of our twentieth-century chemical orgy is to become vegetarian."[52]

Despite their increasing prominence, concerns about antibiotic residues, however, failed to displace other strands of public antibiotic criticism. Discussed in more detail in chapter 7, the late 1950s saw a growing number of British researchers point to agricultural AMR selection as a potential human health hazard. Experts' use of national surveillance data to demolish existing distinctions between AMR on farms and in hospitals led to fierce public debates about who should have access to antibiotics. In contrast to the threshold-oriented thinking of their US colleagues, British researchers deemed low-dosed AGPs particularly likely to select for AMR. Veterinarians in particular highlighted dangers resulting from farmers' unsupervised use of AGPs and therapeutic antibiotics—even though the latter were often sold by veterinarians.[53] Speaking at the 1959 congress of the British Veterinary Association (BVA), the deputy director of the government's Central Veterinary Laboratory (CVL), E. L. Taylor, warned that AGPs eliminated competing microorganisms and enabled resistant pathogens to spread.[54] In early 1960, Britain's Agricultural Research Council (ARC) suggested a general review of medical feed additives. The government subsequently launched a joint inquiry into agricultural antibiotic use. Chaired by the recently retired NFU president James Turner—now Lord Netherthorpe—the committee sat between 1960 and 1962.[55] Although medical journals like the *Lancet* continued to print complacent and even positive reviews of agricultural antibiotic use,[56] the rise of powerful British AMR concerns stood in contrast to Americans' ongoing focus on residues.

Animal welfare concerns constituted a third distinct strand of public antibiotic criticism. In Britain, animal protection had a long and illustrious history. Starting in the late eighteenth century, an increasing number of Britons campaigned for the improved treatment of animals. Founded in 1824, the Royal Society for the Prevention of Cruelty to Animals (RSPCA) was the first organized body for animal protection. During the second half of the nineteenth century, anti-vivisectionist and animal protection campaigns for horses and other animals commanded considerable public support and resulted in legislative reforms. In 1911, the Protection of Animals Act was passed to prevent willful cruelty to animals in public spaces. Their interest in animals not only set Britons apart from other nations but also led to patriotic self-descriptions as being a "nation of animal lovers." This alleged national trait was reinforced during the two world wars with officials and campaigners emphasizing British compassion in opposition to German cruelty.[57]

After 1945, notions of being a nation of animal lovers increasingly clashed with farmyard realities. Concerned about new production methods, a growing number of activists and journalists attacked antibiotics as facilitators of systematic welfare abuse. Although intensive farming was not as widespread as activists claimed (chapter 6), the so-called factory farm soon turned into the main target of rising welfare criticism.[58] In 1959, *Observer* journalist Clifford Selly described the "highly artificial conditions" in which modern "ill-fated chickens" lived.[59] Never seeing daylight, broilers were "heavily drugged to keep them alive" and were victims of a system "more akin to the factory than the farm."[60] Over the next two weeks, Selly's article provoked passionate reader responses both in favour and against intensive farming. G. B. Houston accused the "poor, deluded city dweller" of facilitating the production of "drugged and misused broiler fowls"[61] while F. A. Dorris Smith recommended visits to broiler houses by women's organizations to "bring this abomination to an end."[62] Another reader specifically blamed antibiotics for enabling harmful practices.[63] Although establishment organizations like the RSPCA initially distanced themselves from "factory farm" criticism,[64] antibiotics' association with welfare neglect posed a serious threat to the drugs' public image.

By the early 1960s, three distinct strands of antibiotic criticism were thus gaining ground in Britain. Readers of conservative and liberal newspapers were regularly learning about the hazards of antibiotic residues as well as about AMR selection on farms. At the same time, the nation of animal lovers was coming to terms with antibiotics' ambivalent capacity to provide animal welfare and enable its absence. However, without a common reform agenda to unite them, the distinct strands of public antibiotic criticism remained too disjointed to challenge growing antibiotic infrastructures in agriculture or ongoing antibiotic optimism in large parts of the media.

A Fusion of Concerns

A more holistic framing of antibiotic risks occurred between 1964 and 1966: British research on "infectious resistance" and the publication of Ruth Harrison's bestselling *Animal Machines*[65] led to a fusion of preexisting AMR and welfare warnings around the potent symbol of the "factory farm." Resulting calls for systemic antibiotic reform were amplified by fatal outbreaks of resistant gastroenteritis among newborns.

The 1964 publication of *Animal Machines* was the first event to shift public discourse. A Quaker and vegetarian, the book's author, Ruth Harrison, descended from a family with close ties to the avant-garde Whitechapel Boys and author George Bernard Shaw. She had served in the Friends Ambulance Unit during World War II and then studied at the Royal Academy of Dramatic Art. In 1960, a leaflet against animal cruelty on "factory farms" made Harrison take on the cause of animal welfare and write *Animal Machines*. In her book, Harrison combined easy-to-read summaries of scientific findings with vivid descriptions of animals' plight in factory-like production systems. She also linked animal welfare concerns with a more general consumer and environmentalist critique of factory farming. Appearing one year after the British publication of Rachel Carson's *Silent Spring*, *Animal Machines* not only contained a foreword by Carson but had in fact been edited by Carson herself. The book's authority was further strengthened by a preface from Sydney Jennings, a former president of the British Veterinary Association (BVA).[66]

Claiming that meat eating had become an ethical and health hazard, *Animal Machines* explicitly associated modern farmers' antibiotic-dependency with animal cruelty, drug residues, and AMR.[67] The book underlined its claims by using shocking pictures of conditions in "factory farms" and by referencing *Silent Spring*, the recent milk scandal, and medical AMR warnings. For Harrison, it was "ironic to think that while authorities are steadily urging that antibiotics be used only with great discrimination on the grounds of dangerous resistance building up, the agricultural authorities are encouraging even wider use. Perhaps, these two should get together some time to discuss the matters, before it is too late."[68] According to *Animal Machines*, agricultural antibiotics and "factory farming" were synonymous with health, environmental, and ethical problems. This fusion of concerns was reinforced by Rachel Carson's foreword. For Carson, the days of pastoral agriculture were over. Instead of animals wandering over green fields, producers had erected "factorylike buildings in which animals live out their wretched existence."[69] As a biologist, Carson found it inconceivable that such animals could produce healthy food. Establishments were regularly swept through with diseases, and were "kept going only by the continuous administration of antibiotics."[70] Previously associated with promoting health and spreading prosperity, agricultural antibiotics were now portrayed as upholding an unhealthy and "unnatural" system of dark satanic mills.

While protracted negotiations prevented its publication in the United States,[71] the public attention paid to *Animal Machines* in Britain was impressive. This was in part due to a pre-publishing campaign in the left-leaning *Observer*. Titled "Inside the Animal Factories"[72] and "Fed to Death,"[73] articles by Harrison introduced readers to the main claims of her book. In her first article, Harrison accused the "factory farmer and the agri-industrial world behind him"[74] of acknowledging cruelty only when profitability ceased. As long as animal growth remained stable, rearing systems were not questioned. Antibiotics were "incorporated in [animals'] feed and heavier doses of drugs [were] given at the least sign of flagging."[75] Focusing on poultry, Harrison claimed that young birds suffering from respiratory diseases or cancer often ended up on consumers' tables—the birds' ill health masked by antibiotics.[76] In her second article, Harrison focused on the intensive rearing of calves in darkened sties. Calves' diets consisted almost "exclusively of barley, with added minerals and vitamins, antibiotics, tranquilisers and hormones."[77] Living in these conditions, some calves became blind and many suffered from liver-damage and pneumonia: "their muscles become flabby and they put on weight rapidly, *but they are not healthy*."[78] Using more antibiotics to keep animals alive, farmers and veterinarians contributed to a race "between disease and new drugs."[79] Quoting veterinary practitioners and the first Netherthorpe report, Harrison warned about AMR and residue-laden "tasteless meat"[80] from factory farms.

Reactions to Harrison's claims ranged from furious denial to emphatic support. Seven days after publishing the second article, the *Observer* had received around 320 letters.[81] Many expressed outrage: one reader compared animals' suffering to child labor;[82] a second reader demanded labeling products from intensive farms;[83] a third reader urged compatriots to imagine pets incarcerated in factory farms.[84] While RSPCA Chief Secretary John Hall praised Harrison,[85] animal health lecturer David Sainsbury accused her of presenting a "grossly distorted picture of what is *actually* happening."[86] In Wales, the dean of Llandaff rehashed the "nation of animal lovers" theme by comparing factory farms to Nazi concentration camps—thereby "othering" intensive farming as barbaric and anti-British: in a speech covered by both the *Daily Mirror* and *Guardian*, the dean also warned his congregation about antibiotic and hormone residues.[87] The trope of antibiotic-abuse on "farm Belsens"[88] was powerful in both left-leaning and conservative circles. After publishing a misogynist attack on the "fertile mind" of the "housewife, mother, and vegetarian"[89] Ruth Harrison, the *Daily Mail* was inundated by letters condemning its supposed endorsement of materialism and un-British barbarism on KZ-like farms.[90] Subsequent reporting in the *Mail* was notably more subdued.[91]

Although Karen Sayer and Abigail Woods have shown that intensive indoor farming was by no means ubiquitous,[92] debates about factory farming soon penetrated national politics. In Parliament, Labour MP Joyce Butler launched an inquiry into the agricultural use of chemicals and residues in food.[93] Public calls for the labeling of "factory farmed" food were examined by the British Food Standards Committee.[94] When opening the 1965 Royal Dairy Show, Prince Philip was handed a copy of *Animal Machines*.[95] Cross-party pressure ultimately forced a reluctant government to launch a review of animal welfare led by medical scientist Francis W. Rogers Brambell.[96] Published in 1965, the influential Brambell report stated that animal welfare was more than the absence of physical pain and also comprised mental and behavioral aspects. Animals should have the freedom to stand up, lie down, turn around, groom themselves, and stretch their limbs. Although the drugs were not included in the committee's brief, the Brambell report commented on antibiotics' beneficial role in protecting animals from disease but found itself unable to assess long-term effects on public health.[97] This risk assessment was about to change.

In the same year that *Animal Machines* linked AMR and cruelty allegations, Britain's last major typhoid outbreak in Aberdeen brought home the microbial hazards of industrialized food production. The responsible *Salmonella*

Typhi strain had spread via contaminated Argentinian meat but had proven susceptible to antibiotic treatment with chloramphenicol (chloromycetin).[98] However, experts worried that future episodes might prove chloramphenicol resistant. Resistant typhoid outbreaks with higher fatality rates were already being observed in India, West Africa, Greece, and the Middle East.[99] One of the concerned experts was Ephraim Saul (E. S.) Anderson. As director of the Public Health Laboratory Service's (PHLS) enteric reference laboratory, Anderson had provided expertise to the recent Netherthorpe review of agricultural antibiotics and advised during the Aberdeen typhoid outbreak.[100] Anderson's ability to draw on extensive bacteriological resources and growing skepticism regarding agricultural antibiotic use would play a decisive role in creating widespread public support for AMR-focused antibiotic reform.

In 1965, Anderson and geneticist Naomi Datta published a paper titled "Resistance to Pencillins and Its Transfer in Enterobacteriaceae" in the *Lancet*. In their paper, Anderson and Datta reported the discovery of transferable AMR against multiple antibiotics in *Salmonella typhimurium* isolates from humans and a pig. *S. typhimurium* was a close relative of typhoid and a leading cause of food poisoning. Isolated plasmids (chapter 1) had the same AMR patterns and were likely related to each other. Given pigs' role as *S. typhimurium* reservoirs, the authors speculated whether transferable AMR might have first arisen on farms and then spread to human settings. In the laboratory, Anderson and Datta had also managed to transfer multiple-resistance from wild-type *S. typhimurium* isolates to sensitive *Escherichia coli* strains, which in turn transferred multiple-resistance to sensitive *S. typhimurium* cultures.[101] The fact that bacteria could exchange genetic information was not new. Joshua Lederberg and Edward Tatum had observed bacterial conjugation in 1946, and Tsutomu Watanabe had shown that plasmids could encode resistance to multiple antibiotics during the late 1950s.[102] What was new about Anderson and Datta's 1965 paper was that horizontal resistance selection and transfer also occurred in nonhuman settings and could cross over to bacteria in human populations. Popularizing the dangers of horizontal gene transfer and nonhuman AMR reservoirs, Anderson and Datta warned: "Many of the drug-resistant strains of *S. typhimurium* causing human infection may originate in livestock."[103]

Three months later, Anderson published a second paper together with M. J. Lewis in *Nature*.[104] Reporting a dramatic rise of AMR in *S. typhimurium* phage Type 29, the authors linked the spread of type 29 to calf transports and warned against the "infective hazards of intensive farming."[105] By the end of the year, Anderson published an even more direct attack on agricultural antibiotics in the *British Medical Journal* (*BMJ*): between December 1964 and November 1965, Anderson had collected over 1,200 animal (mainly calf) and 500 human samples of type 29 *S. typhimurium*. Of these, 97.6 percent were drug-resistant. In contrast to earlier papers, Anderson was able to provide concrete evidence of AMR transfer from animal to human bacteria: human and animal *S. typhimurium* samples showed similar resistance to furazolidone, a drug used exclusively in veterinary medicine. Anderson was certain that of the analyzed samples "most human infections of undetermined source were bovine in origin."[106] Anderson's data prompted the *BMJ*'s editorial to wonder whether the risks of veterinary ampicillin use—the other antibiotic of choice against typhoid—could be so great that "it may perhaps be thought advisable to abandon this form of treatment."[107]

The public impact of Anderson's AMR warnings was impressive. Published only one year after *Animal Machines*, scenarios of "infective resistance" underlined Ruth Harrison's criticism of intensive farming and were seemingly corroborated by other international studies showing rising AMR in *E. coli*, *Salmonella*, and *Staphylococci*.[108] In popular discourse, the "factory farm" became firmly connoted as a place of dubious welfare and dangerous bacteria. In February 1965, a *Times* report suggested "that antibiotics should be kept well away from livestock food."[109] Two months later, the *Daily Mail* reported on a medical conference at the Yorkshire Institute of Agriculture. A panel on "chemical farming"[110] had called for residue curbs and AMR-focused restrictions of medically relevant antibiotics. By June 1965, the Labour government was facing calls from its own MPs to investigate AGPs' potential hazards.[111] In November, the *Observer* blamed "super-farms"[112] for AMR. According to the *Times*, Anderson's findings and new research by veterinary bacteriologists necessitated a wider "reappraisal of the use of antibiotics."[113] Meanwhile, another *Observer* article explicitly warned against "factory farm bacteria."[114] Whereas AMR warnings did not resonate strongly in the residue-focused United States (chapter 2), the combined force of *Animal Machines* and "infective resistance" led to the formation of an AMR-focused risk episteme in Britain.

This risk episteme also influenced Britain's organic sector. By the late 1960s, publications like *The Wholefood Finder* referred to AMR and the "cruelties of factory farming"[115] before detailing the Soil Association's new definition of organic food. Relying heavily on *Animal Machines* and going beyond US publications' focus on residues, the Soil Association's "wholefood standards" required that animals be fed organic feeds and kept in a free environment defined according to Brambell-inspired comfort regulations. Organic products had to be free of antibiotics and other chemicals. On farms, antibiotics and other drugs should not be used routinely but only "in an extreme emergency."[116]

British officials similarly revaluated existing antibiotic policies. In spring 1965, the British government reacted to Anderson's AMR warnings by reconvening its Netherthorpe committee on AGPs (chapter 7). In January 1966, the Netherthorpe committee called for a new committee to reevaluate agricultural antibiotics in general.[117] While the government was slow to respond to this demand, new studies, ongoing media reports, and parliamentary inquiries kept the issue of "infective resistance" alive.[118] In 1966, veterinary researcher Herbert Williams Smith published a study showing that transferable plasmid-mediated resistance against multiple antibiotics was common in pathogenic *E. coli* isolated from both humans and pigs, calves, and fowls receiving antibiotics: "The high incidence of resistant strains to a large number of drugs and the complex resistance patterns of some of the strains was a disquieting feature of this survey, particularly as the diseases caused are acute and severe to the extent that they may terminate fatally if the drug with which they are first treated is not active against the infecting strain; the result of sensitivity tests cannot be awaited before commencing treatment."[119] The *Guardian*'s Anthony Tucker and Bernard Dixon from the *New Scientist*—both allies of E. S. Anderson—used scientific warnings to press for antibiotic reform.[120] Dixon in particular attacked "the irritating British habit of seeking expert guidance on a technical matter and then pigeon-holing the advice when it comes."[121] Citing Anderson and Williams Smith, Dixon referred to the danger of multi-resistant *E. coli* strains causing neonatal diarrhea in babies.[122] By December 1967, these warnings sounded tragically prophetic. Described by Robert Bud in chilling detail, multi-resistant *E. coli* 0119 and 0128 caused a severe outbreak of gastroenteritis among infants in the northeastern town of Middlesbrough. Poor hospital hygiene and transferring infected infants to other hospitals spread the infection. Fifteen infants died.[123]

Although there were no proven links, preconditioned British readers connected the multi-resistant Middlesbrough strains to intensive farming and agricultural antibiotic use. It seemed as though there was no end of dangers accruing from factory farms. Following heated exchanges between veterinary science lecturers and organic campaigners,[124] an article in the *London Illustrated News* linked the Teesside epidemic to agricultural antibiotic use: "one cannot help wondering why man should take the chance of placing himself in danger of returning to conditions of the pre-antibiotic era when, for example, the death of fourteen babies from gastro-enteritis would certainly not have made news headlines."[125] In February 1968, a *BMJ* review of the Middlesbrough outbreak by E. S. Anderson poured further oil onto the fire of speculation. Although he noted that it was not possible to distinguish between R-factors of human and animal origins, Anderson warned that the transferable AMR of the Middlesbrough strains might well have originated in nonpathogenic *E. coli* on farms:

> The risk to man that arises from the too free use of antibiotics in human medicine is obvious. . . . It should be remembered, however, that they could also have arisen in livestock as the result of antibiotic malpractice . . . , which has made it practically certain that not only multiple-drug-resistant pathogens such

as *Salmonella typhimurium* but also multiresistant non-pathogenic *E. coli* are transmitted from livestock to man, the latter on a large-scale. . . . It is pointed out that restricting of the use of antibiotics in man and animals is overdue.[126]

The Middlesbrough epidemic put immense pressure on the British government to implement the reconvened Netherthorpe committee's suggestions and launch a wider review of agricultural antibiotic use.[127] Appointed in July 1968 and announcing its findings in November 1969, the Joint Committee on the Use of Antibiotics in Animal Husbandry and Veterinary Medicine—the so-called Swann committee—divided antimicrobial substances into therapeutic and nontherapeutic antibiotics.[128] While therapeutic antibiotics were relevant to human medicine, nontherapeutic antibiotics were considered medically irrelevant. Only nontherapeutic antibiotics below certain doses were to be allowed in growth promoter feeds. Medically relevant penicillin, chlortetracycline, and oxytetracycline were to be banned from AGPs. The Swann committee, however, left many other areas of agricultural antibiotic use untouched: it merely cautioned against the use of chloramphenicol in veterinary medicine and did not address ongoing veterinary prescriptions of now restricted medically relevant antibiotics (chapter 7).[129]

Facing substantial public pressure—not least because Michael Swann, the committee's head, had noted that the Middlesbrough strains might have originated on farms—[130] the British government hastily committed to implementing the Swann report. Despite protests by US manufacturers,[131] the Swann recommendations were endorsed by nearly all segments of the British media. Most commentators explicitly noted that the proposed AGP restrictions were based on precautionary risk predictions rather than hard evidence of AMR-related harm. They were also aware that the Swann proposals left most other aspects of agricultural antibiotic use unchanged. While a *Times* editorial lauded the decision to limit laypersons' access to therapeutic substances,[132] agricultural correspondent Leonard Amey noted that a more ambitious complete ban of antibiotics would have been politically unfeasible and could have jeopardized British intensive animal production.[133] This continued tolerance of many other forms of agricultural antibiotic use was criticized by the *Guardian*.[134] However, overall, even critical voices were satisfied that agricultural AMR selection had now been successfully addressed in Britain.

On both sides of the Atlantic, the 1960s thus saw initial public enthusiasm about agricultural antibiotics become ambivalent. However, antibiotic perceptions were not the same in the United States and the United Kingdom. Economic constraints and accusations of US profits accruing from "British science" had already led to a different tone of British reporting during the 1950s. But it was in the field of risk perceptions that transatlantic views of agricultural antibiotic use differed most. In the United States, deep-seated concerns about food adulteration led to a framing of antibiotic risk nearly exclusively in terms of anti-biotics' potential presence as invisible residues (chapter 2). A very different risk episteme emerged in the United Kingdom when traditional animal welfare concerns fused with new "infective resistance" warnings under the dystopian umbrella of the factory farm. This episteme framed antibiotic risk as AMR spreading from cruel factory farms to humans. On both sides of the Atlantic, interpretations of antibiotic risk were thus influenced by their alignment with already existing deep-seated cultural risk narratives. The resulting national risk epistemes exerted a powerful influence on agricultural and political decision-making.

6

Bigger, Better, Faster

Antibiotics and British Farming

The evolving antibiotic risk episteme had profound implications for British agriculture. In contrast to their market-driven introduction to US agriculture, postwar constraints and powerful veterinary interests led to a more gradual top-down introduction of antibiotics in Britain. The different structure of British livestock husbandry and mixed feed trials also meant that farmers were more hesitant to embrace AGPs following their licensing in 1953. Concerted pro-antibiotic campaigning by British officials, industry representatives, and agricultural experts only gradually overcame this hesitancy. By 1960, many British farmers were regularly using antibiotics but—with the exception of the poultry sector—husbandry systems remained less intensive and more diverse than in the United States. Meanwhile, rising public antibiotic criticism triggered disputes between British farmers and veterinarians. Veterinarians attempted to gain greater control over the animal drug market and promote preventive health schemes. Although many farmers shared concerns about drug residues and new production systems, agricultural experts and organizations like the NFU opposed losing direct antibiotic access. Compromise-oriented corporatist politics subsequently played an important role in shaping British antibiotic reform without jeopardizing antibiotic access.

The Urge to E-X-P-A-N-D

During the early 1930s, the outlook for British farming had been bleak: as inhabitants of the largest agricultural free-trade market in the world, farmers were exposed to sinking food prices and a flood of cheap imports. Unable to compete, employment in the agricultural sector fell and productivity decreased until 60 percent of British food had to be imported. Although it undertook brief forays of agricultural support during the 1920s, the British government only gradually abandoned *laissez-faire* policies: the Agricultural Acts of 1931 and 1933 created tariff walls and marketing boards for various farm products.[1] Reacting to worrying developments on the European continent, the United Kingdom established a Food Department in 1936 and began stockpiling food and agricultural supplies. By 1939, British officials were propagating agricultural expansion to provide additional calories in the likely event of war.[2]

Following the outbreak of hostilities, the nascent alliance between producers and officials developed into a corporatist system of decision-making. Corporatism is a system in which representative groups assume some responsibility for the self-regulation of their constituency in return for privileges, a close relationship with the government, and the ability to reach bargained agreements with state agencies. Corporatism was particularly pronounced in the British health and agricultural sectors.[3] In the case of agriculture, farmers were integrated into War Agricultural Executive Committees, which were controlled by the Ministries of Agriculture and Food. Staffed by officials and farmers, committees enforced ministry directives at the local level but also advised and graded farmers' productivity. Unproductive or recalcitrant farmers could have their land expropriated. Attempting to maximize caloric output, wartime administrators prioritized plant production and introduced guaranteed prices by purchasing farmers' produce.[4] While pig and poultry stocks plummeted and feedstuffs were rationed, British farmers increased caloric output by 50 percent.[5] By the end of the war, farmers were celebrated for feeding the nation during the Nazi onslaught.

Producers' integration into corporatist structures had advantages for officials and farmers alike. Enabling vital government control over wartime food production, it also fostered self-organization of agricultural interests.[6] Despite occasional expropriations and diminished individual freedom,[7] the majority of British farmers and their lobby, the National Farmers' Union (NFU), were eager to continue the profitable corporatist alliance with the state after 1945. Corporatist prospects were promising. In contrast to the interwar years, most farmers' coffers had been flushed by fixed wartime prices and subsidized rural development. The postwar economic situation made the Labour government equally willing to continue the corporatist alliance. In August 1945, the United States's termination of the Lend-Lease agreement necessitated the repayment of wartime loans and left Britain critically short of foreign currency. Attempting to reduce expensive imports, the government embarked on a program of subsidized agricultural expansion with the Agricultural Act of 1947. Perpetuating annual price reviews and intervention purchases, the act was designed to give farmers and farm workers fair returns and stimulate agricultural investment.[8] Agricultural productivity was to be improved by the newly founded National Agricultural Advisory Service, improvement grants, and corporatist marketing boards for agricultural produce.[9] Farm organizations were integrated into decision-making both at the national and regional levels. Eager to control government expenditure, officials soon relied on organizations like the NFU to self-coordinate support by a "myriad of small independent producers"[10] for expansion programs, disease eradication, and farm improvement. In return, the NFU gained an exclusive and powerful place at the table when it came to determining price supports and the regulation of new technologies.[11] According to Michael Winter, the NFU soon became so influential that membership appeared "almost 'compulsory' for many farmers."[12]

Corporatist agricultural arrangements remained stable even after the end of postwar currency shortages and rationing. Ignoring accusations of featherbedding farmers, successive governments supported a system of deficiency payments, which replaced direct intervention purchases: once market prices fell below guaranteed official prices defined by annual reviews, the state paid farmers the difference between guaranteed and real prices. Similar to the United States, sinking international food prices and domestic surpluses soon made state expenditure rise dramatically in areas like beef and dairy production. Attempting to curb expenditure by improving agricultural efficiency, the Conservative government's 1957 Agricultural Act allowed limited annual reductions of price guarantees and shifted the emphasis from subsidies to improvement grants, which by 1960 constituted about 40 percent of agricultural support. The government also abolished powers of supervision and eviction. However, the underlying corporatist consensus on further increasing production remained intact.[13]

During the 1950s, many farmers invested wartime earnings and borrowed heavily to expand production and productivity.[14] Agricultural magazines like *Farmers Weekly* and the NFU's *British Farmer* were full of advice on better husbandry methods, disease eradication, and basic economics for expanding farmers.[15] Presenting technological sophistication as a badge of pride,[16] agricultural boosters intermixed the trope of having "fed the nation at war" with scenarios of global overpopulation.[17] For younger farmers, the rule was "never farm backwards."[18] One article claimed that while "nature intended a bird to lay only 24 eggs a season," scientific nutrition and husbandry meant that "there [was] no reason why she should not reach the 300 mark."[19] Frequently reminded to treat an animal "as a manufacturing unit,"[20] "the urge to E-X-P-A-N-D"[21] was said to be particularly pronounced in young ambitious and modern farmers.

Farmers heeded the urge. While Britain produced 762,000 tons of meat in 1947, it produced 1,713,000 tons in 1960.[22] Meanwhile, the number of workers on British farms declined from 843,000 in 1950 to 645,000 in 1960 with total factor productivity increasing from 67.5 in 1953 to 83.4 in 1963 (1973 = index 100).[23] However, not every farm intensified production. While Andrew Godley and Bridget Williams have shown that vertically integrated intensive rearing systems soon became common in the poultry sector,[24] Karen Sayer and Abigail Woods have noted that intensive and confined production methods were adopted unevenly in the pig, cattle, and egg production sectors. Contradicting public notions of universal factory farming, much of British livestock production remained characterized by a diversity of indoor and outdoor production systems.[25] In part, this diversity had cultural reasons. Some farmers feared that improved efficiency would increase social divides between a shrinking number of farmers and the general public while others feared technological alienation from animals and healthy nature.[26]

From Wariness to Routine

Postwar economics, corporatism, and the diverse structure of livestock production also impacted antibiotics' introduction to British farms. Veterinarians and farmers were already familiar with therapeutic antibiotic use for individual animals. However, in contrast to their rapid demand-driven introduction in the United States, nontherapeutic antibiotic applications had to be actively "sold" to initially hesitant British livestock farmers.

Similar to the United States, the interwar period saw a growing number of drugs and supplemented feeds used on British farms. However, effective remedies for infectious disease remained rare. As late as 1938, manuals like the *Handbook of Modern Pig Farming* had only been able to recommend a combination of preventative measures and culling for most diseases.[27] Advertised in British trade journals from 1938 onward, new sulfonamides significantly enhanced farmers' and veterinarians' ability to treat disease in individual animals. The new drugs later also featured in James Alfred Wight's (alias James Herriot) popular semi-autobiographic books on his time as a veterinary surgeon in Yorkshire between the 1930s and 1950s.[28] In *Vet in Harness* (1974), Wight's character diagnoses calves with pneumonia and convinces the farmer to use M&B 693 (sulphapyridine). The recovery is miraculous: "I didn't know it at the time but I had witnessed the beginning of the revolution.... The long rows of ornate glass bottles with their carved stoppers and Latin inscriptions would not stand on the dispensary shelves much longer .... At last we had something to work with, at last we could use drugs which we knew were going to do something."[29]

Initially, sulfonamides were obtainable without a veterinary prescription. Although interwar officials were beginning to substitute self-regulation with

FIGURE 6.1 An early antibiotic advertisement, M&B 693, *Veterinary Record*, August 1940.

what Stuart Anderson has called "a medical system of control through the writing of prescriptions,"[30] legislation remained fragmented. In 1920 and 1925, the Dangerous Drugs and Therapeutic Substances Acts restricted narcotics and mandated the licensing of producers of biologicals like vaccines and sera. However, so-called poisons remained available over the counter in licensed pharmacies. Meanwhile, the Pharmaceutical Society of Great Britain's power to determine whether a substance was a poison and who could sell it meant that most new drugs—including the azo dye derived sulfonamides—continued to be categorized as poisons.[31] The 1933 Pharmacy and Poisons Act's "trade or business relaxation" enabled farmers to purchase sulfonamides by signing the Poisons Register of a pharmacist, who was an Authorized Seller of Poisons.[32] In 1941, sulfonamides' legal status as poisons was both restricted and confirmed by the Pharmacy and Medicines Act, which listed a range of conditions that should only be treated by medical doctors.

Throughout the early 1940s, actual drug availability was, however, severely constrained by wartime shortages. This was also true for the new biological antibiotics. Produced by biological organisms, antibiotics like penicillin were regulated not as poisons but as therapeutic substances. Their distribution was controlled by the wartime Defence Regulations. Because of AMR concerns and limited supplies, prescription requirements for biological antibiotics were enshrined in the 1947 Penicillin Act, which was extended to encompass chlortetracycline and chloramphenicol in 1951.[33] Access to sulfonamides was also restricted. Although full-time farmers could still purchase sulfonamides directly from pharmacists, the 4th Schedule of the 1933 Pharmacy and Poisons Act now exempted sulfonamides from prescription requirements only when they were "contained in ointments and surgical dressings or in the preparations for the prevention or treatment of diseases in poultry."[34]

Relatively unaffected by wartime cuts of animal production, dairy farmers were among the vanguard of 1940s British antibiotic users. Government efforts to increase milk output and reduce labor inputs had led to a national drive against mastitis. Following veterinary diagnosis, farmers were given access to subsidized sulphanilamide to treat infected udders. The strategic importance of wartime milk production is also illustrated by the fact that precious penicillin, which was still being recycled from human urine, was donated by Howard Florey for experimental use against udder infections in 1941. Although doses remained low (about 10,000 units per treatment), veterinarians studying mastitis were granted preferential penicillin access in 1943. After 1945, penicillin trials against mastitis caused by *Streptococcus agalactiae* in commercial herds fostered optimism about wider disease eradication.[35]

British pharmaceutical producers catered to rising veterinary demand for chemotherapy against mastitis. In 1946, Wellcome introduced collapsible single-dose mastitis tubes with penicillin G in oil-wax suspension, which did not require refrigeration and were less irritating than preparations containing sulfonamides or gramicidin.[36] Sulfonamides and biological antibiotics were also used to treat pneumonia, footrot, and calf diphtheria.[37] However, supply problems prevented trials of US broad-spectrum antibiotics and streptomycin until the early 1950s.[38] As late as 1951, British officials complained about the "absolutely prohibitive"[39] cost of aureomycin treatments.

Rising agricultural demand for antibiotic prescriptions soon led to discussions about what constituted "rational" drug use. Strengthened by their wartime integration into disease control programs and 1940s prescription requirements, British veterinarians were in a far more powerful position to influence antibiotic use than their US colleagues.[40] However, veterinarians' ability to prescribe and sell antibiotics also created a conflict of interest. While national veterinary organizations promoted the image of veterinarians as reliable antibiotic stewards, local vets were incentivized to use antibiotic prescriptions to maximize income and secure the loyalty of paying customers. As early as 1948, articles in the *Veterinary Record*, organ of the British Veterinary Association (BVA), debated responsible prescription practices: "many practitioners are receiving requests from their clients to leave a supply with them, to make provision for the immediate treatment of new cases or to be used for control purposes."[41] Because farmers knew how easy it was to use antibiotic tubes, many were unwilling to pay veterinary fees for every single treatment and demanded bulk prescriptions for self-use. The situation presented a quandary for veterinarians, who depended on local farmers' good-will for their income: "Some, influenced perhaps by ethical considerations or by the sight of boxes of penicillin tubes left lying about on the window ledges of cowsheds, might wish to read into the wording of the [Penicillin] Act the inference that on no account should a penicillin preparation be left for the farmer to administer himself. Some might go to the other extreme and take the view that penicillin may be freely supplied to the client."[42] Veterinarians should try to strike a balance between stewardship and an "unnecessary restrictive"[43] prescription approach.

In practice, veterinary journals' exhortations for "rational" prescriptions could be trumped by what some condemned as "rank commercialism."[44] Veterinarians' control over the increasingly lucrative drug market also exacerbated tensions with pharmacists.[45] Writing to the *Veterinary Record* in 1953, a Cheshire pharmacist noted that little had changed since the 1947 Penicillin Act: "the simple fact is . . . the veterinary surgeon who supplies drugs in bulk for the farmers' use is just as much keeping open shop as any retail pharmacist, and as such should not grumble if he meets competition from the pharmacist."[46] Just how easy it was to obtain restricted antibiotics from some veterinarians is illustrated by the 1956 prosecution of three Yorkshire practitioners. After an undercover inspector had purchased antibiotics from an unqualified assistant, it emerged that it was common practice for local farmers to either come in person, send their sons, or ask for antibiotics on the phone. Drugs were promptly sold without an accompanying herd inspection. According to the accused vets, "it would be ludicrous to run out every time a farmer telephoned to say he had another outbreak of mastitis."[47]

While British farmers were happy to buy prescribed antibiotics to treat and prevent disease, they were initially more skeptical about nontherapeutic applications. Despite closely following developments in the United States,[48] subdued early perceptions of antibiotic growth promotion were caused by disappointing feed trials, the different composition of British animal feeds, and the late derationing of commercial feeds in August 1953.[49]

Similar to the United States, British researchers had been interested in the nutritional value of antibiotic fermentation wastes since the 1948 equation of vitamin B_{12} with APF (chapter 3). Collaborating with Glaxo Laboratories at Reading's National Institute for Research in Dairying, nutritionist Raphael Braude started feeding streptomycin liquor residues to piglets in May 1949. Unaware of the parallel discovery of the antibiotic growth effect, Braude and his colleagues tested different combinations' impact on weight gain and B_{12} concentrations in livers. Initial results were disappointing: no growth promotion was observed and pigs fed streptomycin liquor did less well than controls fed iron supplements.[50] Following the 1950 announcement of the antibiotic growth effect, Britain's Agricultural Research Council (ARC) commissioned a new round of antibiotic feed trials on pigs and poultry in research laboratories in Reading and Northern Ireland as well as by manufacturers like the Distillers Company, Glaxo Laboratories, and the British Oil and Cake Mills (BOCM). However, once again, trials with domestic procaine penicillin, streptomycin, and detergents proved inconclusive. In the case of pigs, antibiotics seemingly promoted the growth of animals fed Midwest-style vegetable protein diets but not of animals fed common British diets, which were rich in animal protein. According to Reading-based researchers, "the effect of the addition of antibiotic supplements [to domestic feeds] is unlikely to be of commercial importance."[51] Positive results were, however, achieved by adding antibiotics to the diets of runts and pigs suffering from scours.[52] Trials on pullets at the Rowett Research institute in Aberdeen produced mixed results.[53]

Following a meeting in October 1951, the ARC decided that initial data was insufficient to justify amending the 1947 Penicillin Act and commissioned further feed trials. This cautious approach was criticized by British antibiotic manufacturers, who warned about impeding commercial progress in "the birthplace of antibiotics" (chapter 5).[54] Focusing on pigs, the second round of ARC-sponsored trials tested US aureomycin supplements alongside British-manufactured procaine penicillin solutions and B_{12}. One group of pigs was fed an antibiotic supplemented diet containing vegetable and animal protein (white fish meal), which was considered typical for British agriculture, while the other group was fed an American "vegetable protein" diet. This time, both penicillin and aureomycin were found to promote animals' weight gain by between 1.4 to 40.2 percent over control animals. Live weight gains were greater on vegetable diets with US aureomycin slightly outperforming British penicillin. A second set of experiments with antibiotic creep feeds for suckling pigs led to similar results. However, strong variations between individual trials made British experts shy away from uniformly endorsing AGPs. In 1953, the ARC cautiously concluded that the average farmer "should be able to derive a more positive commercial benefit from the improvement of food conversion."[55] Despite ongoing uncertainty about feeds' action and performance, it was thought that AGPs led to 8 to 10 percent speedier growth and about 5 percent better food conversion in pigs and poultry.[56]

A second important reason for British farmers' lukewarm endorsement of AGPs may well have been the ample domestic supplies of vitamin B_{12}. As described in chapter 3, an oversupply of cheap vegetable protein and a coinciding scarcity of animal protein supplements partially explained US farmers' 1950s embrace of B_{12}-rich aureomycin fermentation wastes. This was not the case in Britain where the country's highly organized fishing industry created an oversupply of animal protein and commercial compounders were legally required to add animal protein to feeds for young animals.[57] Even before restrictions on commercial feeds were relaxed in 1953,[58] British farmers had had access to relatively cheap animal protein via off-ration fish solubles. This access to cheap animal protein stripped early AGP/APF feeds of half their commercial appeal—especially since some commentators seemed to think that APF and AGPs were the same thing.[59] British antibiotic manufacturers explicitly referred to domestic B_{12} supplies in their campaign to ward off US competition. Writing for *Manufacturing Chemist* in 1952, Distillers Company researcher J. A. Wakelam advocated using *British* procaine penicillin in animal feeds: procaine penicillin production was based on the full extraction of penicillin mold and did not result in viable APF wastes. For British farmers, this was allegedly an advantage since supplementing domestic feeds with pure penicillin would avoid inefficient US AGP/APF "'blunderbuss' treatment[s]."[60] Other researchers soon countered that aureomycin was more able to resist the water and heat necessary for compounding feeds than penicillin.[61]

In view of varying feed trials, uncertainty about AGPs' stability, and new experiments indicating no growth promotion in germ-free chickens,[62] it is perhaps unsurprising that British farmers and their lobby, the National Farmers' Union (NFU), viewed antibiotic growth promotion very cautiously. Ahead of antibiotic feeds' licensing by the 1953 Therapeutic Substances Act, the NFU was less concerned about pushing for the rapid licensing of AGPs than about securing guaranteed minimum antibiotic concentrations in feeds and official guidelines for safe and efficient antibiotic use.[63] Lacking internal antibiotics expertise, the NFU relied heavily on information supplied by the state. As a consequence, government experts played a crucial role in convincing initially cautious farm organizations to promote subtherapeutic antibiotic use. Following a 1953 meeting, the NFU representative thanked officials: "The subject was one about which he and many other farmers were relatively ignorant and he was grateful for the information and advice given. He was in general agreement . . . , but felt that caution in propaganda and in the use of antibiotics was necessary."[64]

Even after Britain's Treasury and the Ministry of Agriculture, Fisheries and Food (MAFF) decided to license penicillin and aureomycin feeds with effect from September 1, 1953 (oxytetracycline AGPs were licensed in 1954), initial AGP sales were disappointing. At a symposium held shortly after AGPs' licensing, it was noted that given "the tremendous amount of publicity and propaganda that originated largely in America, . . . it was thought that there would be a tremendous demand for them in this country."[65] However, "the demand had been considerably less than . . . expected."[66] Antibiotic uptake also varied between livestock sectors.

In the case of pig husbandry, AGPs' popularity was diminished by ongoing uncertainty about feeds' efficacy and farmers' disappointment when weight gains turned out to be lower than advertised.[67] The prevalence of low-intensity outdoor systems may have been an additional factor. Following the slaughtering off of British pigs in 1940, postwar production gains had mostly been achieved by small-scale outdoor producers. Although individual officials and the Pig Industry Development Authority (PIDA) promoted more intensive husbandry systems during the 1950s, their efforts were only partially successful in the East of England. In other areas of the United Kingdom, inconsistent government support and overproduction led to a greater persistence of outdoor farming and a more piecemeal adoption of antibiotics than in the United States.[68]

Pig husbandry manuals were also skeptical of AGPs. While contemporary US manuals routinely recommended AGPs for indoor and outdoor herds (chapter 3), early 1950s British manuals followed prewar traditions by advocating sunlight and cod-liver oil and fish meal supplements as "natural" preventatives. Antibiotics featured alongside serum therapy as prescribed therapeutics but not as nontherapeutic feed additives.[69] In 1945, *Pigs: Their Breeding, Feeding, and Management* had not mentioned therapeutic antibiotics or sulfonamides.[70] Eight years later, its revised edition discussed "experimental" AGP trials on runts but cautioned: "healthy pigs on a [non-vegetable] ration do not make an economical response to the use of antibiotics."[71] In 1956, a new version of the manual maintained that feeding antibiotics to healthy pigs on "good" standard British rations might be unprofitable.[72] Similar concerns were voiced at British feedstuff conferences and in the first edition of *Farm Animals in Health and Disease* from 1954.[73]

Pig farmers' AGP skepticism was only gradually overcome by pharmaceutical marketing and expert endorsement. In 1957, the Rowett Institute averaged outcomes of recent experiments to estimate that feeding AGPs to "baconers" could result in a 5s. profit per pig. Although procaine penicillin was effective at lower doses, broad-spectrum AGPs produced more consistent results.[74] While these figures were bad news for British manufacturers, who tried to compete against US feeds by commissioning patriotic commercials,[75] American companies celebrated the success of their products in the farming press. In 1955, American Cyanamid's British branch boasted: "Last year, 1 in every 10 pigs in the United Kingdom had AUROFAC 2A Feed Supplement throughout its life . . . . This year, 1 in every 7 pigs in the United Kingdom is being fed on AUROFAC 2A Feed Supplement from birth to slaughter."[76] According to the manager of Pfizer's oxytetracycline plant in Kent, AGPs not only enabled British farmers to market pigs three weeks sooner but also saved enough feed to free about 300,000 acres for other crops.[77] Similar views were expressed by commentators in Britain's farming press. According to the popular magazine *Farmers Weekly,* antibiotics were changing the biological rhythms of British husbandry: instead of weaning piglets 56 days after birth, farmers were now advised to wean 24 to 28 hour old piglets with penicillin-enriched milk powder. This way, even runts would survive and piglets would already weigh about 40 pounds at their traditional weaning age.[78] Invoking an ideal of optimized nature, the magazine described antibiotics as a "boon to mankind."[79] Another article, titled "Our Debt to the Chemist,"[80] listed antibiotics, hormones, pesticides and insecticides among the great triumphs of twentieth-century science. By 1958, veterinary researcher Herbert Williams Smith estimated: "about 50 percent of all the pigs in Britain are [fed AGPs] and that nearly all unweaned piglets have access to food containing tetracyclines."[81]

The adoption of AGPs was more straightforward in Britain's poultry sector. British poultry production intensified rapidly following the establishment of the first broiler chicken farm by Geoffrey Sykes in 1953 and Sainsbury's marketing of frozen chickens in self-service supermarkets.[82] Rapid intensification also occurred in turkey and game bird production.[83] Similar to the United States, the increasing size of flocks was enabled by cheap chemotherapeutics and in turn boosted antibiotic sales. British poultry farmers had trialed preventive and therapeutic sulfonamide treatments since 1947. By 1950, popular US products like sulfaquinoxaline were also being manufactured in Britain.[84] Four years later, R. F. Gordon from the Animal Health Trust estimated that about 35 percent of British poultry were housed intensively and routinely received therapeutic and nontherapeutic antibiotics.[85]

Antibiotics' popularity was reflected in British poultry manuals. In 1946, the third edition of *Poultry World's Practical Poultry Keeping* still targeted small backyard producers when it recommended a mix of household remedies and hygiene for the prevention and treatment of disease.[86] By 1952, *Poultry Keeping for Profit* explicitly targeted farmers thinking about expanding production and recommended vaccinations and sulfonamide solutions to maintain birds' health.[87] In 1955, the revised fifth edition of *Practical Poultry Keeping* contained a chapter on feeds with an entire section devoted to antibiotics. Indicating how popular AGPs had become, poultry producers were cautioned

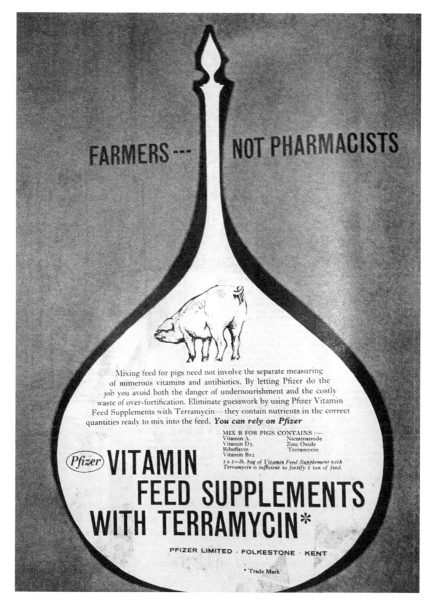

FIGURE 6.2 AGPs were presented as important parts of normal feed rations. Pfizer advertisement, *Farmers Weekly*, 1960.

that antibiotics "cannot be considered as food, but their use does have an effect on the way chicks are able to utilise [feeds]."[88] The manual also warned: "[antibiotics] do not help egg production or breeding and as a rule should not be fed to adult birds, although some breeders claim that they help birds through the moult."[89] Similar to the United States, many manuals' joint advocacy of

FIGURE 6.3 Antibiotic feeds were also sold as disease insurance. Cyanamid advertisement, *Farmers Weekly*, 1964.

routine antibiotic growth promotion, prophylaxis, and treatment meant that legal boundaries between nontherapeutic and therapeutic antibiotic use quickly blurred in practice.[90]

Producers' increasing adoption of antibiotics led to a boom of British drug sales. In 1954, an estimated 69,439 tons of antibiotic supplemented feeds were sold to farmers. By 1959, the number had grown by over 600 percent to 445,706 tons.[91] Feeds for growing animals could include up to 100 parts per million of

Look out now for these signs of coccidiosis . . .

The Farmers Weekly, March 30, 1951   27

1 Chicks may eat ravenously when first infected, but loss of appetite soon becomes pronounced.

2 Listlessness and depression follow, with some birds in a state of near-collapse.

3 Dead chicks — sometimes found without other visible symptoms to give warning of the outbreak.

4 *Blood in the droppings. This is the usual danger signal, and should be acted upon immediately.*

Add 4 tablespoonfuls of 'Sulphamezathine' to each gallon of drinking water. This will rapidly check the outbreak. Moreover, treated chicks are immune for life from further attack. Ask for free I.C.I. Farm Health Books Nos. 8 and 10.

'Sulphamezathine'

quickly controls Coccidiosis

'Sulphamezathine' is controlled as a poison and is obtainable from your chemist, who will advise you as to the regulations. In case of difficulty write to IMPERIAL CHEMICAL (PHARMACEUTICALS) LIMITED, FULSHAW HALL, WILMSLOW, MANCHESTER *A subsidiary company of Imperial Chemical Industries Ltd*

FIGURE 6.4 Poultry producers rapidly adopted antibiotics. ICI advertisement, *Farmers Weekly*, 1951.

penicillin, chlortetracycline, oxytetracycline, and bacitracin. In practice, about 25 parts per million of procaine penicillin or bacitracin were usually added to poultry feeds. In pig production, creep feeds normally contained up to 30 parts per million of broad-spectrum antibiotics while feeds for older growing pigs contained up to 10 parts per million of the same antibiotics.[92] In contrast to their increasing public association with "factory farms" (chapter 5), antibiotics were popular in both intensive and nonintensive settings. In 1962, one observer noted that early weaning, a focus on pork production, and "rough and ready" housing had led to the "greatest employment [of dietary antibiotics] . . . in the feeding of [British] pigs."[93] As indicated in manuals, AGPs were being used for both growth promotion and therapeutic purposes. In 1957, a former industry representative warned that farmers "frequently feed antibiotic-containing feeding-stuff supplements to their stock at levels so high as clearly to be exerting a therapeutic effect. Such action is taken deliberately, and it completely defeats one of the main objects of [the Therapeutic Substances Act], to say nothing of its effect on the incomes of veterinary surgeons."[94] Pharmaceutical revenues grew accordingly. Between 1948 and 1963, adjusted UK veterinary medicines output grew from £2,305,000 to £19,585,000 in current money whereas the output in animal and poultry foods grew from £50.3 million to £466.7 million. Meanwhile, veterinary medicines' share as a proportion of total UK pharmaceutical output increased from 3 percent to 8.3 percent.[95]

FIGURE 6.5 Antibiotics were said to increase efficiency. Cyanamid advertisement, *Farmers Weekly*, 1964.

Despite farmers' initial hesitancy, uneven adoption of drugs, and ongoing recourse to other disease control strategies,[96] an antibiotic infrastructure of easy-to-use growth promoters, therapeutics, and prophylactics was becoming established on British farms. Similar to the United States, drug use would continue to rise over the next decade. In 1960, about 40,000 kilograms of pure antibiotics were annually estimated to be used in British pig and poultry husbandry (about 29 to 35 percent of total antibiotic consumption). By the mid-1960s, the percentage of total British antibiotic output devoted to animals had increased to about 41 to 44 percent.[97]

Early Concerns

Antibiotics' popularity did not mean that British agricultural commentators were unaware of negative side effects. Similar to the United States, the early 1950s saw farming magazines warn producers about penicillin residues' effect on cheese production and mass-poisoning resulting from off-label sulfonamide use.[98] However, in contrast to the United States, AMR soon emerged as an equally prominent issue and divided opinions. Resulting agricultural debates centered on the risks and benefits of uncontrolled as opposed to "rational" veterinary antibiotic use. Whereas animal nutritionists and NFU representatives followed their US counterparts by presenting AMR as a problem of inefficient drug use, veterinarians and public health experts used AMR to assert control over antibiotics. At stake was not antibiotic use per se but antibiotic access.

Coinciding with veterinary debates about "rational" drug use, initial AMR concerns centered on the overuse of therapeutic mastitis treatments.[99] Around 1950, British veterinarians began to warn farmers about a "changing 'clinical picture' which might follow the extensive use of penicillin."[100] Observed changes were due to the replacement of sensitive *Streptococcus agalactiae* with resistant hemolytic *Staphylococci*. Having analyzed 500 udder isolates between 1951 and 1953, researchers from the Boots Pure Drugs Division reported an increase of infections caused by resistant *Staphylococci* but noted that organisms remained sensitive to high doses of penicillin.[101] According to *Farmers Weekly*, the percentage of mastitis outbreaks caused by *Staphylococci* rose from 10 to 30 percent between 1944 and 1955.[102] A further increase of penicillin resistance from 9 to 37 percent was detected in *Staphylococci* isolated from herd milk between 1954 and 1957.[103] In reaction, British veterinarians attempted to shore up antibiotic control by lambasting what they saw as "indiscriminate" US-style drug use "without any veterinary supervision."[104]

Although most agricultural commentators initially agreed that a certain degree of veterinary supervision was necessary to control AMR in the case of therapeutic antibiotic use, opinions were more divided when it came to nontherapeutic AGPs. Ahead of AGPs' licensing in 1953, British veterinarians wondered whether they should agree to exempt low-dosed feeds from prescription requirements. In 1953, a letter in the *Veterinary Record* cited Starr and Reynold's 1951 study (chapter 3) to warn about feeds' selection for AMR on farms.[105] However, other commentators argued that antibiotic dosages and blood stream adsorption were too low to exert selection pressure: "We are assured by bacteriologists [that AMR] is not likely because the action of these antibiotics appears to be a local one in the gut on the microflora there, and . . . the bacterial population is being constantly expelled by normal bowel evacuation and being replaced by a [susceptible] fresh one, ... Secondly, should the organism actually become resistant . . . it would be a simple matter to attack these supposedly resistant organisms by switching to another antibiotic."[106] The threshold-influenced distinction between higher therapeutic and lower nontherapeutic dosages initially reconciled many British veterinarians with AGPs.

This reconciliation was short lived. By the mid-1950s, new AMR data, booming AGP sales, and farmers' therapeutic and preventive use of increasingly higher-dosed feeds to substitute expensive veterinary health care led to clashes within Britain's agricultural community. The BVA's 1956 Conference on Supplements and Additives in Animal Feedingstuffs exposed nascent tensions. Although most speakers praised AGPs, AMR assessments varied significantly between industry representatives, pharmacists, and nutritionists on the one side and veterinary bacteriologists on the other side. According to Reading-based nutritionist Marie E. Coates, who had recently discovered the absence of antibiotic growth promotion in germ-free chickens, AMR fears were "groundless."[107] This view was seconded by British-born American Cyanamid biochemist Robert White-Stevens. Promoting higher-dosed AGPs, White-Stevens claimed that antibiotics were a natural companion of intensifying food production: "it was impossible to keep contagion down [in large intensive poultry flocks], even by the best methods of management; furthermore, the best sanitary management cost a lot of money, generally more than a comparable result from use of antibiotics."[108] Britain's leading AGP expert, nutritionist Raphael Braude, expressed irritation at contemporary veterinary AMR warnings. In a sign of growing professional tensions over antibiotic control, Braude "refrained from drawing analogies involving veterinary prescriptions, but as far as feed additives are concerned, . . . the sole criterion as to whether they should be freely allowed on the market should be . . . that if judiciously used they are harmless to the health of the animal."[109] Market forces would do the rest: farmers would phase out AGPs if AMR diminished their efficacy.

Veterinary bacteriologists begged to differ. Using sensitivity tests and a technology called bacteriophage-typing to differentiate between bacteria strains (chapter 7), the Animal Health Trust's Herbert Williams Smith had analyzed the microbial effects of antibiotic use on farms. His findings were worrying. Responding to a paper on AGPs in pig nutrition, Williams Smith noted that *E. coli* causing calf scours (diarrhea) had become resistant after antibiotic exposure. In the case of pigs, feeding AGPs had caused a sharp rise of tetracycline resistance in *E. coli* isolates. Resistant strains could spread to humans: 36 percent of *E. coli* isolates from pigs entering a bacon factory had been tetracycline resistant.[110]

In a preview of conflicts to come, Williams Smith's intervention led to an acrimonious discussion. Nutritionist Raphael Braude responded by explicitly accusing veterinarians of using AMR to regain control over antibiotics: "the fact that the resistant bacteria existed was meaningless, unless it could be proved that the resistant bacteria present, and possibly growing, were pathogenic organisms, which would cause disease. . . . Speaking bluntly, he thought it was high time that the authorities of the veterinary profession should accept the fact that there was plenty of work to be done without worrying how the antibiotics acted."[111] Attempting to keep the peace, Animal Health Trust founder W. R. Wooldridge warned that Braude "had raised a controversial point by implying that the veterinary profession was endeavoring to stop the use of antibiotics in feeding-stuffs."[112] Wooldridge upheld veterinarians' duty to protect animals and the public from resistant strains like *Staphylococcus* 80/81 (chapter 2) but maintained that it was up to agriculturalists how to use antibiotics on a daily basis. Attending officials followed a similar compromise strategy. Speaking for Britain's Ministry of Health, Senior Medical Officer John Marshall Ross warned about AMR in human medicine but shied away from linking it to agricultural antibiotic use. Instead, he focused on drug residues and antibiotic preservatives' potential selection for AMR among food-borne pathogens.[113] By contrast, Royal Veterinary College microbiologist Reginald Lovell doubted that there was any difference between the risks of therapeutic and nontherapeutic antibiotic use on farms: "We condemn [antibiotics'] indiscriminate use as therapeutic agents and subject them to some control, if we are logical, ought we not to condemn their use as dietary supplements whereby they extend their influence to a wider sphere?"[114]

Within three years of being legalized, AGPs had turned into a divisive issue for Britain's agricultural community. Keenly aware of AMR on farms and in medicine, all sides advocated "rational" antibiotic use for therapeutic purposes but were divided when it came to reinstating medical control over nontherapeutic AGPs. Ironically, the low doses that had made AGPs seem harmless in 1953 were increasingly viewed as particularly dangerous because they allowed bacteria to survive and adapt to antibiotic exposure. This allegation gained additional weight when veterinary bacteriologist Herbert Williams Smith and his collaborator W. E. Crabb published their AMR data shortly after the 1956 BVA conference. Studying *E. coli* in pig and chicken feces, the authors reported a strong correlation between antibiotic use and AMR rates. AMR selection in animals "would undoubtedly have an impact on the treatment of *bact. Coli* infection in those animals and possibly other species, including man, with which they come in contact. It is apparent that considerations of this nature should be given very serious thought before any chemotherapeutic agent is permitted to be used in such a widespread manner as the tetracyclines have been used in pig nutrition."[115]

Battles for Control

The emerging inner-agricultural conflict over antibiotic control would escalate over the next decade. During the 1960s, AMR concerns, the international cost-price squeeze, and an increasingly aggressive veterinary drive for preventive health care led to a wider revaluation of British agricultural antibiotic use and access.

Economically, the 1960s undermined many postwar promises of rural prosperity. With overall agricultural employment and government price guarantees falling, those staying in agriculture tended to work on fewer and larger farms.[116] In livestock production, developments once again varied between different sectors. While UK cattle production stagnated, poultry production increased. Pig production also increased and became more intensive during the second half of the decade.[117]

The direction of agricultural development was not uncontroversial among British farmers. Although many commentators continued to propagate expansion,[118] a growing number of articles in the farming press warned that many smaller producers would not survive the cost-price squeeze.[119] Pointing to parallel trajectories in the United States, a 1962 article in *Farmers Weekly* predicted an "end in sight for the family farms."[120] Two years later, delegates at the NFU's annual meeting clashed over a resolution to limit the size of farms. Mirroring public attacks on factory farms, the resolution called on the NFU to "ensur[e] that production of agricultural commodities remains with the farming industry" and "draw a line between [agricultural factories] and what is traditional agriculture."[121] However, opponents argued that "the resolution was in direct opposition to progress. Hens did not need green fields to run in these days. It was important that some products be produced intensively."[122] After a heated discussion, what would have been a small revolution for British farming was defeated by 174 to 128 votes.

Similar to the United States, economic pressure for increased and more efficient production coincided with controversies over rising antibiotic use on British farms. In 1960, the ARC's decision to review AGPs in view of AMR warnings took many agricultural observers by surprise. According to *Farmers Weekly*, the review "condemns those willing to take risks for what it admits can be considerable gains."[123] Closely integrated into government decision-making, British farmers could, however, be confident that their perspective would be taken into account—especially since the antibiotic review would be headed by former NFU president Jim Turner, now Lord Netherthorpe. While it was unlikely that antibiotic use per se would be challenged, the Netherthorpe review fanned inner-agricultural battles over control of antibiotic access.

Intensifying antibiotic conflicts took place against a backdrop of veterinary campaigns for government subsidized preventive animal health care schemes.

Described by historian Abigail Woods, organizations like the BVA were becoming concerned that successful disease eradication, productivity-oriented notions of animal health, and easy drug access were eroding traditional sources of veterinary income. By lobbying for official veterinary preventive care schemes, they hoped to convert "fire brigade approaches" in which veterinarians were only called to treat acute disease into a system of subsidized health checks and management advice for farmers. The added benefits of establishing "rational" medical oversight of on-farm antibiotic use and curbing AMR selection became a central theme of veterinary campaigning.[124]

British veterinarians' push for greater antibiotic control coincided with a push by livestock producers for unrestricted on-farm access to penicillin and other prescription-only medicines (chapter 7). Confident in their ability to diagnose and use drugs appropriately, a growing number of producers were turning to feed merchants rather than veterinarians for advice. In 1960, *Farmers Weekly* advised a farmer facing resistant coccidiosis to "complain to your feed merchants of the poor results you are getting and perhaps change to some other kind of medicated food."[125] By contrast, the BVA's *Veterinary Record* criticized "unthinking demands" for unsupervised access to "drugs and vaccines, many of them dangerous except in skilled hands and most of them too expensive to be used wastefully."[126] Instead of giving into farmers' antibiotic demands for fear of losing their custom, veterinarians should encourage "enlightened farmers"[127] to seek preventive health advice. Producers disagreed. Doubting the necessity of paying veterinarians to treat diseases that skilled farmers could also handle with appropriate therapeutics, the NFU's Gloucestershire section repeated demands for expanded access to prescription-only medicine in 1962.[128]

In the midst of battles over drug control, hardly anybody considered reducing overall antibiotic use. Throughout the early 1960s most agricultural commentators and manuals remained confident that AMR hazards were an acceptable price to pay for the many benefits of "rational" antibiotic use and continued to advocate it. Although British publications mentioned hazards much more frequently than contemporary US publications (chapter 3), they also mostly presented AMR as a management problem, which could be overcome by improving hygiene and using different drugs and drug combinations.[129] The 1962 Netherthorpe report strengthened this view. Shaped by fierce struggles between veterinarians, public health officials, and farming representatives, the corporatist report marked a victory for farmers and nutritionists by endorsing existing AGP regulations and calling for a legalization of AGPs for calves. However, it also strengthened British veterinarians by calling for an automatic restriction of future medically relevant antibiotics (chapter 7).

The hard-won Netherthorpe compromise on antibiotic access stood in contrast to rising public criticism of modern farming's overall use of agricultural chemicals. Although they quickly endorsed residue controls for milk in the wake of the 1963 penicillin scandal (chapter 5),[130] British farming and veterinary representatives faced increasing pressure from environmentalists and animal welfare activists throughout the 1960s.

Strategies of responding to criticism varied. In the case of rising environmentalist concerns about chemical use, British agricultural commentators frequently downplayed domestic criticism by pointing to more extensive American chemical use and blaming consumer tastes for driving intensive farming.[131] In 1963, *Farmers Weekly* reacted to the British publication of *Silent Spring* by claiming that Carson's warnings were valid but had few implications for British agriculture.[132] Deflecting attention to the US was not possible in the case of home-grown animal welfare criticism. British activists and farmers had two opposing concepts of welfare: while most farmers defined welfare in terms of thrift or physical productivity, campaigners extended definitions of welfare to encompass animals' mental well-being.[133] Since around 1960, British agricultural commentators had reacted to growing public welfare criticism by portraying campaigners as irrational.[134] In comparison to the "pot-bellied" prewar animals "with staring coats, housed in filthy hovels,"[135] scientifically designed modern intensive systems offered animals a much better life. Tensions increased significantly following the publication of Ruth Harrison's *Animal Machines* in 1964 (chapter 5). Reacting to Harrison's *Observer* articles, the NFU's *British Farmer* complained that the newspaper had joined the "anti-land lobby" by presenting a "grossly distorted picture of British agriculture."[136] *Farmers Weekly* bemoaned the Nazi imagery of welfare criticism: "Townspeople . . . have been given a horrifying picture of the 'animal factories' . . . . They are given a chilling picture of broiler house concentration camps and packing station Ausschwitzen [*sic*], of pig 'sweat-boxes'; of darkened torture-chambers for calves, and of animals going blind in intensive beef lots."[137] If animals were truly suffering, they would not produce income.

However, Britain's agricultural community soon found that frontal opposition would not sway domestic welfare criticism. According to Andrew Godley and Bridget Williams, "agriculture was pursuing intensification more energetically in Britain than anywhere else outside North America"[138] but had much less public and economic significance. Failing to prevent the official installation of the 1964 Brambell committee on animal welfare, both the NFU and the farming press chose not to jeopardize corporatist ties with Whitehall and Westminster and began to moderate their rhetoric. Stressing the necessity of an "informed climate,"[139] agricultural documentaries titled "Press Button Farms"[140] and "Look to the Land"[141] instead began to stress the quality, ethical soundness, and safety of intensive food production in Britain.

With initial outrage cooling, farming magazines even began to print occasional criticism of modern production methods' effects on animals and the hazards of technologies like antibiotics.[142] In the *Countryman*, Herbert Sinclair, director of Oxford's Laboratory of Human Nutrition, acknowledged allegations that AGPs were illegally being fed to laying poultry.[143] Although he complained about limited urbanite farming knowledge based on "comics of jolly pigs in trousers,"[144] the chairman of the NFU's Lincolnshire branch similarly cautioned that much still had to be learned about safe drug use while *Farmers Weekly* referenced diseased carcasses "sodden with antibiotics"[145] in abattoirs.

Infective Resistance

While self-criticism in the farming press mostly centered on illegal antibiotic use and resulting residue problems, British veterinarians used public concerns about welfare and AMR to reignite the battle for antibiotic control. Giving evidence to the Brambell committee in June 1964, the BVA called for further limitations on non-nutritive feed additives.[146] These calls were soon significantly strengthened by high-profile warnings of "infective resistance" spreading on farms by Britain's PHLS (chapter 5).

In contrast to the United States, the new horizontal AMR scenarios had an immediate impact on British public and agricultural debate. Whereas most British commentators had previously presented AMR as a spatially limited risk, which could be overcome by better management, hygiene, and drug combinations, horizontal gene transfer challenged the entire rhetoric of sustainable antibiotic use. Because every dose could select for uncontainable AMR genes, producers and practitioners had to ask themselves whether they were squandering precious antibiotic resources. Answers differed. While most British farming manuals and articles continued to endorse "rational" therapeutic and nontherapeutic antibiotic use,[147] commentators also acknowledged the need for some kind of reform. In 1965, the NFU's *British Farmer* warned: "Too many doctors and farmers are dosing human beings, pigs, calves and poultry with antibiotics for minor illnesses or as animal food additives. . . . This can mean that human beings and livestock are less easily treated for more serious epidemics, including typhoid in human beings. In short, the use of antibiotics has been overdone."[148] Ahead of the publication of the second Netherthorpe report (chapter 7), *Farmers Weekly* printed long reports on drug overuse, transferable AMR in calves and humans, alternative therapies, and improved management. The overall agricultural message was clear: AMR was serious and British farmers should self-limit antibiotic use.[149]

Whether certain products should be banned was another question. Having already abandoned the 1962 Netherthorpe compromise on AGP access, British veterinarians pressed for sweeping AGP restrictions. In 1965, the *Veterinary Record* devoted two lead articles to "infective resistance": "No reasonable person would now doubt the need for [a reexamination of agricultural antibiotic use]. They can justly point out to both the medical profession and the agricultural industry that they have always been publicly opposed to the widespread and indiscriminate use of antibiotics in our livestock population."[150] Veterinarians' cause was strengthened when Herbert Williams Smith and Sheila Halls published a study of AMR in *E. coli* in the *Veterinary Record* in 1966. Well over half of fecal isolates from humans, calves, pigs, and fowls were resistant to at least one antibiotic with many displaying multiple resistance.[151] In the case of multiple resistance, the complete AMR pattern was often transferable between nonpathogenic and pathogenic *E. coli* strains and several *Salmonella* serovars. The authors had also transferred AMR from freshly isolated strains and deep-frozen nonpathogenic *E. coli* from 1956 to human pathogens. In most cases, transferred AMR was stably integrated into recipient strains. According to the authors, "infective resistance is probably the most common form of drug resistance among the *E. coli* that inhabit the alimentary tract of human beings, calves, pigs, and fowls in Britain."[152] Findings also indicated that the emergence of new infective AMR was not rare. The authors concluded that their study "greatly strengthened"[153] the case against drug overuse.

Some veterinarians also saw their own profession as partially responsible for AMR problems. According to veterinary pharmacologist Peter Eyre, AMR "in strictly 'veterinary pathogens' can no longer be regarded as a purely veterinary problem!"[154] As a consequence, Eyre endorsed medical calls for a complete review of both therapeutic and nontherapeutic antibiotic use: "the veterinary profession may have felt disquiet about the use of antibiotics as food additives and growth stimulants, but perhaps has never had enough scientific evidence. This new situation suggests that the time may well have arrived for joint action by the veterinary and medical professions immediately. There must be common responsibility to instigate radical changes in the total use of antibacterial agents in treating and preventing all aspects of human and animal disease."[155] While many vets were comfortable with restricting AGPs, Eyre's suggestion of also subjecting veterinary antibiotic use to medical review proved controversial. According to veterinarian John R. Walton, it was not overall antibiotic use but specific aspects like low-dosed AGPs in combination with animal transports and bad hygiene that were driving AMR in rural settings. Containing AMR therefore required an expansion and not a review and potential reduction of veterinary antibiotic oversight.[156]

"Infective resistance" concerns also affected Britain's pharmaceutical market. In the farming media and advice leaflets, US manufacturers continued to promote their products for growth promotion, prophylaxis, and animal welfare.[157] Titled "Have Aureomycin—Will Travel,"[158] a series of Cyanamid advertisements from 1965 depicted calves and pigs in front of small crates and praised aureomycin for reducing transport-induced scouring and mortality. While American manufacturers tried to shore up trust in their popular products, their European competitors tried to profit from public concerns and reports of penicillin and tetracycline AMR by marketing "safe" ersatz products. Risk was always opportunity. In 1963, Glaxo had already reacted to contemporary residues scandals (chapter 5) by printing full-page ads in *Farmers Weekly* for its "residue-safe" Q(uick).R(elease). mastitis treatments.[159] Mirroring what Christoph Gradmann has shown for Bayer's antibiotic strategy in human medicine,[160] concerns about agricultural AMR selection also boosted sales of allegedly medically irrelevant AGPs and of "resistance-proof" semi-synthetic antibiotics for therapeutic purposes. During the meetings of the first Netherthorpe committee, Bayer had already promoted its nontherapeutic virginiamycin as a safe ersatz for existing AGPs.[161] Sold as Eskalin by Beecham in Britain, virginiamycin was subsequently praised by *British Farmer* for answering "criticisms that continuous low level feeding of an antibiotic . . . can induce bacterial resistance."[162] Beecham also used concerns about resistant staphylococci to market its semi-synthetic Orbenin (cloxacillin). Company brochures praised Orbenin's efficacy against resistant mastitis: "Orbenin meets the demand, it kills penicillin resistant and sensitive staphylococci and all streptococci."[163] Glaxo also used AMR concerns to market combination treatments of novobiocin and penicillin G against mastitis.[164] For the mostly European producers of semi-synthetic and nontherapeutic antibiotics, the message was clear: AMR did not mean that antibiotic use had to be restrained. It simply meant that older (American) products had to be substituted with newer (European) ones.

While pharmaceutical manufacturers and large parts of Britain's agricultural establishment continued to endorse routine antibiotic use on farms despite disagreeing on issues of access and antibiotic type, a small minority of producers opposed antibiotic use per se. This hard core of antibiotic critics were often members of the organic Soil Association. Founded in 1946, the Soil Association was initially headed by Eve Balfour (niece of former Conservative Prime Minister Arthur Balfour) and had an often elite membership interested in the health-giving properties of food produced on "living soil." The organization's early outlook was right-wing. Editing the Soil Association's journal *Mother Earth* until his death in 1963, Jorian Jenks had formerly been the agricultural advisor of Edward Mosley's British Union of Fascists.[165] Although its member- and readership remained small compared to the US Rodale Press, the Soil Association's ability to draw on influential supporters often gave it a voice that was out of proportion to its actual size.[166]

The Soil Association had been critical of nontherapeutic antibiotic use from the beginning. Ahead of the 1953 Therapeutic Substances Act, *Mother Earth* printed a long article on antibiotics in animal husbandry. Acknowledging their potential to increase productivity, the article wondered whether antibiotics' popularity was not rather "a measure of deficiencies" in existing production: "there is a grave danger that antibiotic feed supplements will be used, as so many other discoveries have been, to make bad husbandry possible and profitable instead of making good husbandry still more efficient."[167] While artificial rearing might be tolerated for animals "on the grounds that death will intervene before any cumulative consequences become serious," there remained the "subtle question" of its "ultimate effect on ourselves as consumers of [animals'] flesh."[168] Published in 1957, Hugh Corley's *Organic Farming* repeated warnings about "unnatural" all-meal diets and antibiotic additives: "The more varied a pig's diet is, the happier and healthier he will be. To try to provide this variety by putting penicillin in his pigmeal is about as reasonable as feeding people on white bread plus laxatives to cancel the effects of the white bread."[169] However, Corley did not consider all antibiotics bad. Similar to contemporary opinion in the US Rodale Press (chapter 3), Corley claimed that a pig eating normal swill "probably gets all the antibiotics known to science and several that are unknown, for Penicilliums (of various species) are among the commonest mildews on stale food."[170] Corley also endorsed using penicillin and other antibiotics as "quick, harmless and easy"[171] treatments for mastitis.

British organic assessments of agricultural antibiotic use grew less ambiguous during the 1960s. With prominent activists like Ruth Harrison joining its board, the Soil Association became more welfarist and environmentalist in its outlook and more outspoken about agricultural antibiotics.[172] In 1968, *Mother Earth* reported on food poisoning in fifty-nine people in Sussex caused by the now notorious resistant farm-associated *S. typhimurium* Type 29 (chapter 5). In another case, a farmer had allegedly been killed by resistant pathogens carried by his animals. According to *Mother Earth*, restricting antibiotics would not only protect consumers from dangerous resistant strains but also incentivize more "natural" farming practices.[173] Another article cited "the complex and obscure problem of infectious drug resistance"[174] as an example of conventional experts' neglect of new ecological threats. Although the Soil Association continued to endorse antibiotic treatments of sick individual animals,[175] it was clear that industry attempts to protect production systems by switching to new or nontherapeutic antibiotics would not satisfy Britain's organic community.[176]

Following the 1967 Netherthorpe report's unexpected call for a wider antibiotic review that would also encompass therapeutic antibiotics (chapter 7), it seemed as though some form of official antibiotic restriction might actually occur. The question was whether Britain's third antibiotic review within a decade would challenge agriculture's growing antibiotic infrastructure or try to fix it. Agricultural opinions varied.

Unsurprisingly, British veterinary organizations intensified campaigning for bans of prescription-free antibiotic access. Presenting evidence to the new Swann committee, the Royal College of Veterinary Surgeons called for stricter

AGP controls, the Veterinarians' Union advocated a ban of all antibiotic feed supplements, and the BVA supported a ban of chloramphenicol, tylosin, and broad-spectrum AGPs.[177] Repeating demands from the early 1960s, organizations linked campaigning for enhanced control over antibiotic access with calls for a subsidized preventive health service. In October 1969, outgoing BVA president Peter Storie-Pugh looked forward to a time "when his profession could offer farmers an advisory service which could cost far less than a shelffull of drugs."[178] Such implicit criticism of "irrational" antibiotic use by farmers was, however, seldom self-reflexive. Concerned about a loss of status, British veterinarians staunchly resisted any attempt to control their own prescription practices. In 1969, incoming BVA president John Parsons did not mention regulation of veterinary prescription practices when he endorsed more state control over pharmaceuticals and AGPs.[179]

In contrast to veterinarians, most feedstuff and pharmaceutical companies were extremely wary about pending regulatory encroachment on Britain's antibiotic market. Trying to shore up public and agricultural support against restrictions, manufacturers repeated old claims according to which AGPs' ongoing efficacy proved that AMR was merely a theoretical hazard.[180] Ahead of the 1969 Swann report, the industry-sponsored Office of Health Economics (OHE) published a skillful defense of antibiotic use. Picturing quaint pastoral scenes on its cover, the OHE's booklet not only stressed antibiotics' economic benefits but also cited studies downplaying agriculture's contribution to AMR. Hastily banning valuable antibiotics without commissioning more research was reckless. In the case of some disease outbreaks, more aggressive antibiotic use might have prevented harm: "The pharmaceutical manufacturers in particular would welcome the results of studies on these matters, and would willingly cooperate with them. Nothing is more frustrating than being permitted under existing regulations to do something and then being criticised by sincere scientists as being irresponsible for doing it."[181] According to the OHE, there remained "a lack of balanced scientific appraisal" regarding AMR: "this situation can and should be corrected over a period of time, within a framework of flexible regulations intelligently applied."[182]

Similar to 1960, most farming representatives also opposed ceding antibiotic access to officials and veterinarians. Although *Farmers Weekly* cautioned that "confident guesses rule out many antibiotics now used"[183] in October 1969, another article staunchly defended AGPs: if preventing the spread of resistant bacteria to consumers was the main concern, then bacterial transmission via meat and eggs was the problem—and not AMR selection on farms. Better hygiene would be far more useful than AGP bans.[184] After "inspired leaks"[185] about proposed AGP bans emerged in November 1969, commentators complained that the "talk of a 'new peril in food' is an exaggeration of the scientific problems presented by the increased use of these generally beneficial substances."[186] Farmers felt "harassed a bit too much" about methods "which have not yet been proved to be seriously at fault."[187] Concurring, the NFU's *British Farmer* claimed that potential bans were based on "little convincing evidence"[188] and might cost farmers up to £10 million. Referring to the Manchester and Teesside outbreaks of resistant gastroenteritis, another article reaffirmed that there was no evidence linking the respective bacterial strains to farms.[189]

Significantly, however, British agricultural complaints about "purely circumstantial evidence"[190] and pronounced tensions over antibiotic control did not escalate into a wider questioning of the Swann committee's overall authority to propose new regulations. The reasons for this were twofold. First, in contrast to the United States, there was widespread agricultural AMR awareness as a result of sustained reporting in the national, agricultural, and veterinary press. As a consequence, most discussions in Britain's agricultural sphere centered on the degree of threat posed by AMR and not on whether a threat actually existed. Second, the agricultural community's close corporatist integration into expert and political decision-making created greater trust in the soundness of resulting verdicts. British farmers and veterinarians were well aware that the Swann committee's nine members had been carefully chosen to represent a balance between medical, veterinary, and agricultural interests and could be confident that officials would "impose a reasonable measure of control"[191] rather than sweeping bans. Although *British Farmer* warned that the Swann report might also influence the pending regulation of other embattled substances like DDT,[192] agricultural representatives knew that favorable compromise solutions were more likely to occur in discreet corporatist committees than during polarizing public hearings and debates as had recently occurred in the wake of *Animal Machines* and would soon occur in the US (chapter 8). There was also the danger that overly aggressive attacks on corporatist compromise solutions could lead to a political exclusion from future decision-making.

Following its publication in November 1969, British agricultural commentators were thus relieved to find little radicalism in the Swann report.[193] Experts and officials had seemingly mastered the feet of satisfying nearly all sides: restricting penicillin and tetracycline AGPs fulfilled a key public health demand as well as veterinary aspirations for greater control over the animal health market. However, the report also legitimized the wider existing antibiotic infrastructure by leaving prescription-only therapeutic antibiotic use unaddressed and enabling farmers to switch to allegedly nontherapeutic prescription-free ersatz AGPs like virginiamycin (chapter 7).[194]

The respective communities reacted accordingly. Emphasizing the Swann report's calls for preventive health care, the BVA welcomed the additional responsibility vested in British veterinarians and predicted an increase of the "veterinary profession's contribution to productivity in the farming industry."[195]

The Swann committee's alleged inability to find evidence of veterinary malpractice also meant that veterinarians "need not, then, be ashamed of [their] record in using antibiotics."[196] The farming press also reported on veterinarians' increased control over antibiotics. While *British Farmer* joked that the "rows of bottles on some farm office shelves will be seriously depleted,"[197] *Farmers Weekly* printed a comment by farmer G. Armstrong, who drily observed: "My vet seems more pleased to sell products himself. I feel it is not in farmers' best interests for a 'closed shop' to develop."[198] However, nobody questioned the overall wisdom of the Swann report. Even though it lobbied for financial compensation, *Farmers Weekly* admitted, "no sensible farmer would wish to [continue] using a drug which . . . could be a later risk to public health."[199] Others concurred: "By mass use of low-dose antibiotics in farm animals we are creating a reservoir of drug-resistant bacteria. . . . The range of useful antibiotics is limited: we cannot afford to devalue them."[200] Rapidly adapting to the new rules, British livestock organizations soon tried to turn the Swann report into a sales advantage by marketing British poultry and other meat products as "the best and safest in the world."[201] With the exception of US broad-spectrum producers (chapters 7 and 13), most pharmaceutical manufacturers also endorsed the Swann report. In December 1969, the development committee of Wellcome-owned chemical manufacturer Cooper, McDougall and Robertson noted that "the recommendations and conclusions of the Swann committee could not be faulted and were as expected."[202] In the long term, the Swann report might even be a sales advantage for British products. For most parties, it was clear that the specific drugs used on farms would change but that the overall antibiotic infrastructure aiding disease management and animal productivity would remain intact.

The 1969 Swann report marked the end of the pioneering phase of British agricultural antibiotic use. During the early postwar years, structural and economic constraints had led to a comparatively slow adoption of antibiotics on farms. However, by the late 1950s, the drugs had become common in both intensive and nonintensive settings. Although UK and US antibiotic infrastructures gradually converged, agricultural risk perceptions diverged. Influenced by the AMR-centered British risk episteme and concerns about veterinary incomes, agricultural factions fought fierce battles over antibiotic control: while livestock producers and nutritionists pushed for drug access to increase productivity and sidestep veterinary fees, veterinarians used AMR concerns to call for "rational" oversight of both therapeutic and nontherapeutic antibiotic use. Following a brief *détente* after the 1962 Netherthorpe report, battles for control were reignited by mid-1960s concerns about animal welfare and "infective" AMR. Once again, Britain's corporatist system of decision-making played an important role in making resulting reforms acceptable to agricultural parties.

Staffed by medical, veterinary, and agricultural interests, the Swann committee's 1969 call for a precautionary ban of therapeutic AGPs boosted veterinary control over the lucrative antibiotic market but also legitimized ongoing unsupervised access to nontherapeutic AGPs. It also weakened American and strengthened European drug manufacturers. Satisfied by the Swann compromise, nearly all agricultural camps in Britain believed that antibiotic use on farms was now reformed and safe.

7

Typing Resistance

Antibiotic Regulation in Britain

This chapter explores how British officials first promoted agricultural antibiotic use and then tried to balance demands for ongoing antibiotic use with public concerns about animal welfare and AMR. Focusing on the emergence of a national risk episteme, it emphasizes the role that enterprising public health officials and a technology called bacteriophage-typing had in linking agricultural AMR selection with health hazards and driving legislative action. It also highlights the marginalization of residue concerns not only in the public but also in official circles. The chapter then uses ministerial records to trace the evolution of British corporatist decision-making during the three antibiotic reviews of the 1960s. Although the 1969 Swann report pioneered precautionary antibiotic restrictions, a close examination of the report's origins reveals how contingent power struggles between different ministries and professions resulted in a compromise, which had been devised prior to widespread knowledge of horizontal resistance transfer.[1]

Licensing Agricultural Antibiotics

In contrast to the United States, AMR concerns strongly influenced the postwar regulation of antibiotics in Britain. In 1947, fears that AMR would reduce antibiotics' efficacy had led to a restriction of penicillin and other new antibiotics by the Penicillin Act (chapter 6). In Parliament, Labour's Minister of

Health Aneurin Bevan had explicitly resisted calls for antibiotic deregulation: "it seemed to us to be highly undesirable that this very valuable substance should be the plaything of quacks with all sorts of ways of advertising it and selling it—penicillin lipstick, penicillin rouge, penicillin powder, and even penicillin waistbands . . . . It would be an appalling thing if, as a consequence of its misuse, the population might, in a period of years, receive no advantage at all from it because it would have developed resistance in those who take it."[2]

The topic of agricultural antibiotic use only reappeared on British officials' agenda following the 1950 announcement of the antibiotic growth effect. Although Whitehall was interested in US experiences with AGPs and commissioned British feed trials, officials remained ambivalent about nontherapeutic antibiotic use and did not immediately loosen existing restrictions. According to contemporaries, AMR concerns were the "chief reason"[3] for this licensing delay.[4] In 1952, concerned Ministry of Health (MH) officials warned about antibiotic allergies and AMR resulting from unsupervised farm access to antibiotics: "the whole purpose of the Penicillin Act was to prevent penicillin and other antibiotics being used indiscriminately with a consequent danger of producing penicillin resistant strains of pathogens."[5] Such concerns were fiercely opposed by antibiotic supporters consulted by the Ministry of Agriculture, Fisheries and Food (MAFF), who argued that AGPs were too low-dosed to select for AMR. Mirroring contemporary agricultural debates (chapter 6), consulted experts claimed that any "risk to health was negligible."[6] Although it was unclear whether antibiotics were "improving nutrition or curing disease,"[7] presenting AGPs as nutritional and nontherapeutic allowed AGP supporters to diffuse medical criticism (chapter 6). In 1957, chemist J. A. Wakelam remembered:

> The Ministry [of Agriculture] were at great pains to establish that the addition of antibiotics to animal feeds was not intended as therapy, but was a method of improving normal healthy growth. As such it was termed nutritional. Those of us who had carried out work on [penicillin AGPs], were satisfied that these were incapable of producing bloodlevels which could be considered therapeutic, ... If this were not so the Penicillin Act could not have been amended, and the fears of [the veterinary] profession concerning the dangers of developing resistant strains of pathogenic organisms could not have been stilled."[8]

Presenting AGPs as a lucrative nutritional sales outlet for domestic pharmaceutical companies and a means to boost agricultural productivity and reduce expensive feed imports also proved an effective argument in light of Britain's contemporary balance of payments crisis. By the end of the ARC's antibiotic feed trials in late 1952, economic incentives had trumped medical concerns.[9]

Ahead of the 1953 parliamentary debates on licensing AGPs, Britain's MAFF was thus primarily concerned not about AMR but that agricultural demand for broad-spectrum AGPs might exceed domestic supplies. As a consequence, British officials approached US manufacturers to ensure sufficient stocks of antibiotics. In response, American Cyanamid offered free aureomycin magnasol cake to bridge projected shortages ahead of the opening of Cyanamid production facilities in Britain. Lederle Laboratories' director hoped that this would "be the beginning of an association which will be of mutual benefit."[10] However, despite AGP endorsements by veterinary organizations,[11] certain doubts about their safety and efficacy remained. In July 1953—two months ahead of AGPs' licensing—Conservative Minister for Agriculture Thomas Dugdale told NFU president Sir James Turner—later Lord Netherthorpe— that he considered AGPs to be a regulatory experiment: "Our knowledge of antibiotics in feedingstuffs is still comparatively in infancy, but the regulations we are proposing to make will . . . enable all of us including both farmers and manufacturers to increase our practical experience."[12]

Flying Blind

Coming into effect in September 1953, the Therapeutic Substances (Prevention of Misuse) Act (TSA) exempted ready-mixed penicillin and chlortetracycline feeds and self-mix supplements for pigs and poultry from scheduling. Although initial AGP uptake was underwhelming, promotional efforts by officials, pharmaceutical representatives, and agricultural experts soon led to a sustained expansion of antibiotic consumption (chapter 6).

However, similar to the United States, officials were already realizing that they had few tools with which to control antibiotic use on farms. In Britain, regulators lacked sufficient facilities for assaying (testing) antibiotic concentrations in feed, food, or milk.[13] For assay methods, they relied on academic publications and foreign regulatory agencies—most notably the American FDA. Meanwhile, antibiotic enforcement remained confined to the retail level.[14] With no control over the actual use of legally purchased and prescribed antibiotics, British officials had to trust that corporatist self-policing by farmers, veterinarians, and pharmacists would ensure compliance. The limits of this approach became evident during the early 1950s. Reacting to US warnings, British researchers studied penicillin residues' effect on cheese production in 1951. Although no residues were detected in 1,082 bulked milk samples, researchers reported that between 1.4 to 2.8 percent of milk churns arriving at London and Shropshire dairy plants were contaminated with penicillin. In one case, residues had destroyed a starter culture mixed with 800 gallons of milk. Two years later, 3.2 percent of analyzed churn milk samples were contaminated with penicillin.[15] However, in contrast to the United States, where similar detections eventually led to FDA residue monitoring (chapter 4), British residue findings failed to trigger public outrage or official action.

Official complacency about noncompliance with withdrawal times for therapeutic antibiotics was matched by relative complacency about the use of newly licensed AGPs on farms. The reformed 1956 TSA did not improve the situation. While Part I of the TSA dealt with the licensing, manufacture, and importation of medications to ensure their purity, Part II again exempted low-dosed AGPs from scheduling and prescription requirements.[16] Remarkably, the absence of mandatory scheduling for new substances also meant that recently discovered drugs like tylosin, a supposedly medically irrelevant antibiotic, could be sold without any veterinary or official supervision.[17] Relying on corporatist self-policing, officials remained sanguine about this large loophole: as Glaxo's former chief executive scientific officer Alfred Louis Bacharach put it, a "gentleman's agreement"[18] between manufacturers and the MAFF would prevent misuse. Until it was replaced by comprehensive legislation in 1968, an aptly named voluntary Veterinary Products Safety Precautions Scheme established nonbinding guidelines for unscheduled substances.[19] According to Bacharach, a similar "gentleman's agreement"[20] initially also governed voluntary restrictions of antibiotic use in plant protection: "the various [MAFF] officials who have to cope with one aspect or another of the [antibiotic] situation do so in an enlightened and cooperative way and show great skill in not discouraging the enterprising entrepreneur and in simultaneously protecting the public."[21] In reality, the absence of substantial monitoring facilities and statutory powers meant that British officials could do little else than cooperate with industry.

While antibiotic enforcement withered, expert committees bloomed: because new antibiotic applications transcended traditional bureaucratic responsibilities, a veritable jungle of committees became concerned with their use. Originally, the Medical and Agricultural Research Councils (MRC and ARC) had been responsible for advising ministers on agricultural antibiotics. Soon, further committees became involved. Among them were the Preservatives Sub-Committee of the Food Standards Committee and the Scientific Sub-Committee of the Advisory Committee on Poisonous Substances Used in Agriculture and Food Storage. These committees founded an additional joint Antibiotics Panel in 1956.[22] The numerous committees vied for influence and frequently disagreed. As a result, departmental and expert responsibilities blurred and there was no guiding principle driving British antibiotic policy. In 1967 one official complained: "I have been quite unable to understand the relationship between these bodies."[23] Another official admitted: "The situation is now so complicated that it is almost un-understandable."[24]

Meanwhile, antibiotic licensing increased. In 1954, the Therapeutic Substances (Supply of Oxytetracycline for Agricultural Purposes) Regulations legalized oxytetracycline (terramycin) AGPs.[25] Streptomycin and oxytetracycline sprays and paints for plants were licensed four years later.[26] US manufacturers also pressed officials to license antibiotic food preservation and sponsored trials aboard the distant water government trawler and future *Rainbow Warrior*, the *Sir William Hardy* (chapter 5).[27] The merits of antibiotic cheese, poultry, fish, whale, and beer preservation were subsequently reviewed by Britain's Joint Antibiotics Panel. Following extensive testing, Britain first licensed nisin for the preservation of canned foods and cheeses and tetracyclines in ice and dipping solutions for fish in 1964.[28]

Although US technologies were often imported to Britain alongside respective FDA regulations, concerned individual officials at times modified British rulings. In the case of poultry, warnings about the absence of spoilage-indicating bacteria and AMR selection in food-borne pathogens led to a ban of antibiotic preservatives. The decision was made easier by the fact that many British birds were not eviscerated prior to sale, which reduced the efficacy of antibiotic preservation.[29] AMR concerns also delayed the licensing of antibiotic plant sprays. Despite being endorsed by Nobel laureate and penicillin developer Sir Howard Florey,[30] sprays were opposed by MRC researcher Brandon Lush, who feared that residues might alter the human gut flora and select for AMR.[31] Concerned about residues rather than AMR, the new Antibiotics Panel subsequently debated whether farm workers' tough skin would make them less sensitive to antibiotic allergies than soft-skinned nurses.[32] Ultimately, official equanimity prevailed. In 1958, MAFF's proposed labels for antibiotic sprays and paints only recommended washing contaminated skin while Murphy's, the manufacturer applying for the spray's licensing, recommended full-body cover and face-shields for workers.[33]

The fragmentation of public antibiotic concerns meant that *laissez-faire* regulations faced little opposition throughout the 1950s (chapter 5). Trusting in corporatist partners' adherence to often voluntary guidelines, officials expressed a certain pride in the low cost of Britain's food security apparatus. In 1956, British regulators justified their refusal to establish US-style residue limits to the Western European Union Sub-Committee on Health Control of Foodstuffs: "The United Kingdom feels that the problem of consumer hazard can be tackled in more than one way. . . . The successful application of the American system is dependent upon the existence of the necessary governmental machinery. . . . The United Kingdom delegation feels that cost and scientific management problems make it impossible for them to advocate a system of control of residues on prescribed tolerances."[34] Without monitoring or statutory rules forcing their hand, British officials adopted a wait-and see attitude and looked to data from other countries for signs of problems. In 1956, some officials took a positive view of US residue scandals (chapter 4): "In view of the enormous amount of uncooked milk consumed daily by the American population and the propensity of penicillin to produce allergic reactions, it would appear that they have here a large scale experiment already completed."[35]

Members of Britain's Antibiotic Panel subsequently argued that the lack of proven fatal American reactions to penicillin residues indirectly proved the safety of antibiotic food preservatives. Closely involved in corporatist decision-making, pharmaceutical manufacturers agreed. According to Glaxo's Alfred Louis Bacharach: "The number of these involuntary and unconscious penicillin eaters must by now run into many hundreds of thousands."[36] However, none of the "somewhat sinister warnings"[37] about AMR and toxicity had materialized.

British officials' unwillingness to impose further regulations on corporatist partners is all the more remarkable given worrying contemporary residue data. Between 1954 and 1956, spot tests of over 5,000 milk samples revealed that penicillin was present in 3 to 4 percent of samples. An additional 2.2 percent of samples contained antimicrobial substances, which could not be identified: "It is therefore obvious that, in contravention of statutory enactments, a considerable quantity of milk containing penicillin and probably other antibiotics is being marketed for consumption by the general public."[38] However, confidence in industry self-regulation prevailed. In 1959, a Ministry of Health memo maintained that "it has not yet been considered necessary to take action by way of statutory regulations to prevent antibiotics entry into milk as a result of the treatment of mastitis."[39] Instead of investing in US style monitoring, officials asked British farmers to inform milk collecting centers of treatment and only use one dose of an antibiotic without veterinary supervision.[40] Britain's corporatist Milk Marketing Board (MMB) also asked pharmaceutical manufacturers to print cautions on antibiotic tubes.[41]

The glacial pace of corporatist reform quickened only after data on penicillin residues in 14 percent of English and 11.6 percent of Scottish milk caused public outrage in 1963 (chapter 5). The 1961 survey of the Milk and Milk Products Technical Advisory Committee had been available to officials since mid-1962.[42] However, it was only when British newspapers picked up the story that officials took action. Reinforced by a critical WHO report, public outrage triggered the rapid abandonment of corporatist self-policing in favor of US-style penalties and a formal zero-tolerance policy for penicillin in milk.[43] By November 1963, 10 percent of British milk was being tested at twenty-one dairies across the country. Warning letters had been sent to 226 producers and local authorities were also testing supplies. After hastily introducing the 48-hour withdrawal periods already recommended in 1959, the MMB assured MAFF: "We are all aware of the fact that we have a problem that has to be tackled with the utmost despatch and . . . you may rest assured that the industry will use its best endeavours to support you in your in your determination to reduce the risk to an absolute minimum."[44] Following the adaptation of US inhibition tests, the MMB announced that residue offenders would receive price deductions starting in March 1964.[45] Antibiotics' potential presence in British meat was not discussed.

FIGURE 7.1 New bulked milk collection exacerbated residue problems. Gordon Craddock, Bulk Milk Collection, 1963. ©Museum of English Rural Life, University of Reading.

Typing Resistance

British officials' relative complacency regarding antibiotic residues in food and milk stood in contrast to their growing concerns about agricultural AMR selection. This selective focus on antibiotic risk was in part caused by public anxieties (chapter 5) and in part by data supplied by Britain's Public Health Laboratory Service (PHLS). Although corporatist decision-making would protect producers from sweeping bans, PHLS researchers' ability to regularly confront officials with robust AMR data played a crucial role in establishing an AMR-focused British risk episteme.

The monitoring capabilities of the PHLS rested on a technology called bacteriophage-typing. During the Second World War, workers of the Emergency Public Health Laboratory Service (EPHLS) had established a network of bacteriological laboratories across Britain. Tasked with warding off bacteriological attack and epidemics, EPHLS researchers had trialed new surveillance methods. One such method was bacteriophage-typing, which classified and identified (typed) individual bacteria strains with the help of bacteria-infecting viruses called bacteriophages (phages).[46] Originally developed in Germany, phage-typing was adopted in Britain by EPHLS bacteriologist Arthur Felix.[47] During the war, EPHLS workers successfully used centralized phage-typing services to trace the sources of typhoid outbreaks and developed new phage-typing sets for *Salmonella paratyphi* (paratyphoid), *Salmonella typhimurium,* and *Staphylococcus aureus.*[48]

In the decades after 1945, the renamed PHLS's centralized laboratory network and phage-typing capabilities provided unparalleled insight into the evolving microbial environment. Headquartered in London Colindale, PHLS phage-typers were among the first to discern the international threat posed by rising AMR. In the case of human medicine, they uncovered the global "chains of infection"[49] behind the first identified resistant pandemic (*Staphylococcus* phage type 80/81) in 1954. In the case of agriculture, close cooperation between the PHLS and veterinary bacteriologists revealed the threat posed by nonhuman AMR selection.[50] One of the first veterinary researchers to embrace phage-typing was Herbert Williams Smith. During the 1940s, Williams Smith had briefly worked for the PHLS where he shared a laboratory bench with E. S. Anderson—future director of the PHLS's Enteric Reference Laboratory. Later based at the Animal Health Trust in Stock, Williams Smith adapted PHLS phage-typing sets for animal isolates of *Staphylococci* and *Escherichia coli.* He soon noticed that the strains he was typing were becoming increasingly resistant to antibiotics. Seemingly unaware of Starr and Reynolds' 1951 American studies of AMR in poultry (chapter 3), Williams Smith independently identified low-dosed AGPs as major AMR selectors.[51] After warning about AGPs at the 1956 BVA conference (chapter 6), Williams Smith used phage-typing to trace the spread of resistant strains from animals to humans. In a 1960 paper, he compared staphylococcal isolates of 160 pigs fed tetracycline AGPs to an AGP-free control group. Of the pigs fed tetracyclines, 67 percent carried tetracycline-resistant *S. aureus* strains. Of fifty attendants caring for tetracycline- and penicillin-fed chickens, 30 percent carried penicillin-resistant *S. aureus,* 14 percent tetracycline-resistant *S. aureus,* and 4 percent penicillin- and tetracycline-resistant *S. aureus.* Phage-typing showed that resistant human and animal *S. aureus* isolates were mostly identical.[52] Williams Smith's warnings about nonhuman AMR selection were echoed by parallel Canadian studies of AMR in pigs, Cambridge research on antibiotic preservatives, and PHLS surveys of *S. typhimurium* isolates from poultry.[53] In 1960, PHLS workers also detected higher nasal carriage-rates of penicillin-resistant *Staphylococci* in military recruits from rural than from urban backgrounds, which suggested a link to unpasteurized milk and frequent animal contact.[54]

Coinciding with increasing public attacks on intensive farming and the veterinary push for preventive health schemes (chapters 5 and 6), PHLS phage-typers effectively ended official complacency about agricultural AMR selection. In the summer of 1959, Britain's Agricultural Research Council (ARC) referred to still unpublished PHLS data when it demanded a general reassessment of AGPs' safety.[55] Taken aback, the Medical Research Council (MRC) marveled:

"In fact they are seriously considering withdrawing approval of the adding of antibiotics; in other words, they are considering putting the clock back."[56]

Corporatist Compromises

Tasked with undertaking a comprehensive review of AGPs but not of therapeutic antibiotic use, a joint ARC/MRC committee started work in April 1960. The so-called Netherthorpe committee's main body was chaired by former NFU president James Turner—now Lord Netherthorpe—and only met twice. During its first meeting in 1960, it installed a scientific subcommittee headed by Arthur Ashley Miles, director of the Lister Institute of Preventive Medicine. Two years later, it endorsed the subcommittee's report.[57] The subcommittee met five times between 1960 and 1962 and was carefully staffed to reflect competing veterinary, medical, and farming interests.

Although the subcommittee's staffing clearly aimed at producing a consensus-based corporatist fix of AMR problems, it soon became apparent that a fundamental rift divided members. While a medical faction consisting of physicians and veterinarians attacked AGPs on the grounds of AMR, an economic faction consisting of agricultural scientists and officials pressed for an expansion of antibiotic use on the grounds of productivity and unproven harm. Mirroring agricultural struggles over antibiotic control (chapter 6), nearly every subcommittee meeting was characterized by clashes between Robert Fraser Gordon (veterinarian, Houghton Poultry Trust) and Raphael Braude (animal nutritionist, NIRD). Giving evidence in June 1960, Herbert Williams Smith presented new data on the spread of AMR from animals to workers: in one survey, 88.3 percent of *Staph aureus* strains isolated from the noses of veterinary surgeons and 21.1 percent from farmers' noses were penicillin-resistant—14.7 percent of isolates from veterinarians and 2.6 percent from farmers were also resistant to chloramphenicol.[58] Williams Smith warned that even the lowest level of antibiotic use could select for AMR.[59] In response, Braude asked for conclusive evidence of harm resulting from resistant strains. Williams Smith conceded that he was unable to supply such proof. With researchers unable to specify whether AMR resulted primarily from AGPs or therapeutic antibiotic use, the subcommittee therefore compromised on the following statement: "therapeutic uses of antibiotics could lead to the production of resistant strains, . . . the dangers of uncontrolled therapeutic use should be born in mind."[60] Remarkably, evidence submitted by the NFU showed that uncontrolled antibiotic use was indeed taking place. The NFU submission contained three farmers' statements: one farmer confessed having illegally fed antibiotics to breeding pigs, a second farmer stated that he used penicillin but had ignored "fashionable and extravagant claims of the broad-spectrum manufacturers,"[61] and a third farmer reported "certain instances where high-level doses of antibiotics have been used in an attempt to offset bad husbandry practices."[62] The subcommittee's minutes explicitly noted "the difference of opinion between the farming members of the Joint Committee and the farmers whose opinion had been put forward as representative by the NFU."[63]

In view of the division between medical and agricultural members, Professor James Howie from the University of Glasgow—and future PHLS director—presented subcommittee members with three choices:

Complete prohibition of the addition of antibiotics to feedingstuffs (i.e., a reversion to the earlier situation, which would be very difficult)

Maintenance of the present position (on the ground that the conflicting evidence did not provide any basis for a change)

General permission to add antibiotics to feedingstuffs (on the ground that there was insufficient evidence to justify the withholding of such permission)[64]

Howie's phrasing was significant. By presenting only three choices—two of which were extremes—he transformed the status quo ante into an acceptable compromise. Both factions could subsequently tell their supporters to have prevented worse.

Yielding to Braude's objections, the subcommittee agreed that there was insufficient evidence to restrict already licensed AGPs. Acknowledging a common practice, it also recommended licensing AGPs for calves but did not endorse AGPs for layer birds and adult stock. Both sides called for further research. Significantly, the medical faction managed to push through a recommendation that new AGPs should be licensed on the basis of their irrelevance to human and animal therapy.[65] Devised without knowledge of horizontal AMR transfer, the suggested distinction between medically relevant and irrelevant antibiotics was not new: the Antibiotics Panel of the Advisory Committee on Poisonous Substances had discussed such a separation as early as 1956.[66] However, by inserting two-tier licensing into the subcommittee's report, the medical faction scored a long-term victory. Changed licensing would promote the development of medically irrelevant AGPs. Once established, these AGPs would make penicillin and tetracycline AGPs expendable. Following contemporary concepts of vertical AMR proliferation, reduced selection pressure would also lead to a gradual return of bacterial sensitivity.

The resulting 1962 Netherthorpe report "fixed" AMR by shoring up existing legal boundaries between nontherapeutic AGPs and higher-dosed therapeutic antibiotics as well as by creating new boundaries between allegedly medically relevant and irrelevant drugs. In a perfect example of corporatist compromise, the demands of all sides were partially met: veterinarians and public health experts could claim greater control over future antibiotics while agricultural representatives could proclaim business as usual. This insulated mode of consensus-oriented consultation had the advantage of avoiding polarizing public clashes over AMR. However, in a sign of things to come, it also incentivized the creation of narrow reforms via incremental rearrangements of legal terminology over systematic reforms of antibiotic infrastructures.

The 1962 Netherthorpe compromise was short-lived. Although the compromise was endorsed by a 1963 WHO report,[67] the milk residue scandal (1963), *Animal Machines* (1964), and "infective" AMR warnings (1965) soon reignited antibiotic conflicts (chapters 5 and 6). One of the main drivers of the resulting second and third rounds of official AMR reviews was Ephraim Saul "Andy" Anderson. Anderson was the son of Estonian immigrants and had joined the PHLS after being stationed in Cairo during the war. In 1954, he succeeded Arthur Felix as head of the PHLS Enteric Reference Laboratory and took an active interest in the environmental spread of bacterial strains and AMR. His experience in reworking PHLS phage-typing to account for lysogeny (the ability of "weak phages" to integrate their DNA into the host bacterium's genome) also made him appreciate the implications of recently published Japanese research on plasmid-mediated (horizontal) AMR transfer for antibiotic stewardship (chapter 1). In addition to studying AMR on British farms, Anderson cultivated useful friendships with journalists like the *Guardian's* Anthony Tucker and Bernard Dixon, future editor of the *New Scientist*. While his robust AMR studies and media interventions would successfully challenge the Netherthorpe compromise,[68] Anderson's abrasive personality and public partisanship for sweeping antibiotic restrictions would effectively preclude him from participating in the ensuing corporatist redrawing of "safe" antibiotic use.[69]

Anderson had been concerned about agricultural AMR selection since the 1950s. However, it was the combination of Japanese R-factor research and the 1964 Aberdeen typhoid outbreak that turned him into an activist scientist. Concerned that agricultural antibiotic use might foster multiple-resistance in typhoid, he began to look for mobile AMR on farms. In 1965, his first coauthored paper in the *Lancet* reported *in vitro* AMR transfer from ampicillin-resistant *S. typhimurium* to sensitive *S. typhimurium* and *E. coli*. The paper presented compelling evidence of AMR transfer taking place in nonhuman settings and indicated that plasmids could also be transferred to bacteria in human populations (chapter 5). Anderson's next two coauthored papers used the PHLS's unique surveillance capabilities to highlight rising AMR in *S. typhimurium* Phage Type 29 isolates from across the country and presented evidence of furazolidone resistance transfer (a drug used exclusively in animals) from bacteria in animals to bacteria isolated from humans.[70]

Anderson's 1965 papers triggered an important shift toward a more environmental understanding of AMR hazards and led to a recall of the Netherthorpe committee amid rising public pressure in spring 1965. Asked to give evidence and later accused of instigating the whole committee,[71] Anderson argued for immediate antibiotic bans. His team had identified a substantial rise of resistant salmonellosis outbreaks caused by *S. typhimurium* Type 29 since 1961. Type 29 had first been identified in China in 1959 and had been isolated in Britain in 1961. Whereas 16.7 percent of Type 29 isolates were antibiotic resistant in November 1964, the proportion of resistant strains had risen to 59.8 percent in April 1965. Worryingly, Type 29's resistance spectrum had also increased: in 1963, Anderson's team had discovered Type 29 strains with resistance against sulfonamides and streptomycin. Tetracycline resistance was detected in early 1964. By June 1964, most Type 29 cultures were resistant to all three drugs. Ampicillin resistance appeared three months later. Significantly, Type 29's streptomycin, sulfonamide, and ampicillin resistance was plasmid-encoded and transferable. Using any antibiotic on Type 29's resistance spectrum would automatically select for resistance against all of the other antibiotics.[72] Writing in support of Anderson, Herbert Williams Smith reported that his team had analyzed two Type 29 outbreaks that had also proven resistant to neomycin and furazolidone—drugs initially not included in Anderson's AMR survey.[73]

PHLS investigations linked Type 29 outbreaks to the intensive rearing of male calves. These calves were no longer slaughtered after birth in the dairy regions of the West Country but were being sold for fattening in the grain-rich eastern part of England. The calves were often less than one week old and particularly susceptible to salmonella infections. Many were treated prophylactically with a barrage of antibiotics. Remarkably, PHLS phage-typing traced most British Type 29 outbreaks to the premises of a Sussex calf dealer. This dealer had marketed calves together with six-week feeding kits containing many of the antibiotics to which Type 29 had developed resistance.[74] It was likely that this practice had caused severe and occasionally fatal human and animal outbreaks of resistant salmonellosis in thirty-seven counties.[75] Officials, however, lacked the statutory powers to stop the recalcitrant dealer from overusing legally obtained—often prescribed—antibiotics and reform his livestock operation. Reporting on a Type 29 outbreak, which had infected cows, an 18-month-old, and a 6-month-old, the local Medical Officer of Health complained: "The most frustrating aspect of Salmonellosis in cattle is that, having detected the infection and having prevented its spread to the public via the milk, one has no power to eradicate the disease from the herd concerned."[76]

Officials' impotence in the face of Type 29 frustrated Anderson, who increasingly saw AMR as an environmental hazard. Speaking to the reconvened Netherthorpe committee in 1965, he attacked the committee's 1962 report, which had acknowledged agricultural AMR selection but had thought that AMR selection was limited to the antibiotic in use. This was not true: "multiple resistance is now the rule . . . I have already pointed out that the use of any drug in a multiple spectrum of resistance protects the entire spectrum, so that, for example, resistances to ampicillin, streptomycin, sulphonamides, neomycin, kanamycin, and furazolidone will all flourish under the umbrella of tetracycline if they are associated with resistance to that drug."[77] The continuous feeding of low-dosed AGPs was creating the perfect conditions for AMR proliferation. Challenging 1950s scenarios of an inevitable return to bacterial susceptibility once antibiotic use ceased, Anderson also warned that it was unclear whether AMR would "die out in an animal population if the use of antibiotics was discontinued."[78] Since AMR developed in jumps, it was possible that transferable resistance to chloramphenicol—a vital drug against typhoid—might soon emerge. In a crucial statement, Anderson noted: "the position was that R [resistance]-factors now have an epidemiology of their own, covering transfer between strains and species and also between hosts."[79] Containing AMR was no longer just about containing specific bacteria but about containing genes.

Asked what steps he would take if he were given dictatorial powers, Anderson called for a ban of AGPs and antibiotic preservatives, therapeutic antibiotic use only after bacteriological diagnosis, and a reservation of certain antibiotics solely for humans. He also called for enhanced AMR surveillance, antibiotic advertising bans, and improved welfare standards.[80] Unsurprisingly, these recommendations divided the Netherthorpe scientific subcommittee. In a clever move, agricultural committee members referred to the committee's brief of assessing only AGPs and argued that "the use of most of the antibiotics discussed [in calves] was [still] illegal without a veterinary prescription."[81] Because of its focus on calves and forms of antibiotic use, which were either prescription only or illegal, Anderson's AMR data was thus beyond the committee's purview. In June 1965, the reconvened full committee could thus merely call for more research and reconfirm existing regulations—including the planned licensing of AGPs for calves.[82]

For Anderson and his supporters, this status quo was unacceptable. In December 1965, a further recall of the Netherthorpe committee was prompted when Anderson's superior, PHLS director James Howie, withdrew "his concurrence in the recommendations of the [June 1965] draft report."[83] New research by Anderson had strengthened the case both against low-dosed AGPs *and* high-dosed antibiotic prophylaxis. Despite identifying a lab strain of Type 29 whose tetracycline resistance was encoded on a transferable R-factor— designated as T″—Anderson's team had previously thought that transferable tetracycline resistance was rare. Recently, however, they had found an R-factor indistinguishable from T″ in a wild *S. typhimurium* Type 29 strain from pigs. Researchers already knew that the lab-based T″ could transfer resistance to "almost 100 percent of recipient cells of either *S. typhimurium* or *Escherichia Coli* K12 . . . after overnight contact with a donor strain."[84] Although the wild porcine Type 29 strain with T" had not spread outside the pig herd, it was possible that the wild T" R-factor might independently escape into the wider environment by transferring to *E. coli*, which were not being monitored for. Anderson's findings suggested that both low- and high-dosed selection for tetracycline resistance in all livestock sectors was more important for AMR in "*S. typhimurium* than was realized earlier."[85]

With Netherthorpe subcommittee members continuing to disagree about AMR risks resulting from AGPs, Anderson's renewed push for a holistic review of antibiotic use had an unforeseen consequence. In its final report, the subcommittee shifted its focus from prescription-free AGPs to prescription-only therapeutic antibiotics and called for a new investigation of antibiotic use in veterinary *and* human medicine. Further recommendations included a retraction of the subcommittee's previous endorsement of AGPs for calves, a rationalization of British antibiotic committees, more research on therapeutic and prophylactic antibiotic use, and turning salmonellosis into a notifiable disease.[86]

The subcommittee's proposed review of therapeutic antibiotic use proved contentious: not only would it infringe on the jealously guarded legislative boundaries between the Ministries of Agriculture and Health, it also threatened to delay the almost finalized 1968 Medicines Act, which would unify fragmented drug regulations and establish a formalized category of Prescription Only Medicines.[87] In view of the fragile situation, MH officials and the influential antibiotic expert Lawrence Paul Garrod pressed for a deletion of all references to human medicine during the main Netherthorpe committee's final meeting in April 1966.[88] The final Netherthorpe report merely recommended a review of "the use of antibiotics in animal husbandry and veterinary medicine and its implications in the field of public health."[89] However, the subcommittee's attached report stressed that evidence for AGP restrictions was inadequate.[90] In sum, the only area to be reviewed was veterinary medicine.

Unsurprisingly, veterinarians did not take kindly to this proposal. During meetings with the PHLS, MAFF's Animal Health Division resisted pressure to ban prophylactic antibiotic use. Veterinary officials highlighted the difficulties of intensive animal husbandry, stressed the need for differentiating between different forms of antibiotic use, and emphasized educational reform as well as new codes of practice.[91] Complaining about its one-sided focus on veterinarians, the ARC blocked the second Netherthorpe report's publication in January 1967.[92] Opinions within MAFF were more nuanced: while one official downplayed the report as a "storm in a teacup,"[93] others anticipated "a first-class row with the Royal College and the BVA."[94] Although they agreed that withholding publication could inflame public opinion, officials were powerless to override the ARC.[95] As a consequence, MAFF lobbied the MH to reextend the proposed review to both agricultural *and* medical aspects of antibiotic use. In doing so, officials cited a report by MAFF's Scientific Advisory Panel (SAP).

Apparently anticipating problems with the Netherthorpe committee, MAFF had commissioned the SAP with a separate review in 1965. The SAP review was headed by Alastair Frazer, a food additives expert with ties to industry,[96] and advised by penicillin co-discoverer Sir Ernst Boris Chain. Citing declining salmonellosis rates, it endorsed current antibiotic use but recommended further research and a review of control measures.[97] Summing up the SAP report, a MAFF official noted: "In other words, nothing we should do should impede the use of antibiotics in agriculture or food, though of course they must be used with reasonable safeguards."[98]

Following further delays, rising public pressure made Labour's Minister of Agriculture Frederick Peart become personally involved in negotiations about the new antibiotic review in July 1967. During a meeting with senior advisers, Peart agreed that the Netherthorpe report "created some unnecessary alarm, and that [it] picked out veterinarians."[99] However, attempts to extend the planned review to human medicine only increased interministerial tensions. Referring to allegations of antibiotic overuse, a MAFF bureaucrat complained: "there has been a good deal of sniping from certain medical quarters . . . , although I seem to recall something about 'people who live in glass houses.'"[100] In September 1967, all involved parties published a joint press statement accepting most of the Netherthorpe recommendations but rejecting national monitoring plans for resistant salmonellosis.[101] However, despite pressure from the PHLS and MRC, an actual antibiotic review remained unforthcoming.[102]

Similar to the establishment of penicillin monitoring for milk, it took a scandal in the form of the tragic Teesside deaths to break the corporatist stalemate and quicken the pace of reform (chapter 5). Official concern about rising public anger first arose when the BBC's *Twenty-four Hours* linked fatalities to antibiotic overuse in agriculture ahead of Christmas 1967.[103] Previously postponed by an outbreak of foot-and-mouth disease, an intraministerial meeting was hastily scheduled for February 13, 1968. According to an internal letter, "ministers are becoming increasingly vulnerable in this business and we ought quickly to settle our lines on Netherthorpe."[104] A MAFF minute warned: "[MH] have been preparing for the 'battle.'"[105] This prediction was true. With the support of the PHLS,[106] MH representatives argued that medical antibiotic use was the sole concern of their ministry and referred to the 1967 joint press statement's endorsement of the second Netherthorpe report's terms.[107] MAFF officials later complained that the MH had treated the new review's agriculture-focused terms of reference as "a sacred cow which would not be sacrificed at any cost."[108]

The next question to settle was the future antibiotic review's membership. Feeling that the Netherthorpe committee had been "over-weighted scientifically on the medical and para medical sides,"[109] MAFF was keen to rebalance the new committee in favor of agricultural interests. Another point of contention was E. S. Anderson's role: should he be a committee member, or should he function as an adviser? Both ministries were aware of Anderson's public and scientific standing but equally wary of his vocal support of antibiotic bans and his temperamental character. To control Anderson, the MH suggested co-nominating PHLS director Sir James Howie.[110] However, in his eagerness to be appointed, Anderson overshot his goal: in April 1968, he publicly announced that he would refuse to give evidence should he not be appointed to the committee.[111] Anderson's attempt to pressure his way into Britain's confidential world of compromise-oriented corporatist expert consultation backfired. MAFF could now argue that Anderson would endanger the committee's recommendations: "If the committee's conclusions were in line with Dr. Anderson's views, there would be the charge that we had biased it with prejudiced members; if it went the other way, Dr. Anderson would no doubt issue a minority report."[112] Even James Howie, who had previously refused to accept a committee position without Anderson's co-nomination, now changed his mind. In May 1968, a minute by Jock Carnochan, assistant secretary of MAFF's Animal Health Division, triumphantly noted: "Dr. Howie has become impatient of the Prima Donna approach of . . . Dr. Anderson and is no longer prepared to support him."[113]

By late May 1968, all membership decisions had been made: Anderson had been substituted with a Medical Officer of Health from Birmingham and the molecular biologist and University of Edinburgh vice-chancellor Michael Swann had accepted chairmanship of the committee. Fearing attacks by Anderson, MAFF had, however, withdrawn its nomination of Alastair Frazer. In a smart move, agricultural officials convinced the MH to nominate two veterinarians in Frazer's stead. Comprising two agriculturalists, three veterinarians, and two medical scientists, the review committee was now weighted slightly in favor of agricultural interests.[114]

The Swann Report

Eighteen months after the submission of the second Netherthorpe report and three years after Anderson's initial warnings, the Joint Committee on the Use of Antibiotics in Animal Husbandry and Veterinary Medicine started its work in July 1968. The tasks facing the corporatist Swann committee were daunting: it had to strike a compromise between opposing agricultural and veterinary interests, substantial medical AMR concerns, and the provisions of the new 1968 Medicines Act. In contrast to previous committees, public pressure meant that delaying action by calling for more research was no longer an option. Between December 1968 and April 1969, calls for antibiotic reform increased further when thirty babies died from antibiotic resistant gastroenteritis in Manchester in a seemingly grotesque repeat of the Teesside outbreak.[115]

Differences of opinion on antibiotic reform became amply clear during the Swann Committee's evidence gathering phase: farmers and nutritionists defended antibiotic access; veterinary organizations called for prescription-based controls of medicated feeds (chapter 6); medical and public health representatives attacked antibiotic use by both farmers and veterinarians;[116] and pharmaceutical lobbyists used the opportunity to launch a concerted campaign against statutory bans.

The records of the British Beecham company allow a detailed reconstruction of industry campaigning. Concerned about its Vitamealo AGPs, Beecham assessed individual members' voting preferences within days of the Swann committee's nomination: the committee's three animal nutritionists and agricultural experts would support AGP use; the two public health experts would support antibiotic restrictions; and the three veterinarians would support expanded veterinary control over antibiotics.[117] Beecham next mobilized influential supporters to defend its products. One of these supporters was Nobel laureate Ernst Boris Chain, who had helped Beecham develop semi-synthetic penicillins and now directed Imperial College's biochemistry department.[118] Speaking to supermarket representatives in 1967, Chain had already claimed that AMR could be controlled by new drugs and temporary restrictions: "It takes as little as 30 minutes for a bacterium to divide, and for this reason it is impossible to shift the equilibrium of the bacterial population of a world scale. If a shift does occur within a confined area such as a hospital ward or an animal rearing unit, all one has to do is to stop the selection pressure . . . , open doors and windows widely and wait for a week or two; . . . and everything is exactly as it was before."[119] Agricultural antibiotics' ongoing efficacy meant that *in vitro* research on transferable AMR did not apply to field conditions: "if we consider the invaluable benefit which agriculture has derived from the use of antibiotics, it would be outright folly to ban their use . . . acting only on [an] entirely hypothetical chain of events."[120] Chain also accused E. S. Anderson and the PHLS of fueling an environmentalist anti-chemical crusade: "It is never risky to issue warnings. . . . the supercautious public warners do the opposite to performing a public service . . . . The campaign against the use of antibiotics is, of course, part of the general campaign against the use of chemicals altogether in food production. . . . it is the duty of those concerned with food production . . . to resist the entirely irrational and emotion based campaign against [chemicals] use in any form."[121] Beecham adopted this rhetoric by claiming that there was little evidence of harm resulting from agricultural AMR selection: the human reservoir of resistant bacteria might "in theory" be influenced by agricultural antibiotics but in practice seemed "to be fully capable of self-proliferation."[122] Antibiotics should remain available to treat clinical and "sub-clinical" conditions, which posed infectious hazards and limited animal performance.

In May 1969, Beecham joined other members of the Association of the British Pharmaceutical Industry (ABPI) for a confidential meeting with the Swann committee. During the meeting, representatives from Beecham, Cyanamid, Elanco, Smith Kline and French Laboratories, Pfizer, Parke Davis, and the industry-sponsored OHE defended antibiotic use. According to Cyanamid's Keith Grainger, farmers were already shifting away from nontherapeutic growth promotion. Much less concentrate was sold for home-mixing and poultry producers were replacing penicillin with higher-dosed broad-spectrum, copper, and arsenical products. In general, available prescription-free antibiotic feeds were used less for growth promotion and more as prophylactics against stress.[123] Banning AGPs would thus harm animal health. Beecham's representative added "that transfer of resistance was not a new phenomenon and that it could well have always been the main type of resistance, and besides bacterial populations were dynamic and in a perpetual state of change."[124] Considering it "very obvious"[125] that the Swann committee was debating restrictions, ABPI members strongly opposed banning AGPs or antibiotic advertisements to farmers: "[AGP bans] would turn the veterinary surgeon into a pharmacist.... Each material should be considered on its merits and the range of antibiotics available for animal therapy should not be restricted so that the veterinary surgeon's armamentarium was different to the doctor's for treating human disease."[126] Manufacturers also warned that AGP bans could harm about £97.3 million of British antibiotic exports, raise the price of antibiotics for human medicine, and "might induce a manufacturer to choose an alternative place of manufacture on the Continent of Europe."[127]

Faced with a plethora of contradictory demands, Swann committee members struggled to find an acceptable compromise on antibiotic reform. The question was which common denominator the committee's individual factions would be able to agree on. Submitted in November 1969, the Swann report clearly acknowledged that agricultural antibiotics contributed to AMR and had "caused some difficulties in veterinary practice and [had] caused harm to human health."[128] *En bloc* AMR transfer increased the likelihood of a "massive and rapid propagation of antibiotic resistant organisms"[129] as had already happened in the case of *S. typhimurium* Type 29. Although it claimed that horizontal AMR transfer was only a problem in Enterobacteriaceae, the report refused to accept that twenty years of use had proven agricultural antibiotics safe. There was sufficient evidence to take action without allowing the "cry for more research" to "hold up implementation of our recommendations."[130] Echoing previous reports, the Swann committee advised the government to cut the number of advisory bodies, install a permanent committee on all aspects of antibiotic use, ban antibiotic advertising to laypersons, and fund further AMR research and preventive veterinary epidemiology at universities.[131] Most significantly, the report recommended extending the 1962 distinction between medically relevant and irrelevant antibiotics to already licensed AGPs: antibiotics should be used as growth promoters only if they were of economic value, had little or no application as therapeutic agents, and did not impair therapeutic antibiotics. This latter recommendation also included scenarios in which resistance to AGPs was "part of a multiple resistance pattern transferable *en bloc*."[132]

In concrete terms, the Swann report advocated a ban of penicillin and tetracycline AGPs. It also advocated assigning prescription-only status to the still unscheduled macrolide tylosin because of cross-resistance to erythromycin, which was used in human medicine. Prescription-free sulfonamide and nitrofuran feeds should also be restricted due to their selection for multiple-resistant organisms. The availability of nontherapeutic AGPs like bambermycin, virginiamycin, and zinc bacitracin meant that the proposed feed restrictions would not cause economic harm.[133] Taken by itself, the first part of the Swann report marked a milestone in the history of precautionary substance regulation. Its recommendations were based on the strong likelihood—but not quantifiable evidence—of harm.

However, when it came to addressing the wider antibiotic infrastructure of British agriculture, its corporatist roots made the Swann report far less ambitious. In the case of prescribed therapeutic antibiotics, it shied away from challenging veterinary antibiotic use. The Swann report criticized veterinary attempts to control salmonellosis with antibiotics and struggled to "find any excuse in logic or theory" for the "prophylactic treatment of farm animals in the absence of infection."[134] However, it refrained from recommending statutory reform and instead emphasized "that the veterinary surgeon and practitioner, like his medical counterparty, treasures the freedom . . . to prescribe as he thinks best in the interest of his patient."[135] Veterinarians were merely cautioned to "temper credulity with a more critical analysis of the advantages and disadvantages of the antibiotics they prescribe."[136] It was clear that this tame criticism had been controversial within the Swann committee. In the case of chloramphenicol, the report noted that its decision not to call for an end of veterinary prescription rights "may prove to have been mistaken."[137] One official later acknowledged that "issues other than the purely scientific"[138] had influenced debates.

By giving into veterinary pressure, the Swann committee missed its chance to break the narrow AGP-focused mold of the Netherthorpe reports and challenge wider antibiotic infrastructures. Whether limited Swann-style bans would solve AMR was doubtful. Conceived of in the late 1950s, partial antibiotic restrictions were supposed to contain the vertical proliferation of resistant organisms but offered little protection against what E. S. Anderson had already described as the "umbrella-like" selection and ecological proliferation of mobile resistance genes. What would prevent a medically irrelevant AGP from selecting for resistance against medically relevant antibiotics if resistance against both drugs was encoded on the same plasmid? Why was the higher-dosed use of a drug in a prescribed feed safer than the marginally lower-dosed use of the same drug in a prescription-free feed? British decision-makers had attempted to "fix" AMR—but in a way that would not alienate corporatist stakeholders. The result was a report that was revolutionary in its precautionary ambition but tame in its outcomes.

The publication of the 1969 Swann report ended the first era of antibiotic reform. Although AMR dominated British concerns from the 1940s onward, postwar pressures and the belief that low-dosed drugs would not select for resistance resulted in policies, which prioritized antibiotics' economic benefits over potential risks. Relying on loose gentlemen's agreements and corporatist self-regulation, British officials were slow to respond to emerging side effects of expanding antibiotic use. Unlike the residue-focused FDA, Britain only mandated milk surveillance after a public scandal in 1963—meat escaped regulations and would only be monitored from the late 1970s onward (chapter 13). Gradualism also characterized official responses to spreading AMR. Although insulated corporatist decision-making was capable of addressing a systemic challenge like AMR without allowing conflicts of interest to spill into the public arena, its compromise-oriented nature produced conservative outcomes. Under pressure from public concerns about factory farming and medical AMR warnings, the 1960s thus saw two carefully staffed expert committees lay the groundwork for the next three decades of both British and European antibiotic regulation. Differentiating between high- and low-dosed antibiotic use and between medically relevant and irrelevant drugs, the Netherthorpe and Swann committees legitimized public concerns about agricultural AMR selection but constructed and reinforced narrow legal boundaries to contain resulting hazards. Wider antibiotic infrastructures remained unaddressed.

Part III

USA

The Problem of Plenty, 1967–2013

The third part of this book explores the widening transatlantic divergence of agricultural antibiotic use and regulations following the 1960s. In contrast to Britain, AMR selection on farms never turned into a common denominator of public risk perceptions in the United States. Chapter 7 explores how the 1970s "toxicity crisis" and the fracturing of bipartisan support for environmentalism sidetracked calls for AMR-focused antibiotic restrictions. Reform efforts were further undermined by sophisticated industrial counter-science and lobbying. Despite a resurgence of AMR concerns during the 1990s, public pressure remained insufficient to trigger statutory reform and was further fragmented by the boom of organic and antibiotic-free products. Chapter 8 studies how US farmers reacted to public concerns about antibiotics. It shows how ongoing public criticism, economic pressure, and bans of DDT and DES fostered an agricultural fortress mentality against antibiotic restrictions. This opposition to statutory measures persists despite the partial "greening" of agricultural discourse and methods since the 1980s. Chapter 9 analyses US regulators' attempts to curb AMR and residues after 1967. During the 1970s, the FDA attempted to implement Swann-style AGP restrictions three times. Its efforts failed because of lacking public and political support, internal divisions, and the US legal system's increasing insistence on proof of harm and cost-benefit analyses rather than precautionary reasoning. Following prolonged stagnation, a resurgence of antibiotic reform under the Obama administration ultimately failed to implement statutory measures in 2013.

8

Marketplace Environmentalism

Antibiotics, Public Concerns, and Consumer Solutions

This chapter traces the evolution of agricultural antibiotics' public image in the United States from the 1960s to the present. In contrast to Britain, concerns about residues, AMR, and welfare never fused in the US public sphere. Instead, the "toxicity crisis" of the early 1970s prolonged the residue-focused antibiotic risk episteme. It was only toward the late 1970s that a significant number of public commentators began to express concern about agricultural AMR selection. However, an increasing partisan divide on "green" issues and attacks on over-regulation soon diluted calls for wider antibiotic reform. The market stepped in where the state stopped. Concerned about food purity rather than AMR, a growing number of US consumers turned to antibiotic-free or organic produce. By the 1990s, organic food had become mainstream. The effects of this development were ambivalent. While concerned consumers could switch to antibiotic-free food, the creation of a market niche for "safe food" fragmented pressure for systemic antibiotic reform. Green consumerism's emphasis on purity also reinforced the existing US focus on residues over AMR hazards. Meanwhile, higher prices for organic produce led to a class-based distribution of risk whereby wealthier consumers could purchase a sense of safety that was not available to poorer citizens.

Epistemic Divides

The second half of the 1960s did not change Americans' cultural prioritization of residue over AMR hazards. Following the recommendations of the 1966 ad hoc committee on veterinary and nonveterinary antibiotics, the FDA acknowledged potential risks posed by "infective resistance" but focused on residue-oriented goals by banning antibiotic preservatives and establishing a national residue-monitoring program for meat (chapter 4).

The FDA's policy of only partially acknowledging the risks of horizontal AMR transfer mirrored divided assessments in the public sphere. With regard to human medicine, all major US newspapers supported campaigns for "rational" antibiotic use and expressed concern about antibiotic resistant infections.[1] However, the extent of AMR threats remained contested. In the *New York Times*, Vernon Knight of Baylor University claimed that there "was no compelling evidence that a dark age of medicine, bereft of antibiotics, lies ahead."[2] By contrast, R-factor discoverer Tsutomu Watanabe predicted a "pre-antibiotic"[3] era in the *Washington Post*. Unsurprisingly, commentators were even more divided in their assessment of AMR selection on farms. While experts like Watanabe and David Smith from Boston's Children's Hospital attacked agricultural antibiotic use,[4] others called for concrete proof of harm, differentiated between AMR in human and animal populations, and continued to praise antibiotics' economic benefits in food production.[5] After initially equivocating between US and European risk assessments,[6] the *New York Times* was the only major US paper to firmly condemn AMR as an environmental threat to "Spaceship Earth"[7] and endorse Britain's Swann report in 1969. Without a broad coalition of supporters from different communities, the issue of AMR-focused reforms of US agriculture languished in the US public sphere.

The Toxicity Crisis

Americans' relative complacency about nonhuman AMR selection stood in contrast to ongoing concerns about "unnatural" antibiotic residues in food (chapter 2).[8] The late 1960s and 1970s saw a "toxicity crisis"[9] exacerbate this prioritization. The crisis was triggered by scandals and high-level proceedings against toxic and carcinogenic substances like chloramphenicol, DDT, and DES, which damaged trust in regulatory agencies. It also coincided with rising popular environmentalism, the high point of the US consumer protection movement, and the countercultural rejuvenation of the organic movement.[10] Politicians on both sides of the aisle were keen to capitalize on the toxicity crisis and a coinciding wave of what Lizabeth Cohen has termed "Third-Wave Consumerism."[11] Attempting to appeal to a public that was being divided into lifestyle and value segments by polling experts, Democrat and Republican administrations strengthened regulatory science and consumer protection, withdrew hazardous substances, and created new federal agencies like the Environmental Protection Agency (1970). During its heyday between 1967 and 1973, it seemed as though the combined effects of the toxicity crisis and third-wave consumerism might trigger a value-driven environmentalist decade.

American journalists played an important role in reinforcing growing consumer concerns about antibiotic residues and other chemical hazards in agriculture and the environment. Employing the subjective narration of New Journalism, reporters revealed gaps in US food and pharmaceutical controls and attacked industry attempts to water down existing regulations.[12] In July 1967, the *Washington Post* warned that the "meat lobby" was attempting "to sidetrack or modify a bill providing for the inspection of meat, . . . which has been peddled off on unsuspecting housewives for a good many years."[13] "Peddling" was possible because federal inspectors did not target meat sold within state borders where officials were working "hand in glove with the meat packing interests." "[Meat] has been processed with such extras as hog's blood . . . . Even eyeballs, lungs, and chopped hides have been used in processed ham to increase its protein content. Detergents have also been used to freshen up the meat, while such antibiotics as aureomycin have been injected as a substitute for sanitation."[14] Allegations of kowtowing also affected the FDA. In 1967, the *Washington Post*'s investigative journalist Morton Mintz bemoaned the "thundering silence of drug consumers," which was enabling pharmaceutical companies to subvert consumer protection: "Who speaks for the fetus, whose concern with chemicals extends beyond drugs to food additives and pesticides among other things?"[15] Mintz was especially concerned about the antibiotic chloramphenicol (chloromycetin).[16] Although the link had been known since the 1950s,[17] new data suggested that aplastic anemia occurred ten times more frequently after chloramphenicol use than previously thought.[18] Worryingly, the FDA seemed powerless to effect change. After it failed to ban chloramphenicol in early 1968,[19] FDA commissioner Charles Goddard claimed to be at his wit's end[20] and was widely criticized for failing to protect citizens from a drug that was popular in both medicine and agriculture.

Public trust in the FDA eroded further over the next two years. Succeeding Goddard in July 1968, commissioner Herbert Ley launched withdrawal procedures against forty-nine fixed drug combinations and filed price-fixing charges against major manufacturers.[21] However, in the midst of Ley's battle against inefficacious and overpriced drugs, the FDA was shaken by allegations that it had manipulated data on the carcinogenicity of popular cyclamate sweeteners.[22] Commissioner Ley's perceived hesitancy to proceed against cyclamates infuriated public opinion and resulted in his effective sacking by the new Republican secretary of Health, Education, and Welfare (HEW) Robert Finch in 1969.[23] Shortly afterward, Ley claimed that he had been under "constant, tremendous, sometimes unmerciful pressure," sometimes spending "as many as six hours fending off representatives of the drug industry."[24] Attempting to restore trust, Secretary Finch appointed former Booz Allen Hamilton consultant Charles D. Edwards as FDA commissioner in 1970. Edwards was experienced in public relations and announced a revitalization of the FDA: drug recalls would be accelerated, consumer involvement increased, bad advertising and prescription habits targeted, and regulatory science capabilities strengthened.[25]

Within months of being appointed, Edwards had to deal with a new set of food safety allegations. In June 1970, the *Washington Post* reported the results of the USDA's recently established national residue monitoring program. While randomized monitoring indicated low overall contamination, targeted testing had revealed antibiotic residues of sufficient concentration to trigger allergic reactions in consumers.[26] Officials' response was hardly reassuring. Cornelius Donald (C. D.) Van Houweling, head of the FDA's Bureau of Veterinary Medicine (BVM), claimed: "You can't put an inspector at the shoulder of every farmer, veterinarian and meat packer in the country."[27] However, five months later, even limited existing controls seemed under threat when it emerged that monitoring for antibiotic residues was to be cut by 74 percent and for DES by 50 percent.[28] The planned cuts seemingly confirmed the allegations of a new generation of vocal consumer crusaders. Headed by James Turner, the Ralph Nader Study Group on Food Protection and the FDA published *The Chemical Feast* in 1970. The book contained a scathing review of US food production and consumer protection against drug residues and food-borne pathogens.[29] Nader's group also attacked the trustworthiness of contemporary regulatory science and reliance on expertise provided by external "groups, such as the [NAS]," whose members "are themselves either employed by or consultants to the food industry."[30] Attacks by investigative journalists and Nader's activists were effective in damaging public trust in FDA oversight. In 1971, the *New York Times* noted: "These days [the FDA is] just lurching from crisis to crisis"; "In the last year, headlines have proclaimed mercury in fish, botulism in pizzas, pesticides in turkeys, arsenic in chickens, antibiotics in cheese, hormones in meat, salmonella in soup, cyclamates in soft drinks and DDT in practically everything."[31]

The Organic Moment

The US organic sector was the main profiteer of the toxicity crisis and sinking public trust in food safety and—more importantly—purity. The late 1960s saw the fusion of the established organic sector with younger countercultural and environmentalist movements. Throughout the United States, co-ops and communes with names like the Hip Salvation Army began producing and selling organic or "natural" food.[32] Producers and consumers of "natural" food were united in their concern about the poisoning of food, bodies, and the environment by "unnatural" chemicals—most prominently DDT, DES, and antibiotics. Despite agreeing on this *ex negativo* definition of "pure" and thus "natural" food, organic production philosophies varied: while some producers saw the commercial potential of supplying rising public demand for "natural" food, others fused ideals of organic production with Jeffersonian concepts of self-sufficiency and returning to the land.[33]

The informational needs of the alternative community were served by a growing number of publications. Starting in the late 1960s, new magazines like Stewart Brand's *Whole Earth Catalog* and *Mother Earth News* targeted mostly younger technophile dropouts and back-to the-landers interested in small-holdings or self-sustaining communes.[34] In addition to these newer publications, the Rodale Press continued to target both organic consumers and producers. Under the leadership of Robert Rodale (J. I.'s son), the Rodale Press recast organic food as part of a wider—flexibly defined—green lifestyle. In doing so, it helped forge a form of what Andrew Case has called do-it-yourself "marketplace environmentalism,"[35] which placed responsibility for a healthy and safe life in the hands of paying individuals. Citizen consumer action rather than state regulations and traditional politics were presented as solutions to environmental woes. Sales soared. Between 1968 and 1971, the circulation of Rodale's consumer magazine *Prevention* almost doubled to nearly 1 million, while *Organic Gardening and Farming* had 500,000 subscribers.[36]

The rejection of chemical helpers was common to nearly all countercultural and organic publications. In the case of antibiotics, the focus was very much on food that was pure (residue free) rather than food whose production would not select for AMR. Published in 1972, Rodale's *The Basic Book of Organically Grown Food* noted: "A big change is taking place in the way people think about naturally and organically grown food. No longer are you an oddball or a faddist if you buy or sell an apple that hasn't been sprayed. Everybody is concerned about pollution, and many people now believe that food is one thing that we should be able get in unpolluted form."[37] The Rodale Press remained deeply skeptical of wider FDA and USDA policies and attacked practices like the salvaging of edible parts from condemned carcasses: "We wonder how one removes the part affected by drug residues and passes the rest as clean."[38] If AMR hazards were discussed, they were often conflated with residue hazards. According to *The Basic Book*, using agricultural antibiotics "can build a transferable resistance to the drug into the meat tissues."[39]

Inspired by environmentalist bestsellers like *The Limits of Growth* (1972) and the contemporary oil crisis, supporters emphasized organic agriculture's sustainability. In 1973, Rodale's *The New Food Chain* presented organic agriculture as a solution to the "expansionist, over-chemicalized, over-plasticized, over-additived, over-dumped society."[40] The importance of pure food was also emphasized in countercultural restaurants and cookbooks. In 1973, the *Whole Earth Cookbook* from Santa Cruz, California, warned about antibiotics and promoted sustainable food: "The modern feedlot techniques... are so unhealthy that one cattleman commented to us, 'I raise a private stock for my own use.'"[41] Readers should only buy meats "which are not injected with chemicals"[42] and encourage farmers to readopt "natural" production methods. Citing Frances Moore Lappé's *Diet for a Small Planet* (1971),[43] others argued for meat-free diets by mixing concerns about chemical food with warnings about a global protein crisis.[44]

The rejuvenated organic movement received national press coverage. Publications ranging from *Harper's* to the *The Jewish Press* from Omaha described "natural" foods' alleged health and environmental benefits.[45] With organic restaurants and grocery stores appearing across the United States,[46] the *New York Times* published a *Natural Foods Cookbook* in 1971.[47] Journalists were intrigued not only by organic purity claims but also by the movement's "weirdness." In a 1971 article on "Organic Food Cult" guru J. I. Rodale, the *New York Times* enumerated the many groups "loosely clustered under the organic movement's antichemical umbrella": "food cultists, from old-line vegetarians to youthful Orient-oriented 'macrobiotic' dieters..., plus reactionaries yearning to turn back all clocks, urban dropouts..., ecologists..., Dr. Strangelove paranoids... and, increasingly, rather ordinary [people] to whom pronouncements about [chemical perils]..., have stirred a wariness about all man-made chemicals."[48] In 1973, the newspaper printed a similar feature on 69-year-old nutritionist Adelle Davis, "chief showwoman for health foods, a $1-billion-a-year business catering... to a rapidly growing 'organic nation' of health-food devotees."[49] According to the *New York Times*, Davis's vision of pure food was representative of both "archetypes of South California, the little old lady in tennis shoes and the young, barefoot, bearded ex-radical."[50]

Celebrity endorsements played an equally important role in spreading the message of antibiotic-free purity to middle-class consumers. Starting in the mid-1960s, magazines like *Vogue*, *Cosmopolitan*, and *Life* began to report on fashionable "natural" products. Celebrity organic devotees featured by *Vogue* included Yehudi Menuhin, the Marquess of Londonderry, Habib Bourguiba, and Bruno Walter.[51] Ordering meat only from a farm "where they use no sprays and chemicals," performer Carol Channing noted, "Princess Margaret was dying to have my plain roast lamb..., and I think the Kennedys invited me to the White House just to see what I would bring."[52] Advocating surprising amounts of alcohol as part of a healthy diet, *Cosmopolitan* agreed that organic was not just healthy but tasted better.[53] In 1970, *Life* printed a large feature on "the move to eat natural."[54] According to the magazine, "true devotees" went to "great lengths to ensure, that all their food be grown organically, that is without artificial help of any kind"—"natural" meat could only come "from animals raised without benefit of antibiotics."[55] As arbiters of taste, *Vogue*, *Life*, and *Cosmopolitan* agreed that "natural" diets were rejuvenating, fashionable, and ethical—and defined by the absence of antibiotic residue rather than AMR hazards. They also reinforced the notion that antibiotic reform was a question of individual consumer choice rather than traditional politics and state intervention.

The increasing public visibility of organic agriculture provoked fierce responses from supporters of conventional agriculture. Holding fast to scenarios of overpopulation and famine-induced communism, they defended America's mission to produce plenty with the help of modern technologies—including antibiotics. In 1971, USDA secretary Earl Butz attacked organic producers' chemical-free message: "Without the modern input of chemicals of pesticides, antibiotics, of herbicides, we simply couldn't do the job. . . . Before we go back to an organic agriculture in this country somebody must decide which 50 million Americans we are going to let starve or go hungry."[56] Regional perceptions also remained mixed, with student newspapers from Texas emphasizing that organic was not always safer and the *Desert Sun* from Palm Springs complaining about "neurotic"[57] chemical criticism. Despite mourning the demise of Jeffersonian family farmers,[58] the conservative *National Review* was equally skeptical of efforts to "de-chemicalize" agriculture: "The Topsy-Turvy labors of the Whole Earth Catalog brigade go on and on, with no apparent end in sight. We have become accustomed . . . to the efforts of ecologists and their friends in government to slap every conceivable sort of regulation on American business in the name of preserving the environment. . . . Some of the horror stories previously noted . . . include the ban on leaded gasolines, the holy war against DDT, and a rather improbable attack on penicillin."[59] Influential within the budding neoconservative movement, the National Review's equation of antibiotic criticism with eccentric hippies and leftist regulation-excess did not bode well for parallel FDA attempts to curb agricultural AMR selection.

AMR Complacency

Public interest in organic food production stood in stark contrast to the relative lack of attention paid to nontraditional chemical hazards like AMR. With the toxicity crisis focusing attention on invisible residues, few American commentators reported on early FDA efforts to implement Swann-style AGP bans in the United States.

In 1970, the FDA had commissioned an expert task force to review agricultural antibiotic use (chapter 10). After two years of deliberation, the task force published its report. In January 1972, the FDA announced that it would install a "program that should lead to removing some antibiotics from animal feeds as dangers to human beings."[60] The antibiotics in question were penicillin and the tetracyclines. Amid rumors that the sixteen-member task force had argued bitterly, manufacturers were given two years to prove that drugs were safe. Senior FDA officials also did not sound particularly convinced of planned AGP bans. Speaking at the 1972 press conference, BVM director C. D. Van Houweling announced that FDA deadlines could "be extended depending on the drift of safety research in progress."[61] Commissioner Edwards explicitly stated that "the agency had no information that would warrant calling the feeds an 'imminent' hazard"[62]—which would have necessitated bans.

Officials' ambivalence about British-inspired AGP bans was mirrored in the American media. Reporting on the FDA press conference, the *Washington Post* highlighted the discrepancy between task force recommendations for bans and ambivalent FDA statements.[63] In contrast to its 1969 endorsement of the Swann report, the *New York Times* also waivered about whether to support AGP bans. In February 1972, it quoted the president of the pro-industry Animal Health Institute (AHI), James G. Affleck, who maintained that he did not know of "a single case of untreatable bacterial disease in man"[64] caused by AGPs. Five days later, another article compared agricultural AMR selection to the pandemics conjured in Michael Crichton's 1969 *The Andromeda Strain* but failed to reach a definite verdict on the proposed bans: AGPs' alleged economic benefits of $500 million per year were tangible whereas agricultural AMR hazards remained science-fiction-like and contested.[65] This logic of costs and benefits was echoed by Robert Bleiberg, editor of the Dow Jones owned *Barron's National Business and Financial Weekly*. According to Bleiberg, the FDA task force had exaggerated dangers and mobilized left-wing environmentalists for a general assault on modern agriculture and "scientific progress." "Dollars-and-cents aside, [cracking down on antibiotics and sulfonamides] would turn back the clock in agribusiness by a generation or more, risk the eruption of devastating epidemics among the nation's herds and flocks, escalate the cost of living and, in the end, for the first time since the Industrial Revolution, perhaps go a long way toward making the Malthusian nightmare grim reality."[66]

With the public jury on FDA AGP bans remaining out, US media interest faded fast. By late 1972, articles on infective AMR and antibiotic overuse reverted to simply ignoring agricultural resistance selection.[67] In spring 1973, the announcement of an unspecified delay of FDA antibiotic bans received only scant attention. During the FDA's press conference, C. D. Van Houweling merely announced that data submitted so far "[had] not been developed either to prove or disprove the existence of a serious threat."[68]

More sustained US media interest in the environmental dimensions of infective AMR arose in the wake of the 1975 Asilomar conference on recombinant DNA.[69] Using restriction enzymes and plasmids, researchers were now able to insert foreign DNA—including resistance genes—into bacteria.[70] While transferable resistance turned into a valuable laboratory tool, critics used the public focus on "mutant lab bacteria" to renew attacks on agricultural AMR selection.[71] Reporting on Asilomar in March 1975, the *Washington Post*'s Stuart Auerbach quoted British PHLS researcher E. S. Anderson, who emphasized that the "widespread use of antibiotics in agriculture"[72] was a far more dangerous source of resistant pathogens than lab-based bacteria strains. Anderson's warnings seemed verified three months later when Auerbach reported on an AHI- and Pfizer-funded study by a research team under microbiologist Stuart Levy. Having trained under R-factor discoverer Tsutomu Watanabe,[73] Levy had traced the spread of plasmid-mediated AMR from tetracycline-fed animals to a farm family with biochemical markers. After six months, 31.3 percent of the family's weekly fecal samples had contained substantial amounts of tetracycline-resistant bacteria with some bacteria resistant to three or more antibiotics.[74] Levy's team celebrated the end of the study by barbecuing the experimental animals. Some neighbors had, however, "balked at eating the chickens because they feared they would develop a resistance to antibiotics."[75]

Published between 1975 and 1976, Levy's findings had a profound impact on American AMR debates. In what would become a lifelong campaign for antibiotic stewardship, Levy warned: "The rise in frequency of resistant organisms in our environment is the obvious result of antibiotic usage. The only means to curtail this trend is to control the indiscriminate use of these drugs. . . . These data speak strongly against the unqualified and unlimited use of drug feeds in animal husbandry."[76] For microbiologists and liberal commentators, the situation seemed increasingly clear. AGPs' economic benefits no longer justified the medium- to long-term genetic, microbial, and health costs of their use. The question was whether this relatively narrow and partisan coalition of antibiotic critics would be able to convince others to support their cause. A slowing economy and growing political divides on environmentalist regulation and substance bans meant that this was no easy task.

Resisted Regulation

Amid a fresh burst of AMR reports, the FDA's new commissioner—and former microbiologist—Donald Kennedy announced statutory bans of penicillin and tetracycline AGPs in April 1977 (chapter 10).[77] Kennedy's plans were ambitious. Initial AGP bans "should be viewed as a first step toward FDA's ultimate goal of eliminating, to the extent possible, the nontherapeutic use in animals of any drugs needed to treat disease in man."[78] Although the FDA estimated that its measures would only cost five cents per person per year, it also cautioned that industry and political opposition might delay the implementation of AGP bans. This prediction was correct.

Kennedy's AGP bans not only won him few friends in Washington but also occurred parallel to unpopular withdrawal procedures against saccharin sweeteners, which the FDA had been forced to launch following new carcino-genicity data.[79] Popular support for his initiatives was also mixed. Since the early 1970s, a growing number of media commentators had blamed allegedly excessive FDA regulation for creating a so-called drug lag in terms of US drug availability. Referring to neoliberal economist Milton Friedman, the *National Review* claimed in 1978 that thalidomide and DDT had produced "something akin to hysteria"[80] among FDA regulators. Shared concerns about stagnating economic growth and inflation (stagflation) also made more liberal publications like *Harper's* and *Time* join neoconservative attacks on environmentalist "his-trionics."[81] Industry campaigners skillfully exacerbated rising regulation wari-ness by casting doubt on scientific warnings. Employing a well-tested strategy of "agnogenesis" (the creation of ignorance), lobbyists used carefully curated "neu-tral expertise" to undermine uncomfortable data in areas ranging from climate science to cancer research. The goal of this directed research was to counter or delay regulatory action by creating the impression of scientific uncertainty and the need for more research. The public credibility of counter-science was main-tained by financing "friendly" scientists, creating "neutral" research institutes, and designing studies so that results would foster uncertainty.[82]

In the case of AGPs, industry protest was led by the AHI, livestock producer associations, and the Council for Agricultural Science and Technology (CAST). While pharmaceutical and agricultural organizations lobbied Con-gress and organized letter campaigns against FDA antibiotic restrictions (chap-ter 10), CAST specialized in providing counter science. With two-thirds of its later $265,000 budget consisting of industry donations, CAST organized pan-els of reputable experts, whose reports were submitted to a small group of core staff for final editing and publishing.[83] Since its founding in 1972, this strategy had worked well and CAST was able to convince many scientists that it was providing sound research in the murky field of chemical risk assessment. In 1977, CAST reacted to FDA AGP bans by inviting external experts to assess AGPs. However, experts proved difficult to control and CAST's scientific cred-ibility was severely damaged when the six external microbiologists on its AGP panel discovered that their findings were being used in a biased way. Tensions between CAST's editorial board and its AGP panel turned into a full-blown éclat in 1979 when an edited version of the panel's final report contained mis-leading information that had been added without experts' consent. Alarmed by the misuse of their findings, the six microbiologists resigned from the AGP panel and drew scientific and media attention to CAST's dubious practices. Subsequent high-profile controversies between CAST-member Thomas Jukes and Richard Novick, one of the six microbiologists, further damaged CAST's credibility.[84]

However, by this time, FDA antibiotic reform had already come undone. Unable to mobilize large-scale public support and hampered by drawn-out withdrawal procedures (chapter 10), the FDA was outmaneuvered by antibiotic supporters. In 1978, Congress and the Carter administration reacted to concerted industry protests and pro-antibiotic lawmakers threatening to hold the US budget hostage by imposing a moratorium on FDA bans and calling for more research.[85] American lawmakers next commissioned the NAS, whose experts had previously supported AGPs, with an independent review of antibiotic feeds' costs and benefits. Noting the FDA's loss of momentum, the *Washington Post* reported that "many farm state congressmen" were questioning "the need for any FDA action"[86] at all. The loss of FDA momentum was exacerbated by Kennedy's resignation in June 1979. According to the *New York Times*, the FDA had "lost its best commissioner in a long time."[87] When Kennedy had come to the FDA in 1977, the "agency was torn by internal dissension and charges ... that it had become chummy with the industries it regulates. Morale has been raised and the FDA's reputation is decidedly one of independence."[88] This independence had, however, come at a political price: "[Kennedy] lost some big battles of regulation. Congress refused to let him ban saccharin. It impeded his drive against the indiscriminate use of antibiotics in animal feeds. [HEW] Secretary Califano ... blocked ... efforts to phase out nitrites in meat. Several states ignored his warnings against laetrile.... [Kennedy] was probably right on all these issues."[89]

Fighting on many fronts, Kennedy's FDA had failed to transform the residue-focused US risk episteme into one focusing on AMR. Ten years after the Swann report, agricultural AMR selection not only failed to excite widespread anxiety but was increasingly seen as a partisan issue. The eclipse of third-wave consumerism and rise of stagflation concerns spelled further bad news. Defeated under a Democrat administration, it was unlikely that AGP bans would be viewed more favorably by the increasingly neoconservative Republican party. In 1981, the election of the Reagan administration diminished hopes for state-led antibiotic reform. Shortly afterward, the new FDA leadership proposed to expand the number of licensed agricultural antibiotics for the first time in a decade.[90]

Mirroring wider developments in US environmentalism,[91] a coalition of non-state actors filled the void left by the state's departure from antibiotic reform. Around 1980, new mystery diseases like AIDS and the reemergence of resistant scourges like syphilis and tuberculosis undermined many of the optimistic postwar prognoses of improved disease control. Concerned about regulatory stagnation in the face of rising AMR, environmental NGOs, campaigning scientists, and liberal media outlets attempted to transform the public's microbial anxieties into momentum for renewed antibiotic reform.[92] In August 1981, Stuart Levy organized a multinational conference during which 150 doctors from twenty-five nations warned about AMR proliferation in human and non-human settings. Founded in the same year, Levy's nonprofit Alliance for the

Prudent Use of Antibiotics (APUA) continued the campaign.[93] Two years later, the National Resources Defense Council (NRDC) tried to reignite official antibiotic reform. Signed by 300 governmental and nongovernmental experts, an NRDC petition to Ronald Reagan requested the enactment of the FDA's still stalled 1977 AGP bans.[94]

The NRDC's case was significantly strengthened in 1984 when CDC epidemiologist Scott Holmberg published two papers in *Science* and the *New England Journal of Medicine* (*NEJM*). While Holmberg's *Science* paper presented a medium-term epidemiological analysis of US fatalities caused by resistant *Salmonella* outbreaks,[95] his *NEJM* paper used newly developed molecular plasmid profiling to link agricultural AMR selection to a recent case of human harm.[96] In early 1983, Holmberg's team had identified eighteen persons infected with a multi-resistant strain of *Salmonella newport*. Eleven patients had been hospitalized and one had died. By comparing plasmid profiles of all human *S. newport* isolates from the six-state area and all national isolates from animals for eighteen months, Holmberg's team had linked human infections to hamburgers made from antibiotic-fed beef cattle in South Dakota.[97] To most US observers, this was an important breakthrough because it seemed to prove that agricultural AMR selection not only posed a theoretical threat to health. Writing in the same *NEJM* issue, Stuart Levy renewed calls for AGP bans: "Every animal or person taking an antibiotic . . . becomes a factory producing resistant strains . . . . Since there are two or three time more livestock than people in the United States, the number of animals fed antibiotics at subtherapeutic levels . . . is enormously greater than the number of people taking antibiotics in therapeutic amounts . . . . We must reserve these resources for fighting microbial disease."[98] Holmberg's study was widely reported, temporarily depressed cattle futures at the Chicago Board of Trade, and prompted the NRDC to upgrade its petition and call for an immediate ban of AGPs because of imminent harm to health.[99]

Although the FDA's renamed Center for Veterinary Medicine (CVM) promised to review Holmberg's findings, public attacks on regulatory inaction intensified. Californian journalist Orville Schell emerged as one of the most vocal antibiotic critics. In his 1984 bestseller *Modern Meat*, Schell accused agribusiness, scientists like Thomas Jukes, and the FDA of downplaying infective AMR hazards.[100] Trust in FDA officials' ability to critically evaluate AGPs was further undermined by charges of bias against CVM director Lester Crawford. Crawford had previously worked as a consultant for American Cyanamid but had also supported the NRDC's 1983 petition. Exposing himself to criticism from both sides, Crawford added to the controversy by first condemning AGPs and then claiming, "I suppose I could change my mind [on antibiotics]."[101] The *New York Times* reported that Crawford's predecessor C. D. Van Houweling had also changed his mind on AGPs after becoming a consultant for the

National Pork Producers Council.[102] Between 1984 and 1985, congressional and FDA hearings further polarized American debates over antibiotic bans.[103]

In the end, little changed. With political positions on antibiotic reform reflecting Washington's growing partisan divide, mostly liberal calls for antibiotic bans failed to convince senior regulators. In November 1985, Republican Secretary of Health and Human Services Margaret Heckler rejected the NRDC petition as one of her last actions in office. FDA reviews had failed to reveal an "imminent hazard" requiring "emergency action."[104] Despite formally leaving Kennedy's 1977 AGP withdrawal proposals in place, Heckler's decision effectively ended 1980s hopes for statutory restrictions. Unable to formally prove an imminent hazard, antibiotic critics had failed to mobilize sufficient public support to break the partisan deadlock that had emerged around antibiotic reform. Over the next decade, new data on links between agricultural antibiotic use and AMR would trigger an almost cyclical ebb and flow of US reform debates, cost-benefit analyses, and calls for more research.

Purity Pays

Stalling antibiotic reforms encouraged growing numbers of American consumers to seek private ways of protecting themselves from perceived problems in conventional agriculture. Ignoring expert warnings, most consumers continued to fear drug residues more than agricultural AMR selection. Marketplace environmentalism's promise of "safe" and "pure" food reinforced popular residue concerns. The growing market for antibiotic-free produce had ambivalent effects: despite producers' best intentions, wealthy consumers' ability to selectively opt out of conventional production fragmented societal pressure for statutory antibiotic reform. Resulting increases of organic production were also too small to protect consumers of pure food from more serious ecological AMR hazards. As later noted by Stuart Levy and allied scientists, the residues in meat should be of least concern to most people.[105]

Contemporary surveys strengthen the picture of an ongoing public prioritization of residue over AMR concerns. According to a 1985 survey of 500 US households by the National Live Stock and Meat Board, most consumers stated "either a mild or strong health concern about antibiotics in meat." "[Antibiotics] ranked near the middle of 13 concerns. . . . . Sixty percent of the respondents said they had a strong health concern about antibiotics. However, only 15 percent were concerned about the bacteria developing resistance. . . . 17 percent mentioned no specific concern, 9 percent were concerned about transfer of antibiotics to humans through meat, and 17 percent wanted more information. Only 21 percent reported some familiarity with the issue."[106] The existing residue-focused risk episteme was reinforced by a string of scandals. In 1984, the *Wall Street Journal* reported that sulfonamide levels in US veal exceeded official tolerances by as much as a hundredfold and that 6 percent of US pork samples contained sulfonamide residues. Testing also revealed that American veterinarians were prescribing chloramphenicol although the drug was banned for food-producing animals.[107] In 1988, residue concerns resurfaced amid reports that withdrawal times were being ignored on farms and the detection of antibiotic residues in US milk samples.[108] Using a new test called CHARM II, FDA officials found drugs in over 50 percent of seventy supermarket samples.[109] Many samples were contaminated with sulfamethazine, which had been illegally administered to lactating cows. Problematically, the FDA did not publish this data but decided to reevaluate CHARM II positives with less sensitive high-performance liquid chromatography—thereby negating many CHARM II positives.[110] This retesting of CHARM II results came to light in 1990. Amid parallel outrage about Alar residues on apples, congressional hearings highlighted that existing FDA monitoring for antibiotics in milk was only highly effective for penicillin. Speaking to Congress, FDA chemist Joseph Settepani accused his agency of "[ignoring] reliable tests"[111] for other antibiotics.

Residue scandals' effects on US consumer attitudes were predictable. A 1990 mail survey with 706 respondents from New York State found that 28 percent considered agricultural antibiotics and hormones to pose a serious hazard, 43 percent considered them to pose something of a hazard, and only 9 percent considered them to pose no hazard. While only 44 percent of respondents were willing to pay more for milk produced without antibiotics, 75 percent favored introducing labels for milk produced with antibiotics.[112] Two years later, the Food Marketing Institute's annual trends survey found that only 12 percent of US consumers were completely confident that food was safe: concerns about pesticide and herbicide residues were closely followed by concerns about antibiotic and hormone residues ahead of concerns about nitrites, irradiated foods, preservatives, and artificial coloring.[113] AMR did not feature prominently in either survey.

Residue and health concerns made a growing number of American consumers turn to allegedly safe food. Many turned to "natural" or organic food. During the 1970s, organic food had often been a hip lifestyle choice. By the 1980s and 1990s, many media reports presented going organic as a wise response to contemporary food scares. Even regional papers like the *Healdsburg Tribune*, which had previously defended conventional agriculture, reported positively on the alleged benefits of organic food.[114] With subscriptions of Rodale magazines like *Prevention* and *Organic Gardening* reaching record numbers,[115] rising demand soon stretched American organic food supply chains to their limits.[116] The result was a dramatic expansion of organic production. Between the 1980s and 1990s, now familiar companies like Stonyfield, Celestial, and Whole Foods Market turned into semi-industrial operations for pure food (chapter 9).[117] Although "organic" and "natural" remained ill-defined categories,[118] sales of organic food grew by 250 percent to $3.5 billion between 1990 and 1996.[119] A 1999 survey of 26,000 consumers found that about a third of the US population selectively purchased organic produce, with concerns about personal safety and environmental protection listed as prime motivations.[120]

Ongoing "purity" concerns also made many US consumers react furiously to attempts to water down organic definitions. Released after seven years of deliberation, the USDA's 1997 organic standards permitted the use of biotechnology, irradiation, sewage sludge, and a limited amount of antibiotics; organic labels would also contain no information on production methods.[121] The new standards triggered one of the largest consumer write-ins in US history. During the four-month comment period, the USDA was inundated by about 150,000 letters and cards. The scale of protest forced Secretary of Agriculture Dan Glickman to promise a revision of organic standards.[122] Established in 2000, the National Organic Program banned the use of genetic engineering, irradiation, and sludge in organic production. The new regulations also banned organic livestock from receiving antibiotics of any kind. Should an animal fall sick, it could be treated with antibiotics but could no longer be sold as organic.[123]

With organic produce still accounting for less than 4 percent of total food sales, consumers' impressive reaction to USDA proposals stood in contrast to their relative passivity regarding lax antibiotic regulations for the remaining 96 percent of US food.[124] This was not because of a lack of warnings. Despite the NRDC's 1985 defeat, activists continued to campaign for antibiotic restrictions. During the 1990s, a new wave of media reports and bestsellers like Stuart Levy's *The Antibiotic Paradox* (1992) warned about resistant infections.[125] In August 1992, *Science* featured an Hieronymus Bosch–inspired cover on a "post-therapeutic future" with an accompanying editorial on "Microbial Wars" past and future.[126] Data from the field was worrying. While CDC studies showed rising AMR in community-acquired salmonella infections,[127] a 1993 outbreak of multi-resistant *E. coli* 0157:H7 at Jack-in-the-Box restaurant sickened 600 people and killed four children.[128]

The Jack-in-the-Box outbreak highlighted AMR hazards, hygiene problems, and the need for an overhaul of USDA meat inspection.[129] Resistant outbreaks like the 1993 episode also brought home the dangers of an empty drug development pipeline.[130] Following a "Golden Age" between the 1940s and 1960s, only a few new antibiotic classes had entered the market—the most recent class of the fluoroquinolone antibiotics had been licensed during the 1980s. The dry antibiotic pipeline was not due to a lack of promising compounds but due to pharmaceutical companies' increasing focus on more profitable "lifestyle" drugs.[131] Most of the new antibiotic products licensed during the 1990s and 2000s were modifications of older drugs against which resistance had already developed.[132] With reserve antibiotics failing, predictions of a post-antibiotic age became increasingly common.[133] In 1995, the *Coronado Eagle* from California featured an ad by a group of what can best be described as post-antibiotic preppers: "Scientists say antibiotic-resistant microbes could kill large numbers of people at any time. Learn what you can do to increase health and energy. Join a small group of dedicated individuals seeking solutions and alternatives."[134]

Short-lived outbreaks and apocalyptic warnings, however, failed to translate into a societal antibiotic reform movement. Although potential hazards of agricultural AMR selection were now widely discussed, American consumers' clear prioritization of residue hazards, increasing ability to buy "safe" organic food, and ideological divides over federal regulation continued to fragment the public sphere. In contrast to Europe (chapter 12), agricultural antibiotics did not emerge as a dominant and unifying symbol of modern agriculture's environmental and health hazards.

This lack of societal reform pressure was reflected in official licensing decisions (chapter 10). During the mid-1990s, the FDA licensed the fluoroquinolones sarofloxacin and enrofloxacin (Baytril) against *E. coli* in poultry despite known cross-resistance to the reserve antibiotic vancomycin. Officials announced that AMR would be monitored and fluoroquinolones banned if necessary.[135] The decision was criticized in the liberal media.[136] Fears of a resulting increase of fluoroquinolone resistance were confirmed in 1997 when Minnesota's Department of Health reported that 70 to 90 percent of raw poultry samples from supermarkets were contaminated with *Campylobacter* strains—25 percent of which were resistant to fluoroquinolones.[137] However, no immediate bans of either sarofloxacin or enrofloxacin ensued. US regulators also failed to emulate the European Union's 1998 decision to ban five of eight remaining AGPs because of AMR concerns (chapter 13). Despite media criticism and parallel CDC and FDA initiatives to curb AMR in human medicine,[138] FDA officials merely announced AMR testing requirements for new livestock drug applications from 1999 onwards. In response to ongoing scientific warnings that this would not target already licensed products,[139] Deputy commissioner (ex-CVM head) Lester Crawford later defended the FDA's approach: "We think it's far better to look at the real risk . . . instead of just disallowing a category of uses."[140]

It soon became clear that the FDA's partial reforms would protect neither already licensed nor new drugs. In 2000, the *Washington Post* reported that the efficacy of Synercid (quinupristin-dalfopristin), the new hope against vancomycin-resistant *Enterococci* (VRE), was endangered prior to its licensing because of its close relation to virginiamycin, a popular nontherapeutic AGP used since 1974: as much as 50 percent of US supermarket chicken, turkey, and pork already carried virginiamycin-resistant bacteria strains.[141] One year later, further increases of fluoroquinolone resistance forced FDA officials to reevaluate fluoroquinolone use in poultry production. The ensuing legal battle highlighted the problems of restricting already licensed drugs. While sarofloxacin was withdrawn voluntarily, pharmaceutical manufacturer Bayer opposed restrictions of its enrofloxacin (Baytril).[142] Because of Baytril's similarity to Bayer's reserve antibiotic Ciprofloxacin (Cipro),[143] Baytril restrictions became a matter of national security when a series of letters containing anthrax spores were posted to politicians after 9/11. Described as "Cipromania,"[144] panicked Americans purchased gas masks, vaccines, and the entire national stock of Cipro to ward off mostly illusive anthrax spores.[145] Despite clear AMR data and powerful new calls for bans,[146] FDA officials still needed over four years of legal proceedings to withdraw Baytril.

The Synercid and Baytril episodes clearly highlighted the risks of agricultural AMR selection as well as problematic drug withdrawal powers on the part of the FDA. However, their impact on public opinion was limited. With media outrage following partisan lines and polls indicating that most consumers continued to favor choice-based market solutions such as labels for antibiotic-produced meat, wider drug reform remained unforthcoming.[147] It was only following the 2008 election of the Obama administration that hopes for antibiotic restrictions reemerged. In 2009, Democrat Representative Louise Slaughter proposed legislation to reduce nonhuman antibiotic use.[148] Slaughter's move received outspoken support from the AMA and from media outlets like the *New York Times* and *Scientific American*, whose editors accused agribusiness of protecting "a narrow set of interests over the nation's public health."[149] Despite initial endorsement by the Obama administration,[150] legal objections by industry soon threatened the Preservation of Antibiotics for Medical Treatment Act (PAMTA). In April 2010, former FDA commissioner Donald Kennedy warned *New York Times* readers: "More than 30 years ago, when I was [FDA commissioner], we proposed eliminating the use of penicillin and two other antibiotics to promote growth .... When agribusiness interests persuaded Congress not to approve that regulation, we saw firsthand how strong politics can trump wise policy and good science.... It's 30 years late, but Congress should now pass [PAMTA], ... we don't have the luxury of waiting any longer to protect those at risk of increasing antibiotic resistance."[151] However, in what must have seemed a bizarre repeat of history to Kennedy, antibiotic reformers failed to mobilize enough political and societal support for PAMTA's enactment.

By late June 2010, it became clear that the FDA would not support statutory bans and propose "extremely modest"[152] voluntary reforms. Instead of pursuing enforceable statutory bans, post-Baytril legal concerns about officials' ability to withdraw drugs and fierce opposition from the USDA made the FDA propose guidances for antibiotic use under veterinary supervision.[153] Activists were dismayed. Similar to 1984, the NRDC and a coalition of other NGOs reacted to stalling statutory reform by filing a lawsuit requiring the FDA—and not industry—to prove that AGPs were safe. Although the lawsuit was approved by two district courts,[154] the FDA announced only voluntary AGP bans in

December 2013—a move that was welcomed by industry, livestock producers, and veterinarians.[155] In 2014, a two-to-one decision by the US Second Circuit Court of Appeals in New York strengthened the FDA by stating that it was not required to hold AGP safety hearings. In the same year, President Obama's Executive Order 13676 boosted funding for antibiotic research and stewardship. It also proposed to eliminate medically relevant AGPs—but did not specify statutory bans.[156]

Market Solutions?

Four years after PAMTA, thirty-seven years after Donald Kennedy's failed AGP bans, and over fifty years after the first warnings about infective AMR selection on farms, Washington and the US public remained divided over the need for statutory antibiotic reform. One immediate reason for this was the increasing partisan divide of US politics. Since the 1980s, green topics like AGP bans had either been attacked or ignored in conservative media like the *National Review*, and few Republicans supported precautionary policies.[157] Lengthy and complicated withdrawal procedures as in the case of the 1977 bans and Baytril also played an important role in diluting popular pressure for antibiotic reform in the wake of scandals. However, by themselves, political and regulatory factors are not enough to explain why decades of warnings about agricultural AMR selection did not incite more decisive statutory action in the US.

Asking why modern health concerns did not reduce US pesticide consumption, historian Michelle Mart has claimed that this paradox was driven by four cultural factors: (1) an emphasis on acute poisoning; (2) abstract long-term cancer hazards; (3) the belief that risk was best measured in terms of individual health, and not collectively or environmentally; (4) a rejection of precautionary regulation in favor of cost-benefit assessments.[158] The story of the American public's antibiotic risk episteme has many parallels: from the 1950s onward, deep-seated concerns about toxic residues repeatedly detracted from more abstract AMR hazards. Residue hazards were well suited to short-term technical fixes like monitoring and mandatory withdrawal periods. By contrast, the collective and environmental nature of AMR hazards required sustained precautionary antibiotic reductions or restrictions. The potential long-term benefits of these measures were not always apparent when evaluated through the more short-term lens of economic cost-benefit analyses. Distinct perceptions of AMR in human and nonhuman settings further fragmented the US risk episteme. In February 2017, a Harris Poll asked 1,768 respondents, who knew "at least a little about Superbugs,"[159] which types of antibiotic use are among the most significant causes of drug-resistant superbugs: 65 percent listed doctors prescribing antibiotics inefficaciously, 62 percent listed patients needlessly demanding antibiotics, 44 percent listed antibacterial soaps, and only 40 percent listed farmers selecting for AMR.

For American consumers, who remained concerned about antibiotic residues—and increasingly about agricultural AMR selection—the market has gradually replaced the state as a de facto driver of antibiotic change. Validating Lizabeth Cohen's concept of US democracy as a *Consumer's Republic*,[160] agro-pharmaceutical opposition to state intervention has allowed a highly regulated private market for pure and "secure" food to flourish. Risk is always an economic opportunity. This privatization of antibiotic risk has met with approval from an at first glance unlikely alliance of actors. Historian Andrew Kirk has noted how neoliberals' emphasis on market-driven solutions for environmental and societal problems was ideologically compatible with strong libertarian strands within the US environmentalist and countercultural movement.[161] This was certainly true regarding agricultural antibiotics.

Over the past four decades, both the conventional and organic sectors have benefited from the defeat of state-led reform and the privatization of antibiotic risk. In October 2015, a Harris Poll found that 53 percent of 2,225 surveyed consumers considered the question of whether a product was "antibiotic and hormone free" to be "important" or "very important" when shopping.[162] Consumers' willingness to pay for "pure" and "safe" produce has had an important impact on American food production. Despite ongoing controversies about the appropriate scale of sustainable organic production,[163] the US market for "pure" organic food quintupled between 1997 and 2009.[164] According to the Organic Trade Association, organic sales in the US amounted to over $47 billion (5.3 percent of total food sales) in 2016—with sales of organic meat and poultry accounting for $991 million. This was an increase of 8.8 percent since 2015 and compared favorably to the 0.6 percent growth of overall US food sales.[165] Surveys indicate that most new organic consumers were motivated by personal health rather than ecological concerns. Described by Robin O'Sullivan as a "conventionalization" of organic farming, steady profits have not only made larger corporations like Heinz or Nestle develop or acquire organic product lines of their own[166] but also created lucrative market niches for "safer" versions of conventional food. Since the early 2000s, companies like Tyson Foods, Purdue Farms, Foster Farms, McDonald's, and Chipotle Grill have attempted to attract environmentally and health conscious consumers by phasing out AGPs[167]—or at least claiming to do so given the absence of official definitions of "antibiotic free" other than adherence to drug withdrawal times.[168] For many companies, phasing out AGPs without subscribing to stricter organic rules for therapeutic antibiotic use is an effective way of demanding better prices while avoiding higher production costs.

Growing sales of "antibiotic-free" produce are already helping reduce total US antibiotic consumption (chapter 12). However, they also make it easier to present antibiotic stewardship as a question of individual consumer choice rather than an issue of collective social responsibility. According to recent *Farm Bureau* statements (chapter 9), there is no need for further political intervention since concerned citizens can already buy antibiotic-free produce. Marketplace environmentalism is taking care of antibiotic hazards by providing different products for risk-takers and risk-evaders.[169] Reifying Ulrich Beck's concept of risk as a force for socioeconomic stratification and political fragmentation, marketplace solutions are, however, seldom fair.[170] Not everyone can access safe organic or antibiotic-free produce. Given AMR's ecological dimensions, the efficacy of individualized safety solutions is also questionable. Current antibiotic-free niches like organic food remain small in comparison to conventional livestock production.[171] This means that buying antibiotic-free meat will not necessarily lower personal microbial risks. In 2013, a study of kosher and organic chickens found that levels of resistant *E. coli* on organic and conventional chicken were virtually indistinguishable.[172] When it comes to AMR and our interconnected microbiomes, effective stewardship will require collective and not just individual action.

9

Light-Green Reform

Antibiotic Change on American Farms

This chapter explores the development of antibiotic use and perceptions in US agriculture. During the 1970s, animal concentrations and antibiotic use in American livestock production increased significantly. Perceiving antibiotics as an essential disease insurance in times of diminishing profits, producers opposed external interventions into agricultural practice. Concerns about agricultural AMR selection were far less pronounced than concerns about drug residues. While industry-coordinated farmer protest toppled 1970s AGP restrictions, an economic crisis, ongoing environmentalist criticism, and organic farmers' commercial success led to a "greening" of agricultural discourse from the 1980s onward. This greening was, however, not accompanied by wider reforms of production systems or of antibiotic infrastructures. Although contemporary surveys indicate that many farmers were not as antibiotic dependent as claimed by parts of the national press, the agricultural media and farm lobby organizations joined pharmaceutical companies in calling for a rollback of federal regulations during the 1990s. Outcomes were mixed: while official intervention has been minimized and antibiotic access mostly defended, smaller producers have been forced out of a market increasingly dominated by a small number of vertically integrated corporations.

Planting from Fencerow to Fencerow

Economically, the years around 1970 were a good time for US agriculture: farmers' incomes increased because of rising commodity prices,[1] the 1970 Agriculture Act enshrined federal support, and USDA Secretary Earl Butz secured an unprecedented federal appropriation of $8.1 billion in 1971. Emboldened by neo-Malthusian scenarios of overpopulation and Soviet grain purchases, Butz exhorted US farmers to plant "from fencerow to fencerow."[2] This message was passed on via farm magazines. *Progressive Farmer* speculated that the 1970s would be "the time when American agriculture strikes out on a bold new course of influence and prosperity."[3] Family farmers would, however, only survive if they fully engaged in "a new, aggressive, agricultural capitalism"[4] and became more efficient.

Butz's exhortations for aggressive efficiency increases occurred against a backdrop of dramatic changes that had occurred in US farming and livestock production since around 1960. Between 1959 and 1969, annual cattle and calf sales had increased by about 45.6 percent to over 74 million animals, hog and pig sales by about 10.4 percent to over 89 million animals, and broiler and meat-type chicken sales by about 71.8 percent to over 2.4 billion animals. During the same period, the number of cattle and calf farms had, however, decreased by almost a million to 1,719,403, hog and pig farms by over a million to 686,097, and broiler producers by almost 10,000 to 33,757. Remaining farms were not only producing greater numbers of animals but also faced rising expenses for feedstuffs, wages, and chemicals.[5]

Over the next decade, ongoing pressure to improve the efficiency of production methods and stay in control of expenses would have a significant impact on farm animals, environments, and ownership. According to historian Chris Mayda, American hog production was increasingly dominated by specialized operations: "economic breeding . . . limited the breeds in production, so that hogs, like some crops, developed into monoculture."[6] Attempting to increase feed efficiency and minimize disease, producers created "closed herds" in ventilated buildings. The cost of necessary investments also meant that external investors and vertically integrated larger companies began to make inroads into the pork sector. Volatile markets and shifting consumer preferences also began to transform 1970s beef production with an increasing amount of animals being fattened in concentrated feedlots.[7] Of all US livestock sectors, the poultry sector was by far the most intensive and integrated with large corporations controlling as much as 90 percent of broiler production from the early 1960s onward. Concentrated in the US South, carefully bred and medicated birds were fed cheap soy or corn meal in large-scale confined operations, which would further intensify throughout the 1970s.[8]

Agricultural intensification and expansion caused tensions between US farmers and environmentalists (chapter 7). During the early 1970s, the Nixon administration's attempts to burnish its environmentalist credentials by creating the EPA and banning the insecticide DDT proved particularly contentious.[9] Although some farmers expressed health concerns about chemical production helpers,[10] the wider agricultural community rejected precautionary bans in favor of already familiar cost-benefit evaluations (chapter 3): DDT and other chemicals might be dangerous, but if used responsibly according to industry labels, benefits outweighed risks. In 1970, *Progressive Farmer* claimed, "millions of people now living in good health would be dead or anemic cripples if it were not for DDT."[11] Most agricultural commentators were thus appalled when the EPA banned DDT in 1972.[12] According to Kenneth Hood of the American Farm Bureau, the "disaster lobby" was "working overtime"[13] to deprive up to 50 million Americans of food by banning chemical pesticides, herbicides, fertilizers, and drugs. Others warned that the DDT ban might be a first step toward mandatory organic farming.[14]

Defending Antibiotics

Farmers' perceived defeat over DDT had significant implications for simultaneous FDA attempts to restrict AGPs. Whereas British corporatism relied on agricultural self-policing and moderation in exchange for participation in compromise-oriented problem-solving (chapter 6), US farm organizations lacked incentives to tone down public attacks on "overzealous" regulators. Throughout the 1970s, domino scenarios according to which any substance ban would engender further bans made organizations like the American Farm Bureau support pharmaceutical companies' polarizing battles against AGP bans. With US veterinarians failing to harness AMR concerns to gain control over antibiotic access,[15] the only sustained inner-agricultural criticism of antibiotics came from organic farmers.

Ensuing battles over antibiotic access took place against a background of rapidly expanding consumption on US farms. The contemporary growth of nontherapeutic antibiotic use was particularly dramatic. According to FDA figures from 1978, 4.16 million pounds of antibiotics had been produced in the United States in 1960, of which about 1.2 million pounds (about 29 percent) were added to feeds. Ten years later, about 7.3 million pounds of antibiotics (about 43.1 percent of total production) were added to feeds. By 1975, industry figures indicated that nonmedicinal usage accounted for 48.6 percent of total US antibiotic production.[16] Although data on therapeutic antibiotic use is limited and the NAS published higher historical estimates of nontherapeutic antibiotic use in 1980,[17] it is likely that total nonhuman use overtook US human antibiotic use during the 1970s.

Antibiotic consumption varied between livestock sectors. According to contemporary manuals and USDA data, US broiler rations contained between

4 and 10 grams of antibiotics per ton of feed. Hog rations usually contained one or more antibiotics (usually penicillin, chlor- and oxytetracycline, or bacitracin). Piglets would receive 44 milligrams of antibiotics per kilogram of diet, growing pigs 11 to 22 milligrams per kilogram of diet, and finishing pigs around 11 milligrams per kilogram of diet.[18] A significant increase of antibiotic consumption occurred in US cattle production. Starting in the mid-1950s, intensive feedlots expanded from the grain-rich Corn Belt to locations across the United States. Whereas 7,535,000 cattle were "on feed" in 1960, the number increased by about 75 percent to 13,190,000 in 1970. Feedlot conditions were conducive to antibiotic use. During the 1950s, higher-dosed antibiotic injections were used to treat footrot (infected sores on legs) and other infections in individual animals. However, in a departure from earlier recommendations, the second half of the 1960s saw antibiotics like chlortetracycline, oxytetracycline, and zinc bacitracin routinely included in cattle rations at about 20 to 25 milligrams per 100 kilograms of live weight per day. Antibiotic feeds were said to counteract meat quality problems caused by the popular hormonal growth promoter DES. At concentrations of 70 milligrams per head per day, antibiotics were also used to prevent liver abscesses caused by high-grain diets. Cyanamid's AS-700 (350 milligrams chlortetracycline and 360 milligrams sulfamethazine) proved popular against "shipping fever" and was also used to accustom new cows to feedlot conditions. Antibiotic consumption also increased in replacement herds with most dairy and veal calves receiving medicated milk replacers and calf starters.[19]

Despite declining cattle numbers, antibiotic use increased throughout the 1970s. This was due to a new group of antibiotics called ionophores. The ionophore monensin (Coban/Rumensin) was first approved as a coccidiostat for poultry in 1971 and then as a feed additive against bloat and coccidiosis in cattle in 1975. Ionophore additives increased propionic acid production in cows' rumens and improved the fermentation and conversion of high-roughage and high-grain diets. Within ten years of monensin's licensing, ionophores were fed to over 90 percent of US feedlot cattle.[20]

In all livestock sectors, routine therapeutic and nontherapeutic antibiotic use continued to be promoted by agricultural manuals, farming magazines, and the pharmaceutical industry.[21] According to *Lancaster Farming*, antibiotics were part of a modern farmer's identity: "A farmer is a paradox. He is an 'over-alled' executive with his home his office; a scientist using fertilizer attachments; a purchasing agent in an old straw hat; a personnel director with grease under his fingernails; a dietitian with a passion for alfalfa, animals and antibiotics, a production expert faced with a surplus, and a manager battling the cost-price squeeze."[22] In contrast to regular warnings in British farming publications (chapter 6), US commentators rarely discussed AMR hazards. If AMR was mentioned, it was usually in the context of drug residues in food or the need

FIGURE 9.1 The 1970s saw significant increases of antibiotic use on US cattle feedlots. Cyanamid advertisement, *Farm Journal*, 1979.

for more efficient antibiotic use or drug combinations to control infections.[23] According to mainstream opinion, antibiotics' prime hazard consisted of residues and inefficient use.

Many agricultural commentators were thus shocked when the FDA task force recommended AGP bans in early 1972. Unless they were proven safe and efficacious, low-dosed tetracycline, penicillin, streptomycin, dihydrostreptomycin, and sulfonamide feeds would be banned for poultry on January 1, 1973, and for sheep, cattle, and swine on July 1, 1973. Other medically relevant antibiotics would be restricted after December 31, 1973.[24] Taken aback, *Wallaces Farmer* acknowledged, "Evidence indicates using antibiotics in food-producing animals promotes Salmonella and the development of the R factor (resistant) bacteria."[25] It seemed unlikely that bans could be averted by industry safety trials: "it doesn't seem likely that FDA will back off much."[26] Iowa State nutritionist Vaughn Speer agreed with the FDA: "I think the recommendations that certain antibiotics be reserved for human use is a good one. I can't argue with that."[27]

Initial agricultural shock was soon replaced by preformulated industry protest. Since the 1969 Swann report, companies like American Cyanamid had prepared for potential FDA action by commissioning counter science.[28] During a time of growing public distrust of industry influence (chapter 8), pharmaceutical manufacturers, however, knew that classic political lobbying would only go so far. To be effective in Washington, they also had to mobilize a broader popular alliance among farmers, agricultural organizations, and the agriculture-minded public. Founded in 1941 and funded by companies producing more than 90 percent of US veterinary pharmaceuticals and feed additives, the Animal Health Institute (AHI) soon emerged as the most important voice defending AGPs among rural audiences. In magazines like *Lancaster Farming*, AHI experts "emphatically disagree[d] with most of" the FDA task force report: "If one accepts the Task Force premise that the presence of resistant bacteria in animals constitutes a potential human health hazard . . . then the removal of recognized effective prophylactic use of antibacterial agents creates a real potential human health hazard, since it follows logically that such removal will lead to an increase in incidence of various animal bacterial diseases."[29]

America's agricultural establishment soon joined pharmaceutical companies' pro-antibiotic campaign. "Reaffirming their abiding faith in American constitutional government, the private enterprise system, and man's inalienable right to worship God," the 1972 Farm Bureau assembly voted to oppose "a complete ban on the use of any agricultural drug and chemical unless it can be demonstrated positively . . . that the use of such product represents a clear and present danger to health or that such use would seriously jeopardize our environment."[30] Organized opposition to AGP bans spread quickly. Merging with preexisting criticism of environmentalism, reporting in the US farming press became almost homogenous in its criticism of statutory restrictions. Seven months after endorsing the FDA's AGP bans, nutritionist Vaughn Speer suddenly claimed, "[the] possibility [of AMR transfer] has been thoroughly examined and there is no scientific evidence that resistance is transferred this way."[31] Meanwhile, magazines like *Farmers Weekly Review* encouraged readers to send written objections to the Department of Health, Education, and Welfare (HEW).[32]

AGP bans were also criticized by the USDA (chapter 10). According to USDA Secretary Earl Butz, banning antibiotics and other chemicals would force organic agriculture and starvation on the United States.[33] Responding to the 1972 task force report, the USDA's Agricultural Research Service (ARS) published its own antibiotic review. The ARS review recapitulated AGPs' alleged economic advantages and maintained that it was unclear whether resistant organisms and genes in animals constituted a health threat. Failing to cite landmark British AMR publications, the ARS noted that long-term direct exposure of humans to low-dosed antibiotics had not resulted in harm. It was true that AMR could be transferred between enteric bacteria. However, AGPs' ongoing efficacy and the isolation of resistant organisms from animals unexposed to antibiotics indicated that agricultural AMR was "natural," widespread, and therefore not an important public health hazard.[34]

Despite mounting agro-industrial pressure, it initially seemed as though the FDA would follow through with AGP withdrawals. In July 1972, *Progressive Farmer* warned, "stricter regulation of antibiotics in feed look[s] 99 percent certain."[35] In September, *Farm Bureau* members proposed voluntary drug compliance certification to ward off further federal intervention: "Federal authorities have taken DDT away from your use completely. . . . If livestock producers are to continue benefitting from animal health products, they must use them properly and certify that they are doing so. England's Swann Report and the United States' FDA Task Force Report challenged the ability of farmers and ranchers to carry out this responsibility."[36] However, all was not lost. Following the 1972 DDT ban, agricultural observers were beginning to discern cracks in federal agencies' willingness and ability to impose substance restrictions. Senior FDA officials' lackluster support for AGP bans and indecisive handling of contemporary DES bans played an important role in maintaining agro-industrial morale.[37] Bombarding officials and agricultural-minded members of Congress with letters and critical reviews, antibiotic supporters were relieved when the FDA quietly loosened AGP safety review requirements in 1973 (chapter 10).

Victory over FDA bans coincided with darkening economic outlooks. In 1973, the oil crisis and rising inflation brought a return of the agricultural cost-price squeeze. Between 1973 and 1975, US farmers' net income declined from $34.3 to $25.5 billion. Livestock production was hit hard. Although broiler sales continued to increase, sales of cattle, calves, and pigs declined.[38] Faithful to Earl Butz's motto "get big or get out," producers either participated in a further round of agricultural intensification or left agriculture altogether.[39] Those remaining increased their antibiotic use. In 1978, US livestock producers used about 5.58 million kilograms of antibiotics (excluding sulfonamides) in feeds and for other nontherapeutic applications (about 48 percent of total US antibiotic production).[40] Higher-dosed therapeutics also remained popular. In 1978, veterinary county agent Alan Bair resignedly recounted the case of a young farmer who used several antibiotics at once, did not read labels, and purchased "whatever the feedman thought was best that week."[41] Bair's description of mass-antibiotic use was no exaggeration. In 1977, a *Wallaces Farmer* reader poll found that 67 percent of farmers regularly fed drugs to growing pigs and only 7 percent did not feed drugs at all. In a matter-of-fact statement, one farmer noted: "I need drugs to help with production. Keeping the hogs healthy is the only way we can make a living and provide consumers with meat."[42]

Inner-agricultural criticism of rising antibiotic use remained minimal. Most farming commentators emphasized that antibiotic access was absolutely "vital to [the] production of food."[43] Pharmaceutical lobbyists also continued to mobilize against potential restrictions. In 1974, Philip Connell, president of American Cyanamid's Agricultural Division, warned about threats to the "web of technology"[44] underpinning global food production. AGP bans would force US consumers to pay $2.1 billion more for meat (about $10.26 per person) per year. Farmers would need 1.3 million additional acres to maintain 1970 levels of beef and pork production.[45] USDA Secretary Earl Butz echoed industry warnings about threats to global food security: "Every new regulation that hampers agricultural production . . . drives another nail into the collective coffin of mankind."[46] There was even a certain pride in the amount of antibiotics being consumed in the United States. In 1978, *Lancaster Farmer* described an impressionable West German delegation's visit of Pennsylvania farms. During one tour, "more than a few Germans pointed excitedly at a bucketful antibiotics that they found in the swine producer's service room. 'That's forbidden in Germany,' they announced emphatically."[47] The restrictive nature of German regulations was then elaborated.

Although AMR was increasingly mentioned as a driver of federal intervention, it was rarely acknowledged as a genuine hazard. Mirroring public concerns, agricultural commentators instead focused on residue problems resulting from sulfamethazine contaminated feeds, drugged "downer cattle," and more sensitive USDA testing.[48] Livestock manuals similarly either failed to mention AMR or downplayed hazards.[49] In 1978, the fifth edition of *Stockman's Handbook* noted: "Some folks object to the continued use of low levels of antibiotics in feeds on the ground that resistant pathogenic strains of microorganisms might develop . . . . Although it is true that microbial resistance to antibiotics does occur, there is no scientific evidence that more virulent pathogens have evolved .... As a matter of fact, the evidence shows that resistant strains are nearly always less virulent."[50]

Without an adjustment of risk perceptions, US agricultural circles not only remained complacent about growing antibiotic infrastructures but also supported pharmaceutical campaigning against renewed FDA attempts to ban AGPs in 1977 (chapter 10). Their previous victory over the 1972 task force report had equipped campaigners with an effective set of protest tools. With only few readers and commentators voicing AMR concerns,[51] farming magazines portrayed FDA proposals as an irrational attack on "scientific" production. Commentators warned about antibiotic bans' effect on inflation and global food supplies, criticized inefficacious AGP substitutes, and challenged AMR research.[52] Ignoring that American veterinarians' economic decline had been facilitated by unrestricted drug access, agro-industrial representatives also argued that there were too few veterinarians to prescribe antibiotics after AGP bans. While some called on the FDA to allow "feed manufacturers to fill prescriptions at doses not prescribed by law,"[53] American Cyanamid warned that banning AGPs would harm 210,000 smaller farmers without veterinary service contracts.[54] Similar to 1972, lobbyists maximized pressure by calling on farmers and rural communities to send preformulated protest letters to officials and politicians in Washington.[55]

Victory over FDA measures seemed close after Congress stalled AGP bans in mid-1978. By early 1979, farming magazines were encouraging readers to pressure representatives to oppose all three contemporary FDA proceedings against nitrites, DES, and antibiotics.[56] Despite FDA attempts to convince producers of antibiotic restrictions, the farming media consistently prioritized information supplied by industry think tanks like CAST.[57] According to *Progressive Farmer*, CAST researcher Virgil Hays did not "cotton much to the theoretical possibility that resistant organisms ... may be transferred to man."[58] Even articles addressing the difficulties of treating resistant infections on farms did not problematize the reasons behind AMR proliferation and focused on advocating hygiene and different drugs.[59]

Coordinated agro-industrial resistance was effective both in maintaining agricultural cohesion and in achieving political goals. After two years of intense conflict with the FDA, American farmers only lost the battle over DES (chapter 10). In April 1980, *Farm Journal* announced that a congressionally mandated NAS antibiotic review had concluded that "the test probably doesn't exist that can prove or disprove the safety of using low levels of penicillin or tetracyclines."[60] According to the USDA's *Farmline* journal, "possible legal opposition, and the need for further study"[61] should delay any regulations for at least several years. However, victory came at a price. In contrast to corporatist British decision-making (chapters 6 and 12), aggressive public opposition to FDA restrictions fractured already porous ties between the agricultural community and US regulators and AMR experts. Access to AGPs had been defended by using counter science and personal attacks to undermine data, critical scientists, and the authority of public institutions. By allying with the pharmaceutical industry and propagating a polarizing mix of regulation hostility and distrust in microbiological findings, agricultural organizations, commentators, and producers missed an opportunity to rethink the entwined path-dependencies of intensification and growing antibiotic infrastructures. Chances for future antibiotic compromise and level-headed encounters with critics had diminished. Further conflict seemed likely.

A Crisis of Intensification

Ongoing AGP access did little to alleviate an emerging economic crisis. During the 1980s, overproduction and sinking commodity prices triggered a long agricultural downturn. Despite its antiregulatory agenda, the Reagan administration was repeatedly forced to expand subsidies to alleviate farmers' plight. Between 1980 and 1985, the total cost of the US farm program rose to more than $20 billion. However, deficit purchases and storage programs failed to raise commodity prices and stimulated further overproduction. Being able to decrease production costs by increasing efficiency became a determinant of agricultural survival. In many cases, necessary investments were made by taking on debt. Whereas US farm debt had totaled $60 billion in 1972, it totaled an astonishing $216 billion in 1983.[62] Unable to service debts, many farmers were either forced to leave agriculture or become employees of agribusiness corporations. By the late 1980s, the fabric of US farming had changed: similar to the Dust Bowl era, 42 percent of farmland was operated under rental agreements, and corporations and investors had taken over many farm operations.[63]

In the US livestock sector, the economic crisis catalyzed existing trends toward confined, concentrated, and integrated production. In pig husbandry, so-called hog factories became common. Whereas producers raising 1,000 or more animals per year accounted for about 7 percent of US hog production in 1964, this number had already risen to 40 percent in 1979.[64] By 1986, seven out of ten Iowa hogs were raised in confined systems.[65] New total confinement units mirrored Damon Catron's 1940s fantasies of Fordist animal production: housed according to life stages and function (breeding or fattening), pigs were fed tailored diets and progressed down an assembly line from one specialized building to the next (farrowing, nursery, grower, and finisher) before being transported to the abattoir.[66] Accelerating 1970s trends, expensive new high-tech facilities were often financed and built by large corporations with whom farmers contracted to produce animals they no longer owned.[67] This form of vertical integration was already commonplace in the US poultry sector with the same companies often supplying chicks, specifying housing, producing feeds, and slaughtering and processing animals.[68] By the mid-1980s nearly all US chicken broilers and 80 percent of eggs and turkeys were being produced in vertically integrated operations.[69]

Although accelerating intensification triggered renewed clashes with environmentalists and concerned consumers, aggressive agro-pharmaceutical rhetoric did not always match producers' personal views. During the mid-1970s and 1980s, economic pressure made growing numbers of smaller conventional producers realize that farming's future would be for the few and not for the many. Ignoring ongoing attacks on organic farming by agricultural officials and commentators, they began to pay attention to the booming market for "natural" food (chapter 7). Growing mainstream interest gradually carried over into the agricultural press. In 1975, *Lancaster Farming* from Pennsylvania began to print articles by local farmer—and publisher—Robert Rodale on organic gardening and lifestyle. Initially, these articles were followed by disclaimers distancing *Lancaster Farming* from Rodale's views.[70] With organic farming becoming more conventional regarding its supply chains and business models, the need for such disclaimers diminished. In 1981, *Indiana Prairie Farmer* told readers not to "dismiss organic farming as merely a fad or joke"—organic methods could reduce use of "insecticides, herbicides, growth regulators, and fertilizers."[71]

A 1980 USDA review played a significant role in changing wider conventional attitudes toward organic methods. Despite questioning organic health claims, the review gave a balanced overview of the organic sector. While most organic farms were relatively small (10 to 50 acres), some larger operations (100 to 1,500 acres) had also adopted organic methods. The review debunked conventional clichés by noting that most organic farms were well managed, productive, and had not regressed to prewar standards. In terms of pesticide management and crop rotation, organic farmers often adhered to best management practices, although farms' overall ratio of labor input to yield was worse than in conventional systems.[72] In the case of livestock production, most organic farmers did not feed hormones and only used antibiotics to treat individual animals. The size of an organic farm's grain and hay production usually dictated the scale of animal production. Most notably, organic farmers did not "push" animals to achieve the highest rate of gain and market them in the shortest possible time: "A number of farmers reported that with previous chemical-intensive programs they had often incurred a higher rate of birth mortality, decreased reproductive efficiency, and increased respiratory ailments among their livestock, resulting in lower production, and higher veterinary costs."[73] According to the USDA's reviewers, organic methods warranted serious attention from agricultural officials and conventional producers alike. One year later, an organic research budget was integrated into the 1981 farm bill.[74]

Growing agricultural acceptance of organic farming as a tolerable market niche was paralleled by a plurality of opinions regarding other conventional production methods. In the case of agricultural antibiotics, two 1980s surveys indicate that American conventional farmers' reasons for opposing antibiotic bans and attitudes toward antibiotic use were far more nuanced than the uniform public front of agro-pharmaceutical lobbyism suggests. The first industry-sponsored survey of 1,051 Missouri livestock producers was conducted in 1981: 14 percent of respondents said that they always purchased feed with antibiotics, 48 percent claimed to sometimes do so, and 24 percent said that they never did so (surprisingly, 14 percent did not know whether their feeds contained antibiotics or not). Mirroring official consumption statistics, responses varied according to livestock sector: 14 percent of beef producers always and 50 percent sometimes purchased antibiotic feeds (64 percent total); 19 percent of poultry producers always and 51 percent sometimes purchased antibiotic feeds (70 percent total); 27 percent of pork producers always and 58 percent sometimes purchased antibiotic feeds (85 percent total). Interestingly, large pork producers tended to buy antibiotic feeds less frequently than smaller or medium producers. The survey also asked farmers about the impact AGP bans would have on their business. Answers were very different from the apocalyptic picture being painted by industry lobbyists. Of surveyed farmers, only 24 percent said that bans would have a serious impact on their operations; 47 percent did not think bans would have serious effects; and 29 percent thought that bans would have no effect. Farmers' risk perceptions were difficult to judge. When asked whether they believed that using antibiotics undermined future antibiotic treatments, 48 percent responded "don't know," 32 percent said yes, and only 20 percent said no. Of the 24 percent of farmers who did not purchase antibiotic feeds, 48 percent believed that "more harm than good" resulted from antibiotic use, 19 percent believed that AGPs were not effective, and 29 percent believed that they were too expensive.[75]

A 1982 survey of 642 New York livestock producers confirmed many of these findings. Cornell PhD Gilbert Wayne Gillespie Jr. discovered that trust in AGP efficacy for growth promotion was far smaller than suggested by agro-industrial lobbyists and that farmers purchased AGPs—which had initially been licensed as subtherapeutic additives—as a cheap risk insurance against disease. He also found that American farmers' opposition to AGP bans and support of industry campaigning was primarily ideological: "Ideology appears to be a major source of opposition to government regulation of agricultural chemicals and pharmaceuticals in general, with the important antecedent variables being perceptions of negative side-effects from use of these substances, political liberalism, and an orientation toward accepting economic risk in farming. Farm debt is the only structural variable with a significant effect."[76] Similar to Missouri, New York poultry and swine producers used the most AGPs. Larger swine, dairy, and beef producers tended to use more antibiotics than smaller producers while large poultry producers with modern facilities used less drugs than medium-sized ones. Gillespie also found that antibiotic users were more opposed to government regulation, less liberal, and tended to express more feelings of powerlessness than non-users.[77] When asked why they used antibiotics, swine producers were most likely to cite productivity increases and disease prevention. Poultry producers cited disease prevention. Most livestock producers purchased AGPs as insurance against disease and not for growth promotion. Less than one-tenth of surveyed farmers believed that AGPs did more harm than good—AMR knowledge correlated with an endorsement of bans. Most non-users, however, simply thought that they did not need expensive AGPs.[78]

Although many US farmers did *not* believe that AGP bans would seriously affect business, post-DDT domino theories of substance restrictions and their ideological distaste for statutory interference made them support—or at least fail to oppose—aggressive agro-pharmaceutical defense of unrestricted antibiotic access. Responding to the NRDC petition and Scott Holmberg's 1984 link of resistant salmonellosis to antibiotic feeds, pharmaceutical companies, CAST, the AHI, farm organizations, and supportive USDA officials launched a familiar broadside of economic doomsday warnings and counter science to prevent restrictions (chapter 10).[79] In 1985, *Farm Bureau News* (*FBNews*) questioned Holmberg's link between antibiotic use and human illness: "other factors could have caused the outbreak, . . . a direct link was never shown (The resistant Salmonella apparently came from an adjacent dairy farm, where no antibiotics were used)."[80] In Congress, Farm Bureau representatives claimed that farmers would immediately abandon AGPs if their harmfulness was proven: "If the potential hazard to humans is as great as some people claim, why haven't there been more cases of human illness."[81] Others noted that banning AGPs but leaving therapeutic antibiotic use—primarily in human medicine—unreformed would be ineffective. For *FBNews*, it seemed likely that rising AMR was less due to on-farm use and "probably more due to [antibiotics'] prolific use for treating and preventing human infection."[82] The failure of British AGP bans to curb AMR was similarly mobilized.[83]

Not all conventional and pharmaceutical producers thought that opposing AGP bans was in their best interest. Similar to Britain (chapter 7), companies producing nontherapeutic AGPs had no incentive to protect penicillin and tetracycline AGPs at all costs. In 1984, American Hoechst, the company marketing flavomycin (bambermycin), complained that the medicated feed controversy was giving all antibiotics a bad name.[84] Suffering from consumers' increasing preference for "healthy" poultry meat and less reliant on medically relevant AGPs, cattle producers also attempted to gain a sales advantage by discontinuing the use of antibiotic classes that were not crucial to production. In 1985, the National Cattlemen's Association announced that members would discontinue the feeding of tetracyclines—but not of penicillin—"until it can be resolved whether their use causes health problems in humans."[85] Eliciting hostile responses from other livestock groups,[86] cattle producers, however, stressed that they would continue to oppose federal statutory bans to discourage anti-chemical activists: "If a product can be taken off the market by inference instead of fact, why would anybody invest another $70 million to create a new product?"[87]

Margaret Heckler's 1985 rejection of the NRDC petition prevented further agricultural discord. Although opposition to antibiotic-free market niches had weakened, US farmers and their representatives remained opposed to formal AGP bans. This opposition was less grounded in the belief that bans would harm production and more in a general ideological rejection of statutory interference. Most producers were relatively unconcerned about AMR and primarily viewed AGPs as an insurance against disease. Although trust in actual growth promotion was weakening, antibiotic feeds remained in demand. AHI figures indicate that US farmers used about 11 million pounds of antibacterial compounds in feeds in 1985 with tetracyclines proving the most popular.[88]

A Light-Green Rollback

US agriculture's antibiotic infrastructures faced no further statutory challenge until the late 1990s. During the intervening period, economic pressure and changing consumer preferences nonetheless incentivized a gradual *rapprochement* between conventional farmers and external critics. The result was what Michael Bess has called a "light-green" transition whereby green ideas were partially absorbed into agricultural discourse and practice but also trimmed and jettisoned should they threaten core production models or statutory intervention. Both producers and environmentalists "emerged [neither] wholly satisfied nor utterly dismayed"[89] from this process. By the end of the millennium, conventional farmers were happily marketing antibiotic-free produce but still resisting statutory antibiotic reform.

The "light-greening" of US agriculture coincided with ongoing intensification and integration. Between 1987 and 1997, cattle and calf sales increased by about 2 percent to over 74 million animals, pig and hog sales grew by over 47 percent to over 142 million animals, and meat-type chicken sales increased by over 54 percent to over 6.7 billion animals. Animal increases were accompanied by a further decline in the number of farms producing milk, pork, and poultry and by rising production costs.[90] Change was particularly dramatic in hog production where vertical integration, specialization, and economies of scale accounted for most productivity gains. Between the 1990s and early 2000s, the share of farms with 2,000 animals or more increased from less than 30 to 86 percent while farm numbers fell by over 70 percent from over 240,000 in 1992 to about 71,000 in 2009. Varying US state laws also fostered a geographic concentration of large-scale pig operations in sparsely populated low-income regions.[91]

Low profit margins, health warnings, and changing values made a growing number of conventional farmers partially reevaluate production methods. In 1990, *Wallaces Farmer* surveyed 200 farmers' pesticide use. Of the 85 percent who reported changing pesticide management, 94 percent claimed to have done so for economic reasons, 80 percent because of environmental concerns, and 79 percent because of health concerns.[92] Acknowledging that environmentalist values were becoming embedded in public opinion, farm organizations also launched major efforts to promote an image of US agriculture that was green and responsible.[93] This greening was "light": environmental measures were adopted if they did not threaten core practices or production philosophies. In some cases, agricultural commentators also tried to reframe environmentalism. According to *FBNews*, the term *environmentalist* was often used to describe "someone who favors locking up natural resources and opposes the use of chemicals," however, it also meant "someone who cares about the environment"— "Then certainly you could apply the term to farmers and ranchers."[94]

In the case of antibiotics, light green farming led to a renewed focus on "rational" drug use—but little criticism of wider antibiotic infrastructures. During the early 1990s, magazines like *Successful Farming* reacted to rising AMR warnings (chapter 9) by rehashing 1960s criticism of "irrational" antibiotic overuse as an inefficient waste of resources.[95] Veterinary manuals also promoted improved diagnostics for more targeted antibiotic use.[96] Although some advocated making farming less antibiotic dependent,[97] most commentators focused criticism on individual abusers and saw no reason for wider reform. In 1992, the seventh edition of *The Stockman's Handbook* continued to promote routine therapeutic and nontherapeutic antibiotic use. Claiming that eight of ten US food animals received drugs during their life and failing to mention resulting AMR risks, the manual noted that the list of approved agricultural antibiotics was long "and growing longer."[98] Although manuals targeting smaller producers featured more balanced risk assessments, they also emphasized "rational" drug use by enlightened producers rather than explicit antibiotic reductions.[99]

The light-greening of antibiotic use occurred parallel to ongoing Farm Bureau attacks on "environcrat"[100] regulations and critics. Farmers and their representatives were happy to partially rethink and rebrand production systems but remained ideologically opposed to statutory regulations forcing them to do so. When Newt Gingrich's 1994 "Republican revolution" ended the Democrats' fifty-two-year hold on Congress, the Farm Bureau seized the opportunity to push for a rollback of limited precautionary regulations in favor of "flexible" cost-benefit assessments.[101] Speaking to the Senate Agriculture Committee in 1995, Farm Bureau representatives blamed rural plight on a "federal regulatory juggernaut": "[The Farm Bureau] supports four major regulatory reforms: risk assessment . . . ; cost-benefit analysis . . . ; private property compensation when Congress decides to override private interests . . . ; and redirection of regulatory resources into worthwhile private sector incentives."[102] With Republicans controlling both houses and the Clinton administration favoring deregulation, industry pursued a strategy of shifting risk assessment burdens from officials to "rational" consumers on the private market.

Formidable aspects of US consumer legislation soon began to topple. In 1996, Congress passed the Food Quality Protection Act (FQPA) and abolished the Delaney Clause. Environmentalists and consumer advocates supported the FQPA because it ended the distinction between residues on raw and processed food, established low tolerances of a one-in-a-million cancer risk, introduced right-to-know provisions, and required reviews of existing standards. However, the FQPA also marked a significant victory for industry because it simplified regulatory procedures and ended zero-tolerance policies in favor of negotiable risk-benefit calculi.[103]

In the same year, industry also celebrated the passage of the aptly named Animal Drug Availability Act (ADAA).[104] The ADAA had been heavily influenced by petitions from the AHI and American Veterinary Medical Association. Streamlining regulatory procedures, it redefined and reduced the number of efficacy and safety studies required for new animal drug applications. Although drugs were still prohibited from leaving nontolerated residues in food, the ADAA limited the time FDA officials had to review applications and mandated the creation of new tolerances.[105] Significantly, the ADAA also loosened the extra-label provisions of the 1994 Animal Medicinal Drug Use Clarification Act. Whereas unsupervised extra-label drug use had previously been banned, new Veterinary Feed Directives (VFDs) enabled the extra-label addition of often higher-dosed drugs—like antibiotics—to commercial animal feeds after veterinary consultation. In effect, federal responsibility for defining safe medicated feed use via labels was handed over to individual veterinarians (chapter 10).[106] Whether these veterinarians would always prioritize long-term public health interests over more immediate productivity concerns was unclear. American veterinarians were often employed by integrated feed and meat companies and studies showed that health-centered perspectives learned at school were often diluted in favor of productivity-centered perspectives once practitioners began working on farms.[107] Similar to human medicine, what constituted "rational" veterinary antibiotic use was highly context dependent.

Voluntarist Victories

The loosening of US regulations coincided with new European AGP bans (chapter 13). By the late 1990s, the combination of diverging transatlantic regulations and new national AMR surveillance capabilities had ended the post-1985 status quo of antibiotic regulation in Washington. The result was a further wave of voluntarist light-green agricultural reforms. Although antibiotic consumption continued to increase and producers remained united in their opposition of statutory bans, they also reacted to consumer concerns by expanding antibiotic-free market niches. In public, agro-industrial lobbyists no longer categorically denied AMR hazards but instead pointed to market-driven ways of improving antibiotic stewardship and curbing AMR.

Occurring during a time of mass farm closures, renewed antibiotic conflicts made many smaller producers embrace market niches for antibiotic-free produce. In 1998, *Wallaces Farmer* published a survival guide for pig farmers: "This is a difficult column to write. We've always tried to keep a positive attitude and present ways producers can become more efficient, productive or profitable. But nothing we print will change the fact that the pork industry is going through a critical time. . . . some producers are finding a high-value niche for organically raised, antibiotic-free pork. It's not for everyone, but it may be an idea to consider."[108] *Successful Farming* described how conventional livestock farmers had downscaled and transitioned to "all-natural" and antibiotic-free production.[109] According to the magazine, this transition was not being driven by ideology but occurred "all in the name of survival."[110] A converted dairy farmer emphasized that there was no animosity between conventional and organic farmers: "we're just selling milk here, folks."[111] If it occurred voluntarily, going "natural" was fully accepted within the agricultural community. Despite sales increases, its small size also meant the organic livestock sector did not seriously threaten the much larger conventional sector. By 2011, there were 106,181 certified organic beef cattle, 254,771 organic milk cattle, and 28,644,354 organic broilers in the United States.[112] These numbers paled in comparison to the billions of animals produced on conventional farms.

For the large majority of producers remaining in conventional agriculture, further intensification usually entailed rising antibiotic consumption. However, in contrast to previous decades, data provided by the new National Antimicrobial Resistance Monitoring System (1996) made it increasingly difficult to deny farming's role in selecting for AMR. Public fears of resistant "superbugs" also spread within the agricultural community. In 1999, an illustrated article on rural health in *Successful Farming* cited a CDC expert, who noted that agriculture was partially responsible for creating resistant microbial environments: "The same drugs prescribed for human health are widely used in livestock production. . . . Almost half of the 50 million pounds of US-produced antibiotics is used in animals, with the largest share mixed into feed to promote growth."[113] Faced with hard AMR surveillance data, agro-pharmaceutical lobbyists also shifted gears. Instead of claiming that there was no hazard, they launched a sustained battle for voluntarist instead of statutory solutions. While antibiotic stewardship was important, US agriculture would improve at its own light-green pace. Eight months after printing its AMR article, *Successful Farming* quoted the AHI's Richard Carnevale on the necessity of continued antibiotic use. While Carnevale supported AMR research, he maintained that fifty years of experience had shown that farmers were capable of responsible antibiotic use and that AGPs did not constitute an immediate threat to public health.[114]

The new agro-industrial strategy of voluntarism rested on three pillars. The first pillar consisted of developing technical solutions to AMR. During the late 1990s and 2000s, farm and USDA journals discussed options ranging from AGP substitutes, disease-resistant animal breeds, vaccines, competitive inhibition, engineered antibodies, irradiated feeds and food, and bacteriophage therapies.[115] Reacting to microbial vulnerabilities, concentrated livestock operations also enacted ever stricter biosecurity protocols. The last three decades have seen large integrated corporations remold indoor animal environments as well as the habitats and habits of surrounding human populations. In some areas of the United States, entire post-anthropocentric landscapes now center on maintaining the health of isolated animal monocultures.[116]

In addition to biosecurity and antibiotic substitution, the second and third pillars of industrial voluntarism consisted of stressing "rational" antibiotic use and using counter science. In 2002, *FBNews* noted: "There's been a lot of clucking in recent years that livestock and poultry producers are using antibiotics willy nilly so they can crowd their animals together and farm on the cheap. The fact is antibiotics, like most drugs, aren't cheap. . . . Farmers use [antibiotics] when they're needed, and they should be able to continue doing so."[117] Similar to previous decades, lobbyists stressed the costs and limited benefits of statutory intervention. In 2008, a study in the *Review of Agricultural Economics* noted that AGP bans would marginally reduce feed efficacy but increase morbidity and mortality in pig husbandry and raise treatment costs in suboptimal rearing environments.[118] Agricultural organizations used the study to oppose the Obama-era push for antibiotic restrictions (chapters 8 and 10). In 2009, *FBNews* warned that Danish AGP bans had resulted in greater mortality, morbidity, and antibiotic use for pigs. Significantly, the magazine also claimed that organic food had resolved the need for statutory intervention in the United States: "If, however, a consumer still does not trust food from animals treated with antibiotics, there's already a way to avoid it. To be certified organic under USDA's National Organic Program, animals can't be given antibiotics. . . . if someone just wants to avoid products from animals that have been given antibiotics, they can already do that."[119] Farm Bureau president Bob Stallman even claimed that "the possibility of resistance from antibiotics in livestock is declining."[120]

In the US farming press, commentators' opinion on AMR became more ambivalent: while some continued to deny that agricultural antibiotic use was contributing to hazards, others endorsed cautious reform—but no statutory intervention. In 2001, *Successful Farming* had already reported on the hospitalization of a 12-year-old Nebraskan, whose resistant infection was linked to a salmonellosis outbreak and antibiotic use on his father's cattle farm. The infection had been resistant to Rocephin (ceftriaxone) and was being used by the FDA to argue for a ban of Baytril in poultry production (chapter 10).[121] In 2010, *Farmers Weekly Review* participated in the CDC's antibiotic awareness week for human medicine but downplayed agricultural AMR hazards: "Antibiotics mean healthier livestock. Healthier livestock means higher quality food."[122] *Wallaces Farmer* also advocated "rational" antibiotic use but advised farmers to "keep telling our story": "antibiotic resistance in bacteria is a natural part of the evolutionary process. . . . But we also know that underdosing, incomplete treatment or choosing the wrong antibiotic . . . can increase the rate of resistance."[123] Should farmers find themselves "defending an indefensible position," then it was time to "take a serious look at abandoning that particular practice."[124] Responding to a reader, who feared that AGP bans might "erase my profit margin and force me out"[125] in 2010, the magazine's expert panel was surprisingly relaxed: two experts reminded the reader that therapeutic antibiotic use would remain legal, and the third expert noted that hogs were still "being produced profitably in European countries."[126]

What US farmers themselves thought of AMR hazards is more difficult to say. A 2015 study by the USDA's Economic Research Service indicates ongoing reliance on antibiotics for disease prevention but a gradual phasing out of AGPs, which were now estimated to raise productivity by only 1 to 3 percent. In the hog sector, the number of producers using AGPs to finish animals fell from 52 to 40 percent between 2004 and 2009. Although most producers used antibiotics for prophylaxis and treatment, ignorance about whether antibiotics were being fed to promote growth, however, rose from 7 to 22 percent—perhaps indicating the spread of integrated operations and diminished agency on the part of individual producers. Similar trends were apparent in nursery operations where 62 percent (5 percent ignorance) of producers were using antibiotics for disease prevention, 65 percent (5 percent ignorance) for disease prevention or growth promotion (5 percent ignorance), and only 33 percent (8 percent ignorance) solely for growth promotion. Overall use was lower in integrated operations. By 2009, 83 percent of American nursery hogs and 64 percent of finishing hogs were removed under contract.[127] In the broiler sector, twenty large integrators accounted for 96 percent of US production in 2011. Already skeptical of growth promotion during the 1980s, about 48 percent of broiler operations reported only giving antibiotics to animals when they were sick (up from 33 percent in 2000 and 2 percent in 1995). However, 32 percent of surveyed operations also claimed that they did not know whether they used antibiotics for purposes other than disease treatment. Self-limiting antibiotic use was becoming more common due to rising demand for antibiotic-free poultry.[128]

Larger cattle producers defied trends in the pig and poultry sectors. In 1994 and 2011, more than three-quarters of American feedlots with at least 1,000 head (0.2 percent of producers, 8 percent of overall production in 2012) provided antibiotics in feed or water, mostly for the purpose of growth promotion. More than ninety percent of larger feedlots fed ionophores to promote growth, almost 75 percent added antibiotics other than ionophores and coccidiostats to feed, and 44.7 percent added a coccidiostat. AGPs proved less popular among smaller cattle producers. In 2011, only 28.7 percent of smaller feedlots used ionophores and only 26 percent used other antibiotics. Of cow-calf operations surveyed in 2007/2008, 15.8 percent reported feeding antibiotics to prevent disease and/or promote growth.[129] In dairy farming, 90.1 percent of surveyed producers used antibiotics for disease prevention in 2007.[130]

Although USDA data reveals little about farmers' personal risk perceptions, it indicates wider changes in the agricultural antibiotic market. During the first decade of the new millennium, many producers gradually phased out inefficient AGPs. Some also turned toward the antibiotic-free market. However, most producers still invested in routine prophylactic and therapeutic antibiotic use. With antimicrobial sales for food-producing animals increasing to over 14.859 tons in 2013,[131] the majority of US farm animals were still receiving antibiotics at some point in their lives.

Most conventional farmers thus continued to support agro-industrial campaigning against Obama-era moves for statutory antibiotic restrictions—even if this entailed voluntarily abandoning AGPs. In April 2013, *Wallaces Farmer* reported that the AHI was pressing for compliance with formerly suspect FDA guidances to avert "what happened 10 years ago in the European Union, when the use of [AGPs] was stopped via regulation."[132] Although some commentators compared accepting voluntary antibiotic guidances to Israel's withdrawal from Gaza,[133] the AHI's strategy of voluntarist reform seemed validated by the FDA's abandonment of statutory restrictions in the same year (chapter 10).

Resulting voluntary restrictions' impact on the physical and cultural antibiotic infrastructures on American farms seems to have been light. According to the Pew Charitable Trusts, FDA guidances allowed farmers to administer one-quarter of drugs at the same dosages and with no limits on treatment duration for the prevention and control of disease.[134] Despite a recent decline of total antibiotic use and more radical reductions by large poultry producers like Perdue (chapter 10),[135] open-ended prophylactic use of former AGPs continues to occur in cattle production.[136] Voluntary bans have also failed to weaken organized agro-industrial resistance to statutory reform. During Iowa's 2014 annual swine day, ex-USDA Undersecretary of Food Safety Richard Raymond asserted that knowledge about the causes of bacterial resistance remained in flux and that AMR could not be blamed on agricultural antibiotic use: "there is no proof that low doses are any more likely to cause resistance than high doses of antibiotics."[137] The Farm Bureau also maintains that "rational" antibiotic use is unproblematic. In a 2015 policy statement, it expressed "serious concerns about the effects of removing important antibiotics and classes of antibiotics from the market, which would handicap veterinarians and livestock and poultry producers in their efforts to maintain animal health and protect our nation's food supply."[138] In December 2016, *Scientific American* highlighted significant efforts by the National Chicken Council and the AHI to topple new antibiotic-related legislation. Tasked with deciding over PAMTA's final fate (chapter 8), half of the Health Subcommittee of the House Energy and Commerce Committee had received donations of more than $15,000 by pharmaceutical companies or farming organizations. Larger corporations are also being accused of using biosecurity protocols and fines to keep critical researchers and journalists away from production facilities.[139]

Over the past decades, this dual strategy of resisting statutory reform and stressing voluntarist measures has been remarkably successful in defending US agricultural antibiotic use. Since the 1970s, pharma-led agricultural lobbying defeated multiple statutory interventions and even expanded drug access during the 1990s. Part of this success has rested on compartmentalizing antibiotic problems: beginning with the 1972 task force report, agro-industrial commentators dismissed AMR concerns by claiming that popular products like AGPs remained efficacious, confusing AMR and residue issues, blaming AMR on medical overuse, and focusing on individual aspects rather than wider structures of drug use. Compartmentalization made problems seem reformable via new drugs, antibiotic substitutes, market niches for pure food, improved biosecurity, and narrow restrictions—as in the case of AGPs. It also helped maintain inner-agricultural cohesion and unified opposition against excessive state intervention. Despite internal doubts about AGP efficacy and varying consumption patterns, the US agricultural alliance for broad antibiotic access has remained firm. While voluntary restrictions and antibiotic-free product lines have undeniably reduced total antibiotic use since 2013, control over the shape of future reforms seems to rest in the hands of industry—and not of the officials tasked with regulating it.

From the historical perspective, it is debatable whether further victories over state intervention are in US farmers' best interest. Helping the pharmaceutical industry ward off FDA action has committed organizations like the Farm Bureau to an ideological rear-guard defense of antibiotic use that will continue to face external criticism. Weaker US regulations may also give a long-term advantage to European producers, who have already adapted to production without AGPs and are now preparing for likely restrictions of prophylactic drug use. In the long term, stricter statutory regulations may aid European markets by allowing producers to sell "safer" produce and argue for barriers against American food produced with less stringent standards (chapter 12). Economically, polarized antibiotic battles are moreover contributing to a situation in which American farmers' representatives defend a system of industrial production that is consistently pushing independent producers into industry contracts or out of farming altogether. Since antibiotics' mass introduction to agriculture, the number of US farms has declined by over 60 percent from 5,388,437 in 1950 to 2,109,303 in 2012.[140] Although antibiotics did not cause intensification, they facilitated the rise of a new mode of integrated food production, which is unlikely to increase—or even sustain—existing farmer numbers. Arguing for state support to protect the antibiotic commons via alternative production methods might. At several moments in history, an open agricultural debate on antibiotics' long-term costs and benefits could have stimulated a wider revaluation of expansion and intensification's ambivalent impact on farmers themselves.

10

Statutory Defeat

Voluntarism and the Limits
of FDA Power

This chapter explores why the American FDA struggled to assert authority over antibiotics. As the two previous chapters have shown, the FDA's failure to implement statutory AGP bans in 1972, 1977, and 2013 was in part due to agro-industrial resistance, regulation wariness, partisan divides, and consumers, who feared residues more than AMR. However, the agency's ambivalent record was also caused by official factionalism, ineffective enforcement, and a weakening of the FDA during the 1980s. Amid polarizing public debates about antibiotics' costs and benefits, officials struggled to defend limited bans in courts and Congress. Precautionary reasoning was repeatedly defeated by industry's insistence on proof of harm. From the 1990s onward, officials reacted to the defeat of statutory interventions by attempting to improve antibiotic stewardship via AMR monitoring and voluntary guidelines. Similar to Europe (chapter 13), resulting voluntarist reforms initially narrowly focused on AGPs and left wider antibiotic infrastructures intact.

Institutional Change

Following the 1965 resignation of commissioner George Larrick, heightened responsibilities, complicated drug withdrawal procedures, and growing political interference resulted in a rapid succession of FDA commissioners and bureaucratic turmoil.[1] Characterized as a "wild-eyed crusader with a battle-ax flailing boldly,"[2] Larrick's successor, James Goddard, was soon damaged by the chloramphenicol scandal. Following Goddard's resignation in 1968, the less-outspoken physician and bacteriologist Herbert Ley became the last FDA commissioner to have risen through the agency's ranks.[3] Both Goddard and Ley oversaw a significant rise of FDA responsibilities and capabilities under the consumer-oriented Johnson administration: whereas the FDA had employed 800 people in 1954, Goddard's agency employed 4,441 people. In 1968, the FDA was also incorporated into the HEW's new Consumer Protection and Environmental Health Service (CPEHS), which aimed to create a "super bureau" capable of research-based hazard prevention.

However, the era of consumer-oriented reform was brief. In 1969, the cyclamate scandal (chapter 8) not only led to a sacking of Ley but a sea-change in the FDA's organization and outlook. The newly elected Nixon administration responded to industry and medical demands by separating the FDA from the only recently created CPEHS and by prioritizing substance regulation based on flexible tolerances rather than zero-tolerance rules.[4] A HEW review also recommended a reorganization of the FDA along product rather than function lines. The Bureaus of Science, Medicine, and Compliance were subsequently replaced with the Bureaus of Drugs, Foods, Pesticides, and Product Safety.[5]

Although responsibilities were now divided between three bureaus, the Bureau of Veterinary Medicines (BVM), which had already been made independent of the Bureau of Drugs in 1967, oversaw most aspects of American agricultural antibiotic regulation. The BVM's authority was further strengthened by the 1968 Animal Drug Amendments (ADA).[6] Supposed to redress the "stepchild treatment" of animal drugs, the ADA streamlined licensing procedures by creating the category of a "new animal drug." New animal drug applications (NADAs) could be handled under human drug procedures. In the case of most antibiotic feeds, products no longer needed to be cleared separately as new drugs (Section 505), food additives (Section 409 FDC), and antibiotics (Section 507). NADAs had to prove that a product was safe and efficacious if used according to label directions and had to include assay methods. FDA officials had 180 days to review a NADA or ask for additional information. A further 90 days were allowed to process applications for inclusion of new animal drugs in commercial feeds. The goal was to have new feed additives on the market in nine instead of fourteen months.[7] The ADA was supported by the AHI, the American Veterinary Medical Association, and the feed industry, who argued that more drugs would improve animals' health, increase meat supplies, and reduce consumer bills.[8] In 1970, the FDA's BVM was formally placed in charge of processing NADAs.[9]

The streamlining of drug licensing contrasted with the complexification of drug withdrawals. With the exception of the Delaney Clause's zero-tolerance policy for carcinogens, the main FDA safety standard for additives and animal drugs remained the reasonable certainty of no harm, which was often interpreted as "relative safety."[10] This requirement of "no harm" was, however, replaced with a requirement of "no significant risk of harm" in 1971.[11] If the FDA wanted to withdraw a substance, it now had to provide affected individuals and organizations with 60 days to register protest. Should complaints reveal differences on "matters of substance," the FDA was required to hold formal administrative hearings on substances' safety, which were often lengthy, expensive, and inconclusive. The only way administrative hearings could be avoided was if FDA commissioners declared that a substance posed an "imminent hazard". Because affected parties were entitled to challenge such a move, officials could, however, end up defending substance bans on the grounds of an "imminent hazard" in regular courts. Losing a public trial was not only embarrassing for the FDA but could also lead to very expensive compensation payments to plaintiffs.[12]

The growing difficulty of withdrawing substances incentivized the FDA to avoid polarizing all-out public confrontation with industry. During the 1970s, officials used interim regulations to circumvent formal withdrawals. Interim regulations permitted the ongoing use of a suspect substance while safety questions were being resolved. The FDA justified interim regulations with reference to the increasing amount of substances being challenged and the rising burden of evidence required to prove safety. What exactly "safety" was and whether a substance's benefits outweighed limited risks was not just decided by regulators.[13] In courts and Congress, the authority of FDA science faced rising challenges by industrial or activist counter science.[14] Polarizing battles over proof of harm would almost inevitably result in officials being accused of either inaction or overreaction.

Proof of Harm

Complicated and antagonistic withdrawal procedures did not bode well for US antibiotic reform. Proving harm was already complicated in the case of known carcinogens and toxicants. Implementing precautionary bans of entire substance groups on the basis of abstract AMR scenarios would be exceedingly difficult—especially, since the issue was not high on the public agenda. Unsurprisingly, many 1970s FDA regulators were thus reluctant to endorse calls for British-style AGP bans.

Officials' hesitancy to act on "infective AMR" warnings was exacerbated by their contemporary struggle to restrict a limited number of therapeutic antibiotic products on the grounds of residue hazards. In April 1968, commissioner Goddard had announced FDA proceedings against drugs likely to leave hazardous residues in food or milk. However, in the absence of proof of "imminent harm" resulting from residues, withdrawal proceedings left the FDA on thin legal ice. Following Goddard's departure in mid-1968, industry resistance first led to an extension of the official comment period on withdrawals to 120 days and a subsequent abeyance of FDA proposals for a further two years to wait for results of new industry residue research. The legal stalemate only ended when companies stopped resisting FDA withdrawals following passage of the industry-friendly 1968 ADA.[15] In May 1969, the FDA published new regulations for labeling indicating withdrawal periods, limited several residue-prone antibiotic products to veterinary prescription, and banned bacitracin as well as oral chloramphenicol solutions, which were proving popular as unapproved growth promoters, from use in food animals.[16]

The difficulty of withdrawing products causing measurable residues made BVM director Cornelius Donald Van Houweling reluctant to contemplate further AMR-inspired regulatory action. Appointed in 1967, Van Houweling had previously directed programs in what would turn into the USDA's Animal and Plant Health Inspection Service.[17] Faced with a contemporary breakdown of drug compliance on US farms and feed mills, Van Houweling instead focused limited BVM resources on curbing residue and salmonellosis hazards.[18] In 1967, officials informed Democratic Senator Birch Bayh that "infective" AMR was not an immediate policy priority: "At present there is only one antibiotic available for animals which is not used in humans. Even this antibiotic causes cross-resistance . . . however, there is no definitive evidence linking antibiotic resistant organisms of animal origin to human disease or allergies. Consequently, we are not contemplating any [antibiotics bans]."[19] Echoing the FDA's 1966 ad hoc report, officials maintained that reducing direct human exposure to antibiotic residues was the most effective way of curbing AMR risks. To determine the "*real* hazard to man"[20] [emphasis in the original], the FDA authorized $25,000 for research on "resistant organisms resulting from medicated animal feeds."[21] Initial studies were inconclusive. Pigs and chickens fed chlortetracycline developed a predominantly resistant flora. However, resistant bacteria had already been detected in animal stomachs, feeds, and farm soils prior to the feed experiments.[22]

The BVM's wait-and-see attitude was criticized by other FDA bureaus.[23] In 1968, *FDA Papers* published an attack on AGPs by David Smith from Boston's Children's Hospital. Questioning AGPs' efficacy, Smith warned that resistant *Salmonella* were already causing problems in the United States: "All strains of *S. typhimurium* isolated before 1948, when antibiotics were seldom used on farms, were sensitive to tetracycline; 30 percent of strains isolated from poultry in 1962 were resistant to tetracycline, while 94 percent and 57 percent of strains isolated from cattle and hogs were resistant to tetracycline."[24] Coinciding with the 1969 cyclamate scandal and commissioner Ley's sacking, the publication of Britain's Swann report exacerbated inner-FDA tensions. Asked to summarize the situation in light of media enquiries, Van Houweling stated: "It is reliably estimated that approximately 40 million tons of animal feed containing drugs was consumed in 1968. Also, that almost 80 percent of the meat, milk, and eggs consumed in the United States comes from animals fed medicated feeds. The [AHI] reported that $72.5 million of antibiotics went into animal feed last year."[25] While the FDA's Bureaus of Science and Medicine were "acutely aware" of "the possible ecological effects of using these large amounts of antibiotics in animal feeds," the BVM maintained that there was no evidence that "such resistance has caused difficulties in treating diseases in animals or man in the United States."[26] Two new studies were underway to assess hazards: the FDA was attempting to trace resistant *Salmonella* outbreaks back to three farms but had been unable to "process all the culture received," and William G. Huber at the University of Illinois College of Veterinary Medicine was studying resistant bacteria in animals and people who had contact with farm animals.[27] In the absence of watertight proof of harm, Van Houweling advised against launching legally uncertain AGP withdrawals.

Over the next decade, this situation would repeat itself numerous times: while a reluctant BVM led by Van Houweling would warn about limited proof of harm, external AMR warnings would make other FDA officials push for antibiotic bans. Each of the three 1970s FDA commissioners would use a different approach to solve this dilemma: the first would try to displace internal AGP conflicts by seeking external expert validation of restrictions; the second would try to stage-manage external endorsement of AGP withdrawals; and the third would try to withdraw AGPs unilaterally. All three attempts ended in defeat.

Attempt Number One: Seeking External Endorsement (1970–1974)

Taking over as FDA commissioner shortly after Van Houweling's 1969 AMR memorandum, the trained surgeon and former consultant Charles Edwards faced a difficult situation: he had to restore trust in FDA consumer protection while not antagonizing superiors in the more industry-friendly Nixon administration.[28] There were plenty of fires to put out regarding antibiotic regulation: medicated feedstuff controls had broken down, USDA monitoring was revealing residues in food, and FDA officials had to respond to the British Swann report. Between 1970 and 1972, Edwards decided to use limited BVM resources to tackle residues and displaced the issue of AMR by commissioning an expert review to evaluate AMR hazards and AGPs. In 1972, a resulting opportunity for reform quickly disintegrated in the face of industry opposition, counter science, and FDA factionalism.

Announced in April 1970 and corresponding with a general increase of FDA reliance on external expertise,[29] the FDA Task Force on Antibiotics in Animal Feeds was chaired by Van Houweling and was composed of officials, industry representatives, and experts on infectious disease, microbiology, and veterinary medicine. Consumer representatives were not invited. The task force was charged with undertaking an "in-depth review of the usage and actual value of antibiotics in animal feed"[30] and assessing health hazards. Between 1970 and 1972, the task force met nine times. During its first meeting, members decided "that while there was not enough evidence to indicate an imminent and immediate health hazard, there was sufficient data to assure that there is a potential, if not probable, health hazard associated with feeding of antibiotics to animals."[31] Separate expert groups were formed on antibiotic research, human health problems, animal health problems, and antibiotic effectiveness. The task force also called for a moratorium on AGP licensing for the duration of their review. Over the next year, members gathered evidence across the United States and attended a "Conference on the problems of AMR" in New York in October 1970 where they met European experts and industry scientists like former Cyanamid-employee Thomas Jukes (chapter 13).[32]

In early 1971, eight task force members also embarked on a fact-finding mission to Britain. Between February 1 and 3, the US delegation met with British agricultural and health officials.[33] The meetings exposed significant differences with the British prioritizing precautionary risk scenarios and AMR and the Americans focusing on proof of harm and residues. US delegates were particularly surprised by the lack of post-Swann AMR surveillance in Britain.[34] During frank exchanges, British officials conceded that the Swann report "could be criticized for the absence of references and for its lack of attention to the problem of residues."[35] They even "accepted that some of the conclusions of the Report had been based on inadequate evidence."[36] However, they maintained that AMR scenarios were based on logical extrapolations and that precautionary AGP bans were necessary. The implementation of other British antibiotic reforms seemed surprisingly weak. Although veterinary guidelines had been reformulated according to the principle of "sharp, short and high,"[37] low-dosed antibiotic prescriptions remained legal. There would also be "no 'army'"[38] checking whether guidance was being adhered to.[39]

On February 3, task force members met PHLS director James Howie and E. S. Anderson, the researcher behind the 1965 "infective" resistance warnings about *S. typhimurium*. According to Howie, British AMR concerns were increasingly centering on AMR selection in coliforms rather than in salmonella where infections were usually self-limiting. In a significant statement, he also noted that the Swann committee had been primarily concerned not so much about AGPs but about "farmers prescribing and using antibiotics in therapeutic and near therapeutic levels in all kinds of stock."[40] Agro-pharmaceutical data in favor of using antibiotics against vaguely defined stress had been the "most pitiful scientific evidence presented before Swann."[41] E. S. Anderson warned US colleagues about neglecting AMR—even if detected resistance seemed insignificant. Although "little resistance" developed in response to ampicillin use on farms, this use was probably behind most ampicillin resistance in *Staphylococci* isolated from disease outbreaks.[42] When asked whether Britain's *S. typhimurium* Type 29 epidemic had been unique, Anderson noted that Type 29 had provided a precedent: "Other strains of *S. typhimurium* [like phage types U29 and U163] are transferred from animals to man, and follow this same model but [the] kinetics may not be so explosive."[43] There was no reason to believe that agricultural AMR hazards were unique to *S. typhimurium*— the problem was ecological:

> [Anderson] didn't in the past and doesn't now distinguish between growth prophylactic and therapeutic doses of antibiotics. He cannot tell the difference in numbers of resistant strains arising in man as a result of these three regiments and apparently the bacteria also do not distinguish amongst these three regiments. . . . Dr. Anderson considers the entire problem of transferable resistance as microbial ecology and for example he pointed out that low level transferable ampicillin resistance was first found in his laboratory and was almost certainly of porcine origin.[44]

Questioned about a recent paper by British veterinary researcher and early AGP critic, Herbert Williams Smith, who had swallowed resistant organisms but had not been able to detect a permanent change of his gut flora, Anderson maintained that "transfer and colonization could occur."[45] Despite expressing confidence that AGP bans would reduce AMR, Anderson attacked antibiotic prophylaxis and proposed a complete separation of human and animal antibiotic use. Accompanying Anderson to his laboratory, an anonymous task force member commented on Anderson's forceful personality—which extended to changing the American's cough syrup— but also described the "incredible amount of data" the British had accumulated on AMR:

> The surveillance of Salmonella strains both in animals and in man that is carried out is truly astounding since not only are organisms phage typed and sera typed but a hunt for resistance patterns is routinely followed in all organisms coming to the laboratory's attention. In those instances where an organism in Dr. Anderson's opinion seems a little strange, . . . an enormous amount of work goes on in transferring or attempting to transfer this resistance from one organism to another . . . . It was truly an impressive demonstration which I think is unique.[46]

Interviewed by the task force on February 4, Herbert Williams Smith also regretted that the Swann report had not gone further. Following his recent self-experiment, he "believed that human beings are the major source of R-factors for other human beings"[47] but worried about the large reservoir of mobile resistance genes being selected for in animals. Ongoing selection pressure might eventually result in R-factors that could transfer to humans more readily. Existing R-factors against tetracycline acted as "catalysts" for further AMR by allowing other R-factors and virulence and toxicity genes to "hitchhike."[48] In the United Kingdom, AMR was already leading to tetracycline treatment failures for bacteremia in swine and poultry: "if we don't need to have the animal reservoir of R-factors we are better off without it."[49]

Tasked with summarizing the FDA task force delegation's UK visit, John V. Bennett, head of the CDC's Bacterial Diseases Branch, again highlighted differences between precautionary British reasoning and the US focus on proof of harm. Bennett was particularly frustrated by the "absence of solid, factual data on many issues" underlying British decision-making: "After a review of the available data I am convinced that there is not an imminent human or animal health problem resulting from current animal husbandry and veterinary practices in the United States. Though a non-imminent problem probably does exist, its magnitude has not been sufficiently defined to justify present action that might have substantial economic, political, and professional impacts."[50] Bennett thus advocated commissioning studies to test AGPs' efficacy and their potential to select for AMR against medically relevant drugs. If effective drugs without medical relevance could be found, then only they should be permitted for prescription-free use as AGPs. Should only medically relevant drugs prove efficacious as AGPs, then studies would have to document that their use did not cause therapeutic problems and residues. AMR concerns also made Bennett propose reviewing US veterinary prescription rights for chloramphenicol and gentamicin.[51]

Bennett's proposals proved controversial with both medical and agro-industrial task force factions. Following the delegation's return, the FDA task force drafted an intermediate report in late February and a final report in October 1971.[52] Fierce conflicts characterized both drafting stages. While all sides agreed that medically relevant AGPs could be withdrawn from poultry, restrictions for other species led to significant disagreement. Antibiotic supporters were organized in the task force's subcommittee on antibiotic effectiveness and critics in the subcommittee on human health aspects. The subcommittee on veterinary medical aspects remained divided. In early 1971, the first draft report recommended a two-year phase-out of most AGPs and of prophylactic antibiotic use but allowed "safer" antibiotics to remain on the market until users could switch to other drugs or devise alternative solutions. This proposal failed to satisfy either antibiotic critics or supporters. By April 1971, both sides had compiled minority reports forcing BVM director Van Houweling to redraft the main task force report. Conflicts continued. In November 1971, *Food Chemical News* reported that a preview of the second draft report had "degenerated into a donnybrook after [commissioner Edwards] and his staff departed."[53] A last-minute compromise was eventually brokered by proposing AGP bans but allowing industry to prove products' safety first.[54]

Passed with a thin margin but signed by all members, the final task force report was released in January 1972. It cautioned that the "efficacy and safety of long-term feeding of subtherapeutic levels of antibiotics for animal disease control and prophylaxis [had] not been adequately demonstrated."[55] In 1970, antibiotics had been worth $414,135,000 to livestock producers and pharmaceutical companies had earned $64,030,323 from selling them between 1968 and 1969. Estimating the impact of AGP restrictions was not possible because "some antibiotics will undoubtedly continue to be available for growth promotion purposes."[56] Significantly, the final report concluded that AGPs' selection for AMR posed a "not fully documented" but logical health hazard: "Evidence suggests that the use of certain antibiotics in food-producing animals promotes an increase in the animal reservoir of Salmonella ... the use of some antibiotics in animals produces a marked increase in the prevalence of R-factor containing bacteria which may be transmissible to man's enteric flora."[57] It was "the consensus of the task force that it would be highly desirable that in the future, a group of antibacterial agents be reserved exclusively for human use."[58] The task force advised restricting tetracyclines, dihydrostreptomycin, sulfonamides, penicillins, and "all other approved antibiotics"[59] to prescription-only status if producers failed to prove their safety and efficacy. The task force also endorsed existing feed restrictions for chloramphenicol, semi-synthetic penicillins, gentamicin, and kanamycin. FDA proposals based on the task force report were published in the *Federal Register* on February 1, 1972.[60]

The *Federal Register* announcement did not end controversies. The subsequent publication of contradictory task force minority reports instead triggered significant disagreement between officials and agro-pharmaceutical representatives as well as between government departments and American scientists.[61] At the government level, agricultural regulators were displeased with the FDA's proposed bans. While state officials warned that it would be difficult to enforce AGP restrictions,[62] the USDA's Agricultural Research Service could not find definitive evidence of harm in April 1972 (chapter 9).[63] The review's recommendations for more research were echoed by a cautious USDA working paper on antibiotic bans' economic effects for turkey and broiler production.[64] US academia was also divided over AGP bans. The 1972 annual meeting of the American Society for Microbiology exposed significant scientific and ideological rifts between an older generation of neo-Malthusian experts and younger, more industry critical scientists. While microbiologist Stanley Falkow warned that agricultural antibiotic use might compromise human therapy, American Cyanamid's Gordon Kemp stated that there was "absolutely no evidence"[65] that R-factor transfer was impairing treatment. AGP co-discoverer Thomas Jukes emphasized that potential problems with AGPs would have materialized long ago. Falkow disagreed and pointed to diarrhea outbreaks among swine, which had been caused by *E. coli* made resistant by feeds. This assertion was attacked by animal scientist James Leece, who claimed that outbreaks had been caused by a virus. Proceedings got tense when task force member Arthur K. Saz declared: "It is a microbiological monstrosity to have an ever-replenishing pool of antibiotic-resistant organisms being fed to the population."[66] Jukes responded in style: "This has been my life, improving growth in animals with antibiotics. To feed the world's population we need every trick we know."[67]

Determined to defend AGPs, well-connected older American scientists also used their position as government advisors to attack FDA AMR warnings. In June 1972, the NAS commissioned a hostile review of the FDA task force report. The NAS review was chaired by infectious disease specialist Maxwell Finland. A friend of Thomas Jukes, Finland had already endorsed AGPs at the 1955 antibiotic feeds conference and as a member of the FDA's 1965 ad hoc committee (chapter 4). Despite campaigning against inefficacious antibiotics in human medicine, Finland "remain[ed] unconvinced"[68] that infective AMR scenarios undermined existing agricultural antibiotic use. The NAS Division of Biology and Agriculture had already published a critical position paper on AGP restrictions. However, NAS president Philip Handler wanted the Drug Research Board to review this paper before sending it to the FDA. Asked to conduct this review by Drug Research Board secretary Duke Trexler, Finland was allowed to nominate other reviewers and was supplied with task force files and a hostile letter, which Thomas Jukes had recently written as a member of the President's Science Advisory Committee.[69] After less than four months deliberation, Finland's committee recommended approving the NAS position paper against AGP bans and funding research on AMR and medically irrelevant antibiotics solely for agriculture.[70]

The rushed NAS report attracted scientific criticism not only from former task force members but also from pharmacologists, who had been members of Finland's internal review.[71] In a sign of how polarized academia had become, letters between Jukes and Finland discussed the degree to which opponents were ignoring the "facts." Jukes particularly disliked the "zealot"[72] David H. Smith, who had headed the FDA task force's Human Health Hazards Committee. Finland agreed: "It was only after [Smith] joined the task force that he was 'converted' or perhaps brainwashed."[73] Controlling influential positions within academia, Finland and Jukes may not have been able to prevent critical studies but had sufficient power to undermine their opponents' standing.

Emerging academic and bureaucratic rifts weakened FDA reformers, who were encountering well-organized agro-pharmaceutical opposition (chapter 9). After proposing AGP bans in February 1972, the FDA received over 380 responses within the mandatory 60-day comment period—formal hearings would now be necessary.[74] In August 1972, NAS president Handler exacerbated reformers' loss of political momentum by personally asking commissioner Edwards to delay regulatory action until Finland's ad hoc review had completed its work.[75] Worried by the extent of industry protest and weak public support of AGP bans (chapter 8), Edwards informed Handler that he was willing to defer action and consider the NAS review. Writing to Finland, Drug Research Board secretary Trexler described Edwards' move as a "volte face."[76] Victory over FDA AGP bans was close.

Commissioner Edwards now faced a dilemma. NAS, USDA, and industry opposition had successfully undermined AGP bans' political momentum. Although Edwards had never actively pushed for restrictions himself, the FDA had, however, publicly committed to implementing the 1972 task force proposals in the *Federal Register*. The only way out seemed to be a loophole in the task force report itself. As a result of pro-industry lobbying, manufacturers had been allowed to prove that their products would not: cause a significant increase of resistant pathogenic and multiple resistant bacteria in humans, animals, and feed; prolong the shedding of resistant bacteria or increase their pathogenicity; result in cross-resistance to other therapeutics.[77] However, decisions over the design of appropriate testing protocols had been left to the FDA. The solution to Edwards's dilemma lay in drastically narrowing AGP safety reviews' scope. In December 1972, the FDA's associate commissioner for medical affairs informed Edwards: "The previous documents . . . approach the problem in a broad way, considering that antibiotics *per se* in this particular use might constitute a hazard . . . . [A new BVM draft] narrows this scope considerably by restricting the studies of human hazard (aside from the possibility of salmonella reservoir increase) to drugs which are (1) used in human clinical medicine, and (2) which promote gram negative transferable resistance."[78]

Modified safety reviews were announced in the *Federal Register* in April 1973. According to the FDA, AGPs constituted an imminent hazard if they significantly increased the *Salmonella* reservoir in animals and food.[79] Ignoring British warnings, the FDA claimed that there was "less agreement on the hazard to human health presented by other animal-source bacteria (e.g., coliforms)"[80] and R-factor transfer. It also referenced Finland's NAS review and noted that AGPs appeared "safe under the conditions of use."[81] Manufacturers were given one year to produce "an assessment of the effects of subtherapeutic levels of [tetracyclines, streptomycin, dihydrostreptomycin, the sulfonamides, and penicillin] on the salmonella reservoir."[82] Producers would also have to submit studies "concerning (1) the colonization and R-factor transfer from animals to man and (2) increased pathogenicity due to toxin-linkage with R-factor."[83] Continued AGP use was enabled by an interim regulation that would remain in place until evidence of safety or harm was provided. Proving safety did not "require complete certainty of the absolute harmlessness of a drug, but rather the reasonable certainty in the minds of competent scientists that it is not harmful, *when balanced against the benefits to be obtained from the drug*" [emphasis added].[84] By downplaying AMR's ecological dimensions, narrowing risk assessments to *Salmonella* and allegedly subtherapeutic dosages, and including cost-benefit evaluations in its safety review, the FDA quietly began to abandon its first attempt at precautionary antibiotic restrictions.

The results of industry-led safety reviews were predictable. By 1974, eight manufacturers had submitted twenty-one *in vivo* studies of AGPs' impact on the *Salmonella* reservoir. Behind the scenes, FDA scientists expressed dismay about the quality of industry data: "omissions, deficiencies, or areas which raise questions [existed] in almost every study."[85] An external review of three industry salmonella studies also warned that they would provide no suitable base for regulatory action: "these are the kinds of data for the most part that will say whatever one wishes them to say . . . if the salmonella data are not clear, the other [*E. coli*] studies will only 'muddy the waters' more."[86] Having reached a similar conclusion about four additional extramural studies in March 1974,[87] FDA and Canadian officials traveled to Britain to enquire after the Swann bans' impact. However, the stagnation of British AMR monitoring dashed hopes for supportive findings (chapter 13).[88] Referencing the lack of reliable AMR data and the need for more research, the BVM decided against pursuing AGP bans in August 1974 and formally ended proceedings following the final deadline for industry studies in April 1975.[89]

Attempt Number Two: Stage-Managing Endorsement (1975–1977)

The regulatory respite caused by the abandonment of the 1972 task force proposals was short lived. In 1974, a European WHO working group report recommended reserving medically relevant antibiotics for human use.[90] In the US, two University of Illinois studies also revived pressure for reform. Both studies analyzed AGPs' effects on the human and animal gut flora on 30 farms. According to the first paper from 1974, AGPs' effects on animals' fecal flora were significant with many isolates proving resistant to oxytetracycline, streptomycin, dihydrostreptomycin, and ampicillin. Transferable resistance to oxytetracycline, dihydrostreptomycin, ampicillin, neomycin, chloramphenicol, and cephalothin was detected.[91] The second paper from 1975 compared fecal samples from (1) 22 people working on 16 farms using antibiotics, (2) 20 people residing on 13 farms with no direct exposure to animals, (3) 18 people treated with antibiotics for salmonellosis, (4) untreated people residing with treated individuals, and (5) untreated people with no exposure to farm animals. Unsurprisingly, samples from groups 1 and 3 contained the highest and group 5 the lowest proportion of resistant gram-negative enteric organisms. However, AMR levels were also elevated in groups 2 and 4, which suggested that AMR selection in medicated animals and humans could also alter untreated adjacent enteric flora.[92] Concerns seemed verified one year later by Stuart Levy's research on AMR transfer from farm animals to a farm family (chapter 8).

The new AMR data made FDA officials relaunch AGP withdrawal efforts. Learning from his predecessor's failure, new FDA commissioner, Alexander M. Schmidt, however, decided to use existing advisory committees rather than a further potentially polarizing expert review to legitimize AGP bans. During the 1960s, the FDA had created national advisory committees to enhance transparency and public trust. The committees were composed of officials, experts, and consumer and industry representatives. If staffed "correctly," they offered a covert way to create external support for FDA-led change. In early 1975, FDA general counsel Peter Hutt suggested using the National Advisory Food and Drug Council (NAFDC) to circumvent a further task force–style review.[93] The NAFDC's "neutral evidence" could be used to fend off industry challenges to formal AGP withdrawals. Industry opposition would "come in the form of formal litigation, . . . congressional inquiries and committee hearings."[94] The FDA would staff a "friendly" special NAFDC subcommittee with three NAFDC members and select consultants to bolster its regulatory authority.[95]

Chaired by commissioner Schmidt, the NAFDC established an Antibiotics in Animal Feeds Subcommittee (AAFC) in June 1975.[96] Eighteen months later,[97] the AAFC recommended banning tetracycline and penicillin AGPs. Going beyond the 1972 task force report, it also called for penicillin bans in disease prevention if effective substitutes were available. Tetracycline and sulfaquinoxaline use was to be limited "to those periods of time for which the presence of the drug in the feed . . . is necessary due to the threat of animal disease."[98] Although it did not completely heed internal FDA guidance,[99] the AAFC report seemed to provide FDA officials with robust external endorsement of statutory antibiotic restrictions.

However, in a surprisingly blow, the main NAFDC rejected the AAFC's recommendations. During a one-day session in early 1977, the NAFDC accepted the proposed penicillin and sulfaquinoxaline restrictions but rejected tetracycline restrictions. The committee also called for more research and recommended that the FDA's position be "reevaluated within three years."[100] According to one observer, three people had influenced the NAFDC's break with the AAFC: "(one the Chairman of the Board of a drug firm, another the President of a feedlot) whose sweeping generalities were not based on scientific fact and nevertheless went unchallenged."[101] Microbiologist Rosa Gryder recalled how, ahead of the meeting, some NAFDC members admitted not having read the background material supplied to them while "others did not clearly understand it."[102] Having been consulted by the AAFC, microbiologist Stanley Falkow fumed:

> Without mincing words, to accept the recommendations on the restriction of penicillin and sulfonamide and to table the [AAFC] recommendations on tetracycline simply reflects the ignorance of the full Committee . . . , the action of the full Committee was an insult to [AAFC members], . . . it is no exaggeration to say that the ecology of the enterobacteria, and recently other bacterial groups . . . has been changed by the pattern of antibiotic usage in man and his domestic animals.[103]

For antibiotic reformers, the NAFDC rejection could hardly have come at a worse time. Following Jimmy Carter's 1976 election victory, the FDA was headed by an interim commissioner and resources were strained by the fact that the Delaney Clause was forcing officials to proceed against popular saccharin sweeteners.[104] For a while, it was unclear whether the agency would remain committed to antibiotic restrictions.

Attempt Number Three: Unilateral Restriction (1977–1985)

Taking office in April 1977, the new FDA commissioner Donald Kennedy faced a considerably more hostile climate than his predecessors had. In Congress, fears of "stagflation" and the "drug lag" were tempting an increasing number of politicians to attack FDA regulations (chapter 8).[105] The Carter administration further complicated things by strengthening external oversight and introducing fiscal restraints such as mandatory inflation impact assessments for any regulatory action projected to cost more than $100 million. Some observers worried whether Kennedy, who had no prior political experience, would be able to strike a workable balance between regulatory action and Washington's campaign to reduce the bureaucratic footprint.[106]

Breaking with his predecessors' attempts to gain external expert endorsement, Kennedy's unilateral campaign for AGP bans heightened such concerns. Eleven days after taking office, Kennedy announced that he considered the NAFDC's decision nonbinding and would ban penicillin and tetracycline AGPs. Justifying his course, Kennedy referenced ecologist Garett Hardin's "tragedy of the Commons":[107]

> In short, the evidence indicates that enteric microorganisms in food animals and man, their r-plasmids and human pathogens form a linked ecosystem of their own in which action at any one point can affect every other. . . . the

vulnerability of microorganisms to antibiotics is a kind of "commons"—a resource which if we consume it by the use of antibiotics for non-medical purposes in animals, is diminished in man.[108]

According to Kennedy, the long-term benefits of antibiotic restrictions outweighed short-term economic costs.[109]

Kennedy's disregard of the recent NAFDC AGP endorsements was aided by parallel Government Accountability Office (GAO) criticism of his predecessor's use of the NAFDC to circumvent formal safety reviews.[110] However, his actions divided FDA officials. In a 1977 memorandum, BVM director Van Houweling had warned that banning all nonprescribed uses of penicillin and the tetracyclines would "have the approval of that segment of society represented by the consumer activist, scientists and those members of the medical profession who feel that action should be taken."[111] BVM officials, however, worried that such bans had the "potential for causing the greatest change in US animal food production."[112] Van Houweling reported: "five out of seven scientists from the staff of the BVM Antibiotics in Feeds Group and the Veterinary Research Division are willing to compromise and adopt the recommendations of the [AAFC]."[113] One BVM member supported the NAFDC endorsement of tetracyclines, "while another individual and the FDA Office of Science"[114] preferred strict tetracycline limitations to therapeutic use. Advising that the FDA follow AAFC recommendations to ban penicillin AGPs but permit more restricted prophylactic use of tetracycline AGPs, Van Houweling noted: "Politically, the Subcommittee position lies between the extremes desired by different segments of the American public."[115]

Other officials warned about congressional and industrial insistence on proof of harm. In June 1977, a memorandum cautioned: "we may not have enough [evidence] to avoid a hearing since there may well be substantial and material issues of fact."[116] Future BVM director Lester Crawford later remembered that the agency "simply stated in the [official FDA argument in the 1978] Congressional record that there was a theoretical chance of a compromise of human therapy. In fact we labored for almost a year over the choice of the term 'theoretical,' or the alternative of presenting direct evidence."[117] This lack of direct evidence soon came back to haunt the FDA in the face of impressive agro-pharmaceutical protest (chapter 9). In the American National Archives in College Park, the large moving boxes containing the FDA's general correspondence between 1977 and 1979 are close to bursting with private, industrial, congressional—and occasionally USDA[118]—letters opposing AGP restrictions.[119] In April 1977, a *Cyanamid News Release* claimed that "banning [all] antibiotics"[120] would annually cost consumers a staggering $2.1 billion. Readers were asked to make their "voice heard"[121] and contact their representatives with a list of arguments and writing tips provided by Cyanamid:

Make it known at the start of your letter that you think the proposal is harmful, and that you disagree with it.

Tell how long you have used tetracycline antibiotics on your farm, and the
benefits you have reaped that could not have come from any other
source.

Stress that you have seen no indication of adverse effects, to either animal
or human, from tetracycline use.

Say that you want to keep using tetracycline antibiotics, and what your
operation would be like without them.

When writing your Congressman and Senator, urge him to protect his
constituents' interest.

You should also consider writing to your state Commissioner of Agriculture.... When writing a letter to a Representative, Senator, or any
government official, there are rules of etiquette.... When writing a
Congressman, the envelope is properly addressed to "The
Honorable."[122]

The agro-industrial letter campaign was extremely effective. In June 1977, commissioner Kennedy complained about the volume of hostile correspondence
he was receiving: "The majority originate from a campaign orchestrated by a
major antibiotics producer."[123] According to a contemporary BVM memo, the
scale of opposition was entirely predictable. In 1977, the US AGP market was
worth about $118.1 million. Fierce competition in the pharmaceutical sector
would make "companies such as American Cyanamid Company, Pfizer, Inc.
and others ... vigorously resist any change in use."[124] Similar to farmers (chapter 9), pharmaceutical companies mostly opposed FDA regulations because of
the "belief that this is a prelude to other restrictions and more control by the
government."[125] Companies had spent much "time and money" on "defensive
research" and had anticipated FDA action "since about 1970 when the Antibiotics Task Force was formed"—"what to do when the change comes, has probably been in the planning for several years."[126]

Official withdrawal proposals for penicillin in animal feeds and for subtherapeutic tetracycline use were published in the *Federal Register* four months
after Kennedy's initial announcement in August and October 1977.[127] However, by this time, AGP bans were already showing signs of stalling: Congress
had just indicated its readiness to intervene in FDA affairs by mandating an
eighteen-month "breather" on saccharin regulations;[128] Canada had reneged
on parallel AGP bans;[129] and FDA AMR assessments were under attack from
other government agencies and industry scientists.

Similar to 1972, the increasing prominence of external reviews constrained
the FDA's ability to make authoritative and rapid decisions in the name of public health. In addition to diluting FDA authority, they also tended to deflect

political attention from long-term AMR hazards toward more short-term economic calculations. The result was a battle of often questionable cost-benefit estimates. While American Cyanamid used 1975 estimates to claim that restricting all AGPs would cost $801.7 million,[130] FDA officials pointed to 1976 Office of Planning and Evaluation and in-house estimates that producers using substitute antibiotics would only incur a cost increase of between $65 and $74 million—both well below Carter administration thresholds for a "major" inflationary impact.[131] In a September 1977 paper, the USDA's Economic Research Service (ERS) also reassessed antibiotics' benefits and hazards. Erroneously claiming that "it was not observed until 1955 that bacteria could develop resistance to antibiotics,"[132] the ERS stated that banning AGPs because of theoretical hazards would cause real economic costs. However, it also noted that industry estimates of costs between $1.6 and $1.9 billion resulted from models that assumed static 1970 production levels, a ban of all AGPs, or a significant ensuing production decline. Models assuming partial bans and a dynamic adaptation of production showed that consumer costs were either nonexistent or very low.[133] Competing industry and official cost estimates made the USDA and the Senate Committee on Agriculture, Nutrition and Forestry commission additional reviews of drug and chemical feed additives.[134] In November 1977, FDA officials warned that these new reviews would likely be unfavorable and that formal withdrawal hearings would delay regulatory action "until late 1978 at the earliest."[135]

Aware that time was not working in its favor, the FDA decided to create a "done deal" in December 1977 by establishing a new category of prescription-only restricted medicated animal feeds, which would include penicillin, chlortetracycline, and oxytetracycline feeds—something that was eventually implemented via voluntary guidances after 2013. The new feed category promised to restrict AGP access to veterinary prescription and licensed feed mills well ahead of what were likely to be drawn-out and uncertain statutory withdrawal procedures for subtherapeutic feeds. Although FDA projections showed that AGP restrictions to prescription-only status would only cost $15.6 million—"well below the criterion for a major economic impact"[136]—it was clear that Congress would interpret the so-called Controls Document as an attempt to bypass its authority.

During informal FDA hearings following the publication of the Controls Document in January 1978,[137] prominent critics attacked FDA plans[138]—174 witnesses opposed prescription-only restrictions and only 15 supported them.[139] Congress also reacted quickly. Representative Jamie Whitten, a Mississippi Democrat chairing the House Appropriations Subcommittee on Agriculture and Rural Development, vowed to hold the US budget hostage if AGP restrictions proceeded. A deal was brokered with the Carter administration whereby FDA action would be put on hold until further research provided more clarity on AMR hazards.[140] In February 1978, a resolution was introduced to the Senate directing the FDA to refrain from restrictions pending the outcome of new studies. The FDA first attempted to appease critics by extending the comments period on its Controls Document to June 1978. However, in May, the BVM warned that extensions would not be enough "in view of Congress's increasing tendency to [take] action where administrative agencies have taken unpopular stands on issues having significant public impact."[141] This prediction was correct. In July 1978, following hearings during which twenty-six of twenty-nine witnesses rejected FDA restrictions, Democrat Charles Rose, chairman of the Subcommittee on Dairy and Poultry of the House Committee on Agriculture, proposed a resolution to stall FDA action.[142] The resolution would force the FDA to await the outcome of a new NAS study for which Congressman Whitten's Appropriation Committee had earmarked $250,000 two months earlier.[143] Ahead of the final vote, Rose showed a film highlighting the costs of FDA action and featuring industry and CAST experts claiming that no AMR problems had emerged since the 1950s. Rose also claimed that FDA bans would annually cost approximately $2 billion. His resolution was passed unanimously and endorsed by the Senate in September 1978.[144]

Despite the publication of more favorable external assessments of AGP bans by the Office of Technology Assessment and the USDA,[145] FDA officials were powerless to take further action ahead of the congressionally mandated NAS review. Its narrow terms of reference, pre-selected membership, and limited recommendation options made it very unlikely that this review would support FDA action. Tasked with assessing "the scientific feasibility of additional epidemiological studies"[146] and focusing only on "subtherapeutic" penicillin and tetracycline feeds below 200 grams per ton, NAS reviewers were primed to call for more research. In September 1979, the NAS review's staff director indirectly acknowledged this by stating that reviewers could make only three recommendations: specific further studies are needed, no further studies are needed, or there is no way of devising further studies. Review members had moreover been selected to avoid public charges of bias. As a consequence, both CAST scientists and many renowned experts who had worked on critical FDA, WHO, and OTA panels were omitted from the NAS review.[147]

Published in March 1980, the NAS report predictably prolonged the *status quo* by concluding "that the postulated hazards to human health from the subtherapeutic use of antimicrobials in animal feeds were neither proven nor disproven."[148] Although it proposed additional studies of local AMR transmission, the NAS report cautioned that conclusive data on AMR hazards caused by subtherapeutic antibiotic use alone was unlikely to emerge: "it is not possible to conduct a feasible, comprehensive epidemiological study of the effects on human health arising from the subtherapeutic use of antimicrobials in animal feeds."[149] This assessment was, however, ignored by Congress, which provided

$1.5 million for a "definitive epidemiologic study of the antibiotics in animal feeds issue"[150] for the fiscal year 1981. Ahead of taking any regulatory action, the FDA was to study the effect of *both* therapeutic and subtherapeutic antibiotic use on plasmid-mediated resistance in enteric organisms in animals; compare the enteric flora of vegetarians and meat eaters; study how exposure to animal bacteria affected slaughterhouse workers and their contacts; and analyze whether carriage of resistant fecal flora was associated with increased morbidity or mortality from urinary tract infections.[151] Interim regulations from the 1970s would permit ongoing AGP use until final certainty on AMR hazards emerged.

Not all members of Congress were willing to further postpone FDA action. In June 1980, Democrat Representatives Henry Waxman and John Dingell proposed a bill (H.R. 7285) that would allow the secretary of health to limit antibiotic feeds if they did not meet a "compelling need."[152] The Subcommittee on Health and the Environment held hearings on AGPs. Boycotted by the National Pork Producers Council,[153] the hearings highlighted ongoing differences regarding statutory intervention and AMR hazards. The American Medical Association could not "state at this time that there is sufficient evidence of the transfer of disease-causing antibiotic resistant bacteria from animals to humans to warrant alarm."[154] Having only recently defended physicians' autonomy to prescribe antibiotics,[155] the organization opposed legislative measures, which would reduce regulators' ability to deal with AMR flexibly. Joyce Lashof from the US Office of Technology Assessment disagreed:

> Our [1979] conclusion was that the increasing pool of resistant bacteria is a serious health risk to humans, and that the contribution from low-level antibacterial use in animal feeds played a significant part in increasing the general pool of genetically resistant organisms. . . . We also pointed out that it was not relevant that the therapeutic use in humans of these same antibacterials may be a larger contributor to the development of resistance, as long as animal feed use was in itself a significant contributor to resistance.[156]

The new BVM director Lester Crawford also endorsed banning medically relevant antibiotics from use as AGPs.[157] However, without the support of Tim Lee Carter, ranking Republican member of the subcommittee, H.R. 7285 failed to make it past the "marking up" stage of legislation.[158]

Chances for FDA-led antibiotic restrictions decreased further after Ronald Reagan's 1980 election victory. Between 1981 and 1989, the FDA's workload grew significantly while its budget stagnated and workforce declined. Enforcement dropped from 1,041 annual actions between 1977 and 1980 to 577 actions in 1981.[159] In a significant move, the Reagan administration also changed decision-making within the renamed Department of Health and Human

Services (HHS). Taking over as HHS secretary in 1981, former Republican Senator Richard Schweiker acquired significant powers. Following 1981, FDA regulations had to be personally signed by Schweiker, who was intent on reducing "overregulation" and the "drug lag." The new FDA commissioner Arthur Hayes was equally unlikely to push for AGP bans. Hayes enjoyed a close relationship with pharmaceutical producers and had to resign in 1983 for accepting financial honoraria.[160] Even if the FDA and HHS had approved AGP restrictions, it was unlikely that they would have passed the Office of Management and Budget (OMB). In February 1981, Executive Order 12291 required all federal agencies to submit cost-benefit analyses on regulatory actions for OMB approval. The order increased the possibility of external interference on regulatory decision-making and disincentivized officials from pressing for substance bans.[161]

Weakened FDA officials could do little else than administer the stagnation of American antibiotic reform. Between 1980 and 1984, the FDA neither acted on nor withdrew the 1977 notice of hearings on AGPs. In response to NRDC and industry pressure to drop or pursue bans, officials referred to ongoing AMR research and licensed a limited number of penicillin and tetracycline premixes so as not to disadvantage companies that had not marketed these products prior to the 1973 interim regulations.[162]

The next round of congressionally mandated AMR research was published in August 1984. Despite finding significant overlaps in human and animal *Salmonella* and *Campylobacter* populations, the Seattle-King County Department of Public Health study validated the 1980 NAS report by neither proving nor disproving harm caused by AGPs. It seemed likely that antibiotic supporters would again be able to delay regulatory action by calling for more research. However, within a month, now familiar regulatory delays were unexpectedly jeopardized by Scott Holmberg's CDC study on links between resistant salmonellosis and antibiotic-fed cattle. Responding to Holmberg's findings, the NRDC upgraded its 1983 petition for hearings on stalled FDA bans to a request for the immediate suspension of all penicillin and tetracycline AGPs because of imminent harm (chapter 8).[163]

Despite Holmberg's data and NRDC pressure, chances for statutory AGP restrictions remained slim.[164] During hearings by Al Gore's Subcommittee on Investigations and Oversight in December 1984, the formerly supportive director of the renamed FDA Center for Veterinary Medicine (CVM), Lester Crawford, dampened hopes for rapid FDA action. Battling charges of bias (chapter 8), Crawford cautioned that a suspension of AGP-marketing would still necessitate formal evidentiary hearings on drug withdrawals.[165] Other senior officials were equally hesitant to endorse CDC AMR warnings. Remarkably, the USDA's Donald L. Houston claimed that the USDA had no position on AMR.[166] Gore responded: "I understand why USDA is reluctant to get

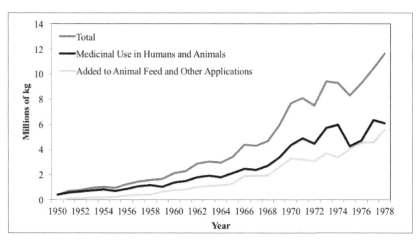

FIGURE 10.1 The 1980 NAS review compiled the first historical antibiotic consumption statistics for the US (excluding sulfonamides). "The Effects on Human Health of Subtherapeutic Use of Antimicrobials in Animal Feeds" (Washington, DC: NAS, 1980).

involved in this. . . . I understand the political pressures that are brought to bear. Believe me, I do."[167] While senior officials failed to support AGP withdrawals, expert witnesses cautioned that Britain's partial AGP bans had failed to reduce AMR or antibiotic use. Having recently traveled to London, one expert noted: "The recommendation of the Swann committee was held with—and actually laughed at there—as being grossly ineffective."[168]

Following public hearings in early 1985, FDA commissioner Frank Young did not find that AGPs' selection for AMR in *Salmonella* posed an imminent harm to health. In October 1985, the FDA recommended rejecting the NRDC petition to HHS secretary Margaret Heckler, who formally rejected the petition in November.[169] Although Donald Kennedy's 1977 notice of hearings formally remained in place, Heckler's announcement effectively ended FDA-led statutory proceedings against AGPs.[170]

Regulation in a Time of Deregulation

Despite cyclical pressure from concerned scientists, the CDC, and individual Democrats, the next ten years would see US antibiotic politics stagnate. In an age of deregulation, there was little political support for new antibiotic rules— or a coordinated defense of existing regulations. Tasked with administering the ongoing status quo, FDA officials focused limited resources on fire brigade responses to microbial and chemical food safety scandals.

The stagnation of official antibiotic reform took place against a backdrop of worsening drug compliance. In December 1985, Democrat Ted Weiss's

Intergovernmental Relations and Human Resources Subcommittee published a scathing review of FDA medicated feed oversight. Internal FDA estimates indicated that "as many as 90 percent or more of the 20,000 to 30,000 new animal drugs estimated to be on the market"[171] had not been approved as safe and effective. Noncompliance encompassed the entire agribusiness community: "Illegal veterinary drug sales are of such magnitude and pervasiveness that they threaten the 'credibility of the veterinary drug approval and regulatory process.'" On one two-week road trip in Iowa, an FDA investigator was able to make 40 illegal buys out of 43 attempts.[172] According to one CVM official, growing political pressure for rapid drug approvals was also leading to rash FDA licensing.[173]

Although a major corruption scandal, supreme court rulings, and a blue-ribbon review allowed the FDA to shake off some of its 1980s shackles, drug compliance and residue problems continued to haunt the agency during the 1990s.[174] Appointed in August 1990, FDA commissioner David Kessler had to contain the emerging scandal about allegedly carcinogenic sulfamethazine residues in milk (chapter 8). The scandal focused public and congressional attention on the joint problem of insensitive drug assays, inadequate official enforcement of rules, and increasing extra-label drug use by veterinarians. In Congress, the already familiar Ted Weiss accused FDA officials of not "really keeping up with scientific advances" and not being "diligent enough nor aggressive enough in pursuing those people who use and prescribe those drugs, unlawfully."[175] In 1992, the US Government Accountability Office (GAO) and external experts similarly warned that monitoring programs for milk were unable to detect many residues: "Fifty-three drugs have been approved by FDA for use in dairy animals; 25 drugs have been reported to be used in an extra-label fashion. Only a small number of drugs are looked for in the milk supply; of these, only 6 drugs have confirmatory procedures."[176]

FDA officials responded by promising improved monitoring but also defended the large US market for over-the-counter drugs and existing extra-label drug guidance: "[The 1968 Animal Drug Amendment] required us . . . to make drugs available to lay persons, if adequate directions for use could be written. So 80 percent of the therapeutic drugs for use in food-producing animals are legally sold over the counter."[177] A voluntary FDA guidance from 1984 dealt with extra-label drug use: veterinarians were advised to prescribe and sell drugs for nonlicensed applications only if an animal was suffering or its life was threatened. Clients should be informed regarding responsible drug use and withdrawal times.[178]

Food safety and residue problems were not limited to milk. Following the 1993 Jack-in-the-Box outbreak of resistant *E. coli* (chapter 8), Washington's narrow focus on drug residues appeared increasingly outdated. In 1994, a GAO report pressed for new risk-based microbial inspections at neuralgic points of the food chain.[179] Shortly afterward, a second GAO report criticized the USDA's National Residue Program for meat. Random monitoring was not random and there was no appropriate risk ranking for priority testing of up to two-thirds of "367 compounds already identified as being of potential concern."[180] The GAO also criticized FDA enforcement. Between 1989 and 1992, the FDA had only investigated about 20 percent of 21,439 reported residue violations. Only one prosecution, 12 injunctions, and 383 warning letters had resulted from investigations.[181]

Passed by Congress in 1994, the Animal Medicinal Drug Use Clarification Act (AMDUCA) attempted to solve compliance and residue problems by defining rules for extra-label drug use within a formal veterinarian-client-patient relationship. Extra-label prescriptions were permitted when the health and life of an animal were threatened or when the animal was suffering. Extra-label use was not allowed to result in dangerous residues or residues above legal tolerances. In a significant move, compounding of extra-label prescriptions in feed was banned and a limited number of drugs like chloramphenicol and the nitrofurans were excluded from extra-label use altogether.[182] However, the tightening of compounding rules proved short lived. Following industry lobbying, the fittingly named 1996 Animal Drug Availability Act (ADAA) loosened recent restrictions by creating the new legal category of Veterinary Feed Directive (VFD) drugs. VFDs nominally increased veterinary supervision over new animal drugs but also relegalized the compounding of extra-label drugs in commercial feed mills (chapter 9).[183] Foreshadowing future US responses to AMR, attempts to impose rules on the veterinary drug market had been thwarted by replacing statutory bans with supervision by often industry-employed veterinarians.

Problems with US drug enforcement and food poisoning were paralleled by ongoing scientific warnings about agricultural AMR selection. In 1987, the FDA had financed further AMR research. Narrowly briefed with focusing on subtherapeutic tetracycline and penicillin AGPs' effects on foodborne pathogens and human health, the resulting report had, however, been able to do little more than call for more research.[184] External pressure for more aggressive action increased during the 1990s. Commissioned by Congress to assess AMR in 1994,[185] the Office of Technology Assessment had already supported the FDA's 1977 AGP bans and its advisory panel included renowned critical microbiologists like Stuart Levy and Nobel-laureate Joshua Lederberg. The 1995 OTA report identified AMR as a serious economic and political hazard. Although studies showing "a direct connection between agricultural use of antibiotics and human illness or death" remained "sparse and difficult to obtain,"[186] the report noted that agricultural antibiotic use could turn farmers into carriers of resistant bacteria and select for environmental AMR. Over 40 percent of the US population already harbored resistant bacteria in their colons. These could cause harm following ingestion of antibiotics for other reasons. Concerned about a further rise of AMR, OTA experts cited CDC warnings against licensing reserve antibiotics like the fluoroquinolones for agricultural use.[187]

However, in a sign of prevailing regulation hostility, calls to increase drug availability proved more powerful than OTA warnings. Facing a concerted agro-industrial campaign to roll back regulatory barriers for farm chemicals, the FDA ignored warnings and licensed sarofloxacin and enrofloxacin (Baytril) in poultry in 1995 and 1996 (chapters 8 and 9). With little public momentum for wider antibiotic reform, American officials felt unable to refuse industry applications highlighting agricultural antibiotics' short-term economic benefits despite the more significant long-term costs of AMR selection.

The Era of Voluntarism

Ongoing US deregulation contrasted with a concerted push for antibiotic restrictions in Europe (chapter 13). Responding to EU policies and AMR warnings, US regulators attempted to improve antibiotic stewardship with the help of voluntary industry compliance. This voluntarist strategy yielded mixed results.

In November 1998, EU AGP bans and domestic warnings about rising fluoroquinolone resistance made FDA regulators propose a cautious framework for "evaluating and assuring the human safety of the microbial effects of antimicrobial New Animal Drugs."[188] Industry applicants for new antibiotic licenses would have to submit data on products' effects on animals' gut flora and human exposure to resistant bacteria, R-factors, and pathogens.[189] Resistance-prone drugs would not be licensed, and licensed drugs would be evaluated via post-licensing surveillance. Critics, however, bemoaned that FDA proposals would not necessarily apply to previously licensed drugs and that post-licensing surveillance had already failed to trigger regulatory action (chapter 8).[190]

Established in 1996, the National Antimicrobial Resistance Monitoring System (NARMS) tested *Salmonella* (1996), *Campylobacter* (1997), *E. coli* and *Enterococci* (2000) as sentinel organisms for susceptibility to seventeen antibiotics.[191] NARMS relied on interagency collaboration. The FDA and CDC monitored retail level isolates and the USDA monitored isolates collected at slaughter, from sick animals, and from a limited number of healthy farm animals.[192] The number of NARMS isolates (about 4,000 in 1998) was, however, insufficient to detect slight changes in antimicrobial susceptibility or to identify rare but important resistant phenotypes. Agricultural AMR monitoring was moreover accused of prioritizing poultry isolates and over-relying on passive surveillance.[193] Even if NARMS revealed emerging AMR problems, complicated drug withdrawal procedures meant that official action to reduce selection pressure was slow and uncertain. Despite acknowledging clear AMR warnings, the FDA licensed fluoroquinolone use for beef cattle in 1998 and needed five years to withdraw the drugs after eventually launching formal procedures in 2000.[194]

There was little hope that statutory drug withdrawals would become easier in the near future. Despite supportive WHO reports, congressional bills aiming to restrict AGPs failed to gain support between 1999 and 2003. Meanwhile, federal stewardship initiatives either ignored or proposed voluntary solutions for nonmedical antibiotic use.[195] Instead of pursuing new uncertain statutory drug withdrawals, FDA officials reacted to past defeats and the Baytril experience by trying to control agricultural antibiotic consumption via a mix of voluntarism, educational programs, and updated licensing procedures. In 2003, the FDA released voluntary Guidance #152 for "Evaluating the Safety of Antimicrobial New Animal Drugs with Regard to Their Microbiological Effects on Bacteria of Human Health Concern." Based on the 1998 framework, Guidance #152 instructed licensing applicants for animal drugs on preparing an AMR hazard characterization of their product. Officials and producers would next co-develop an official risk assessment and the FDA would finally determine appropriate conditions of use.[196] Success was limited: evaluated by the GAO in 2004, the new licensing and surveillance framework failed to address sales of older antibiotics and did not monitor actual antibiotic use. Voluntary measures' impact on AMR was uncertain.[197]

More decisive reforms seemed likely following the 2008 election of the Obama administration and the introduction of the Preservation of Antibiotics for Medical Treatment Act (H.R. 962).[198] Politically, the context for antibiotic restrictions had not been so promising since the 1970s: a Democrat president could count on the support of two Democrat-controlled houses of Congress. Antibiotic restrictions were also endorsed by a 2008 PEW report on Industrial Farm Animal Production, which recommended a licensing stop and phasing out of nontherapeutic antibiotics as well as significant improvements of federal AMR surveillance and the official collection of antibiotic sales data.[199] Congress subsequently required drug sponsors to submit an annual report for each approved antimicrobial drug sold or distributed for use in food-producing animals. From 2010 onwards, officials would know how much antibiotics were being used on US farms.[200] In 2009, updated versions of PAMTA (H.R. 1549; S. 619) contained clauses that would have required US authorities to withdraw, within two years, the approval of any "nontherapeutic use" in food-producing animals of a "critical antimicrobial animal drug"—a category that encompassed any kind of penicillin, tetracycline, macrolide, lincosamide, streptogramin, aminoglycoside, or sulfonamide.[201]

Scarred by three decades of stalled AGP withdrawals, FDA support for PAMTA was mixed. Speaking to Congress in 2008, CVM director Bernadette

Dunham had been hesitant about EU-style AGP bans.[202] One year later, deputy commissioner Joshua Sharfstein stated, "both [commissioner Margaret Hamburg] and I strongly support action to limit the unnecessary use of antibiotics in animals to protect the public health."[203] For Sharfstein, it was clear that "the use of antimicrobials should be limited to those situations where human and animal health are protected"[204]—but not limited further. Sharfstein also called for a restriction of antibiotic use to situations where there was veterinary supervision, evidence of efficacy, and no "reasonable alternative."[205] Although the FDA supported PAMTA's aim of AGP restrictions, it did not explicitly endorse it but instead called on Congress to provide relief for the underlying statutory problem of "burdensome"[206] drug withdrawals.

When this statutory relief did not prove forthcoming, FDA regulators deemed the burden of mass AGP withdrawals too great and broke with PAMTA. Ignoring protest in the scientific and liberal media, FDA officials proposed alternative voluntary policies in June 2010 (chapter 8). A voluntary draft guidance had been co-developed with the USDA to reduce consumption of medically relevant antibiotics on farms but avoid statutory restrictions of "judicious" antibiotic use.[207] Similar to the 1996 abandonment of extra-label compounding restrictions, formal feed restrictions were circumvented by placing veterinarians in charge of both nontherapeutic and therapeutic antibiotic use. In 2011, the FDA also officially dropped Donald Kennedy's 1977 proposal to withdraw approvals for penicillin and tetracycline AGPs.[208] One year later, it formally released Guidance for Industry #209 (GFI #209) on the "Judicious Use of Medically Important Antimicrobial Drugs in Food-Producing Animals."[209] GFI #209 extended the voluntary pre-licensing principles of GFI #152: already licensed products containing medically important antibiotics should now be limited to uses "necessary for assuring animal health" under "veterinary oversight or consultation."[210] Following legal battles (chapter 8), the FDA released Guidance for Industry #213 (GFI #213) in December 2013. GFI #213 recommended voluntary label changes for medicated feeds and water so that products could be used only under veterinary supervision. Claims for increased weight gain or feed efficiency were no longer considered suitable.[211] Officials threatened statutory action if, after three years, "we determine that adequate progress has not been made."[212] All FDA Guidances were to be fully phased in by 2017.[213]

It was at times easy to forget that—in contrast to PAMTA—FDA reforms were voluntary. Similar to the 1969 Swann bans, some "medically irrelevant" AGPs would also remain available over the counter while access to restricted AGPs was possible via veterinary prescription. Speaking at London's Chatham House in 2014, commissioner Hamburg nonetheless claimed that FDA Guidances were a success: twenty-six pharmaceutical companies were in the process of revising labels and thirty individual over-the-counter preparations had already been withdrawn. Aware of considerable skepticism in the audience, Hamburg presented FDA voluntarism as a pragmatic response to decades of regulatory defeat: "Experience has shown us that this in fact is the quickest, most efficient way to reach our collective goal—considerably faster than a mandatory ban that would have required dozens of individual legal proceedings on each product."[214] Trusting in industry cooperation and reinvigorated antibiotic development, Hamburg, however, remained vague on how the FDA would ensure compliance and how therapeutic antibiotic use could be reformed in the absence of uniform, statutory regulations.

The impacts of FDA voluntarism have been mixed. In September 2014, journalists reviewed over 320 "feed tickets" detailing practices in Tyson Foods, Pilgrim's Pride, Perdue Farms, George's, and Koch Foods: "antibiotics were given as standard practice over most of the life of the chickens . . . the doses were at the low levels that scientists say are especially conducive to the growth of so-called superbugs."[215] Continuous "prophylactic" use of former AGPs also occurred in the US cattle industry.[216] Senior politicians pressed for further reforms. In December 2014, Democrat Senators Elizabeth Warren, Kirsten Gillibrand, and Dianne Feinstein warned "that the FDA may lack the authority to ensure veterinarians adhere to the criteria for determining an appropriate preventive use . . . , that the FDA does not have a clear mechanism for collecting the data necessary to evaluate whether its policies effectively reduce the public health threat, and that the administration has no clear metrics or benchmarks that will be used to determine success or a need for future action."[217]

Establishing these benchmarks is not easy. In the case of AMR, US data collection remains patchy. In 2014, the CDC estimated that more than 2 million Americans annually fall ill with resistant infections, with 23,000 resulting fatalities. Excess associated health care costs range between $20 billion and $35 billion per year (2008 US dollars). On a ranking from urgent to serious and concerning threats, AMR in agriculture-associated nontyphoidal *Salmonella*, *Campylobacter*, extended spectrum beta-lactamase (ESBL) producing *E. coli* and *Klebsiella*, as well as vancomycin-resistant *Enterococcus faecium* (VRE) were listed as serious. According to Laura Kahn, US health care costs resulting from resistant nontyphoidal *Salmonella* and *Campylobacter* alone could amount to $2.6 billion.[218] However, gathering more fine-grained AMR data and assessing the microbial effects of voluntary AGP bans is difficult. Despite NARMS, US officials have struggled to track the sources of resistant organisms and genes in the field. Surveillance resources are limited. In 2015, the FDA, USDA, and CDC discussed collecting more on-farm AMR data. However, the FDA received none of the $7.1 million it requested to study AMR in animals in the fiscal year of 2016.[219]

In contrast to AMR, the establishment of mandatory sales reporting in 2008 has made it far easier to measure antibiotic consumption trends on US farms.

At the point of writing, published FDA data covers the period between 2009 and 2017. After consistently rising to 15,577,940 kilograms in 2015, the amount of antimicrobial drugs sold for use in food-producing animals declined by almost 33 percent to 10,933,367 kilograms in 2017. This is an impressive shift of consumption. In contrast to previous years, only 51 percent of sold antibiotics were considered medically important, with cattle (42 percent) and swine production (36 percent) consuming the majority of these antibiotic classes. Driven by industry-led change, FDA guidances, and shifting consumer habits, reductions of antibiotic use have been particularly impressive in the US poultry sector. However, certain caveats remain concerning this data. Because of the removal of production indications from feed labels in 2017, all sales of medically important drugs were officially recorded as being for therapeutic indications only. This is an unlikely increase from 2016 when only 31 percent of medically important antibiotics were sold for therapeutic indications and likely masks ongoing nontherapeutic antibiotic use on farms.[220] Prescribed extra-label drug use is also not adequately captured in current statistics.

Despite the substantial and commendable drop of American antimicrobial usage to levels last seen during the early 2000s, it not only remains to be seen how sustained reductions will be but also whether they will be sufficient to curb AMR levels. Reliable benchmarks for regulatory success remain rare. It is equally unclear how American officials will react if overuse of prescription-only antibiotics emerges as a problem—as occurred after European AGP bans (chapter 13). In 2019, the *New York Times* reported that drug producer Elanco was urging US pig producers to administer antibiotics daily for prophylactic purposes.[221] In contrast to AGPs, controlling veterinary prescription practices without recourse to statutory interventions will be difficult.

Fifty-three years after the FDA ad hoc committee first debated horizontal gene transfer, US regulators are thus still struggling to develop a coherent approach to agricultural AMR selection. During the late 1960s, limited resources, bureaucratic turmoil, and the prioritization of residues slowed FDA responses to "infective" AMR. Regulators only cautiously began to push for narrow, Swann-inspired AGP restrictions during the 1970s. The highpoint of this campaign was Donald Kennedy's 1977 attempt to statutorily withdraw tetracycline and penicillin AGPs. Similar to previous efforts, Kennedy's bans were, however, checked by a mix of counter science and cost-benefit assessments, congressional interference, public regulation wariness, and—perhaps most importantly—proof-of-harm requirements for drug withdrawals, which were unsuited to nonlinear AMR hazards. Another problem was the narrow nature of proposed AGP restrictions, which had already failed to solve problems in Britain. After the defeat of the 1984 NRDC petition, regulators embraced a strategy of voluntarism to curb AMR. Over the past thirty years, this strategy has produced mixed outcomes. Between 1985 and 2015, antibiotic consumption increased significantly while Washington's focus on deregulation led to the licensing of reserve antibiotics and extra-label drug use. Although the FDA's 2013 strategy of voluntary AGP withdrawals has helped to significantly reduce total drug consumption, it still fails to address wider therapeutic antibiotic infrastructures. European experience shows that fragmented reforms of anti-biotic use will not curb AMR (chapter 13). Almost half a century after its 1972 task force report, it is worrying that the FDA still feels unable to implement broad, transparent, statutory regulations covering all aspects of agricultural antibiotic use. Leaving guardianship of the antibiotic commons to consumer preferences and commercial interests is a substantial gamble.

Part IV

Britain

From Gluttony to Fear, 1970–2018

This part explores how British antibiotic perceptions and regulations developed after the 1969 Swann report. Chapter 11 reconstructs how the enactment of limited AGP bans fragmented 1970s antibiotic protest until a series of residue and salmonellosis scandals reignited popular demands for antibiotic reform around 1980. Coinciding with rising demand for antibiotic-free food, public pressure for sweeping reforms of intensive animal production and antibiotic use reached fever pitch after the 1996 bovine spongiform encephalitis (BSE) crisis. Between 1998 and 2006, Britain supported EU-led antibiotic reform and the phasing out of remaining AGPs. Chapter 12 shows that British agricultural antibiotic use was not significantly altered by the Swann bans. Although British corporatism reduced tensions between farmers, regulators, and antibiotic critics, battles over antibiotic control continued to be fought between the veterinary and medical communities. Following a partial "greening" of British agriculture during the 1980s, the 1996 bovine spongiform encephalitis (BSE) crisis fragmented agricultural support for AGPs. While farmers and veterinarians still defend routine therapeutic antibiotic use, opposition to EU AGP restrictions was limited. Chapter 13 reconstructs the watering down of many of the Swann report's recommendations during the 1970s. Although Britain enacted AGP bans and exported its "Swann gospel," officials resisted calls for enhanced monitoring of antibiotic use, resistance, and residues. While EEC pressure eventually led to residue monitoring for meat and Britain committed to AGP bans in the wake of BSE, official impulses for statutory reforms of therapeutic antibiotic use were limited until the EU's 2018 decision to phase out prophylactic use.

11

Between Swann
Patriotism and BSE

Antibiotics in
the Public Sphere

This chapter explores the development of public perceptions of agricultural antibiotic use after 1969. During the 1970s, the AMR- and welfare-focused risk episteme that had driven previous antibiotic reform fractured and gave way to a complacent sense of "Swann patriotism." It was only around 1980 that residue scandals, resistant salmonellosis outbreaks, and concerted reforms in other European countries triggered renewed societal pressure for antibiotic reform. Concerns about agricultural AMR selection were heightened by medical reports about mystery diseases like AIDS and failing reserve antibiotics. Similar to the United States, consumer fears also drove the rapid expansion of antibiotic-free market niches. As a result of the moral panic triggered by the cataclysmic 1996 bovine spongiform encephalopathy (BSE) crisis, antibiotic bans reemerged as a common denominator of national reform calls. Ensuing European AGP restrictions were widely praised in the British media. The persistent focus on nontherapeutic AGPs, however, means that there has been considerably less public pressure for statutory reforms of allegedly therapeutic antibiotic use.

Fragmented Pressure

After the publication of the Swann report in late 1969 and ensuing bans of penicillin and tetracycline AGPs, the unified risk episteme driving late 1960s British antibiotic reform fragmented and gave way to a new form of Swann patriotism. Whereas antibiotics had previously functioned as a common denominator of environmentalist, ethical, public health, and consumer concerns, framings of antibiotic risk once again split along the familiar lines of residues, AMR, and animal welfare.

Fragmented antibiotic concerns were soon displaced by other popular issues. Similar to the United States, the early 1970s were characterized by an explosion of new activism. Dismissive of the "'softly-softly' reformism of the 1960s,"[1] a younger generation of activists often operated outside traditional structures and favored symbolic protest designed to provoke media interest. Although problems relating to modern food production featured prominently, single-issue campaigning fragmented activist agendas.[2] Concerns about limited resources, overpopulation, and technology-focused Western growth models were amplified by a number of British bestsellers like Barbara Ward's *Spaceship Earth* (1966), the *Ecologist's Blueprint for Survival* (1972), and *Small Is Beautiful* (1973). Arguing for a fundamental ecological reform of politics, all three books focused on new models of agriculture but mostly glossed over hazards relating to agricultural antibiotic use.

Diminished interest in antibiotic risk was also apparent in media reporting. Although they defended the Swann report against attacks from US pharmaceutical manufacturers (chapter 13), British newspapers printed only a small number of articles addressing agricultural antibiotic use.[3] Many of these articles expressed national pride in the pioneering nature of the Swann report. Similar to what historian Frank Uekötter has described as "green patriotism" in the case of post-1970s German environmentalism,[4] this Swann patriotism could lead to complacency about the actual state of affairs. Reporting in the *Times* is a good example for complacent Swann patriotism: during the early 1970s, the newspaper warned about AMR selection in human medicine and laboratories and lambasted countries like Ireland or Mexico for failing to implement Swann-style legislation.[5] When typhoid with plasmid-mediated chloramphenicol resistance emerged in India and Mexico in 1972, the *Times* condemned "indiscriminate" Mexican antibiotic use and reminded readers that "few other countries"[6] had introduced Swann standards. The newspaper, however, forgot to mention that prescribed chloramphenicol use remained legal in Britain; it remained sanguine when veterinarians challenged plans to limit its use; and it continued to print advertisements for both human and agricultural antibiotics.[7]

Popular Swann patriotism stood in contrast to reports indicating that British AGP bans were failing to curb either antibiotic consumption or AMR.

Published in *Nature* in 1975, a study of British pigs by Herbert Williams Smith showed that all animals carried resistant organisms with many isolates proving resistant to restricted tetracyclines.[8] Meanwhile, an article in the *BMJ* attacked ongoing "indiscriminate"[9] antibiotic use on farms. In 1976, former Netherthorpe committee member and antibiotic supporter Raphael Braude claimed that the Swann report's "only positive achievement" was its removal of "public anxiety."[10] According to Braude, a temporary post-Swann drop of antibiotic consumption had been reversed by increased use of medically irrelevant nontherapeutic AGPs and lucrative higher-dosed veterinary antibiotic prescriptions. Although his numbers have to be taken with a grain of salt, Braude estimated that prescribed post-Swann therapeutic antibiotic use had annually increased by about 15 percent between 1973 and 1975.[11] Other observers warned that a lucrative black market had sprung up to meet ongoing antibiotic demand on British farms.[12]

Although there were numerous signs that the 1969 Swann report had failed to "fix" British agricultural antibiotic use or AMR hazards, interest in seeing these signs was limited not only among antibiotic supporters (chapter 12), but also among former critics. In contrast to 1960s reporting on systemic AMR hazards, 1970s journalists instead focused on more limited lifestyle-related risk scenarios involving antibiotic residues in food and the ethical implications of eating antibiotic-produced meat. Writing for the *Times* in 1974, French journalist Josée Doyère warned about the dangers of illegal "hawking"[13] of feed additives and hoped for common EEC residue limits. One year later, the British Consumers' Committee called for improved residue monitoring of British milk. Committee members were concerned that official tests were too slow to stop drug-contaminated milk from being sold.[14] In contrast to purity-focused US reporting (chapter 8), British journalists often linked warnings about antibiotic residues with wider ethical concerns about conventional livestock farming. With media interest in animal welfare heightened by contemporary conflicts about new concepts of animal rights,[15] articles on residues also blamed antibiotics for enabling "unnatural development"[16] on factory farms, whose products were held to be both morally and physically tainted.

Nightmarish popular visions of factory farming played an important role in stimulating interest in food that was both "pure" and ethical. Despite *Daily Mail* articles mocking eccentric health nuts and hippies,[17] the 1970s saw an increasing number of British stores, restaurants, cookbooks, and handbooks promote "natural" and ethical food. In 1971, even the Good Housekeeping Institute publishing its own *Wholefoods Cook Book*.[18] The absence of antibiotics and factory methods in animal rearing was usually viewed as a precondition for "natural" and organic food.[19] Although it also mentioned transferable AMR selection, Doris Grant's *Your Daily Food—Recipe for Survival* (1973) primarily described antibiotics as part of a "chemical flood"[20] that was poisoning

British consumer with invisible residues and enabling production on ethically "desensitizing" factory farms. Printed in 1975, the *Eco Cookbook* of Friends of the Earth expressed similar concerns about residues and the "horrors of factory farming": "In the process of intensive feeding, animals take in a great amount of hormones, antibiotics, and other chemicals; they become concentrated in the meat, since animals have no way to pass such substances out of their bodies."[21]

With different publics once again focusing on different aspects of antibiotic risk, the broad 1960s consensus on antibiotic reform had come to an end. In the absence of a unified risk episteme, fragmenting popular risk perceptions failed to create sufficient momentum for renewed societal action around antibiotics.

Reemergent Crises

The post-Swann complacency about British antibiotic use on farms only began to ebb around 1980 when a long series of residue- and AMR-related scandals began to undermine trust in existing regulations. Facilitated by the growing bipartisan endorsement of green values, antibiotic reform gradually reemerged as a prominent issue in the public arena.

In May 1979, the cover of the *Radio Times* showed a friendly piglet lying on straw. While the headline asked, "Should this little piggy go to market?" a second caption read: "Health Warning. Meat and Poultry May Seriously Affect Your Health."[22] The health warning referred to a popular BBC program called *Brass Tacks*, whose upcoming episode was titled "It Shouldn't Happen to a Pig."[23] Asking "whether it is time to choose between safe meat and cheap meat,"[24] *Brass Tacks* featured a Pharmaceutical Society spokesman, who claimed: "there is a substantial black market involving at least £500,000 worth of antibiotics, compared with the estimated £20 million worth used by farmers each year."[25] Three months later, the Government Chemist's annual report seemed to confirm *Brass Tacks*' allegations. According to the *Guardian*'s Anthony Tucker, "itinerant 'con men'"[26] were endangering public health. Often operating out of plain vans, dealers sold pharmaceuticals with forged brand labels. Using mislabeled drugs could result in animals' death, residues in meat, and AMR. In 1978, antibiotics including chloramphenicol had been found in two-thirds of 350 confiscated samples of illegal merchandise. *Guardian* warnings were echoed in the conservative *Spectator*. The magazine accused agricultural officials and veterinarians of turning a blind eye to antibiotic abuse and linked problems to a contemporary surge of resistant salmonellosis: "The abuse of these 'miracle' drugs by farmers, vets and pharmaceutical salesmen is an obvious danger to everyone's health; and, knowing the Ministry mind, it is remarkable that such abuse has not already been made a criminal offence. Perhaps it would comfort the relatives of the next victims if the gravestones could be marked with the words. 'He perished to make British farming more efficient.'"[27]

Over the next years, black market scandals involving antibiotics continued to occur with worrying regularity. In a 1983 interview for the *Daily Mirror*, the head of the Pharmaceutical Society's law department, Gordon Applebe, described the challenges of monitoring Britain's black market with only twenty inspectors and twelve additional staff from MAFF. In Applebe's opinion, British authorities were "probably only scratching the surface of the problem."[28] In 1984, the *Guardian* estimated that the British pharmaceutical black market was worth about £3 million, with the bulk of supplies coming from Ireland.[29] One year later, it emerged that British veterinarians were involved in illegal marketing too. Suspected of extending from the West Country to Cheshire, a drugs ring was accused of flooding farms with "illegal supplies of antibiotics amounting to more than £1,000 a week."[30] According to the Pharmaceutical Society, stopping the drugs ring was the "biggest operation in the society's 140-year history."[31]

Scandals about illegal antibiotic sales occurred parallel to growing concerns about resistant gram-negative infections and heightened tensions between British physicians and veterinarians.[32] In 1980, a *BMJ* paper by PHLS microbiologist Eric John Threlfall analyzed the spread of multi-resistant *S. typhimurium* types 204 and 193 from cattle to humans. In 1979, the strains had caused 290 cases of salmonellosis in Britain and killed an elderly patient and a three-year-old. According to Threlfall, the rise of types 204 and 193 had been facilitated by agricultural antibiotic use. Pharmaceutical advertisements for prescription-only drugs had increased agricultural demand for veterinary prescriptions of therapeutic antibiotics, which had in turn selected for AMR in salmonella. Focusing only on AGPs, "current regulations have failed."[33] An anonymous *BMJ* editorial reinforced these claims and blamed "over-enthusiastic representatives of pharmaceutical firms," "black market operators,"[34] and farmers for resistant salmonellosis. Swann was doomed to fail because it had left British veterinarians and large parts of the antibiotics sales infrastructure unregulated. Unsurprisingly, the *BMJ* articles provoked angry reactions. In addition to veterinary attacks on antibiotic overuse in human medicine (chapter 12), Herbert Williams Smith complained that the Swann report could not be blamed for salmonellosis in bovines since feeding antibiotics to them had never been permitted. As the report's adoption by "many other countries" showed, Britain should "take some pleasure in having initiated it."[35]

Such pride in the Swann bans seemed increasingly out of place during a time of rising concerns about antibiotic resistant threats to British food security. In October 1981, Bernard Rowe, director of the PHLS Enteric Pathogens Division, warned that Britain was threatened by a "food super germ."[36] *S. typhimurium* type 204 had reached "a disgraceful level of drug resistance."[37]

According to the *Times*, notified British salmonellosis cases rose from 10,000 to 17,000 between 1977 and 1983 with resulting deaths rising from 25 to 65 between 1972 and 1982.[38] In public discourse, fears of antibiotic resistant food poisoning, lacking hygiene, and drug residues fused into general uncertainty about British food safety. While the *Telegraph* warned about microbial "death risk[s]"[39] posed by salmonella-tainted hospital food, the *Daily Mail* reported that a police investigation codenamed Operation Meathook had found up to 100 tons of contaminated meat being used to produce convenience food per week. Using "devious plans and odious skullduggery," dealers were selling sick cows pumped full of antibiotics: "only when the animal is killed later that night will the buyer realise what has happened. Often he will not admit what has happened to the meat inspector, but will push it through with the rest."[40] By 1985, the *Guardian* noted: "food additives and residues from pesticides, hormones and antibiotics now rival AIDS as the number one health issue."[41] One year later, Jan Walsh's *The Meat Machine* blamed antibiotic overuse and AMR on "the unnatural conditions of intensive rearing units" where disease spread "like wildfire."[42]

Unsurprisingly, 1980s food security concerns further boosted public interest in allegedly safe and pure food. In 1981, the *Observer* noted that the organic movement was fast discarding its image of "dirndl, beads, sandals and an atmosphere of folkloric guitar strumming"[43] in favor of professional marketing and sales networks. The opening of new organic venues was commented on in the national media. In 1984, the *Daily Mail* published an enthusiastic review of an affordable cash-and-carry center for organic produce.[44] Similar reports on Wholefood Butchers and other organic vendors appeared in the *Guardian*.[45] In British book stores, the message of safe, pure, and ethical food was promoted by a new wave of guides and cookery books ranging from *The Cranks Recipe Book* to publications by Britain's Women's Institute.[46] According to *Good Housekeeping*, organic wholefoods were "not just an upper-class fad . . . wholefood shops are becoming the norm in local high streets and shopping centres and even supermarkets are more aware of the public demand for this type of food."[47] Organic food also received celebrity endorsement from Andrew Lloyd Webber, Prince Charles, and Paul Eddington, star of the popular TV series *Yes, Prime Minister*, who delivered organic food to Margaret Thatcher in 1990.[48]

Similar to the United States, the amount of public attention paid to alternative agriculture bore no resemblance to actual sales of organic food. Although all sectors of organic produce grew rapidly during the 1980s and a survey found that 72 percent of consumers were willing to pay more for organic food, organic sales accounted for only 1 percent of overall British food sales in 1989.[49] This does not mean that media interest in organic food was a mere hype. Instead, 1980s organic reporting was indicative of an important gradual shift of

Britain's wider public risk episteme with regard to *conventional* agriculture. Although only small parts of the public were willing to spend money on selected organic goods, a majority of consumers and media commentators began to expect that conventional agriculture should also abandon unpopular practices and become greener.

Reporting in the *Times* is indicative of this parallelism between organic endorsements and reform demands for conventional agriculture. Starting in the mid-1980s, the newspaper joined the chorus of organic praise after breaking with over thirty years of neo-Malthusian overpopulation warnings (chapter 2). In 1985, agricultural correspondent John Young claimed that an alliance of "doom-mongers"[50] had exaggerated Malthusian scenarios. According to the UN World Food Council, population growth had not outpaced cereal production. Global hunger was "political, not economic."[51] Following this conceptual shift away from productivity-oriented neo-Malthusian theories, *Times* articles not only began to praise organic agriculture but also intensified criticism of conventional technologies like agricultural antibiotics.[52] In 1986, the newspaper printed a positive review of Peter Cox's *Why You Don't Need Meat*.[53] According to the Vegetarian Society's former chief executive, antibiotics had changed the way British agriculture worked. An anonymous veterinarian complained: "once vets were people who looked after the well-being of animals . . . . But now, we just suppress the disease until it's time for the animal to be killed."[54] Focusing on AMR, the *Times* also printed a three-part series on "The Global Overdose"[55] in 1987. Titled "The Bitter Harvest,"[56] the third part of the series showed a piggy bank filled with pills. The newspaper accused physicians and veterinarians of shunning responsibility for AMR rather than jointly improving antibiotic stewardship.

The steady rise of concerns about antibiotic abuse on British farms peaked between 1987 and 1989. In 1987, the *Guardian* reported that British farmers were adding the enzyme penicillinase to milk to obscure illegal penicillin residues. Hard hit by recent EEC quota cuts, a West Country farmer noted, "I can't afford to throw away a 250 gallon tank of milk at 80p a gallon. Enough penicillinase to neutralise the problem only costs me £8."[57] Although the NFU and MMB claimed that penicillinase was harmless, Joe Collier, a pharmacologist at St George's Hospital in London, warned that neutralized penicillin could still trigger allergic reactions.[58] Illegal residues were not only detected in milk. In January 1988, health inspectors found antibiotic residues in sixteen of eighty-eight carcasses at a Bradford abattoir.[59] Trying to maintain customer trust, supermarkets Marks & Spencer and Waitrose announced that they would stop buying meat produced with AGPs.[60] In February 1988, the *Daily Mirror* asked, "What has gone wrong with our food laws?"[61] and referred to resistant pathogens and inadequate hygiene controls before calling for a complete review of food controls.

Food safety concerns reached fever pitch levels ten months later. On December 3, 1988, a large-scale outbreak of multiple resistant *Salmonella enteritidis* P4 prompted Junior Health Minister Edwina Currie to warn British TV viewers to avoid "all raw egg products like mayonnaise, home-made ice cream, and even lightly cooked eggs."[62] Reacting to Currie's announcement, Richard Lacey from the government's Veterinary Products Committee confirmed 450 recent cases of *S. enteritidis*–induced food poisoning.[63] However, following another televised warning by Currie, Lacey corrected the number to approximately 3,000 infections with one resulting fatality every week.[64] Overall, the number of salmonella infections had increased by 152 percent over 1987.[65] The substantial number of antibiotic-resistant infections prompted a national debate on food controls and agricultural drug use.[66] In Parliament, Sir Richard Body, Conservative MP for Holland with Boston deplored "the over-use of antibiotic drugs in hatcheries"[67] and unsupervised antibiotic use by farmers. With egg sales dropping nearly 15 percent ahead of Christmas, the NFU threatened to sue Currie for her alleged alarmism. Dubbed "Eggwina" by the media, industry pressure forced Currie to resign on December 16, 1988.[68]

Currie's resignation did little to curb the mounting wave of public alarm about chemical and microbial food hazards. Although the Thatcher administration promised a new Food Safety Act,[69] media reports on "food danger[s] from 'barbaric' factory farms"[70] continued to appear. In April 1989, the head of the Government's Institute for Food Research warned that "food poisoning in Britain is out of control."[71] With articles titled "Not Even Fit for Our Pigs"[72] and "Cages of Cruelty,"[73] the *Daily Mirror* launched a series of attacks on the methods of modern intensive livestock production : "[animals] are born and reared in the dark and the dirt. They are pumped full of hormones and antibiotics. . . . They are the next potential food poisoning timebomb. . . . And no Tory Government has dared to take on its masters, the agriculture lobby."[74] Food poisoning statistics added to public unease. Despite the culling of infected flocks and reformed hygiene protocols, the total number of all confirmed salmonellosis cases rose from 27,478 in 1988 to 29,998 in 1989. Cases caused by multiple-resistant *S. enteritidis* P4 rose by a further 25 percent to 16,151 in 1990.[75] Actual incidence was likely higher. Experts believed that salmonellosis was underreported by as much as 100 percent.[76]

Resistant salmonellosis and a decade of residue and black market scandals dealt a severe blow to Swann patriotism and undermined trust in British antibiotic regulation. Whereas antibiotic-related concerns had been fragmented for much of the 1970s, 1980s media reports revived the specter of not only ethically but also microbially hazardous factory farms. Similar to the 1967 Teesside outbreak (chapters 5 and 7), a major scandal would soon catalyze and fuse concerns about agricultural antibiotics into a new national reform movement.

BSE and Moral Panics

Since the second half of the 1980s, a new disease called bovine spongiform encephalopathy (BSE or mad cow disease) had been causing growing concern among British veterinarians and public health officials. First examined before Christmas 1984 and officially identified in 1986, BSE is believed to be caused by misfolded proteins—so-called prions—that accumulate as plaque fibers in brain tissues and cause death.[77] Significantly, BSE is transmissible to humans in the form of the equally fatal variant Creutzfeldt-Jakob Disease (vCJD). While there are competing theories on the origins of BSE, the disease was probably spread by feeding meat and bone meal to herbivore cattle.[78] In other words, the spread of BSE was inherently linked to the efficiency-focused (re-) processing logic of intensive agriculture and the "factory farm." With officials only slowly addressing growing scientific concerns,[79] the 1996 confirmation of a link between BSE and vCJD caused a "moral panic" in Britain during which charged public reactions triggered agricultural reforms that were much broader than any specific threat posed by BSE. Although they did not cause BSE, antibiotics' preexisting image as a particularly suspect tool of conventional livestock husbandry meant that they were publicly singled out for stringent regulation.[80]

In British popular culture, associations between antibiotic and BSE risks had been growing since the early 1990s. Writing for the *Guardian* in 1990, Lucy Ellmann complained that intensive farming had "given a new meaning to the term, fast food: the cattle themselves grow unnaturally fast on their diet of pig's blood, sheep offal, decaying chickens, chicken shit, hormones and antibiotics.... Writing this piece has given me such a headache.... Oh, what the hell, might as well finish myself off with a chicken sandwich."[81] Public criticism of factory farming not only mixed with contemporary warnings about superbugs (chapter 8) but also with controversies about American companies' planned introduction of the synthetic growth hormone BST and of the genetically modified "Flavr Savr" tomato to British agriculture.[82] Coinciding with public clashes over "frankenfoods" and artificial AMR, Compassion in World Farming organized a well-publicized conference on factory farming in 1992 during which six veterinarians including a former RCVS president and a former MAFF assistant chief veterinary officer criticized intensive farming's disease risks.[83] Although some commentators soon grew tired of the "media hype" surrounding the "same old story with the 'killer bug,'"[84] Hollywood movies, TV series, and bestsellers like *The Hot Zone* further fueled public anxiety about a coming post-antibiotic era.[85]

With AMR fears increasing and trust in conventional livestock production wearing thin, agricultural antibiotics were already re-emerging as a common denominator of different protest camps: physicians blamed agricultural antibiotics for fatal infections; tabloids attacked them for facilitating animal abuse; and organic farmers, who were beginning to sell their produce in major supermarkets, pointed to antibiotics' absence as a marker of quality and safety. In contrast to the United States, this criticism did not follow partisan lines. Similar to the 1960s, antibiotics' status as a common denominator of diverse strands of agricultural criticism made them vulnerable to wider moral panics about British farming—even if problems had nothing to do with residues or AMR.

Such a moral panic occurred on March 20, 1996, when the British government confirmed a possible link between BSE and human vCJD. The following weeks and months saw officials and farmers face unprecedented outrage and embargoes that threatened to destroy Britain's beef sector.[86] Writing for the *Guardian*, Patrick Holden, president of the organic Soil Association, described BSE as "testimony to the breathtaking arrogance of 20th century western agricultural science."[87] Significantly, Holden's subsequent criticism immediately targeted agricultural antibiotics: "When, inevitably, the animals get sick, farmers use antibiotics to prevent infectious diseases taking hold. This is like trying to put a cork in a bottle that is actively fermenting—it cannot possibly work for very long."[88] The *Observer* also interpreted BSE as a systemic failure of conventional agriculture and criticized antibiotics' role in facilitating dangerous practices on factory farms.[89] According to the *Daily Mirror*, Britons had "mad farming disease."[90] BSE was described as the tip of an agro-industrial iceberg kept afloat by antibiotics. Having only recently criticized "Euro-law" for raising the cost of veterinary antibiotic use,[91] the *Daily Mail* now expressed understanding for continental embargoes of British animals "awash in antibiotics, additives and hormones."[92] In the *Times*, Clive Aslet, editor of *Country Life*, noted that BSE justified consumers in turning to antibiotic-free food. From the ruins of BSE, "Britain must build a system of agriculture that is acknowledged as the safest and most humane in the world."[93] Faced with a widespread moral panic about British agriculture and parallel outbreaks of resistant *E. coli* 0157, staphylococci, and salmonellosis,[94] commentators throughout the political spectrum agreed that BSE necessitated antibiotic reform.

Reform

This message was not heard by the Conservative government. Trying to lift the EU boycott of British beef, it resisted contemporary Scandinavian-led EU initiatives to ban AGPs in European agriculture (chapter 13). British officials' defense of alleged national interests not only underestimated public opinion but also failed to convince EU partners.

In December 1996, Britain's representative in the EU's animal food committee referred to lacking evidence of harm before voting against a proposed ban of the popular growth promoter avoparcin. The vote on avoparcin had been filed by EU Commissioner for Agriculture Franz Fischler after European data indicated that the drug could select for resistance against the reserve antibiotic vancomycin (chapter 13). Highlighting Britain's increasing political isolation in European matters of food security, the vote ended fourteen to one against Britain and an EU ban of avoparcin was announced in February 1997.[95]

The restriction of avoparcin was the opening salvo of a new ambitious Scandinavian-led round of EU antibiotic reform. Having banned all AGPs in 1986,[96] Sweden had negotiated a three-year exemption from mandatory antibiotic compliance following its 1995 EU accession. With the three-year exemption about to expire, the Swedish government announced that it would not lift its domestic AGP ban.[97] Linking AMR and BSE hazards, senior Swedish officials actively promoted AGP bans with newspaper articles, letters, and advertisements in other EU member states.[98] Although Swedish letters failed to convince Britain's embattled Conservative Party, they struck a chord with Tony Blair's Labour Party. Winning a landslide victory in May 1997 and mentioning the BSE crisis three times in its election manifesto, the New Labour government was not only keen to avoid further damaging agricultural conflicts within Britain but also eager to regain the trust of its EU partners.[99]

The Blair government's resolve for reform was almost immediately tested when German, Danish, and Finnish ministers expressed grave concerns over AGPs during a meeting of the EU Council of Agriculture Ministers in November 1997.[100] Coinciding with domestic reports on vancomycin resistant MRSA, allegedly pan-resistant *Pseudomonas aeruginosa*, and multiple-resistant foodborne pathogens, precautionary bans of low-dosed growth promoters were widely supported in the British media.[101] In early 1998, a unilateral Danish ban of virginiamycin and a voluntary phasing out of all AGPs until 1999 by Danish livestock associations increased pressure for British support of far-reaching EU restrictions. In Britain, influential reports on antibiotics and AMR by committees from both Houses of Parliament also endorsed bans (chapter 13).[102] Following an international conference in Copenhagen in September 1998, twelve of the EU's fifteen agricultural ministers decided to ban four of the remaining eight AGPs. Two years after voting against the avoparcin ban, Britain now supported much more sweeping AGP bans.[103] Following US precedent, EU governments also agreed to establish a European-wide AMR surveillance system (chapter 13). Sensing a permanent shift of antibiotic politics, the *Financial Times* began to print investment advice for companies producing probiotic substitute feeds.[104]

Although Pfizer's chairman criticized them as a return to "the Dark Ages, where witchcraft and sorcery are prevailing,"[105] the 1998 AGP bans bore testament both to the BSE crisis and to European consumers' increasing power over agricultural policy. By the late 1990s, consumers' ability to vote against controversial production methods was impacting both the organic and conventional sector. Profiting from unified labeling and nearly two decades of food scandals, Britain's organic food sector grew from £40 million in 1987 to £267 million in 1997. Sales were projected to grow to over £1 billion by 2000.[106] Although they still only accounted for about 1 percent of total British food and drink expenditure in 2000,[107] organic products' cultural and indirect economic influence was significant. In the *Times*, radio presenter Libby Purves smugly reflected, "In about 1992, we had John Gummer, then Agriculture Minister . . . laughing charmingly and pooh-poohing our organic attitudes."[108] Seven years later, there were "signs of a genuine popular rebellion against the culture of ghastly farming and ghastly food."[109]

Consumers' rebellion was also transforming the conventional sector. Shocked by BSE, under pressure from popular green values, and trying to stay ahead on the competitive food market, major British supermarkets like Tesco, Sainsbury's, and Asda proactively announced that they were prohibiting suppliers from using AGPs in April 1998—eight months before the EU formally decided to ban them.[110] Other companies followed suit. Probably anticipating further AGP bans, the Grampian Country Food Group, the United Kingdom's biggest chicken producer, announced that it would stop using AGPs in September 1999. Secret trials had indicated that AGP withdrawals would not lead to any price increases.[111] According to the *Guardian,* Grampian's initiative could "signal the biggest revolution in years in the way that animals are reared."[112] Having abandoned its first organic line in the mid-1980s, Marks & Spencer announced that it too would now restock "chemical free" produce and ban all poultry products produced with AGPs. Large multinational food conglomerates like Mars further strengthened the domestic trend toward greener production methods by developing their own antibiotic-free and organic product lines.[113]

Boosted by EU courts' rejection of industry lawsuits against the 1998 antibiotic bans,[114] critics soon began to call for restrictions of the four remaining AGPs. Their calls were strengthened by a series of antibiotic scandals and official reports on AMR. In July 1999, British and Irish authorities cracked down on illegal Irish drug imports. According to the *Sunday Times*, price differences meant that "Irish pharmacies [were] being bombarded with requests for antibiotics from farmers in Britain."[115] The newspaper also referred to a recent BBC *CountryFile* episode in which an undercover team had purchased therapeutic antibiotics over the counter in Britain and via mail order from Ireland. According to the organic Soil Association, "as many as 10,000 farms in Britain"[116] could be using antibiotics illegally. Reacting to a 1999 antibiotics report by Britain's Advisory Committee on the Microbiological Safety of Food (chapter 13), the *Daily Mail* attacked the "grotesque irresponsibility" of intensive farming: "Many farmers have [reacted to the 1998 bans by switching] to another drug, avilamycin. . . . The trouble is that avilamycin is almost identical to a 'vitally important' drug now on trial in Britain's hospitals [Synercid]. . . . The demand for cheap meat is one thing. But if it reduces our ability to fight disease, it may yet prove the most ruinously expensive choice we could have made."[117]

The new millennium did not alter now entrenched public hostility toward AGP use on British farms. In 2000, the *Daily Mail* warned that *Salmonella enteritidis* had "effectively become unbeatable."[118] WHO director Gro Harlem Brundtland warned about a "pre-antibiotic age,"[119] and *Times* commentators reckoned that opinion had turned for good "against destructive industrial farming."[120] Meanwhile, the popularity of antibiotic-free produce continued to grow. While the Soil Association offered "Food for Life" packs to parents concerned about conventional school food, tabloids endorsed organic food and a new generation of celebrity chefs like Jamie Oliver advised using organic—if possible.[121] In politics, Conservative and Labour MPs showed progressive credentials by conspicuously eating antibiotic-free organic produce.[122] Unsurprisingly, there was little protest in 2002 when EU Commissioner for Health and Consumer Protection David Bryne announced that he planned to phase out remaining AGPs by 2006.[123]

Higher-Dosed Problems

While the phasing-out of low-dosed AGPs between 1998 and 2006 fulfilled a major public health demand, reforms of higher-dosed therapeutic antibiotic use remained limited. A significant reason for this was waning public attention. Similar to the early 1970s, the seeming success of pioneering EU AGP bans made antibiotics lose their status as a common denominator of public protest against conventional agriculture. To many commentators, it seemed that the gaps of the 1969 Swann report had now been fixed. This was despite evidence pointing to similar AMR selection by higher dosed products and the fluid boundaries between antibiotic growth promotion, prophylaxis, and therapy on farms (chapter 13).

In a repeat of 1970s Swann patriotism, many British newspapers once again asserted that Britain had done its homework and accused other countries of antibiotic misuse. Although black market and residue scandals continued to occur in Britain,[124] observers increasingly focused on weaker standards in other parts of the world. In 1999, the conservative *Daily Mail* allied with British ministers and the NFU for a "Just Say Non" campaign. The campaign protested French boycotts of British beef and the alleged use of sewage, hormones, and antibiotic feeds in continental agriculture. Producers were encouraged to "stamp a Union Jack logo on all home-produced food to help shoppers fly the flag."[125] In Parliament, Conservative MPs forgot about their 1996 defense of avoparcin feeds and demanded import bans of AGP-fed poultry.[126] Throughout the

2000s, other reports criticized US AGP use, residues in Chinese honey, and resistant bacteria on South American food imports without conducting detailed analyses of similar problems in Europe.[127]

Media reports on the routine use of higher-dosed therapeutic antibiotics on British farms remained rare.[128] In 2003, the *Daily Mail* published a scathing review of Grampian's and Tesco's return to using "rooster boosters"[129] like avilamcyin for disease prevention. According to the newspaper, one in five of the producers registered with Britain's Assured Chicken scheme had reverted to using growth-promoting antibiotics. Others attacked rising prophylactic antibiotic use. A 2003 *Observer* article titled, "If Max eats up all his chicken, he'll grow to be a big, strong boy. Unless it kills him first."[130] The article reconstructed Max's contamination with resistant pathogens: "As the chicken oozes unappetisingly on the top shelf of your fridge, . . . blood drips on to the cheddar cheese below . . . . Making yourself a cheese sandwich next day, you don't notice the bacteriological accompaniment—but you have inadvertently eaten uncooked enterococci."[131] According to the *Observer*, the 43 tons of AGPs annually used on British farms were "only the tip" of a wider 463-ton agricultural "antibiotic iceberg."[132] Banning AGPs without reforming therapeutic antibiotic use and conventional husbandry systems would have little effect on overall AMR. In a seeming repeat of history, these warnings and new reports on new livestock-associated pathogens like tetracycline-resistant "non-typeable-MRSA" (NT-MRSA), however, failed to trigger wider public concern (chapter 13).[133] In 2004, a twenty-point *Guardian* list of ways to "cut out chemicals" mentioned antibiotics only in place 16—AMR was not mentioned.[134]

The renewed lull of British antibiotic risk awareness was only gradually overcome as a result of dramatic 2013 warnings about a looming post-antibiotic apocalypse by the United Kingdom's Chief Medical Officer (CMO) Sally Davies and the publication of the 2016 O'Neill AMR Review (chapter 13). However, despite these warnings, popular calls for renewed antibiotic reform remain limited. In 2018, the EU Parliament's vote to restrict veterinary antibiotic use attracted only minimal media attention.[135] Historically, this is unsurprising. In Britain, national antibiotic protest emerged following moral panics and well-publicized health tragedies. It also tended to focus on easily identifiable—and vilifiable—practices like AGP use by amorphous farmers. Focusing public attention on the complex infrastructures driving and supplying wider antibiotic use has proven more difficult. Despite medical warnings about a resistant apocalypse, there is no obvious symbol technology—like AGPs—around which to unite public concerns. Industry voluntarism has further fragmented protest potential: concerned consumers can now buy antibiotic-free products in all major supermarkets, and agricultural organizations have successfully presented routine therapeutic antibiotic use as important for animal welfare (chapter 12). Similar to the 1970s, there has also been a tendency to publicly rest on the laurels of the 2006 AGP bans and recent voluntarist reductions of overall usage or blame rising AMR on antibiotic overuse in other parts of the world. That current nonhuman antibiotic regulations are not the result of British but of Scandinavian and EU leadership is often forgotten. What will happen to transnational stewardship efforts after Britain's pending departure from the EU is rarely discussed.

12

Persistent Infrastructures

Antibiotic Reform and
British Farming

This chapter explores how antibiotic perceptions and use evolved on British farms after the 1969 Swann report. Similar to public commentators, most agricultural observers initially believed that the corporatist report had "fixed" AMR problems by differentiating between medically relevant and irrelevant antibiotics and by entrusting the former to veterinarians. Despite a brief dip in the wake of the 1971 AGP bans, wider antibiotic infrastructures remained culturally and physically intact. By the mid-1970s, antibiotic consumption had recovered as a result of veterinary prescriptions and "nontherapeutic" AGPs. Although antibiotic usage remained common, Britain's corporatist structures reduced polarizing conflicts between farmers and public critics once new AMR data challenged the Swann compromise. Reacting to the 1980s economic crisis, residue and salmonellosis problems, as well as European reform pressure, British farmers and veterinarians engaged in a "light-green" restructuring of their industry. Similar to the United States, farm organizations like the NFU supported "antibiotic free" ventures but resisted statutory interventions. Organized British resistance to EU-led antibiotic restrictions crumbled following the 1996 BSE crisis. During the late 1990s, many farmers and agricultural commentators ignored calls by the NFU and the pharmaceutical industry to defend AGPs. However, this did not mean that public and agricultural risk perceptions fully merged. Despite achieving impressive reductions of antibiotic usage following Dame Sally Davies' 2013 AMR warnings, British veterinarians and farmers have repeatedly opposed statutory restrictions of therapeutic antibiotic use.

Business as Usual

The decade following the 1969 Swann report brought major changes to British agriculture. Throughout the 1970s, farmers faced the dual challenge of surviving lower profit margins and justifying rising subsidies. This situation strained postwar corporatist arrangements between farmers and government officials. With many smaller farms struggling to survive an ongoing cost-price squeeze, the NFU and officials of Britain's Ministry of Agriculture, Fisheries and Food (MAFF) had to reconcile competing interests of larger and smaller producers. Problems for the NFU were compounded by challenges to its monopoly on agricultural representation from regional farming organizations and the relegation of national decision-making powers to Brussels following Britain's 1973 accession to the EEC's Common Agricultural Policy (CAP).[1] Meanwhile, agricultural surpluses continued to grow: between 1970 and 1980, total meat output from British cattle, pigs, and sheep increased by over 9.7 percent to 2,305,000 tons.[2] Although British farmers initially profited from CAP membership, the saturation of European meat markets gradually depressed farm incomes and forced CAP member states to make intervention purchases to shore up commodity prices.[3] Throughout the EEC, notorious state-acquired "butter and meat mountains" grew rapidly. By 1983, Britain alone was storing 177,000 tons of intervention butter stocks.[4]

With farmers trapped in a buyer's market, risk-minimizing technologies like antibiotics remained popular. Similar to the United States (chapter 9), 1970s British farmers seem to have purchased antibiotics primarily for disease prevention rather than alleged growth-promoting effects. According to a 1970 NFU survey of 1,200 farmers, less than 50 percent fed antibiotics to pigs and poultry for growth promotion alone. However, over 50 percent used antibiotics to promote growth *and* combat stress. Significantly, surveyed farmers' primary source of antibiotic information came from industry commercials followed by advice from veterinarians and extension officials.[5] Increasing meat production led to a rise of antibiotic use. In 1978, a paper by NIRD nutritionist Raphael Braude showed that Britain's 1971 bans of medically relevant AGPs had failed to curb drug consumption. Despite Braude's background as a vocal antibiotic supporter (chapter 7), his findings match wider sales trends.[6] According to Braude, AGP bans had actually increased the dosages of still popular antibiotic feeds: tetracycline AGPs with dosages of 10 to 20 parts per million had been replaced by feeds with dosages of 100 to 600 parts per million. Although AGP bans had briefly halved the tonnage of agricultural antibiotic sales, sales of restricted "therapeutic" antibiotics had annually increased by about 15 percent between 1972 and 1975. The use of allegedly medically irrelevant *ersatz* AGPs (like flavomycin, virginiamycin, and zinc bacitracin) had also increased. With the exception of oxytetracycline, pigs received the largest proportion of antibiotics, followed by poultry and calves. Similar to the 1970 NFU survey, only a minority of livestock farmers used antibiotics solely for growth promotion. All in all, there had been a slight increase of overall antibiotic use since the Swann bans. Braude estimated that in 1975 26,500 kilograms of penicillins and tetracyclines and 55,000 kilograms of other antibiotics had been used as feed additives on British farms.[7]

Braude's data did not encompass rising illegal antibiotic use. In 1971, the *Veterinary Record* reprinted complaints by the Pharmaceutical Society's solicitor that the Society's fourteen inspectors only had the power to conduct farm-level inspections in the case of synthetic sulfonamides but not when it came to biological antibiotics. Offenders could tell inspectors: "You have no power to search or make inquiry. . . . You have only power to institute proceedings. Go away."[8] Meanwhile, Customs and Excise data indicated "substantial smuggling of Dutch and other continental preparations along the Norfolk coast, and of animal medicines from the Republic of Ireland."[9] A limited number of prosecutions had occurred when veterinarians had convinced farmers who had obtained substandard drugs illegally to make formal complaints. Illegal sales networks were sophisticated and hard to uncover because the majority of sales occurred "from market huts, and more frequently, direct to farms by itinerant sellers."[10] In 1971, the Society had uncovered that two companies with annual turnovers in excess of £100,000 had illegally distributed antibiotics with the help of a full-time team of three or four traveling salesmen. Confiscated drugs had been produced abroad and were substandard. The Pharmaceutical Society also admonished British veterinarians for fostering lackluster antibiotic stewardship: "A veterinary surgeon . . . supplying antibiotics for a 250-mile distant herd of cows which he has never seen (instances of which have occurred . . .) produces an equally casual attitude towards antibiotics from the lay public."[11]

However, such outspoken criticism of inappropriate therapeutic antibiotic use and sales remained rare in Britain's agricultural and veterinary press. Throughout the 1970s, the tenor of agricultural opinion was rather that the Swann report had "fixed" earlier antibiotic problems. Although they differed from US publications by emphasizing AMR hazards, British farming manuals continued to advocate nontherapeutic and therapeutic antibiotic use.[12] From the early 1970s onward, the NFU, BVA, and MMB also launched what would become a "five-point plan" for mastitis. The plan consisted of teat dipping in disinfectant, antibiotic treatment of mastitis, dry cow therapy ahead of calving (antibiotic prophylaxis), the culling of chronically infected animals, testing of milking machines, and monthly bulk milk cell counts.[13] Promoting the

FIGURE 12.1 Antibiotic infrastructures were hard to reform. Cyanamid advertisement, *British Farmer*, 1969.

Cyfac pigs are never a burden

Cyfac* is the unique 3-in-1 feed supplement that gets pigs off to the right start—they grow faster, market sooner. Cyfac protects pigs against stress and scours—and profits against loss. With Cyfac in your creep feed and weaner rations there is everything to gain. During the 4-week post weaning period it can add up to a 12½ lb increase in live-weight gain with a 23% overall improvement in feed conversion efficiency. Leaving out Cyfac could mean trouble; adding Cyfac is avoiding it.

CYANAMID OF GREAT BRITAIN LIMITED, BUSH HOUSE, LONDON W.C.2.
*Regd. Trade Mark

CYANAMID
RESEARCH SERVES BRITAIN

80

BRITISH FARMER, SEPTEMBER 9, 1969

FIGURE 12.2 Cyanamid advertisement, *British Farmer*, 1969.

five-point plan boosted veterinary antibiotic sales. In 1972, it was estimated that over 10 million single-dose antibiotic tubes were annually prescribed and sold for the treatment of around 3 million animals.[14] Routine antibiotic prophylaxis was also promoted in the farming media. In 1975, *Farmers Weekly* published an educational song titled "Mastitis, yeh, yeh, yeh" [*sic*]:

> Treat all your udders with a dry cow tube
> And smile, smile, smile,
> Maybe you think the cost is pretty rude,
> But it really is worthwhile.
> Bugs can cause mastitis,
> They always run so wild,
> So treat all your udders with a dry cow tube
> And smile, smile, smile.[15]

Routine antibiotic use also continued to be fostered by pharmaceutical manufacturers. Reacting to the 1969 Swann report, British companies like Beecham and May & Baker increased advertisements for "safe" nontherapeutic AGPs as well as for new therapeutic antibiotic classes to treat and prevent disease—often with considerable discounts for bulk orders.[16] Meanwhile, US companies like Pfizer and Cyanamid attempted to defend market shares by stressing the prophylactic qualities of their now prescription-only tetracycline feeds. Ahead of the enactment of Britain's AGP bans, Cyanamid purchased a full-page ad in *Farmers Weekly* to provide farmers with "an important reassurance":

> Many farmers consider that after 1st March 1971 CYFAC and AUROFAC will no longer be available. This is not the case. CYFAC as CYFAC 25 and new CYFAC PELLETS and AUROFAC will be available **on the recommendation of your Veterinary Surgeon.** If you feel that the condition of your stock or poultry is—or is about to be—such that you need the help that these supplements have given you in the past, ring your veterinary surgeon at once.[17]

The minimal physical implications of the Swann bans for British agriculture stood in sharp contrast to their symbolic value for warding off foreign competition and critics. Following AGP bans' enactment in 1971, British farmers repeatedly appealed to popular Swann patriotism (chapter 11) to ban imports produced with therapeutic AGPs. In 1973, the British Poultry Federation pressured supermarkets to reject consignments of Dutch poultry fed therapeutic antibiotics.[18] One year later, French AGP use featured prominently in a trade war between British and French egg producers.[19] AMR concerns were not at the center of these conflicts. Putting it bluntly, *Farmers Weekly* commented:

"The point at issue is not that French eggs are a health hazard to consumers. It is that the French have a way of shipping unprofitability to Britain in an unfair trading package."[20] With emotions running high, the UK Egg Producers' Association gathered money for legal action and farmers organized pickets and boycotts against French produce.[21] While the increasing harmonization of EEC rules eventually cooled Franco-British egg wars (chapter 13), it created new problems for dairy farmers struggling with high rates of mastitis and antibiotic residues.[22] In this situation, Swann patriotism re-emerged as a way to undermine stricter EEC residue testing plans. Playing AMR concerns off against residue concerns, a 1978 article in *Country Life* contrasted British Swann rulings with "fussy" continental concerns about antibiotic and hormone residues in food: "There is the great fear, especially in the Latin races, that masculinity could be at risk through the careless use of oestrogens."[23] "Swann put Britain in the clear" with regard to AMR and British residue rules were "as good as anywhere in the world, if that is any comfort. Which it surely should be."[24]

Unfounded claims of stricter British regulations were paralleled by mostly uncritical views of wider post-Swann antibiotic use. Expert contributions to the University of Nottingham's 1977 Easter School highlight widespread agricultural and veterinary confidence in the post-Swann state of affairs. According to most agricultural attendees, British antibiotic use was necessary, efficient, and safe.[25] This assessment also included veterinary antibiotic use despite the fact that "fire brigade" approaches to antibiotic prescribing remained common on farms and Swann-supported preventive veterinary health services had failed to materialize. Mostly called to treat already sick animals, veterinarians seldom had time to conduct the sensitivity tests and diagnoses necessary for "rational" antibiotic use and continued to rely on drug sales to bolster incomes.[26] At the 1977 Easter School, a Scottish veterinarian noted that farmers still preferred convenient antibiotics to routine veterinary advice—even if this meant paying markup prices for drugs: "Antibiotics, therefore, may remain an indispensable means of protecting the productivity and profitability of many a livestock enterprise."[27]

While many practicing veterinarians remained confident that "rational" drug use in the form of sensitivity tests, high doses, and drug combinations would check AMR,[28] microbiologists disagreed. At the 1977 Easter School, bacteriologists Alan Linton and Herbert Williams Smith warned that limited AGP bans had failed to reduce AMR. Linton had analyzed *E. coli* from animal, human, and environmental sources. High levels of resistance against prescription-only drugs had been detected in all isolates and serotypes of resistant isolates from humans and animals were indistinguishable. AMR was often plasmid-mediated with plasmids transferring multiple-resistance.[29] While AMR was more common in the rich and diverse gut flora of animals, which had been "penetrated widely"[30] by resistance-conferring plasmids, it was likely that AMR in the human gut flora could have originated in animals. In abattoirs, high levels of resistant *E. coli* had been found on poultry, pig, and beef carcasses—probably as a result of fecal contamination. Humans in contact with the carcasses had also been colonized. Higher-dosed therapeutic antibiotics were just as good at creating AMR reservoirs as lower-dosed AGPs had been: "Any cross-infection ultimately derived from farm animals will be additive to that already cycling by cross-infection in man himself. . . . The evidence to date confirms the views of the Swann Committee prohibiting the use of antibiotics of clinical importance for growth promotion . . . and possibly this prohibition should be extended to their use for prophylaxis and generally to reduce their use for all nontherapeutic purposes."[31]

Herbert Williams Smith also called for further antibiotic reform. Although AMR transfer varied between organisms, constant antibiotic exposure made it easy for resistant organisms to become dominant and increased chances of successful resistance transfer.[32] The Swann committee had believed that curbing tetracycline use via AGP restrictions would reduce corresponding resistance in animals' gut flora. However, studies in British pigs had shown that this was not the case. The failure of the Swann bans to reduce AMR could be caused by the fact that resistant strains no longer had an evolutionary disadvantage in the absence of selection pressure as well as by higher-dosed veterinary antibiotic use and co-selection. Williams Smith ended his paper by warning that restrictions of already licensed antibiotics would not easily reverse AMR. New antibiotics had to be restricted immediately before widespread resistance could develop.[33]

Bacteriologists' AMR warnings failed to elicit widespread agricultural concern. Similar to Britain's public sphere (chapter 11), there remained a strong belief that the Swann report had fixed problems. Faced with rapid structural change, many also saw antibiotics as an important tool to survive relentless pressure for productivity increases. Trends toward intensification were similar across British livestock sectors. Between 1967 and 1980, the number of British broiler flocks decreased from 3,700 to 2,200 while the average flock size increased from 9,800 to 26,500 birds—95 percent of which were now raised in confinement systems.[34] Between 1970 and 1980, the average dairy herd size increased from 33 to 56 cows, producer numbers fell from 80,265 to 47,169, and productivity increased from 825 to 1,037 gallons of milk per cow.[35] Although US-style feedlots did not become common, overproduction and sinking demand triggered the first decline of British beef cattle in nearly 100 years.[36] Pig production was affected by similar trends. With subsidies declining after Britain's EEC accession, producers had to rely on feed efficacy and productivity to beat shrinking profit margins. Although pig farming remained more diverse than in the United States, surviving producers tended to concentrate in the north and east of the county and tried to cut costs by concentrating animals and investing in labor-saving technologies and new breeds.[37]

As indicated by the parallel rise of 1970s drug usage, antibiotics were used to facilitate herd growth and productivity. On British farms, the Swann bans had neither reduced antibiotic access nor changed antibiotic mentalities. Aside from the scale of operations and veterinarians' control over medically relevant drugs, there remained many similarities between antibiotic infrastructures on both sides of the Atlantic (chapter 9). While some commentators reflected on rising chemical use and small farms' demise as two sides of the same coin,[38] there was little appetite for further antibiotic reform.

Thatcherism, Farming, and Limited Reforms

The post-Swann status quo of British agricultural antibiotic use began to be challenged during the 1980s. Although the agricultural community often reacted with hostility to rising public antibiotic criticism (chapter 11), diminished political support, changing EEC policies, economic hardship, and scandals led to self-reform. Similar to the United States, this reform was light-green.

For many British livestock farmers, the 1980s were a period of crisis. With average real income falling from £12,058 in 1973 to a nadir of £4,894 in 1980,[39] many farmers struggled to survive. In the farming press, agricultural commentators often displaced blame for British farming's woes on continental protectionism and foreign overproduction.[40] Angry about French import bans in 1980, British farmers attempted to deliver "a British lamb to the firmly closed French embassy" while singing "jingle jangle, Giscard dangle."[41] Generating political support for a further subsidized increase of domestic production proved difficult. Whereas previous decades had been characterized by a relatively insular mode of corporatist decision-making, the 1980s saw ties between British farmers and Whitehall fray. The weakening of corporatist bonds was caused by rising subsidy costs, neoliberal policy reviews, and an increasingly fragmented agricultural policy landscape shaped by environmentalist and animal welfare concerns.[42] In contrast to the Reagan administration's reluctant expansion of agricultural subsidies (chapter 9), Britain's new Thatcher government did little to redress farmers' situation. In 1980, it "axed" school meals[43] and remained committed to overvaluing the "green pound," one of a number of artificial EEC currencies created to determine CAP prices in relation to national currencies, despite halving British CAP contributions in 1984. While consumer prices were kept low, CAP payments to British farmers were worth less.[44]

Dismayed by government inaction, many conventional livestock producers tried to survive the cost-price squeeze by expanding production and increasing productivity. Antibiotics facilitated this process. A 1982 article by G. H. Yeoman, head of clinical studies at Beecham's Animal Health Division, illustrates contemporary trends. According to Yeoman, recent years had seen substantial shifts of British animal production and antibiotic use. More animals were housed in confined buildings, and new farming systems were creating novel disease challenges: "These are not new pathogens: rather they are old pathogens exploiting new ecosystems provided by the new methods and thus appearing as modified, if not new, syndromes."[45] Because new housing systems were mostly designed for maximum production, veterinarians had to "somehow do [their] best to ameliorate the bad effects on animal health and to a great extent this means resorting to medication whether therapeutically or pre-emptively. Indeed, there is a danger . . . that the use of antibiotics may take on the status of a routine input, regarded little differently from that of food, fuel and water."[46] Vaccines, better animal management, and all-in-all-out systems had taken care of many diseases. However, a "short" list of bacterial diseases still generated "the critical battleground where massive exposure to antibiotics occurs."[47] These "diseases of intensive farming"[48] were caused by *E. coli*, *Streptococci*, *Staphylococci*, *Salmonella dublin*, *Salmonella typhimurium*, the recently discovered *Campylobacter*, *Haemophilus*, *Pasteurella*, and *Klebsiella*. According to Yeoman, it was important to remember that the factors influencing veterinary antibiotic prescriptions were fundamentally different from human medicine: "The farmer is concerned with cost-effectiveness, the vet with providing treatment at a price that will ensure that his bill is likely to be paid."[49] Popular antibiotics like penicillins, tetracyclines, macrolides, and sulfonamides were cheap and easy to mass-administer via feed or water. "The managerial pressures of factory farming" could easily bias veterinary prescription habits: "[The] veterinary adviser [of a large farmer] may well have provided prescriptions in advance and his compounder may hold ready-medicated rations: thus within hours antibiotic treatment of the whole house can be implemented and any potential set-back pre-empted. . . . If his veterinary surgeon has reservations about this, there is fairly ready access to grey-market or black-market supplies."[50]

While inner-agricultural criticism of antibiotic-intensive production remained minimal, British farmers and veterinarians had thin-skinned reactions to reemerging public criticism of animal welfare, AMR, environmentalist, and residue issues.[51] In 1979, farm organizations reacted furiously to *Brass Tacks*'s attack on antibiotic use (chapter 11). The NFU considered taking out an injunction against the *Radio Times* and promised to send "hot missiles" to the BBC's chairman and director-general,[52] the former of whom was none other than Michael Swann. In the same year, the *Veterinary Record* responded to allegations of indiscriminate veterinary prescriptions. The journal conceded that the recent spread of multiple-resistant *S. typhimurium* 204 through human and cattle populations had been facilitated by excessive "and sometimes illicit"[53] antibiotic use. However, AMR could not be blamed solely on veterinarians.

Pharmaceutical companies and physicians could equally be accused of facilitating antibiotic overuse.[54] Reporting on the 1980 BVA congress, *BFS* noted that "despite an attack by the medical profession represented by [PHLS] Dr [John] Threlfall," the "use of antibiotics in agriculture received strong support"[55] from veterinarians. According to veterinarian John Walton, "the medicos [were] wrong."[56] Further bans would not prevent resistant salmonellosis. Two years later, significant opposition from veterinarians, producers, and industry representatives toppled proposed restrictions of oral chloramphenicol preparations.[57]

Controversies about legal antibiotic use were paralleled by revelations about substantial black and grey markets for antibiotics on farms. During the early 1980s, significant regulatory gaps continued to enable semi-legal and illegal drug access on British farms (chapter 13). In 1983, Britain's animal health black market was estimated to be worth £1 to £2 million per year.[58] Corporatist attempts to solve problems in cooperation with officials proved complicated. In 1979, the British government tried to reduce semi-legal drug sales by reforming the Pharmacy and Merchants' List (PML). Drugs listed as PML, which also included certain antibiotics, could be sold by veterinarians, pharmacists, and listed agricultural merchants.[59] The new PML regulations attempted to ban dubious itinerant sales by mandating that only listed "stationary" vendors could sell PML products. The goal was to control illegal sales without restricting agricultural drug access. However, it soon became clear that many listed merchants were no better than van salesmen. Of the 1,800 names and 3,000 premises officially listed in 1981 only 500 were believed to be upright merchants. Many listed premises did not keep records or demand to see prescriptions prior to selling drugs. In one case, a newspaper agent was fined £1,000 after it emerged that he was illegally supplying local farmers with restricted drugs.[60] Although new rules required a pharmacist to supervise all PML sales and introduced a mandatory code of practice in 1984,[61] farmers could still legally import foreign feeds and milk replacers containing restricted drugs.[62]

The pharmaceutical industry was happy to supply rising antibiotic demand. Between 1979 and 1983, Britain's animal health market grew from £71 to £114 million—making it the sixth-largest in the world. Prescription-only medicines accounted for about 44 percent of British sales—with antimicrobials constituting 35 percent of prescription-only sales (14.5 percent of total market). Despite the 1971 AGP bans, antibiotic (9 percent) and coccidiostat feed additives (5 percent) made up another 14 percent (£16 million) of total sales. Antibiotic consumption varied according to livestock sector. Prescription-only and PML antibiotics constituted about 60 percent of the £6 million feed additive market for pigs but only 40 percent of the poultry market. PML additives included AGPs like virginiamycin (Smith Kline), flavomycin (Hoechst), avoparcin (Cyanamid), and monensin (Elanco). Writing in 1984, industry analysts also noted that sinking incomes were intensifying veterinary competition for farm-ers' custom when it came to selling prescription-only drugs with markup charges of between 25 and 50 percent. Competition was also increasing between veterinarians and registered pharmacists. Charging a 25 to 30 percent markup for PMLs, many veterinarians were trying to outcompete pharmacies by bulk buying and storing medicines.[63]

Residues resulting from legal, semi-legal, and illegal antibiotic use increas-ingly caused problems. In November 1979, more sensitive MMB penicillin monitoring showed that 900 to 1,000 of 47,000 farmers regularly produced milk with excessive residues.[64] While some blamed the "odd cow" getting "milked by mistake,"[65] the NFU's *BFS* warned: "What is disturbing about these figures is that the incidence of test failures in the United Kingdom is 20 times that in other countries, apart from Eire, despite the fact that most use a more sensitive test: And equally most (again excluding Eire) impose more severe pen-alties."[66] Reacting to problems, the MMB increased residue penalties: first-time offenders would be fined 5p per liter, second-time offenders 7p per liter, and third-time offenders would have to pay a "swingeing rate"[67] of 9p per liter. However, penalty increases were unsuccessful. Receiving 11p for every liter of uncontaminated milk, farmers continued to sell contaminated milk because the chance of incurring a fine was less problematic than foregoing earnings com-pletely.[68] Contradicting popular claims of higher British standards, British milk was found to contain the highest residue levels in Europe in 1982. Accord-ing to the *British Veterinary Journal*, many farmers believed that antibiotics in the milk of one cow could be diluted below detection limits although trials had shown that one cow treated with 200 milligrams of penicillin G could con-taminate the milk of 8,000 other cows.[69] In accordance with new EEC rulings, testing sensitivity was raised to 0.01 international units of penicillin per milliliter in 1986. This time, better recording, higher penalties, random test-ing, and awareness campaigns succeeded in reducing violation rates to below 0.5 percent.[70]

Mounting problems with residues, AMR, and black market drugs coincided with a new round of agricultural reform. With overproduction and subsidy costs exploding, the European Community introduced dairy quotas (1984), a co-responsibility levy for cereals (1986), and forced farmers to let land lie fal-low (1986). In 1987, EEC Regulation 1760/87 introduced a voluntary scheme for farmers to reduce the output of cereals, beef, and wine for three to five years. Although measures failed to solve overproduction, they marked the beginning of a sea-change by weakening postwar interventionist modes of agricultural expansion in favor of diversification and an emphasis on small farmers and the environment.[71] The sea-change of agricultural policies mirrored an emerging change of agricultural attitudes. Worn down by the 1980s economic crisis, pub-lic criticism, and weakened ties to Whitehall, British farm organizations became more willing to consider alternative production methods. Following its 1984 annual general meeting, the NFU acknowledged: "it seems right to conclude that we are now at a watershed and that the era when agricultural expansion was widely accepted as a desirable goal has passed."[72]

Similar to the United States (chapter 9), solutions proposed by the agricultural community were mostly voluntary and light-green. For struggling smaller producers, organic or "natural" forms of agriculture became increasingly attractive. During the 1970s, members of the still relatively small Soil Association and other organic organizations had published a growing number of manuals and advice on how to farm and live sustainably.[73] Emphasizing environmentalist themes, early publications' rhetoric had, however, often been aggressively anticonventional.[74] This changed during the mid-1970s. Although there was no publication comparable to Rodale's US journals, Britain's organic community began to moderate its rhetoric and cater to the needs of new audiences like conventional farmers looking to transition to organic production or organic farmers looking to expand production.[75] The ensuing professionalization, commercialization, and integration of Britain's organic movement coincided with a generational shift. Two new producer groups—the Organic Growers Association and British Organic Farmers—were formed in 1981 and 1983. In 1985, the Soil Association's new leadership moved headquarters from the countryside to Bristol.[76] Rejuvenated organizations were more media-savvy and commercially oriented. They also profited from new access to UK supermarkets. In 1981, Safeway became the first British supermarket to stock organic produce, followed by Waitrose, Sainsbury's, Tesco, and a short-lived experiment by Marks and Spencer in the second half of the 1980s.[77]

Its professionalization and integration into established supply chains made organic production more palatable to the wider farming community. Previously lambasting organic farming,[78] conventional representatives began to attend Soil Association conferences while organic methods and profits drew favorable comments in the farming press.[79] From the mid-1980s onward, even bastions of conventional agriculture like *Farmers Weekly* began to advise struggling cattle farmers to sell "natural" drug-free produce at a premium price.[80] Organic producers also received official recognition with the creation of the UK Register of Organic Food Standards in 1987 and the 1989 organic standards.[81] The absence of antibiotics remained a central element of organic purity and safety claims. According to the 1989 organic standards, livestock producers were only allowed to use antibiotics to save an animal life, prevent suffering, or treat conditions with no alternative treatment or management practice—AGPs remained taboo.[82]

Although most British farmers did not transition to organic agriculture, their increasing acceptance of alternative production methods and acknowledgement of shifting public values led to a gradual "greening" of conventional production and rhetoric. When officials challenged an EEC hormonal growth promoter ban in 1986, Britain's beef industry cautioned that unilateral action might provoke import bans and stoke consumer fears: "Privately, they believe it might be better to face the ban."[83] According to a 1986 article in *Farmers Weekly*, environmentalism and intensive farming were not mutually exclusive: "there is no reason why we should not compete in the world's agricultural markets, . . . and still have a country fit for Robin Hood or Rupert Bear."[84] The magazine also printed complaints by "suburban housewife" Audrey Curran: "I am fed up of being told, as a consumer, that it is my fault if animals are being reared in these intensive units to supply me with cheap food. I don't want it and I don't know of anyone who does when made aware of what is involved. And, who asked me if it was OK to stuff them with antibiotics?"[85]

Unwilling to risk profitable production systems, conventional producers were, however, unsure how far "green" reforms could go. In 1987, *Farmers Weekly* commentator Robert Gair described the fundamental dilemma he shared with many other farmers: criticizing attacks by "Greenpeacers" and "the anti-farming, anti-chemical brigade," Gair confessed that he too had "no desire to see a countryside without birds, mammals, frogs, butterflies, orchids, and the rest."[86] In order to arrest "detrimental changes in the environment," all parties should engage in a "rational examination" of factors likely to disturb the "balance of nature."[87] The "light-green" dilemma described by Gair was similar regarding agricultural antibiotics and resulted in a curious parallelism of antibiotic endorsements and cautious reform initiatives.

On the one hand, many 1980s agricultural commentators and manuals continued to routinely advocate "rational" therapeutic and nontherapeutic antibiotic use alongside other preventive health measures.[88] Similar to the United States (chapter 9), articles in the British farming press defended antibiotics and reacted to external criticism by attacking the "inane agitation of the lunatic fringe of the animal welfare movement."[89] Hailing the 1987 results of the national meat surveillance scheme as proof of safe drug use on farms, *Farmers Weekly* launched a scathing attack on testing systems "hell-bent on proving that wholesome food is positively dangerous" and capable "of sniffing down to parts per billion."[90] Farmers had "to spread the gospel [of meat safety] before political pressures remove yet more useful pharmaceuticals from the market and restrict research."[91] British veterinarians also attacked "excessive" consumer fears. According to John R. Walton, there was a risk that more sensitive residue testing methods could cause unnecessary alarm about safe drug use. Regarding AMR, it was important not "to confuse the use of antibiotics in agriculture with outbreaks of human disease particularly caused by antibiotic-resistant *Salmonella* organisms."[92] Contemporary salmonellosis outbreaks would be solved not by antibiotic regulation but by better hygiene and infection control. The BVA similarly condemned rising antibiotic criticism as "emotive": "There is every reason to believe that problems of antibiotic resistance encountered in human medicine are overwhelmingly due to the way in which they are used in man."[93] The British pharmaceutical industry supported agricultural antibiotic defense. In 1986, it founded the National Office of Animal Health—with the compelling acronym NOAH. Tasked with "giving a more effective voice"[94] to the drugs industry and its allies, NOAH would become an agro-industrial bulwark defending on-farm antibiotic access.

On the other hand, the 1980s also saw a growing amount of inner-agricultural skepticism contradict mostly benign NFU, BVA, and NOAH antibiotic assessments. Although it recommended antibiotics for various diseases, the 1981 *Pigkeeper's Guide* criticized units in which animals were "continually stuffed with antibiotics to keep down some disease or other, when all that is needed is a bit of space and fresh air."[95] In the same year, the new edition of *The Agricultural Notebook*—one of the United Kingdom's standard agricultural textbooks—questioned the efficacy of AGPs and of low-dosed therapeutic antibiotic use.[96] The *Notebook* stressed that there were alternatives to intensive production: "[animal welfare criticism] has forced us to enquire whether it is necessary to house animals using methods which involve a high degree of restriction on their movement."[97] Although critical commentators did not follow organic farmers and completely reject antibiotics, they were keenly aware that AMR, residue scandals, and changing consumer preferences would eventually change production systems: "the public will buy what it wants, and not what some scientist thinks it should buy."[98] In *Farmers Weekly*, articles began to advertise "no-additive feed"[99] and warned about AMR selection by feeding antibiotic-laden milk to calves.[100] In the *Veterinary Record*, commentaries by continental veterinarians also started to question senior British veterinarians' defense of post-Swann arrangements.[101]

Meandering between defensiveness and premonition, 1980s controversies about antibiotics, welfare, and environmentalism primed early agricultural reactions to BSE. In October 1987, MAFF veterinarians announced that BSE "is not of epidemic proportions, . . . and is not very significant when compared with losses from other nervous disorders."[102] *Farmers Weekly* joked that "BSE thrives on rumours": "Thank goodness witchcraft is out of fashion, otherwise the old lady who lives in the cottage down the lane with a black cat for company would be accused and ducked in the village pond."[103] However, behind the façade of prescribed calm, there was growing awareness that the vortex of public insecurity created by BSE, salmonellosis, and residue scares was beginning to pose a substantial challenge to existing agricultural practices (chapter 11). With environmentalist and consumer power rising and trust in food safety eroding, conventional British agriculture was about to face systemic change.

A Decade of Crises

The 1990s were a hard decade for British farmers. With economic pressure, food scares, and subsidy reforms further reducing farm numbers, remaining producers had the choice of either expanding production or of diversifying into tourism or "green" market niches.[104] In the case of antibiotics, NFU, NOAH, and BVA strategies of voluntarist self-regulation were shattered by the 1996 BSE crisis. BSE not only triggered a reform of post-Swann antibiotic rules but also highlighted rifts between large intensive and smaller producers.

The decade started inauspiciously. At the European level, the 1992 Mac-Sharry reforms marked the most significant modification of the CAP since its inception.[105] Instead of subsidizing prices, commissioner Ray MacSharry attacked overproduction and subsidy costs with quotas, set-aside schemes, and direct payments linked to a farm's size and animals' age.[106] Together with the 1992 EU Flora-Fauna-Habitat guidelines, the MacSharry reforms also embedded environmentalist principles in EU agricultural policy-making.[107] Although the reforms benefitted intensive producers by lowering grain prices and providing greater income security, British farmers were simultaneously hit hard by falling consumer trust, lower commodity prices, and a strengthening pound, which reduced EU payments. Politically, the Conservative government exacerbated the situation by stripping once powerful bastions of corporatist decision-making like the MMB of its powers in 1994 and ending annual price reviews in 1995.[108]

Economic pressures, new regulations, and a relative loss of political influence had a significant impact on British livestock production. In the pig sector, decreasing profitability and currency fluctuations led to a decline of animal numbers from around eight to under five million by the end of the 1990s. Remaining herds were usually larger and more concentrated. While the majority of animals were now housed in confined intensive settings, lower capital and maintenance costs also led to a revival of outdoor systems in the South and East of England.[109] Cattle numbers also declined—with the 1996 BSE crisis severely impacting the beef sector. Although poultry numbers continued to increase, birds were being produced in larger units on fewer farms. By contrast, welfare regulations, salmonellosis scares, and changing consumer preferences led to a parallel boost of free-range systems in egg production. In all livestock sectors, total factor productivity rose as a result of labor substitution and more productive breeds.[110] Remaining producers also found themselves exposed to an unprecedented scale of monitoring. The *Salmonella*-inspired 1990 Food Safety Act and new slaughtering clauses allowed inspections of farm animals for forbidden substances, forced farmers to keep detailed medication records, and created enormous pressure to keep herds disease-free.[111]

Weakened by its declining membership and fraying corporatist influence, the NFU responded to contemporary food scandals and rising demands for reforms of production practices (chapter 11) with a light-green strategy of voluntarist change. In addition to intensifying reporting on "green topics" in *British Farmer*,[112] NFU president David Naish announced a new program called Farming for the Environment ahead of the 1992 general elections.[113] The goal was to signal and inspire a gradual shift of agricultural values and practices before more radical change was thrust on farmers via statutory measures. Although some commentators remained hostile toward "the greenies,"[114] major magazines like *Farmers Weekly* also fostered "green values" among farmers and their families.[115]

Light-green image campaigns occurred against a backdrop of steady sales increases of antibiotic-free food. In 1992, total organic food sales amounted to £92.5 million and sales of meat and dairy products to about £13.9 million. Over the next five years, sales of organic meat increased by a further 50 percent.[116] With major supermarkets also establishing premium lines for ethical and antibiotic-free conventional produce,[117] breaking with standard antibiotic use was becoming a sales advantage. This message was propagated by the farming press. In 1991, the NFU's *British Farmer* announced that East Anglian Dalehead Foods was looking for "pigs from 'welfare-conscious' systems" raised on cereal-based feeds with "no antibiotic growth promoters or probiotics."[118] Delivering "green pig" products to a "southern-based supermarket chain," Dalehead offered suppliers a "generous premium."[119] The magazine also printed an advertisement for Daisy Hill Feeds' Headstart Challenge. Targeting conventional farmers, the company claimed that its antibiotic-free feeds were just as good as AGPs, answered consumer criticism, and would help farmers transition to a coming antibiotic-free era: "Please your customers and [get] ahead of any ministry or EC legislation, ... Give the Headstart range of piglet diets a trial against your existing supplies— ... once you have removed the fear factor of not using antibiotic growth promoters you will have the confidence to remove them from your other pig feed diets. In our opinion you will not be disappointed and you will be helping dislodge an area of criticism and concern levelled at the British Pig Industry."[120] Targeting wealthy urbanites and large land-owners rather than small farmers, *Country Life* also published articles referring to Prince Charles's campaigning for organic farming, homeopathy, and profitable antibiotic-free production systems.[121]

Similar to the United States (chapter 9), there were, however, clear limits to inner-agricultural reform efforts. Despite supporting "rational" antibiotic use and individual farmers' switch to antibiotic-free production, most 1990s commentators did not endorse statutory restrictions. In contemporary manuals, AMR and residue warnings continued to feature alongside instructions for routine antibiotic use.[122] In 1995, the nineteenth edition of the *Agricultural*

Notebook recommended therapeutic and prophylactic drug use for numerous diseases. The manual also provided lists of AGPs while simultaneously cautioning farmers about pressure by "anti-factory-farming lobbies,"[123] AMR, residues, and potential drug restrictions.

With only a minority of farmers switching to antibiotic-free production, green voluntarist rhetoric had little actual impact on antibiotic infrastructures and usage. According to Britain's Veterinary Medicines Directorate (VMD), sales of therapeutic antimicrobials in food animals increased by about 40 percent from 392 to 533 tons between 1993 and 1996. Tetracyclines, sulfonamides, and β-lactam antibiotics accounted for 72 to 81 percent of sales. Drugs were usually administered via feeds. Antimicrobial sales were highest for pigs with poultry sales slightly outcompeting those for cattle. Similar to the 1970s, nontherapeutic AGPs accounted for only 83 tons of total antimicrobial sales in 1993 (about 17.5 percent) and 96 tons in 1996 (15.3 percent). Coccidiostat sales remained relatively constant.[124] The numbers validated Raphael Braude's 1978 assessment that the AGP-focused Swann bans had achieved little—except raise the dosage of medically relevant antibiotics in feeds. The moral panic triggered by the 1996 BSE crisis (chapter 11) would soon end the post-1969 stasis of British agricultural antibiotic infrastructures.

BSE and AGPs

Throughout the 1990s, the shadow of BSE had loomed ever larger over British agriculture. In 1992, *British Farmer* reported that "a billion pounds [had been] wiped off the value of the nation's cattle"[125] following the BSE-related death of a Bristol cat in 1990. However, the journal remained optimistic that the "crescendo" of "unjustified public anxiety"[126] would ebb. Four years later, the official announcement of a possible link between BSE and vCJD on March 20, 1996, dashed hopes for a recovery of public trust. Within 24 hours, several EU countries issued unilateral bans on British beef and refused to lift them despite diplomatic action by the British government. On March 22, the Consumer Association recommended removing beef from personal diets. At this point, some voices began calling for a complete cull of the national cattle herd.[127] Whereas domestic beef consumption fell by 50 percent in the first week after the announcement, it recovered to 25 percent below average in the second week. The loss of export markets resulted in a further 30 percent drop of sales.[128]

Agricultural reactions to the moral panic triggered by BSE ranged from shock to anti-European outrage. NFU president Sir David Naish assured farmers that he was "deeply aware of the immense uncertainties and anxieties facing you and your families."[129] According to Naish, "the NFU [would] not rest in its efforts to restore our customers' confidence in our product"[130] and attacked calls for untargeted mass culls. Agricultural magazines reported that Britain did not have enough incinerators to cope with the proposed cull and advertised suicide helplines for struggling farmers.[131] Farmers' reactions were mixed. While John Pidsley from Cheshire blamed "media hysteria" for the unnecessary "wholesale slaughter of complete herds,"[132] a "worried farmer from Gloucestershire" thought that "feed-makers" were the real villains: "They included the meat and bone meal in the rations. We did not ask for it. Now they must pay for the damage suffered. . . . Just like the oil disasters, Baring Bank, lead in feed and thalidomide, the firms involved should be made to pay the price and suffer the consequences."[133] According to Anthony Carter from West Sussex, "BSE must teach us all that current perceptions of safe are wrong."[134] Instead of relying on technological artifice, farmers should accept that "nature works very well on its own."[135] Farmer John Newman predicted that BSE would boost sales of "safe" and traceable organic products.[136] Meanwhile, members of the Soil Association called for an increase in organic subsidies and reduced CAP support for conventional production methods.[137]

With consumers turning away from British beef, a window for inner-agricultural reform opened. Mirroring the attention paid to antibiotics in the national press (chapter 11), critical commentators used the BSE crisis to question other controversial aspects of conventional production like AGPs. Changing agricultural attitudes are best exemplified in the context of the EU's 1997 avoparcin ban. Up to 1997, avoparcin AGPs had been used by about 80 percent of British poultry producers and about 30 percent of pig and cattle producers.[138] Despite its popularity and attempts by the Conservative government to prevent its ban (chapter 11), farm media protests against EU withdrawal were surprisingly muted. In 1997, *Farmers Weekly* not only failed to criticize the pending ban but also reassured producers that Swedish farmers had been able to phase out AGPs with improved diets, hygiene, and all-in-all-out housing.[139] Post-BSE manuals like *Growth of Farm Animals* predicted further welfare- and drug-related reforms: "unresponsive farmers may find themselves unable to continue in stock farming and much as they might regard such measures as undue interference . . . , they will perhaps find very little sympathy from the general public."[140]

Pharmaceutical and feed companies were concerned about developments. Following the 1997 general elections, they attempted to whip up popular agricultural protest against further AGP bans. NOAH warned that critics were "confused over the facts behind farming's role in foodborne disease, antibiotic resistance and growth promoters."[141] According to NOAH director Roger Cook, antibiotics increased food safety and were "a major factor in reducing salmonella."[142] Demanding that "all sides of the argument" be "represented accurately,"[143] NOAH also mobilized counter-expertise. During a press briefing, ex-BVA president Karl Linklater claimed that antibiotics brought "significant economic benefits," made "enormous contributions to animal welfare," and had been used "for 40 years without difficulty."[144] NOAH chairman Bill Hird accused Scandinavian reformers of exporting AGP bans to maintain "their own high cost agricultural production."[145] His colleague Roger Cook used identical arguments against the Soil Association: "the Soil Association represents organic farmers who, for years, have sought to justify the high prices they demand for their products . . . . They have a vested interest in maintaining public anxiety about British food."[146] At the EU level, the industry-sponsored Federation of Animal Health (FEDESA) accused powerful consumer groups of displacing blame for AMR from human medicine onto AGPs.[147]

Pharmaceutical lobbying divided Britain's agricultural community. Favoring the interests of larger intensive producers, organizations like the NFU, the Meat and Livestock Commission, and the British Pig Association joined NOAH's pro-AGP campaigning.[148] Others begged to differ. According to Jim Reed, director of the UK Agricultural Supply Trade Association, it was time to start acknowledging consumer demands: "And if that mean[s] a ban on certain in-feed antibiotics then so be it."[149] *Country Life* slammed agricultural antibiotics for threatening a return to a "pre antibiotic era."[150] While doctors were also to blame for AMR, it was problematic that "all a farmer needs to collect a tonne of 'growth promoter' is a fork-lift truck."[151] The NFU was making things worse by "playing for time, and doing itself no favours with a deeply distrusting public after the BSE scandal. . . . Instead of constantly retrenching in defence of intensive farming, it must direct its efforts to campaigning for higher-cost, higher-value foods, educating the public in the process."[152] In response, NFU president Ben Gill asserted that his organization was merely demanding a revaluation of AGPs on the basis of "good science and good knowledge of on-farm practices."[153] According to Gill, there was "no widespread misuse of antibiotics on our farms. There is use, and declining use, within the rules imposed by the British and European authorities."[154]

NFU and NOAH appeals for voluntarist alternatives to AGP bans failed to inspire significant agricultural protest. Despite attempts to rebrand AGPs as environmentally friendly "digestive enhancers,"[155] the majority of British farmers and agricultural commentators chose not to emulate contemporary US opposition to statutory interventions (chapter 9). Two years after the BSE crisis, there was little appetite to defend a small group of substances whose efficacy was doubtful and whose use could easily be substituted. With major supermarkets like Waitrose and Tesco beginning to demand animals produced without any growth promoters, popular magazines like *Farmers Weekly* focused on preparing farmers for an AGP-free future.[156]

In December 1998, the EU banned virginiamycin, tylosin, zinc bacitracin, and spiramycin AGPs. Together, the substances accounted for about 80 percent of AGPs used in EU pig and poultry rations.[157] In contrast to industry predictions, British farmers adapted quickly. On farms, animal nutritionists estimated that AGP bans would cost between 50p and £1 per pig but hoped "to claw at least half of that back"[158] with better management. Jasper Renold, pig unit manager on *Farmers Weekly*'s experimental farm, questioned the entire economic logic behind AGPs: "I think we see them as necessary to safeguard performance, particularly in weaners. But if you were to ask me how much benefit they give, I couldn't tell you. . . . I think we're continuing to use them because they're seen as a relatively cheap form of insurance."[159] Not only did Renold's statement contradict NOAH claims, it also revealed how credulous many had been regarding AGP efficacy claims. *Farmers Weekly* veterinarian Richard Potter noted that he "wouldn't be at all surprised if there was no dip in grower performance following AGP removal, given the right management and hygiene."[160] Management and hygiene were not the only ways to replace AGPs. Similar to the 1971 Swann bans, it was also possible to substitute lower-dosed AGPs with feeds containing higher-dosages of prescribed antibiotics. Between 1998 and 2000, British AGP sales dropped from 141 to 44 tons of active ingredient while veterinary sales rose from 433 to 465 tons of active ingredient—despite a parallel drop of pig and cattle numbers.[161]

Despite legal rearguard action by pharmaceutical companies,[162] most agricultural observers freely acknowledged that remaining AGPs would not stay on the market for long.[163] Speaking at the 1999 Pig and Poultry Fair, Tesco's agricultural manager announced: "it's no longer a question of if there's a total ban on use of AGPs for pig production but when."[164] Suppliers were encouraged to "remove prophylactic use of AGPs," "the quicker the better."[165] Tesco's prediction came true. In April 2002, the EU Commission proposed phasing out the four remaining AGPs (monensin sodium, salinomycin, avilamycin and flavophospholipol)—coccidiostats would remain available.[166] In 2003, EU agriculture ministers confirmed the phasing-out of remaining AGPs by January 1, 2006.[167] Reflecting on AGPs' imminent end, the 2002 edition of *Growth of Farm Animals* noted:

> The classical antibiotic era produced an almost miraculous facilitation of growth. Its corollaries were equally extraordinary in terms of the intensification which it allowed and the improvements in efficiency and productivity. However, there are also causes for regret in that it seems, with hindsight, that other improvements remained undeveloped. These lost opportunities are once again beckoning. We now have improved understanding of the physiology of the digestive tract and the role of gut bacteria in health and disease combined with the ever inventive mind of the farming world and those associated with it. This will ensure that the future for the industry may not be as bleak as some have feared, and indeed, in a world without routine antibiotics, the future comfort and well-being of farmed animals may in fact be improved and sustained profitability achieved again.[168]

Antibiotics Reformed?

In late 2005, the end of AGPs attracted barely any mention in Britain's farming press. Most agricultural commentators believed that the trajectory of antibiotic reform that had started with the 1962 Netherthorpe and 1969 Swann reports had now been fulfilled: prescription-free antibiotic access had been restricted and veterinary oversight established. Despite data indicating that narrow AGP bans alone might not reduce antibiotic use or AMR, hardly anybody thought that further statutory reforms were necessary.[169]

Consumption data shows that agricultural antibiotic infrastructures were not severely impacted by the 2006 AGP bans. While the phasing out of AGPs reduced overall antibiotic use by about 10 percent, the period between 1998 and 2013 saw higher-dosed therapeutic antibiotic use in the United Kingdom increase by 41 percent. The use of coccidiostats and antiparasitic agents also increased by 45 percent after 2006.[170] Similar to the 1990s, tetracyclines, sulfonamides, and β-lactams remained the most popular antibiotics with macrolides also gaining in popularity. Pigs continued to receive the most antibiotics (40 to 45 percent), and drugs were mostly administered via medicated feedstuffs or water.[171]

Therapeutic and prophylactic antibiotic use continued to be advocated in agricultural manuals. Despite also containing chapters on transitioning to organic farming, the 2003 version of the *Agriculture Notebook* provided detailed advice on how to maintain the health and productivity of livestock. According to the manual, larger production facilities made preventive infection control essential. Although alternative measures were described, prophylactic antibiotic use remained an important method of disease control.[172] Whether this use was always "rational" is questionable. In 2004, *Farmers Weekly* offered a glimpse into the reality of prescription-only therapeutic antibiotic use when American veterinarian Sam Leadley comically addressed common mistakes on farms:

Pickup-itis
When after purchase, antibiotic remains in the pickup and was never given to the sick animals.
Too-much-water-itis
Directions for reconstituting a powder were not followed—allowing treatment of three calves instead of two. But each injection then carries too little active drug to do the job.
Store-the-syringe-in-the bottle-itis
You always need a needle handy, so just stab the contaminated needle back into the bottle. . . .
Under-dosing-itis . . .

Windowsill-itis
Exposure to strong sunlight and heat destroyed much of the antibiotic's
potency when it was left on the barn windowsill.
Quit-treating-too-soon-itis . . . [173]

Other "itis"-types included "one-drug-fits-all-itis" and "virus-it is,"[174] which
meant using antibiotics against viral infections.

Pharmaceutical companies did their best to promote antibiotic trust in both
the veterinary and farming press. In 2001, Schering-Plough sponsored a prize
quiz in *Farmers Weekly*. Winners were awarded £1,250 worth of weighing equipment.[175] In its three quiz articles, Schering-Plough stressed that farmers should
treat calf pneumonia early and "trust an antibiotic that is effective against all
three main pneumonia-causing bacteria."[176] Schering-Plough's Nuflor (florfenicol) was just such a "proven first-line antibiotic for pneumonia," "effective against
all major bacterial causes of pneumonia," with "no recorded resistance"[177]
and "now available in extra-value 250ml bottles."[178] Winners could use their
weighing kit to "monitor how well cattle recover after treatment with Nuflor."[179]
In 2005, pharmaceutical companies also sponsored "Farmers Weekly Academy," which "educated" about treatments against mastitis, metritis, and other
conditions. Antibiotics produced by the sponsor were conveniently mentioned
below the article.[180] Although the EU forced Britain to restrict direct antibiotic
commercials to farmers in 2011 (chapter 13), similar advertisements remain
common in the veterinary press.

The narrow scope of AGP reforms and ongoing popularity of routine higher-
dosed use of similar antibiotics meant that AMR problems continued. During the early 2000s, Veterinary Laboratories Agency surveys revealed static
AMR among many *Salmonella* isolates but a sharp post-1980s increase of AMR
in *S. typhimurium* against all tested antimicrobials.[181] Surveys also revealed
increasing *E. coli* resistance against tetracyclines, trimethoprim/sulfonamide,
and fluoroquinolones. In 2004, the agency isolated the first animal strains of
extended-spectrum beta-lactamase (ESBL) producing *E. coli* from Welsh calves.
Another outbreak involving the death of several calves occurred in 2006, and
the ESBL-encoding gene (CTX-M-15) was detected on affected farms. In
humans, ESBL *E. coli* were simultaneously causing a rise of resistant urinary
tract infections (UTIs) and blood poisoning. AMR in farm strains likely originated in humans but had been selected for and amplified by agricultural antibiotic use.[182] On farms, ESBL selection pressure was being exerted by rising use of
third and fourth generation cephalosporins, which were listed as critically important antibiotics (CIAs) by the WHO in 2007. Between 1999 and 2006, British
veterinary use of cephalosporins had more than tripled from 220 to 680 kilograms of active ingredient. Prescriptions of other CIAs like fluoroquinolones

had increased by more than 50 percent from 1,230 to 1,951 kilograms between 2000 and 2007.[183]

The effects of rising reserve antibiotic use were predictable. By 2012, 82.8 percent of calves on one UK dairy farm tested positive for ESBL *E. coli* at one day of age. Observers also warned that ESBL proliferation could be driven by feeding antibiotic-tainted waste milk to calves.[184] This practice had been condemned as unacceptable by MAFF researchers in 1990 but had continued due to loopholes in EU regulations.[185] Contaminated waste milk accrued after routine mastitis treatment. In 2010-11, a survey of 557 British dairy farms revealed that 93 percent used antibiotic inframammary tubes to treat mastitis and 96 percent used antibiotic tubes to dry off cows: 29 percent of tubes for lactating cows contained cefquinome and 43 percent of dry cow tubes contained cefalonium. Both drugs were cephalosporins. Over 90 percent of farms fed resulting waste milk to calves.[186]

Improved AMR surveillance also showed that agricultural AMR selection could threaten the health of farmers and veterinarians. A 2008 study of 272 attendees of a Copenhagen pig health conference found that 12.5 percent carried strains of community-acquired methicillin resistant *Staphylococcus aureus* (ST398 MRSA). Carriage among attending veterinarians and farmers was significantly higher than in the average population (about 0.03 percent to 3 percent).[187] Although it only slowly spread to Britain, livestock-associated MRSA was detected on UK farms and food in 2015.[188]

More fine-grained AMR surveillance and improved usage data made it increasingly difficult to resist further statutory regulations for therapeutic antibiotic use in agriculture. Boosted by new government campaigns to curb AMR (chapter 13), critical medical, environmentalist, consumer, and agricultural organizations like the Soil Association have proposed numerous ways to curb AMR hazards.[189] While voluntary reform proposals mostly focus on farmer and veterinary education, statutory proposals include: banning the use of CIAs in agriculture, decoupling veterinary prescription and sales privileges as occurred in Denmark in 1994, introducing annual antibiotic allowances as occurred in Denmark in 2010, phasing out prophylactic antibiotic use, requiring mandatory reports for metaphylactic antibiotic use, and establishing national antibiotic reduction targets. In 2014, the European Consumer Organisation (BEUC) also called for bans of off-label use and the so-called cascade, whereby veterinarians can prescribe drugs listed in the Table of Allowed Substances in Commission Regulation (EU No37/2010) for unauthorized uses.[190]

Reacting to these calls, Britain's conventional farmers and veterinarians have for the most part pursued a strategy of emphasizing voluntarist over statutory reform. The Responsible Use of Medicines in Agriculture (RUMA) Alliance has played an important role in this strategy. RUMA comprises not only important producer and veterinary organizations like the NFU and BVA but also pharmaceutical manufacturers gathered in NOAH. Similar to industry-affiliated think tanks in the US (chapters 8 and 10), RUMA describes itself as an independent "farm to fork" advisory organization with a strong focus on antibiotic policy.[191] Since its foundation in the wake of the 1997 avoparcin ban, RUMA has repeatedly called for more research, challenged statutory reform initiatives such as bans and reduction targets, and developed guidelines for best-practice conventional antibiotic use[192]—the first of which was published in 1999 and immediately accused by the Soil Association and Compassion in World Farming of "'putting a gloss' on Britain's reluctance to reduce the use of drugs."[193]

RUMA's "light green" voluntarism was mirrored by partner organizations like the BVA. Since the late 1990s, the BVA has organized antibiotic awareness campaigns, promoted targeted "rational" drug use, issued an eight-point plan to limit AMR, and developed biomass-based calculations to limit over- and underdosing.[194] Referring to these initiatives, veterinarians reacted angrily to ongoing allegations of antibiotic overuse. Similar to the early 1980s, letters in the *Veterinary Record* complained that physicians were equally complicit regarding antibiotic overuse and justified drug use with references to global population growth.[195] After the United Kingdom's CMO Sir Liam Donaldson called for a ban of quinolone and cephalosporin use in animals in 2008, RUMA asserted that "veterinarians need the full range of antibiotics to help protect the health and welfare of Britain's farm animals, and to help ensure the safety of food derived from those animals."[196] In 2012, RUMA also challenged an EU Parliament resolution calling on member states to reduce the "uncontrolled prophylactic use of antimicrobials in animal husbandry."[197] According to RUMA, the resolution and a similar one in the House of Commons repeated "some of the myths on the use of antibiotics in agriculture and the impact this has on antibiotic resistance in humans . . . . There is, however, a scientific consensus that use of antimicrobials in human medicine rather than antibiotic use in the veterinary sector is the driving force for antibiotic-resistant human infections."[198] In 2014, antibiotic prophylaxis was again defended by BVA president Robin Hargreaves on animal welfare grounds and by NOAH and RUMA in front of the House of Commons Science and Technology Committee.[199] Similar concerns about "overly restrictive" bans and impacts on animal welfare were also put forward by the National Pig Association in response to the EU Parliament's 2018 decision to ban prophylactic antibiotic use.[200]

Social sciences research is providing more detailed insight into how veterinarians and farmers perceive antibiotics and their risks. In 2014, the BVA found that 90 percent of veterinarians were concerned about AMR. Poor client compliance, ignorance, and pressure for antibiotic prescriptions were listed as significant problems.[201] A parallel study found that most farmers and veterinarians believed that better housing, vaccinations, biosecurity, and other environmental improvements could reduce antibiotic use. Similar to the 1970s, antibiotics' cost was described as a significant factor influencing prescriptions. While veterinarians denied that they prescribed antibiotics for reasons of profitability, farmers also denied that that they were pushing their veterinarians to prescribe drugs. Both sides thought that they used antibiotics prudently, opposed statutory restrictions, denied being influenced by industry advertisements, and defended antibiotic prophylaxis. While farmers considered veterinarians to have overall responsibility for stewardship, veterinarians listed NOAH's compendium on drugs as their most trusted source of antibiotic information. Many farmers and veterinarians also thought that there was insufficient evidence to prove a decisive link between antimicrobial use in food-producing animals and AMR problems in humans.[202]

In Britain, the last five years have seen doubts about agricultural AMR hazards come under fire. Improved AMR surveillance, antibiotic reduction targets, and campaigning by CMO Sally Davies have significantly reduced antibiotic use (chapter 13). Pressure for British reforms was increased by aggressive reduction campaigns in other EU countries like Denmark and the Netherlands. Between 2012 and 2017, British antibiotic use in animals (including horses and companion animals) decreased by over 50 percent from 464 to 228 tons—extra-label drug use was not measured. The EU also introduced more reliable measurement units. Antibiotic use is now correlated to the technical population correlation unit (PCU), which equals 1 kilogram of animal treated. Between 2012 and 2016, British antibiotic use almost halved from 66 to 37 milligram per PCU. Use of reserve antibiotics like the fluoroquinolones and cephalosporin also decreased from 2015 onward. Overall antibiotic consumption declined significantly in both pig and poultry production. Reductions proved more difficult in the dairy and cattle sectors.[203]

This rapid drop of British antibiotic use is impressive. Its semi-voluntarist (nonstatutory) nature may also signal an important shift toward bottom-up value-driven antibiotic stewardship in agricultural circles.[204] Cynical observers might, however, note how quick agricultural change occurred once external pressure and the likelihood of further statutory interventions were high enough. Historically, what is remarkable about British antibiotic infrastructures is how physically and culturally robust they have been. Even after EU AGP bans, the proportion of British antibiotics used in animals (about 45 percent of total use in 2013) remained roughly the same as prior to the 1969 Swann report (about 40 to 44 percent). Overall agricultural antibiotic use in 2017 (228 tons) was still 60 tons higher than in 1967 (168 tons).[205] Antibiotic infrastructures' resilience has in part been due to producers' and veterinarians' belief in antibiotics' safety and in part due to the narrow nature of external and internal reforms. Empowered by the Swann report, British veterinarians tended

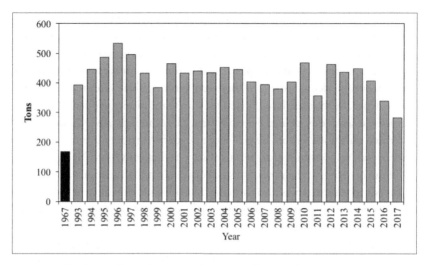

FIGURE 12.3 Total annual sale of active antibiotic ingredients in the UK. Source: "Joint Committee on the Use of Antibiotics in Animal Husbandry and Veterinary Medicine" (London: HMSO, 1969); "UK Veterinary Antibiotic Resistance and Sales Surveillance Report" (UK-VARSS 2017) (New Haw: VMD, 2018).

to blame rising AMR on physicians rather than address economic conflicts of interest leading to antibiotic overuse and jeopardize prescription rights. Despite employing "green" rhetoric and tolerating antibiotic-free market niches, farm organizations repeatedly defended production models in which antibiotics were a routine and—more importantly—cost-effective solution to disease pressure. This was despite evidence highlighting the failure of narrow AGP restrictions from the late 1970s onward. While changing consumer preferences, fraying corporatism, and the 1996 BSE crisis eventually fractured internal support for AGPs, voluntarist impulses for systemic reforms of therapeutic antibiotic use remained relatively ineffective until recent British and EU (semi) statutory pressure.

Resulting antibiotic reductions *are* commendable. The question is how long the current semi-voluntarist push for reform will last. Until 2013, the biggest reduction of British antibiotic use occurred not as a result of voluntarist guidelines but as a result of statutory intervention. Interventions occurred within a unified European market and were the result of continental pressure, initiatives by individual British governments, and increasingly integrated AMR surveillance. What will happen to antibiotic standards and industry's willingness to reform should British agriculture become decoupled from the EU's common market is unclear. At the point of writing, Britain has failed to commit to the EU Parliament's path-breaking 2018 decision to limit metaphylactic and prophylactic antibiotic use.[206]

13

Swann Song

British Antibiotic Policy
After 1969

This chapter explores the development of British antibiotic policy after the 1969 Swann report. Despite Britain's role in pioneering precautionary antibiotic restrictions, subsequent decades were characterized less by British and more by European leadership on antibiotic policy. During the 1970s, officials watered down many recommendations of the original Swann report. Although Britain banned medically relevant AGPs in 1971, bureaucratic rivalries and industry pressure led to an abandonment of other recommendations like improved surveillance, advertising bans, and a strong joint committee on all forms of antibiotic use. Similar to public Swann patriotism, the Swann report turned into an unquestioned official "gospel" to be spread to other nations. It was only during the 1980s that EEC pressure and *Salmonella* scandals forced British officials to establish stricter residue and microbial food controls. With AMR-oriented reforms stagnating under the Thatcher and Major governments, continental Europeans also played a decisive role in pushing for further AGP bans. Britain resisted early reforms despite finding itself isolated on relevant EU committees. In the end, British support for EU AGP bans resulted from the combined pressure of Scandinavian countries, the 1996 BSE crisis, and the election of the New Labour government. While EU bans fulfilled the narrow AGP-focused policy trajectories of the 1960s, statutory reforms of prescription-only antibiotic use continue to divide British decision-makers. Despite the marked recent success of voluntarist antibiotic reductions, it remains unclear whether post-Brexit Britain will enact EU bans of prophylactic antibiotic use by 2022.

Implementation

On November 20, 1969, public pressure and an upcoming election had forced Cledwyn Hughes, Labour's minister of agriculture, to publicly commit his ministry to the Swann report's implementation. The initial promise was that penicillin and tetracycline AGPs would be banned sooner rather than later. As a consequence, MAFF officials were surprised when three weeks ahead of the 1970 general elections a minute suddenly announced that the original plan to ban AGPs by July 1 was "off."[1]

MAFF officials had spent the past months negotiating the July deadline with all interested parties. Moving speedily, they had authorized the use of European "nontherapeutic" antibiotics like Zinc bacitracin (AL Pharma, Norway), virginiamycin (Beecham), and flavomycin (Hoechst) in feeds in 1970.[2] However, behind closed doors, industry opposition to elements of the Swann report remained strong. Although most members of the Association of the British Pharmaceutical Industry (ABPI) eventually accepted the report's recommendations, the NFU continued to insist on the "need to use [therapeutic] antibiotics for stress"[3]—thereby paving the way for low-dosed prescriptions of banned AGPs. American producers also protested the Swann report. As one of the companies most affected by bans of medically relevant AGPs, American Cyanamid launched a systematic campaign to prevent bans of its broadspectrum AGPs. Writing to Parliamentary Secretary John Mackie in May 1970, Cyanamid International president Ernest Hesse warned that Britain had taken on a "heavy responsibility in introducing legislative controls."[4] Having "pioneered"[5] agricultural antibiotic use, Cyanamid considered AMR hazards insignificant.

In contrast to other ABPI members, Cyanamid not only lobbied British officials but also decided to abandon corporatist etiquette and adopt an antagonistic American style of public campaigning. In 1970, Cyanamid sponsored a symposium at the Royal Society for Medicine and flew in two ex-employees from the United States. Having cut their teeth in campaigns against Rachel Carson's *Silent Spring*,[6] Thomas Jukes and Rutgers biochemist Robert White-Stevens cast doubt on AMR hazards. White-Stevens claimed to speak for large parts of "the scientific community" when he portrayed the Swann report as the most recent manifestation of a "tendency to provoke pessimism over scientific progress" and "loudly [bewail] the usually quite insignificant side-effects of technology."[7] Convinced that "scientific agriculture must 'hold a finger in the dike' against starvation," White-Stevens urged authorities to "maintain meat production at its highest level."[8] Thomas Jukes "flatly rejected"[9] Swann. Referring to "an exploding human population," Jukes claimed that the report was not based on "facts":

1) Antibiotics have retained their effectiveness for the production of growth of farm animals after continuous use for nearly eighteen years.
2) There is no evidence that the use of antibiotics in animal feeds has led to an increase in resistance either in animal or human pathogens.[10]

Another familiar speaker at Cyanamid's symposium was Nobel laureate Ernst Boris Chain. Having only recently advised Beecham on how to oppose antibiotic restrictions (chapter 7), Chain attacked the Swann report: "The Swann report has changed nothing. No one has banned the use of antibiotics in veterinary praxis. The farmers can get hold of exactly the same antibiotics as before, only it is more expensive now because you need a vet's prescription. Of course, in all probability more antibiotics will be used, ..."[11] Chain also noted that AMR in animals was less hazardous than claimed: "I thought the main point which Dr. [E. S.] Anderson was making was that there was a danger that [*S. typhimurium* resistance] could be transferred to typhoid. . . . Where is the evidence for that? So far that transfer from typhimurium to typhoid has never occurred and I do not think it will occur."[12] Congratulating Chain on his "quite devastating contribution on the Swann recommendations,"[13] Cyanamid's W. E. McAlister hoped that Chain could also speak at an upcoming NFU meeting. Chain replied by noting that many British companies hoped that the Swann bans would increase sales of their own products. His personal grievance with the Swann report lay in its alleged alarmism: "I am alarmed by the ease with which public opinion can be stirred up . . . and the readiness of the politicians to pass legislation under the pressure of such statements."[14]

Cyanamid's public relations campaign against AGP bans backfired. Trying to foment US-style public opposition to regulatory science (chapter 10), the company and its media advisors underestimated growing British Swann patriotism (chapter 11) and Whitehall's abhorrence of public controversy. Similar to E. S. Anderson's 1960s exclusion from expert consultation (chapter 7), Cyanamid representatives soon discovered that they too could be ostracized from confidential corporatist decision-making. Having attended Cyanamid's 1970 briefing, a MAFF official reported that the "press representatives present" had been "surprisingly hostile."[15] Journalists had wanted to know "why they should believe what [Jukes and White-Stevens] had said in preference to Swann's report."[16] In the end, "only the *Guardian* . . . covered [the] story."[17] MAFF's Animal Health Division was relieved to note that while manufacturers also considered the "Swann proposals too sweeping," they were "not prepared to use the publicity methods adopted by Cyanamid."[18] Learning from its mistakes,

Cyanamid subsequently adopted more indirect forms of lobbying. In November 1970, the company invited Ernst Boris Chain to comment on plans to invite "on an international scale" "a select group of the larger research-based firms engaged in human, animal and plant health work." "The object is to provide a more direct and more effective communications pathway with key governments. . . . recent examples of hasty adoption of ill-judged regulations by various Western parliaments indicates a great lack of understanding of the wider implications."[19]

Despite the failure of Cyanamid's PR campaign, MAFF's position remained complicated. Without evidence of direct harm resulting from AMR selection by low-dosed AGPs, the precautionary Swann bans remained based on a theoretical argument, which was difficult to communicate. MAFF officials complained, "It is not possible to produce conclusive scientific evidence to justify fully either accepting the proposals or rejecting them."[20] Wary of MAFF's public commitment to the Swann report, an internal memo warned, "In view of the uncertainties we cannot afford to wait until [the debate over hazards is over]—if ever."[21] Meanwhile, officials remained committed to maintaining good relations with corporatist partners in farming and industry and tried to offer "as smooth a transition as possible"[22] even if this led to implementation delays. Within nine months of accepting the Swann report, MAFF twice postponed deadlines for penicillin and tetracycline bans, first to October 1970 and then to January 1971.[23]

Following the Conservatives' victory in the July 1970 general elections, implementing the Labour-commissioned Swann report became even more complicated. The new Conservative minister of agriculture, James Prior, initially "agreed that failure to implement the Swann recommendations would be very difficult to defend politically."[24] However, despite officials' warning that "too many deadlines had been breached in the past,"[25] Prior soon began to waver. Six days after a meeting between Prior and his officials in August 1970, MAFF further postponed the implementation date of the Swann bans to March and September 1971.[26] Three days before the August meeting, Prior had received a letter from Cyanamid Britain's Keith Grainger. The intimate tone of Grainger's letter to Prior is striking. After congratulating "dear Jim" for "[getting] off to a very good start!!"[27] Grainger immediately broached the topic of the Swann bans: "Obviously, I would be considered to be biased, but there is little doubt that this report caused considerable comment in scientific circles and some outstanding figures have taken issue with Professor Swann."[28] Grainger warned that the "practical problems and the cost of fully implementing 'Swann' would be immense."[29] Casting doubt on AMR hazards, Grainger was sure that Prior would not "wish farmers and veterinary surgeons to be made the scapegoats for a subject which has much wider implications."[30] While Grainger appreciated that "Jim" had "inherited this particular 'hot potato' from [his] predecessor,"[31] he was eager to supply "proper" information. Signing with his first name, Keith invited MAFF to send representatives to an upcoming symposium on the "The Problems of Drug Resistant Pathogenic Bacteria"[32] at New York's Waldorf-Astoria Hotel. MAFF decided to send A. B. Paterson, director of the Central Veterinary Laboratory (CVL) in Weybridge: "First, and this is as it were a public relations reason, in view of the strong attack which has been made on the Swann Committee recommendations we ought to make it abundantly clear that we are prepared to listen to all the views which are being put forward. . . . Second, . . . we ought in fact make sure that we are in touch with the latest developments."[33]

Similar to the 1967 NAS meeting on drugs in feeds (chapter 4), the New York symposium exposed divisions between European and US attendees. Contradicting European warnings, scientists associated with the US pharmaceutical industry downplayed horizontal AMR risks. While the omnipresent Thomas Jukes repeated familiar arguments,[34] Harold Jarolmen from Cyanamid's Agricultural Division claimed that R-factor transfer in live animals was negligible. In the rare cases that *in vivo* transfer did occur, bacterial strains supposedly lost their "virulence" and were put "at a competitive disadvantage."[35] The symposium also highlighted differing official risk assessments. In his presentation, FDA BVM director C. D. Van Houweling noted: "There are important differences in the uses of antibiotics in animals in Great Britain and in the United States. We believe that through our new drug approvals . . . we have controls that they do not have in Great Britain. However, we do recognize that the continuous or prolonged use of antibiotics in feed does cause gramnegative organisms to develop resistance."[36] Writing to Van Houweling one week later, A.B. Paterson from Britain's Central Veterinary Laboratory looked forward to reading the upcoming FDA task force report on antibiotics in feeds (chapter 10), "which can perhaps be described as the 'U.S. Swann.'"[37] Having learned "a good deal" in New York, Paterson promised to send British material on AMR selection under "feedlot conditions."[38] However, Paterson also warned Van Houweling that combating AMR entailed more than curbing residues in milk and meat: "At the Conference itself I felt that quite unwittingly we were talking rather at cross-purposes in that the FDA has concerned itself very largely with the problem of residues and the possible effect of these residues on the human population, whilst Swann is almost entirely concerned with the possible development of antibiotic resistant strains and their significance in outbreaks of disease in both animals and humans."[39]

Although the 1970 symposium did not resolve transatlantic differences, MAFF was unable to renege on its public commitment to the 1969 Swann report. In March 1971, Britain banned penicillin and tetracycline AGPs. In September, further controls of tylosin, sulfonamides, furazolidone, and furaltadone were published. While the bans looked tough on paper, MAFF had considerably eased the transition for British farmers. In addition to delaying the implementation of AGP bans by over a year, officials provided access to numerous alternatives: farmers could now feed zinc bacitracin (at 100 parts per million to pigs and poultry and 20 parts per million for calves), virginiamycin (at 5 parts per million), flavomycin (at 20 parts per million), and nitrovin. They could also still feed sulphaquinoxaline (at up to 125 parts per million) and sulphanitrin (at up to 300 parts per million) without veterinary prescriptions.[40] Meanwhile, legal access to therapeutic antibiotics remained a prescription away. With agricultural antibiotic infrastructures remaining unchallenged (chapter 12), Ernst Chain's prediction of rising antibiotic use soon came true.

Dilution

In contrast to AGP bans, there was significantly less public pressure on MAFF to implement other Swann recommendations like improved AMR surveillance, antibiotic advertising bans, and a new expert committee on all forms of antibiotic use. Implementation of these additional proposals was further complicated by bureaucratic rivalries and opposition from corporatist partners in farming and the pharmaceutical industry.

In the case of the proposed expert committee on antibiotics, internal power struggles repeatedly delayed official action. It had already been clear in 1969 that the new antibiotics committee would sit uncomfortably between the Veterinary Products Committee (VPC) and the Committee on Safety of Medicines (CSM), which had both been established by the 1968 Medicines Act to review drug applications.[41] Despite pressure from MAFF and the Department of Health and Social Services (DHSS),[42] internal power struggles delayed the formation of the new antibiotics committee. Following the dissolution of the already existing Antibiotics Panel in 1970,[43] it took involved parties two years to compromise on a joint advisory committee, whose advice would be nonbinding to both the VPC and CSM.[44] In the absence of an official body capable of assessing industry applications, British veterinary antibiotic licensing ground to a halt. Bureaucratic power struggles were only fully resolved as a result of pressure by pharmaceutical companies on the VPC.[45] In March 1973, a VPC official informed the CSM: "because of the need for expert advice on EEC matters and also because of the number of antibiotic applications awaiting scrutiny ... I should be most grateful if everything possible could be done ... to save any further embarrassment."[46]

The Joint Sub-Committee on Antimicrobial Substances (JSC) finally started work on July 2, 1973 and was chaired by former PHLS director and Swann member James Howie.[47] The JSC also included other prominent members: officially reconciled with MAFF after his 1960s fallout with the ministry, PHLS researcher E. S. Anderson served alongside University of Liverpool veterinarian

John R. Walton and Houghton veterinary bacteriologist Herbert Williams Smith.[48] Despite their scientific standing, members of the JSC, however, soon found that their committee lacked real political power: pharmaceutical manufacturers repeatedly refused to provide basic sales data and wider policy decisions on antibiotics were made without JSC input.[49] Relations with the CSM and VPC also proved difficult. Because the CSM preferred to consult its own experts, JSC members gradually turned into VPC antibiotic consultants without access to confidential licensing information. In September 1979, members sent a list of grievances to the VPC and CSM and called for a strengthening and reform of their committee. While some distinguished members had simply stopped attending JSC meetings, remaining members were frustrated by their inability to properly evaluate often poorly submitted antibiotic licensing applications. JSC members "not unreasonably consider[ed] that they are too often being invited [to VPC meetings] merely to hazard a guess about the value or safety of the products under consideration."[50] Referring to parallel reports on black market antibiotic sales and "uncritical prescribing" by veterinarians and physicians, the JSC noted that its attempts to "secure rational use of antimicrobial substances"[51] had failed.

For a while, the fate of the dysfunctional JSC hung in the balance. Although some officials considered it unwise to disband the committee due to the "emotive area"[52] it dealt with, the VPC and CSM were both unwilling to devolve power and were facing parallel pressure to simplify drug licensing procedures.[53] According to the CSM, "little would be lost if [the JSC] were disbanded."[54] Unwilling to fund the JSC by itself, the VPC agreed to disband the committee. Although JSC members deplored their dismissal, the JSC was disbanded on December 31, 1980, and Britain returned to the pre-1969 separation of official responsibilities for agricultural and medical antibiotic use.[55]

The JSC was not the only Swann recommendation to suffer from lackluster official support. MAFF and DHSS were also reluctant to commit funds to AMR monitoring. Although the Central Veterinary Laboratory in Weybridge routinely screened *Salmonella* isolates for resistance from 1971 onward,[56] this passive surveillance was not necessarily representative of overall AMR. Pathogen isolates sent to Weybridge came from clinical infections and provided no overview of farms' wider microbial ecology. Meanwhile, routine PHLS surveillance mostly only covered samples received from medical practitioners.[57] To gain a better understanding of AMR on farms, food, and the environment, the PHLS and CVL wanted to conduct an active "in depth [study] of the drug resistance in enterobacteria."[58]

MAFF officials feared that improved AMR surveillance might lead to inconvenient truths. Responding to protectionist agricultural demands for AMR surveillance of meat imports in 1970,[59] officials initially worried that resistance detections might necessitate the "drastic step"[60] of import rejections. According to MAFF, monitoring would surely reveal "major difficulties"[61] with continental and Irish imports. Attempting to avoid a diplomatic incident but appease farmers,[62] officials proposed a limited pilot survey of imports. The AMR survey was to be evaluated by an interdepartmental committee and conducted by E. S. Anderson, who—with characteristic bluntness—had already stated that it was "eyewash."[63] Unperturbed by Anderson's views, MAFF planned to use the pilot survey and official inquiries to foreign governments to appease domestic farming organizations and dissuade the use of banned AGPs for British-bound animals.[64] Britain's program of "spreading the Swann gospel"[65] was, however, soon put on hold.[66] Because of renewed EEC membership negotiations following the 1970 general election, MAFF argued that Britain should put its "own house in order"[67] instead of antagonizing EEC members.

Behind the scenes, E. S. Anderson was nonetheless commissioned to go ahead with a limited import survey. Confident in the efficacy of the Swann bans and planning ahead to the time following Britain's EEC accession, MAFF officials wanted "to confront countries with a vested interest in antibiotics with scientific facts but we would prefer to keep quiet until such data are available."[68] Although the EEC had agreed to introduce Swann-style restrictions in November 1970, EEC Directive 70 524 allowed the use of some substances that were prohibited in the United Kingdom.[69] MAFF's plan was to use AMR data to push for EEC bans of these substances and "general acceptance in the Community of the Swann philosophy"[70] prior to the end of Britain's exemption from full EEC compliance in 1978.

Reporting in April 1972, Anderson seemingly confirmed MAFF suspicions of lax foreign standards. Between February and October 1971, veterinarians had collected samples of Irish beef and pork and of US sheep and lamb livers at British ports.[71] Twenty-nine of thirty-two US lamb livers and two of seven US sheep livers carried resistant *E. coli*.[72] Despite cleansing by high-pressure hosing, 57.8 percent of Irish beef samples and 75.6 percent of Irish pork samples were contaminated with *E. coli*; 25.2 percent of isolated *E. coli* from Irish beef and 94.9 percent of isolated *E. coli* from Irish pork were antibiotic resistant.[73] It was "relatively common" for Irish *E. coli* to be "resistant to combinations of the restricted 'therapeutic' drugs ampicillin, chloramphenicol, neomycin-kanamycin, streptomycin and sulphonamides."[74] According to Anderson, this was unsurprising, "since the Irish farmer has free access to all therapeutic antibiotics."[75] Resistance "was transferable or mobilizable"[76] from 94.8 percent of resistant strains from Irish beef, from 99.5 percent of resistant strains from Irish pork, and from 88 percent of resistant strains from US sheep livers. The confidential study was received with interest by British regulators and the American FDA with whom Britain shared the results.[77] In May 1972, a MAFF official hoped that Anderson's data would pressure the Republic of Ireland to reform antibiotic regulations and block illegal antibiotic sales into Northern Ireland.[78] One month later, the Working Group on the Monitoring of Imported Meat for Antibiotic Resistant Enterobacteria pressed for further studies of French and Dutch imports, as the latter country had "a vested interest in the use of antibiotics in animal husbandry."[79]

To their dismay, British officials, however, soon found that they were throwing stones in a glass house. In January 1973, the planned expansion of AMR monitoring was halted when spotlight surveys indicated ongoing problems within British agriculture. Between 1971 and 1972, streptomycin resistance among domestic *E. coli* isolates had risen from 47 to 50 percent, tetracycline resistance from 44 to 50 percent, and ampicillin resistance from 49 to 56 percent.[80] Although AMR levels were lower than in other EEC countries, the results contradicted projected Swann outcomes and undermined Britain's attempt to promote Swann-style AGP restrictions within the EEC. A MAFF official warned that previously planned comparative AMR surveys could "only be done on the basis of a free exchange of information": "this might rob us of some of our advantage during the 5-year derogation from EEC practice . . . in the course of which we hope that the [EEC's antibiotic policy] will align with us."[81] While he was confident that results would not "invalidate the Swann doctrine,"[82] it might take longer than five years for positive data to emerge. For fear of jeopardizing the "Swann gospel," no new systematic AMR surveys were commissioned. Potential studies would likely have undermined limited AGP restrictions: when British farmers protested against egg imports from French hens fed antibiotics in 1975 (chapter 12), a MAFF survey found no difference regarding AMR on French and British produce.[83]

With British officials stopping AMR monitoring rather than endangering an ossified gospel of partial restrictions, the Swann report's broader goal of curbing AMR was effectively forsaken within four years of its publication. Instead of critically revaluating and expanding existing regulations, short-term economic considerations reduced chances for an effective response to rising AMR. The lack of robust British monitoring data also impacted parallel FDA attempts to restrict AGPs (chapter 10). In 1974, a joint US/Canadian fact-finding mission noted that the limited Swann bans had not resulted in significant economic problems for British farmers but also found no evidence with which to support plans for North American AGP restrictions.[84] In the same year, PHLS and CVL representatives warned that passive surveillance of British *E. coli* isolates showed a rise of chloramphenicol resistance—probably as a result of ongoing veterinary chloramphenicol use against *Salmonella*. With the exception of declining streptomycin resistance, AMR levels in British *Salmonella* isolates had remained constant. AMR patterns were similar to those in Ireland where AGPs had not been banned.[85]

In addition to weakening the JSC and stalling AMR monitoring, officials also failed to enact the Swann report's proposed ban of antibiotic advertisements to farmers. In 1972, MAFF's Animal Health Division noted that the BVA, the Royal College of Veterinary Surgeons, and the British Pharmaceutical Society all favored advertising bans: "They argue that [veterinarians'] . . . task should not be made more difficult by uninformed pressures from clients responding to advertising. Also they think that . . . there is a danger that some clients whose interest has been aroused will obtain supplies illicitly."[86] Unsurprisingly, the ABPI, NFU, and British Poultry Federation opposed bans. The latter faction argued that restrictions gave further "control to veterinarians and that the Government should not 'molly-coddle' them."[87] Following several inconclusive meetings, MAFF officials noted that in the absence of a corporatist compromise between the two factions the "proper course would be to accept the logic of the Swann recommendation" and announce "the intention to make regulations in the absence of an effective voluntary [restriction] scheme."[88] However, similar to his 1971 delay of AGP bans, Conservative minister of agriculture James Prior again failed to heed his officials' advice. Despite appeals by junior health minister Lord Aberdare, Prior referred the matter to the still nonexistent JSC. In a draft letter to Aberdare, Prior stated: "I do not subscribe to the view that the veterinary profession is not strong enough to resist pressure from its farmer clients; and although the drug manufacturers are obviously keen to sell I am not sure that advertising necessarily increases overall demand."[89]

Following Prior's promotion to leader of the House of Commons, the issue of antibiotic advertising bans resurfaced two years later when both the JSC and VPC endorsed restrictions.[90] However, Prior's successor failed to address the matter ahead of the 1974 general election. Following the Conservatives' defeat, MAFF officials duly resubmitted the proposed restrictions to their new Labour ministers.[91] A frustrated official noted: "My own view is that, while in matters of this sort there is often much to be said for leaving well alone [sic], in this case we cannot ignore the advice of 3 official committees nor swallow the assertion of the manufacturers that they do not seek to enlist the support of the farmer in building up an even bigger market in therapeutic antibiotics, regardless of true need."[92] Half a decade after their initial proposal by the Swann report, the opportune moment for advertisement restrictions had, however, passed. Despite ministerial approval, proposed restrictions were quietly dropped after technicalities further delayed official action in 1975.[93]

By 1980, many important proposals of the 1969 Swann report had thus either failed to materialize or been watered down. Although limited AGP bans had been implemented in 1971, there was no systematic British AMR monitoring, no JSC, and no ban of antibiotic advertising. The absence of these accompanying measures facilitated an increasing stagnation of British antibiotic policy.

Leading British experts were well aware of the problematic state of affairs. In 1982, microbiologist Alan Linton gave an overview of British agricultural antibiotic use and AMR. On farms, medically relevant antibiotics were still regularly used for growth promotion, prophylaxis, and treatment. Ongoing direct agricultural access to tylosin was particularly worrying because of AMR selection against macrolides. Overall, AMR hazards remained difficult to judge. Whereas studies indicated that AMR in *E. coli* had surged and plasmids were being exchanged between bacteria in humans and animals, there was no evidence that resistant *E. coli* of animal origin had compromised human therapy. Most cases of now notifiable salmonellosis remained relatively antibiotic sensitive. However, since 1974 there had been a surge of cattle-associated outbreaks of new multiple-resistant *S. typhimurium* strains.[94] Between 1977 and 1982, 611 human infections—three of them fatal—had been caused by *S. typhimurium* phage types 204, 193, 204a, and 204c. Many more had probably not been reported.[95] Linton nonetheless considered therapeutic antibiotic restrictions impractical. Although veterinary antibiotic use was often "lax and empiric," restrictions of therapeutic antibiotics would come too late to curb AMR "unless set at such a level as to make a major [economic] impact, which would be unacceptable."[96] According to Linton, Britain's 1971 AGP bans could not have curbed AMR "in the absence of simultaneous restrictions on the prophylactic and therapeutic use of antibiotics in both animals and man."[97] This bleak assessment was seconded by former Swann member and ex-PHLS and JSC director James Howie:

> the [Swann] Committee failed to recognize certain factors which ... ensured that farmers could get direct access to as much antibiotics as they wanted. . . . The [JSC] ... soon became aware that the problems with *S. typhimurium* in calves were recurring. . . . It had no powers to control the black market in antibiotics or authority to command information from industry, and it was ultimately disbanded. The result is that we now have a free-for-all, and there is a real danger that the usefulness of antibiotics may be compromised by this.[98]

Residues and Meat Hygiene

The situation was little better regarding antibiotic residues in food. Established following the 1963 milk scandal (chapter 7), Britain's only formal residue monitoring program had reduced violation rates of penicillin in milk to about 1 percent in the mid-1970s.[99] However, in contrast to the United States (chapter 10), demands for an extension of monitoring to meat fell on deaf ears. Meat inspection was comparatively weak, and there were no systematic veterinary or hygiene controls in many abattoirs.[100] There were also no standardized official assays for antibiotic residues in meat.[101]

With no monitoring data forcing authorities to act, residue rules remained based on the 1955 Food and Drugs Act and the Preservatives in Food Regulations, which prohibited adding antibiotics to food.[102] However, there was no legislation outlawing the sale of food in which antibiotic residues were "present" as opposed to deliberately added.[103] Until the 1980s, British consumers' main protection from drug residues in food lay in the wording of Section Two of the 1955 Food and Drugs Act according to which it was an offence "to sell to the prejudice of the purchaser any food which is not of the nature, substance or quality demanded."[104] In a few cases, local authorities used Section Two to exert pressure on producers and MAFF to reduce residues in culturally sensitive products like eggs.[105]

From the mid-1970s onward, the rising international emphasis on laboratory-based monitoring of food safety placed lax British policies under pressure.[106] Pushing for a ban of antibiotic residue-tainted EEC shipments,[107] West Germans rejected three consignments of residue-laden British meat in 1975 and 1976.[108] Following further rejections by US, Dutch, and Scandinavian authorities, MAFF scrambled to restore trust in £150 million of annual meat exports: "If we are to avoid placing our export meat trade in jeopardy, and one could argue that it is already on the brink, . . . it is imperative that a more positive policy on 'residues in meat' be formulated."[109] In March 1976, a Sub-Group on Antibiotic Residues was tasked with establishing limited residue monitoring for meat exports.[110] Progress was slow. Denied statutory access to abattoirs and unable to obtain confidential assay information from industry or the VPC,[111] it took until 1978 for officials to publish a first survey: between July 1977 and March 1978, 933 voluntary samples from 23 export abattoirs had been analyzed using new EEC bacterial inhibition tests. Although positive test incidence was below 0.3 percent, experts warned that results were compromised by decaying samples of both domestic and imported meat and the absence of key test organs. In the case of 153 samples from London's Smithfield abattoir, "many liver and a proportion of the kidney and beef samples were of extremely poor quality . . . . Certain livers were green and strong smelling, and really should not have been tested."[112] MAFF nonetheless used initial results to reassure trade partners. Reporting on a meeting with US and Canadian delegates, an official noted: "It clearly came as something of a surprise to the American and Canadian delegates to realise the extent to which a combination of administrative and legal provisions could be effective. They had clearly heard . . . that the so-called loopholes in the law, plugged only by administrative recommendations, were less than effective."[113]

European partners like West Germany and Denmark were more skeptical and pressed for unified EEC residue monitoring, which would analyze a fixed percentage of the national meat trade.[114] Its lack of monitoring capabilities made Britain oppose fixed percentages (around 5,200 samples per year) and favor a

US-style system, which used a smaller number of random samples (around 1,800 per year) to extrapolate overall contamination levels. Should residues be found, more intensive local sampling would in theory reveal offenders.[115] Britain's proposal caused conflicts. In 1978, the British delegate to an EEC meeting on residues noted that random sampling had led to a "good deal of acrimonious discussion with the German representative proving the most vocal."[116] Continental opponents had talked "a good deal of nonsense ... about the willingness of consumers to pay for extra protection."[117] British resistance to percentage-based monitoring ultimately weakened proposed EEC rules. Member states agreed to trial randomized testing, which would indicate whether more extensive sampling was necessary. Countries like Germany, Italy, and the Netherlands were, however, allowed to monitor more extensively within their borders.[118]

Coinciding with media reports about black market antibiotic sales (chapter 11), business-friendly British pilot testing began in 1980. In 1982, the first national residue report claimed with 95 percent certitude that less than 1 percent of British cattle, calf, sheep, and pig products contained antimicrobial agents above tolerances. Sulphadimidine was probably present in less than 4 percent of British meat. Poultry products were not tested.[119] However, once again, official data had to be taken with a grain of salt: voluntarily participating slaughterhouses could theoretically preselect uncontaminated samples, state-employed veterinarians threatened to manipulate sampling rather than antagonize clients, and the sensitivity of EEC residue tests varied substantially for different antibiotic classes.[120] A contemporary survey of more extensively monitored German-bound exports found that 31.2 percent of meat samples tested positive for antibiotics.[121]

British residue sampling programs gradually improved as a result of new European directives. By early 1985, the EEC mandated traceable samples, follow-up sampling of offenders, maximum residue limits, and outlawing the slaughtering of residue-laden animals.[122] Although Britain delayed their introduction for four years, the new EEC regulations forced officials to expand surveillance to 28,000 samples in 1989.[123] The provision of samples was now mandatory, and veterinary surgeons and environmental health officers were empowered to routinely enter and inspect slaughterhouses. In 1988, EC Directive 88/409 also required inspections of domestic slaughterhouses to be brought in line with stricter ones for export abattoirs.[124] Although critics bemoaned its financing by industry, a new Veterinary Medicines Directorate (VMD) was placed in charge of veterinary drug licensing, enforcement, and residue surveillance in 1989.[125] Between 1990 and 1995, salmonellosis outbreaks and BSE concerns also led to mandatory ante- and post-mortem hygiene inspections across Britain.[126] Ending decades of voluntarist policies, integration into Europe's common market had forced officials to address pharmaceutical and microbial food contamination with statutory surveillance and regulations.

Reform

A similar process of domestic scandals, external European pressure, and ensuing British reforms occurred in the case of AMR. Despite national alarm about resistant *S. typhimurium* (chapter 11), British AMR policies had stagnated throughout the 1980s. Economic problems and its aim to reduce regulatory burdens made the neoliberal Thatcher government reluctant to consider further antibiotic restrictions or a reform of passive *Salmonella*-focused AMR surveillance.[127] A new round of precautionary AMR-oriented antibiotic reform was initiated as a result of continental pressure and the 1988 salmonella scandal. Whereas a comparative paucity of data had made it easy for British officials to downplay problems during the 1980s, the 1990s saw improved AMR surveillance increase the political cost of regulatory inaction: the more one knew, the more one had to do. Initially opposing rather than shaping European antibiotic reform, British officials found themselves outmaneuvered by more proactive EU countries from 1995 onward.

British officials' hesitancy to reform antibiotic use was not because of a lack of warnings. Amid BSE and salmonellosis concerns, MAFF had convened an Expert Group on Animal Feedingstuffs in February 1991. Headed by University of Nottingham animal physiologist George Eric Lamming, the so-called Lamming committee identified "gaps in [feed industry] legislation and its enforcement."[128] Regarding residues, the Lamming committee criticized insufficient official oversight and lacking assay methods for prescription-only medicines.[129] The committee also warned about rising AMR. Between 1981 and 1990, multiple resistance among *S. typhimurium* isolates from British cattle had risen from 15 to 66 percent and from 2 to 8 percent in poultry isolates.[130] The committee recommended expanding routine AMR monitoring to human *E. coli* isolates, discouraging the "prophylactic use of antibiotics with cross-resistance to those used in human medicine,"[131] and changing rules allowing manufacturers to send diluted antibiotic substrates to farmers for home-mixing.[132]

Published in 1992, the Lamming report, however, failed to provoke a rethinking of British antibiotic policy. Similar to contemporary FDA decision-making (chapter 10), British officials ignored the Lamming committee's warnings about cross-resistance selection on farms and approved agricultural fluoroquinolone use (Baytril) in 1993.[133] The risks of agricultural fluoroquinolone use were well known in Europe. After fluoroquinolones' introduction to German food production in 1988, fluoroquinolone-resistant variants of *S. typhimurium* DT204c reached "a prevalence of 50 percent"[134] in calf isolates in certain areas. Following Britain's licensing of the drugs, multiple resistant strains of *S. typhimurium* DT104 with reduced susceptibility to ciprofloxacin were soon isolated from domestic animals.[135]

Within the EU, British antibiotic policies caused increasing consternation. Bowing to critical member states, the EU had established a commission to assess agricultural growth promoters in 1992.[136] Two years later, the EU banned the agricultural use of chloramphenicol.[137] Reports that avoparcin AGPs selected for cross-resistance to the reserve antibiotic vancomycin also caused disquiet. While carcinogenicity concerns had prevented their licensing in the United States, avoparcin AGPs had been licensed in the EEC in 1976. In 1986, the first glycopeptide-resistant strain of *Enterococcus faecium*, which could resist high vancomycin doses, was detected in humans. Reports that allegedly nontherapeutic avoparcin could also select for glycopeptide-resistant enterococci appeared in the early 1990s when vancomycin-resistant *Enterococcus faecium* (VRE) was isolated from German, Danish, and British animals. Germany, Denmark, and the Netherlands subsequently vetoed a British request to license avoparcin in European dairy cow feeds. In 1995, VRE concerns led to a voluntary and then statutory Danish ban of avoparcin according to Article 11 (the safeguard clause) of Directive 70/524/EEC. Germany and other EU members followed suit.[138] Britain resisted. Unable to lift continental BSE embargoes, the Conservative government was reluctant to endorse EU-led antibiotic reform ahead of the May 1997 general elections (chapter 11). Although British calls for more research were supported by the EU's Scientific Committee for Animal Nutrition, the EU Committee of Experts on Feed Additives and European Commission considered evidence linking avoparcin and VRE sufficient for precautionary bans.[139] In a far cry from its 1960s leadership on AMR, British resistance to bans was overruled 14:1 and the EU restricted avoparcin on April 1, 1997.[140]

Following its 1997 election victory, the New Labour government tried to restore trust in British food regulations. One of its most important decisions was to end the institutionalized conflict between business and consumer interests within MAFF. Responsible to the Department of Health, an independent Food Standards Agency (FSA) started work in April 2000. In 2001, MAFF itself was dissolved and turned into the Department for Environment, Food and Rural Affairs (DEFRA). While the VMD retained responsibility for residue monitoring, DEFRA was responsible for the control of microbial threats such as food-borne zoonosis.[141]

Institutional changes occurred parallel to a significant reordering of British antibiotic policies. Between 1997 and 1998, a large number of high-profile AMR warnings led to significant pressure on British and European regulators to consider further precautionary AGP bans. In 1997, a WHO meeting on "The Medical Impact of Antimicrobial Use in Food Animals"[142] repeated 1970s calls to exclude antibiotics effective against gram-negative pathogens from animal nutrition and stressed the need for improved AMR surveillance in *Salmonella*, *E. coli*, *Campylobacter*, and *Enterococcus*. In April 1998, a second WHO meeting

expressed general concern about "non-medical uses of antimicrobials."[143] Two
months later, a third WHO meeting on agricultural quinolone use admonished
veterinarians to reduce prescriptions and warned against using quinolones
"for performance enhancement."[144]

In Britain, three additional reports strengthened WHO calls for antibiotic
reform. Published in early 1998, a House of Lords report on AMR explicitly referenced "a continuing threat to human health from imprudent use of
antibiotics in animals."[145] Sections of the report that discussed agricultural
antibiotics bore a strong resemblance to the 1969 Swann report. This was no
coincidence. Veterinarian Lord Soulsby, the committee's leader, had been a close
friend of the late Michael Swann.[146] According to the Lords, Britain had once
"led the world in addressing the threat to human health posed by antibiotic
use in farming practices."[147] However, important parts of the Swann report had
been watered down. Criticizing the JSC's dissolution and inadequate AMR surveillance, the Lords warned, "departmental and agency boundaries must not
be allowed to prevent the Government from getting a grip on the whole of this
issue."[148] The Lords recommended phasing out virginiamycin AGPs because of
cross-resistance to the new antibiotic dalfopristin/quinupristin (Synercid) and
also criticized veterinary antibiotic use: "mass-treatment of herds ... and
flocks ... with [antibiotic] agents cannot be best practice."[149] Remaining loyal to
the Swann report, the Lords, however, urged veterinary "self-regulation in preference to legislation."[150] Published shortly afterwards, a second report by the
House of Commons Select Committee on Agriculture also recommended "a
ban on the use of antibiotics in farming as growth promoters, and tighter restrictions on their use for subtherapeutic or prophylactic purposes."[151] A third report
by Britain's Standing Medical Advisory Committee subsequently reinforced
parliamentary criticism of AGPs and veterinary "fire brigade" treatments.[152]

The wave of high-level warnings left little political room to resist contemporary Scandinavian lobbying for AGP restrictions (chapter 11). At an EU
conference on 'The Microbial Threat' in Copenhagen in September 1998, over
400 participants—including EU CMOs—stressed the necessity of coordinated
action and more reliable data on antibiotic consumption and AMR.[153] Fourteen days after the conference, the EU passed Decision No 2119/98/EC,[154]
which established the European Antimicrobial Resistance Surveillance System
(EARSS) for resistance in humans and animals.[155] On December 17, 1998—
three months after the Copenhagen conference—the EU Commission decided
to ban four of the most popular AGPs (virginiamycin, zinc bacitracin, spiramycin, and tylosin phosphate) with Directive (EC)2821/98. Abandoning three
decades of post-Swann complacency, Britain supported the precautionary
expansion of AGP bans.

With government reports highlighting ongoing stewardship problems
on British farms and Scandinavian countries pressing for additional bans,

remaining AGPs would not stay permitted for long.[156] In 2000, Denmark unilaterally restricted all antibiotics to prescription only status.[157] Spearheaded by the Irish Commissioner of Health and Consumer Protection, David Byrne, the EU decided to emulate Danish measures. On March 25, 2002, the EU Commission proposed to ban all AGPs. Nine months later, regulation (EC) No 1831/2003 commenced the phasing out of AGPs but not of coccidiostats like ionophores by December 31, 2005.[158]

For Britain, the 2006 AGP bans fulfilled a fitful regulatory trajectory that had begun with the 1962 Netherthorpe call to reserve new antibiotics for therapy and the 1969 Swann recommendation to ban medically relevant AGPs. Both reports made important contributions to precautionary drug regulation by basing recommendations on "abstract" AMR hazards. However, their corporatist origins also made them try to "fix" antibiotic infrastructures by redistributing control over drug access. The resulting Swann "gospel" of narrow legal distinctions between relevant and irrelevant drugs as well as between low- and high-dosed uses of the same substance failed to reduce antibiotic consumption and AMR. Predicted by both E. S. Anderson and Ernst Boris Chain, this failure was, however, not officially acknowledged. Afraid of jeopardizing trade interests, British officials did not support comprehensive AMR studies and held fast to popular Swann patriotism. A similar tendency to stall surveillance and reforms also occurred regarding antibiotic residues in meat. In both cases, British resistance to further reforms was only overcome as a result of major food safety scandals (salmonellosis, drug detections, and BSE) (chapter 11), fraying agricultural resistance (chapter 12), and foreign pressure resulting from integration into the European common market. Although the 1996 BSE crisis and its parallel 14:1 defeat over avoparcin arguably left it with no other option, Britain's support of EU AGP bans ended decades of regulatory stagnation.

After AGPs

What would happen after 2006 was not immediately apparent. Consistently accounting for less than 20 percent of total use, it was clear that banning AGPs would only partially reform British agriculture's antibiotic infrastructures. In this situation, improved surveillance capabilities helped to keep official attention focused on AMR selection and problems resulting from therapeutic antibiotic use.

From the beginning, reinvigorated British surveillance efforts were closely coordinated with European partners and the new EARSS. In 1999, the Labour government founded DEFRA Antimicrobial Resistance Coordination (DARC). DARC was tasked with encouraging prudent antibiotic use and reviewing AMR surveillance in select sentinel pathogens and commensal organisms, which Britain had launched following the 1998 Copenhagen conference.[159] In

2003 and 2007, AMR monitoring of *Salmonella* and *Campylobacter* in food and animals became mandatory EU-wide. Britain provided additional AMR data for *E. coli* and *Enterococci*.[160] Readdressing antibiotic residues in 2001, the British government also installed an independent Veterinary Residues Committee (VRC) to advise the VMD and Food Safety Agency. In the same year, the EU's Veterinary Medicinal Products Directive (2001/82/EC) called for harmonized controls for the manufacture, authorization, marketing, and distribution of veterinary medicines. Coming into force in October 2005, the resulting British Veterinary Medicines Regulations consolidated the plethora of controls previously contained in the 1968 Medicines Act and over fifty amending Statutory Instruments.[161] The VMD also compiled antibiotic sales data. Although Britain had collected this data on a voluntary basis since 1999, 2005 statutory requirements produced more detailed insight into antibiotic consumption on British farms. In 2009, data gathering was further standardized and made comparable at the European level with the creation of the European Surveillance of Veterinary Antimicrobial Consumption (ESVAC) by the European Medicines Agency.[162]

By establishing regular and independent monitoring of AMR, residues, and antibiotic usage in coordination with EU partners, the British government has fulfilled key demands of antibiotic critics. It is now far easier to identify and resolve antibiotic-related problems both within agriculture and corresponding bureaucracies. Mirroring what Nicolas Fortané and Frédéric Keck have described in relation to zoonosis monitoring, current AMR surveillance is also indicative of a wider policy shift toward biosecurity. In the EU, real-time surveillance of emerging microbial threats and sentinel organisms has strengthened "ecological" notions of disease and boosted the standing of epidemiologists regarding AMR decision-making.[163] In the case of agricultural antibiotics, the regular publication of national AMR and antibiotic sales data has also been crucial to keeping antibiotic reform on the political agenda. With public concerns ebbing in the decade after the 2006 AGP bans (chapter 11), UK governments were initially reluctant to push for further statutory reforms of agricultural antibiotic use. In 2011, the EU Commission was forced to criticize Britain for infringing Directive 2001/82/EC by continuing to permit direct antibiotic advertising to farmers. While the practice has since been banned,[164] British legislators also failed to support CMO Sir Liam Donaldson's 2008 call for statutory bans of CIAs, endorse European calls for statutory restrictions of extra-label cascade prescribing, or follow Scandinavian countries and divorce veterinary prescription and sales rights (chapter 12). In this situation, neutral reports on antibiotic usage and AMR and EU-wide comparisons helped hold decision-makers in politics and industry to account.

In recent years, CMO Sally Davies has emerged as a further leading force for British antibiotic reform. Appointed in 2011, Davies made national headlines in

2013 by stating that AMR was "as big a risk as terrorism"[165] and should be added to the national register of civil emergencies. In Britain, Davies's warnings triggered a flood of new AMR research and reports. High-profile initiatives under the Conservative Cameron government subsequently attempted to turn antibiotic stewardship into a signature area of British leadership. Since 2013, measures supported by Britain have included the Longitude Prize for rapid diagnostics, high-level international commitments to tackling AMR by the EU, UN, and G20, and the financing of new antibiotic development via public-private initiatives.[166]

Coinciding with WHO initiatives to standardize international AMR surveillance, the British government also commissioned a major review of AMR by Goldman Sachs's former chairman of asset management Jim O'Neill. Published in 2016, the O'Neill report tried to put a number on the global cost of AMR ($100 trillion by 2050 with 10 million annual deaths) and proposed "cost-effective" policies to curb it. The main report's recommendations include: financing awareness campaigns; funding the development of new antibiotics and rapid diagnostics; promoting better sanitation, hygiene, and infection control; supporting the development and use of vaccines and alternative treatments; improving global AMR surveillance; and establishing binding goals for antibiotic reductions.[167] In the case of agriculture, the O'Neill report has called for uniform definitions and a phasing-out of CIAs, more transparency about antibiotic use, and the establishment of clear reduction targets to an agreed level of antibiotic use per kilogram of animal production. It also argues for a new focus on curbing antimicrobial waste in the environment.[168] Although the full O'Neill report stresses that a range of statutory and nonstatutory tools should be considered, its special report on agriculture primarily emphasizes fiscal measures like taxing antibiotics to discourage prophylactic and growth promotional uses.[169]

As a signature initiative of the Cameron government, the O'Neill report highlighted AMR's economic risks and made UK officials commit to cutting agricultural antibiotic use to 50 milligrams per PCU by 2020—something that was already achieved in 2016.[170] However, following David Cameron's resignation in the wake of the June 2016 Brexit referendum, the fate of British reform again hangs in the balance. Co-developed by Britain, a new and bold set of statutory interventions into therapeutic antibiotic use was overwhelmingly approved by the EU Parliament in October 2018. Breaking with over eighty years of voluntarist antibiotic arrangements, measures include a statutory ban of reserve antibiotic use in veterinary medicine, restrictions on prescribed prophylactic antibiotic use, and improved monitoring of veterinarians' antibiotic use.[171] It remains unclear whether Britain will continue to adhere to new EU standards after Brexit. Similar to the 1970s, there is a credible threat that short-term trade incentives may make future British officials decide to weaken antibiotic standards.

Decoupled from the EU, Britain's self-proclaimed leadership regarding AMR remains fragile. Historically, antibiotic reforms have been most successful when they were supported by a broad coalition of actors. Reducing AMR requires individual *and* collective action. The recent upsurge of Anglo-American nationalism threatens to undermine both international risk governance and collective responsibility for antibiotic stewardship.

Conclusion

Antibiotics Unleashed

Roughly eighty years after their first use in agriculture, antibiotics have affected nearly every sector of global food production. In the United States and Europe, no conventional farmer or veterinarian remembers a time when antibiotics were not at hand to treat and prevent disease in animals.

Antibiotic infrastructures are also shaping farming outside of Europe and the United States. Although American livestock producers pioneered concentration, intensification, and integration, their leadership is now being contested by operations in countries like Brazil and China. In the case of Brazil, poultry production increased twenty-fold between 1968 and 1990 and amounted to 12 million tons in 2009. Having overtaken the United States as the world's largest poultry producer, over 90 percent of Brazilian poultry are produced in confined and intensive settings with two mega-companies dominating the market.[1] In China, Western firms were initially allowed to develop industrialized production systems but later ceded ownership to Chinese companies. In 2008, China produced over 450 billion pigs—nearly eightfold the number produced in the United States. In 2013, the Chinese WH Group (then known as the Shuanghui Group) acquired the American Smithfield Foods company and is now the largest pork producer in the world.[2] Driven by middle-class spending, the centers of gravity in global animal production are shifting. However, the production methods being used are more similar than ever before.

Antibiotics have played an important role in facilitating the global spread of standardized animal monocultures. In 2010, global agricultural antibiotic use was estimated conservatively at about 63,151 tons. Average antibiotic use amounted to 45 milligrams per kilogram of cattle, 148 milligrams per kilogram of chicken, and 172 milligrams per kilogram of pork. China already accounted for 23 percent of global agricultural antibiotic use, followed by the United States (13 percent), Brazil (9 percent), India (4 percent), and Germany (3 percent). By 2030, global agricultural antibiotic use is projected to increase to over 100,000 tons per year (67 percent increase). One-third of increased consumption will likely be due to a further shift toward intensive and integrated animal production.[3]

It is clear that this expansion of antibiotic use is not without risks—the past seventy-five years have given us ample warning. Although it is not within the scope of this book—nor the abilities of its author—to comment on agricultural antibiotics' contribution to global AMR, it is important to stress how unpredictable the dynamics of the planetary "resistome" are—and how difficult it is to assess, manage, or reverse any changes to the global burden of AMR.

Recent work by Laura Kahn highlights the complexities of antibiotic stewardship. In Denmark, legislators and industry went to great lengths to reduce agricultural antibiotic use. Denmark's small size and sophisticated AMR surveillance means that it has received a lot of scientific attention. After phasing out AGPs and separating veterinary prescription and sales rights, Denmark introduced a "yellow card" system to further reduce therapeutic antibiotic use in 2010. Pig producers are allotted thresholds for average antibiotic consumption over a ten-month period. Should producers exceed thresholds, injunctions are issued. Farmers are required to pay a fee for each injunction and resulting inspections. However, the effects of Danish leadership on AMR have been difficult to measure. While banning avoparcin in 1995 lowered VRE in livestock, it has since emerged that 70 to 90 percent of animal and community-acquired VRE strains differ. Rather than farm animals, companion animals seem to act as major VRE reservoirs for humans. On farms, enterococcal resistance against streptogramins, avilamycin, and erythromycin fell following 1990s AGP bans— but tetracycline resistance increased. Judging the effects of AGP bans at the European level is even more difficult: after 2006, some forms of AMR seem to have declined—but others increased.[4]

At the microbial level, whole-genome sequencing is currently also recasting the way we understand both the history and current context of AMR selection and proliferation. Recent studies have highlighted that many of the resistance genes found in current bacteria are not new but ancient because of millennia of inter-bacteria competition.[5] In 2014, researchers reported that the first accessioned strain in Britain's National Collection of Type Cultures—a strain of *Shigella flexneri* isolated from the Western Front during World War One— was already resistant to penicillin and erythromycin and also contained a complement of chromosomal AMR genes similar to more recent isolates.[6] Other studies have reconstructed the unpredictable effects of the modern era of mass antibiotic use. In 2017, researchers showed that the 1950s use of first generation β-lactams drove the evolution of MRSA and of ampicillin resistance in *S. typhimurium*—before either methicillin or ampicillin were marketed. These remarkable findings challenge long-established linear chronologies of selection pressure and AMR emergence. They also highlight the potential fallout of widespread agricultural and medical antibiotic use for existing and not yet discovered drugs.[7]

Over half a century after the discovery of horizontal gene transfer, the dynamics of the planetary resistome remain bewildering and frequently defy linear cause and effect models. This is not only challenging from a judicial point of view but also means that final certainty about the costs and benefits of different stewardship models remains elusive. Similar to 1969, the question is whether this inherent uncertainty justifies regulatory inaction or precautionary restrictions? When asked this question by the FDA in 1971, Herbert Williams Smith answered by emphasizing the "bewildering" aspect of AMR: "It is a danger because it can happen. . . . if we don't need to have the animal reservoir of R-factors we are better off without it."[8] In the United States, Congress thought differently. Forty-seven years later, opinions on antibiotic restrictions remain divided. What seems clear is that once bacterial sensitivity to an antibiotic is lost, it is often lost for the foreseeable future. Defying traditional dose-response models, AMR can occur gradually or suddenly. The extent of resulting problems depends on a whole host of further genetic and epidemiological factors. Although AMR selection in medicine poses more immediate threats to human health, decades of agricultural antibiotic use have facilitated the rise and spread of dangerous resistant organisms and genes ranging from *S. typhimurium* Type 29, *S. typhimurium* DT 104, LA-MRSA, to *mcr-1*. Some threats disappeared, others stayed. The ongoing growth and rising interconnectedness of global antibiotic infrastructures in food production increases the likelihood of future outbreaks.

Four Stories

So, is it antibiotics' destiny to be pyrrhic? Will victory over microbes in the present negate future victories? And what can this book contribute to current debates? At first glance, the historical perspective gives little reason for optimism. In Britain and the United States, the history of agricultural antibiotics reveals a seemingly repetitive pattern of overenthusiasm, warnings, technical fixes, complacency, and a rediscovery of earlier warnings. After seventy-five years of debate, no ready solution to AMR hazards seems at hand. However, ending the story here would be too simple. A second, more nuanced glance can also uncover the underlying infrastructures, epistemologies, and path dependencies that have contributed to the current situation. This second glance is useful in fostering awareness not only for the complexity of AMR as a biological but also as a social, economic, and cultural phenomenon. In his history of tuberculosis, Christian McMillen notes that appreciating the cyclical nature and historical contours of major health challenges is to "learn to be humble: people to whom problems they are just encountering seem new and fresh might do well to consider they might not be the first responders . . . history does not march progressively onward; we are not always getting better."[9] Knowing about the rootedness of complex biosocial problems is not only a prophylactic against rediscovery but also a way of averting a reduction of history into "simple moral lessons" about past mistakes and successes.[10] A similar approach is useful when thinking about the history of agricultural antibiotics. What emerges from this book are four stories about short-termism, epistemic fragmentation, infrastructure, and narrow reform.

1. A Story of Short-Termism

The book's first story is the story of a technology that was hastily introduced because of an emphasis on short-term benefits over long-term hazards. The motivations for initial antibiotic experiments differed on both sides of the Atlantic: responding to pressure from industry and the specter of hunger-fueled communism, US regulators licensed antibiotics' mass introduction to agriculture without considering hazards other than acute toxicity. As a relatively unknown group of substances, antibiotics were rapidly integrated into the well-developed but relatively unsupervised supply chains linking pharmaceutical companies to farmers. The ensuing expansion of agricultural antibiotic use was market driven in every sense of the word: pharmaceutical companies generated additional value from existing patents and fermentation waste; farmers used antibiotics to insure themselves against disease, enhance efficiency, avoid veterinary fees, and substitute labor; and consumers drove demand by buying increasing amounts of cheap meat. Struggling to keep up with the antibiotic boom, US regulators retrospectively licensed many already popular practices. By contrast, Britain's antibiotic experiment was initiated from the top down. During the 1940s and 1950s, officials responded to supply and currency pressures by boosting therapeutic antibiotic use and by licensing AGPs to reduce feedstuff imports. While the British poultry and dairy sectors rapidly adopted antibiotics, Whitehall had to "sell" AGPs to the pig sector. In doing so, officials chose to ignore not only AMR and residue warnings but also inconclusive feed trials. National pride in "British" penicillin and concerns about stagnating markets for pharmaceutical companies sweetened licensing decisions.

While the context of antibiotics' introduction varied, the factors making officials want to see new products in a favorable light were similar. In both the United States and United Kingdom, short-term considerations of immediate strategic and economic interest superseded contemporary concerns about hazards and dubious efficacy. As a consequence, the two main Western powers of the postwar period licensed agricultural antibiotics in the full knowledge that nobody knew what the long-term consequences of their introduction would be. Neither country developed surveillance programs with which to assess licensed products or provided authorities with adequate statutory powers to withdraw substances should they prove problematic. Once established for the sulfonamides, penicillin, and the tetracyclines, the short-termist antibiotic licensing matrix was reapplied again and again to substances ranging from chloramphenicol to ampicillin. Avoparcin and the fluoroquinolones were licensed for agricultural use even after AMR assessments became mandatory components of drug applications. While precautionary AMR concerns prompted the restriction of entire antibiotic classes in Europe in the 1970s and 1990s, US regulators remain hamstrung by complicated withdrawal procedures for licensed drugs. Both Britain and the United States have also struggled to systematically address veterinary antibiotic use.

The tendency to focus on short-term benefits at the expense of long-term hazards is by no means limited to antibiotics and has been explored for numerous other substances ranging from DES to BPA.[11] Ultimately, every licensing decision is an experiment based on imperfect information: no technology is without its risk and some risks like horizontal AMR transfer will only become apparent decades after original licensing decisions. What the first story of short-termism highlights is the need for a resilient mode of substance regulation that factors flexibility into every licensing decision. In the case of antibiotics, the past seventy-five years have shown that licensing should never be considered a one-off decision. Regulatory gatekeeping has to be complemented by surveillance mechanisms capable of studying substances' often unpredictable long-term effects and by statutory instruments powerful enough to guarantee speedy withdrawal once risks have been identified.

2. A Story of Epistemic Fragmentation

The second story of this book is one of epistemic fragmentation in which one group of substances was viewed through different cultural lenses and acquired varying meanings on both sides of the Atlantic. Despite recent attempts to rationalize antibiotic policies by "putting a number" on the costs and benefits of regulatory action, cultural interpretations of risk continue to differ not only between but within societies. Effective antibiotic stewardship thus not only depends on sound technocratic decision-making but also on the successful staging of risk in different cultural, social, and national milieus.

In the United States, antibiotics were immediately integrated into a deep-seated risk episteme prioritizing potential drug residues in food over AMR or animal welfare concerns. Both the public and regulators engaged in heated discussions about whether to tolerate or ban antibiotic residues. Toleration depended in part on the cultural valence attached to different foodstuffs. Whereas antibiotics in milk were taboo, this was not the case for residues in meat, poultry, fish, and on plants. Deemed noncarcinogenic, most antibiotics also escaped regulation under the 1958 Delaney Clause. Whereas public residue concerns eventually resulted in pioneering US monitoring, British regulators were under far less pressure to tackle residues despite similar evidence of antibiotics in food and milk. While a Briton's meat could be an American's poison, concerns about AMR and animal welfare were markedly stronger in Britain. It was during moments when health and moral concerns fused that powerful British reform movements emerged. Driven by *Animal Machines* and E. S. Anderson's AMR warnings, a first fusion of diverse antibiotic concerns occurred around the cultural crucible of barbaric "factory farms" around 1965. In 1996, the moral panic triggered by BSE again turned antibiotics into a common denominator of anxieties about intensive farming. In both cases, uncontrolled antibiotic use by farmers rather than veterinarians became the target of public reform demands. A similar fusion of antibiotic concerns around AMR or animal welfare never occurred in the United States.

Evaluations of antibiotics' risks and benefits also differed within societies. In the United States, opinions varied strongly between the public and agricultural spheres. Following an initial phase of general optimism about progressive antibiotic use, 1960s environmentalist and consumer criticism fostered an agricultural fortress mentality. Conventional producers reacted to criticism and economic pressure by reaffirming their commitment to the interwar gospel of efficiency. Ideologically opposed to statutory intervention, most farmers continued to view antibiotics as important components of "progressive" farming. Although agricultural commentators were aware of residue and AMR problems, they tended to present them as "fixable" management and efficiency challenges rather than as systemic risks. Ironically, perceptions of antibiotics as "rational" components of the neo-Malthusian US fight for plenty prevented critical evaluations not only of resulting health hazards but also of AGPs' actual economic efficacy. From the 1970s onward, the vast majority of agricultural commentators and experts endorsed the polarizing public defense of AGPs mounted by the US pharmaceutical industry and the Farm Bureau. Although actual AGP use was less common than publicly presented, there was also no popular rebellion by conventional producers against agro-industrial campaigning.

A similar divorce of public and agricultural risk perceptions occurred in Britain. Although corporatist decision-making prevented a US-style public polarization of conflicts, British agricultural commentators also emphasized the "manageability" of antibiotic problems. Trusting in Britain's compromise-oriented system to produce palatable "antibiotic fixes," the agricultural community saw no need to resist partial AGP bans in 1971 and defended the narrow Swann-compromise until the 1990s. The inner-agricultural consensus on nontherapeutic antibiotic use only broke down in the wake of the 1996 BSE crisis. While AGPs have been nominally phased out on both sides of the Atlantic, British and US commentators continue to defend routine therapeutic antibiotic use by presenting AMR as manageable via "rational" and "efficient" drug use. Voluntarist reform remains preferable to statutory intervention.

Since the 1960s, the widening gap between public perceptions of uncontainable antibiotic risks and the agricultural emphasis on manageable risks has created lucrative opportunities for selling safety. Although definitions varied, popular residue concerns gradually turned antibiotics' absence into a powerful common denominator of most organic and "natural" food. The higher prices paid for "pure" produce not only drove the rapid expansion of the organic sector but have also triggered a diversification of conventional production. Nowadays, companies like McDonalds and Tyson Foods are marketing antibiotic-free produce in an attempt to profit from health and environmental concerns that their production methods helped create. Still mostly driven by residue rather than abstract AMR fears, antibiotic-free sales have been praised for reducing antibiotic use but have also helped displace political responsibility for antibiotic stewardship onto the marketplace.[12] This marketplace is not a level playing field. Antibiotic-free sales follow the rules of class-based risk distribution predicted by Ulrich Beck's *Risk Society*: while wealthy consumers can afford to purchase culturally prized and priced "pure," "ethical," and "safe" food, poorer consumers buy riskier food—and are subsequently blamed for contributing to AMR, obesity, and climate change. Commoditized risk does not make equals.

So what does this Babel of fragmented risk perceptions and markets mean for antibiotic stewardship? A central message of this book is that there is no one universal way to regulate or communicate antibiotic risks. Since the 1940s, divergent risk cultures have emphasized either the residue, the AMR, or the welfare aspect of the "antibiotic problem." On both sides of the Atlantic, antibiotics' cultural connotations often had a stronger influence on national drug policies and markets than "objective" expert evaluations. Fragmented risk understandings are not limited to the Anglo-American context and pose a major challenge for international antibiotic stewardship. Creating a common AMR-focused risk episteme is not straightforward and will require more than top-down appeals for "rational" antibiotic use or cyclical warnings about a pending apocalypse. Because it has never been "natural" to be more concerned about AMR than about residues or animal welfare, stewardship policies have to take into account stakeholders' distinct risk epistemologies. Once we abandon the premise that there is a single compelling rational logic for international action against AMR, we will be more attuned to the many negotiations, translations, trade-offs, and financial commitments that will have to accompany any effective collective action to preserve antibiotics' efficacy.

3. A Story of Infrastructure

The third story of this book is the story of the powerful and salient effects of antibiotic infrastructures. Although it is important to remember that antibiotics alone did not drive the expansion and intensification of twentieth-century livestock production, their use created significant cultural and material path dependencies within global agriculture. From the 1950s onward, easy access to therapeutic and nontherapeutic forms of antibiotic use facilitated the concentration of livestock, aided the adoption of confined housing, and changed relations between farmers, veterinarians, and animals. Once an antibiotic infrastructure had become established, it was difficult to dissolve.

In the United States, booming antibiotic sales cemented already close ties between pharmaceutical manufacturers and farmers. On farms, easy antibiotic access and commercial encouragement to self-diagnose and medicate livestock undercut veterinarians' traditional role and facilitated a shift from labor-intensive individual health care to performance-oriented mass animal health management. By the early 1960s, widespread antibiotic use had begun to create cultural and material path dependencies: increasing production seemed to naturally entail more antibiotic use to curb disease and maintain productivity. Farmers were expected to routinely make "rational" decisions on which drug to use at which concentration to maintain animal health, cost-effective production, and consumer safety. Low antibiotic costs also led to a relative neglect of alternative disease management. Caught in an escalating cost-price squeeze, most producers soon considered abandoning antibiotic infrastructures as too risky and equated external criticism with an attack on their rationality and freedom of choice.

In Britain, antibiotic infrastructures developed in a more fragmented fashion. Whereas poultry and dairy producers rapidly adopted antibiotics, pig producers were slow to embrace AGPs. This hesitancy was in part caused by mixed feed trials, farmer concerns, the less concentrated nature of British animal production, and alternative methods of disease control. It was only gradually overcome by promotional efforts and pressure to increase postwar food production. By the 1960s, most livestock sectors had embraced antibiotics as an important means to combat and prevent disease, reduce labor, and enhance productivity. Similar to the United States, an ongoing cost-price squeeze entrenched antibiotic infrastructures and fostered resentment against critics. Farmers' increasing reliance on antibiotics also triggered inner-agricultural conflicts over drug access. Whereas US veterinarians effectively lost control over antibiotics following the 1951 Durham Humphrey's Amendment, Britain had maintained prescription requirements for a large number of antibiotic products. During the 1960s, British veterinarians tried to consolidate control over the animal health market by calling for drug restrictions and advisory health schemes while the NFU pressed for more antibiotic access. Both sides fought over control rather than antibiotic use per se. Despite its focus on AMR, the resulting 1969 Swann compromise on partial AGP bans redistributed control over medically relevant antibiotics but did not seriously challenge existing antibiotic infrastructures.

Resulting antibiotic infrastructures proved remarkably resilient. On both sides of the Atlantic, they have survived decades of public criticism, organic competition, and several rounds of FDA and EU reforms. Recent work by Clare Chandler, Eleanor Hutchinson, and Coll Hutchinson highlights the importance of paying attention to the deeper societal, economic, and relational pillars that support international antibiotic infrastructures. Seeing these relational pillars allows us to appreciate the "scale at which shifts would be required to make a substantial difference to antimicrobial use."[13] What emerges in the case of agricultural antibiotics is a global entanglement of: farming practices; varying agricultural, pharmaceutical, and veterinary interests; dependencies created by the physical design of production systems; an agricultural knowledge system which has relied on antibiotics to fix microbial problems for over seventy-five years; cultural hostility to external interference; and an ongoing arms race between resistant microbes and antibiotic-wielding producers. A holistic view of the many branches of this antibiotic entanglement challenges simplistic narratives of irresponsible "pharmers" and reveals a complicated story of physical path dependencies, cultural blind spots, lacking incentives for change, and a vicious circle of global hunger for cheap protein and rising antibiotic use. Realizing the scale and breadth of reforms necessary to address antibiotic infrastructures is a crucial requirement for effective responses to rising AMR.

4. A Story of Narrow Reform

The last story of this book is one of narrow reform. While short-termism and fragmented epistemologies facilitated antibiotics' spread, narrow visions of reform played an important role in leaving wider antibiotic infrastructures intact. Narrow reform visions were the result of inter-professional rivalries over antibiotic control and the attempt to manage amorphous risks like "infective" AMR within traditional regulatory frameworks based on linear cause-and-effect models of harm, cost-benefit calculations, and corporatist compromise. The patchwork of narrow reforms emerging from this constellation of competing interests and traditional risk regulation unsystematically targeted individual aspects of antibiotic use and rarely complemented each other. Despite the divergent evolution of national risk epistemes and regulatory interventions, Anglo-American antibiotic infrastructures thus remained remarkably similar until the late 1990s.

Similar to Scott Podolsky's periodization of US medical antibiotic regulations,[14] transatlantic reforms of agricultural antibiotic use can be divided into distinct phases. Following an initial phase of mass licensing and deregulation, the mid-1950s saw officials and public campaigners emphasize "rational" antibiotic use as a way to curb emerging AMR and residue hazards. Stressing cognitive reform was not enough to curb either problem. From the late 1950s onward, residue detections forced officials to abandon voluntary measures and establish statutory residue controls for milk and meat in the United States and for milk in the United Kingdom. In Britain, AMR concerns also made officials increase veterinary power over antibiotic access. In both countries, 1960s reforms were designed to "fix" antibiotic use. They were also the smallest common denominator between conflicting consumer, medical, veterinary, and agricultural demands: physicians wanted to reduce AMR selection on farms but resisted calls for reviews of medical antibiotic use; consumers were concerned about residues, factory farms, and AMR but continued to demand cheap meat; veterinarians wanted to limit farmers' antibiotic access but feared that wider reforms might restrict their own drug access and compromise their standing within the medical community; farmers were concerned about a loss of consumer trust but equally feared jeopardizing antibiotic infrastructures.

The narrow 1960s compromises on antibiotic reform would shape the next thirty years of regulatory intervention. Foreshadowed by the 1962 Netherthorpe report and targeting only a limited number of AGPs, Britain's 1969 Swann report reinforced legal distinctions between subtherapeutic and therapeutic dosages and between medically relevant and irrelevant antibiotics. The Swann compromise proved so popular because it focused precautionary AMR-oriented reforms on a small number of low-dosed antibiotics while leaving wider antibiotic infrastructures unchallenged. Its palatable nature also meant that British officials, farmers, and veterinarians continued to defend the Swann compromise even after data indicated that it had failed to reduce either antibiotic use or AMR. Although the ossified "Swann gospel" was abandoned in the wake of the 1996 BSE crisis, British and EU officials continued to follow narrow 1960s precepts by initially limiting renewed statutory interventions to AGPs. Ahead of the EU's 2018 decision to regulate veterinary antibiotic use, officials reacted to fierce agricultural and veterinary resistance by relying on voluntarist measures to reform the remaining 80 percent of antibiotic use.

Narrow reforms also hampered US responses to AMR. Limiting AMR concerns to antibiotic residues' presence in food, FDA regulators and NAS experts initially created the impression that "infective" AMR was containable. When FDA regulators later pressed for Swann-inspired AGP restrictions, their efforts failed not only because of fragmented risk epistemologies, regulation wariness, and well-financed industrial opposition but also because of the US legal system's insistence on proof of harm. Forced to engage in lengthy formal drug withdrawal procedures, officials found it impossible to prove harm resulting from AGPs' selection for AMR as opposed to harm resulting from AMR selection by therapeutic antibiotic use in agricultural and medical settings. 1970s regulators' inability to provide evidence of harm or imminent harm was not because AGPs were not selecting for dangerous AMR but because it was technologically and judicially impossible—as confirmed by the 1980 NAS report. Ironically, even if the FDA had been able to prove imminent harm, implementing narrow Swann-style AGP bans would likely not have reduced wider AMR. Their narrow focus on nontherapeutic antibiotic use also threatens the long-term success of the most recent 2017 voluntary AGP restrictions.

On both sides of the Atlantic, policies targeting only individual aspects of antibiotic use have failed to reform wider antibiotic infrastructures or AMR. Often representing the smallest common denominator of competing interests, narrow reforms promised quick fixes by blending out the complexities of existing antibiotic infrastructures and the interconnection of environmental, medical, and agricultural AMR selection. More uncomfortable reforms were often avoided by either displacing stewardship decisions onto the "green" marketplace or by relying on industry voluntarism. Globally, the countries that have been most successful in reducing their antibiotic footprint have done so by acknowledging entanglements and interlayering different regulatory approaches. In Denmark, antibiotic reductions were achieved with the help of educational campaigns, detailed surveillance, vaccines, a divorce of veterinary prescription and sales rights, antibiotic quotas, and injections of taxpayer money to help farmers develop alternative production systems.[15] Change was neither cheap nor achieved overnight. In the long term, reducing global antibiotic infrastructures will require broad reforms of all aspects of antibiotic use, creating microbially resilient animal production systems, and winning conventional producers over to a sustained value-driven[16] push for antibiotic-free farming.

Toward Long-Term Resilience

The four stories of short-termism, epistemic fragmentation, infrastructure, and narrow reform highlight the need for sustained, multipronged, global collective action to preserve antibiotics' efficacy.[17] Even the best-intentioned national reforms will not contain the global proliferation of resistance genes and resistant bacteria. While the current resurgence of nationalist politics is not helpful, waiting for new legislators or new antibiotics is unlikely to resolve problems—nor is ongoing interprofessional and international finger-pointing.

In this situation, it is productive to think of antibiotics in analogy to climate change. Driven by global middle-class consumption, carbon dioxide and AMR levels will likely continue to rise in the short to medium term. Any political interventions will only see payoffs in the medium to long term. Successful antibiotic stewardship therefore depends less on short-termist warnings about a pending apocalypse than on (a) long-term commitments to sustained antibiotic reduction with integrated and independent feedback mechanisms, (b) new efforts to communicate AMR risks to stakeholders, (c) and factoring microbial resilience into agricultural and health care planning. The latter point is particularly important. As mentioned in the introduction, AMR is an ancient biological phenomenon that becomes a risk only within antibiotic dependent systems. Over the next decades, rising selection pressure means that bacterial susceptibility to some of our antibiotic work horses may well be lost. Although other antibiotics will remain effective and new drugs may well enter the market, fostering microbial resilience in health care and food production systems minimizes the fallout of resistant disruptions to antibiotic infrastructures.

Improved management of our microbial commons seems the most promising way of moving forward. Since every antimicrobial intervention will eventually produce a microbial reaction, money spent on reopening the antibiotic pipeline will only ever generate limited returns. In the long run, managing and accepting the inevitable presence of microbes seems a more sustainable—and cheaper—option than constantly waging antimicrobial war in hospitals and on farms. In agriculture, options for reducing antibiotic use and increasing microbial resilience are manifold and range from designing environments that foster the competitive inhibition of pathogens, reducing herd densities, improving animal welfare, promoting vaccinations, investing in alternative therapies and preventive health care, taxing unsustainable forms of production, banning drugs, introducing antibiotic quotas, subsidizing antibiotic-free production, enhancing surveillance of antibiotic use and AMR, and so on.[18] Back among his coughing pigs, the central question for the pig producer featured in the first chapter is not so much whether access to antibiotics will be banned but how not to rely on them in the first place.

Acknowledgments

This book is the result of seven years of thinking and writing about antibiotics. Every sentence is rooted in countless discussions and suggestions by friends and colleagues from across the world. Since moving to Oxford in 2012, I have benefited from the generous academic and personal support of the Wellcome Unit for the History of Medicine and the Oxford Martin School. I would particularly like to thank Mark Harrison, Erica Charters, Roderick Bailey, Pietro Corsi, Conrad Keating, Margaret Pelling, Sloan Mahone, Robert Iliffe, Belinda Clark, and members of the Martin School's Programme on Collective Responsibility for Infectious Disease for their comments on the book, help in navigating the cliffs of doctoral and postdoctoral research, and offering many kind words of support—especially when it seemed as though this book was no more than an endless series of drafts.

Over the years, I have also had the immense privilege of being mentored by Scott Podolsky, Christoph Gradmann, Thomas Le Roux, Ulrike Thoms, and the late Mark Finlay—all of whom generously shared their findings and some of whom—like Scott—patiently read multiple drafts of this manuscript. The content and tone of this book are the result of their collegiality, friendship, and invaluable advice. I am also very grateful to Nicolas Fortané, Delphine Berdah, Nathalie Jas, Andrew Singer, Robert Bud, Javier Lezaun, Clare Chandler, Coll Hutchinson, Anne Kveim Lee, Anne Hardy, Alex Bowmer, Avinash Sharma, Frank Uekötter, and Abigail Woods for helpful comments on different chapters and related projects—I look forward to discussing the whole book with them. My editor at Rutgers, Peter Mickulas, has been a friendly but firm force pushing me forward and not allowing me to get lost "in the weeds." Kendra Smith-Howard must be singled out for particular praise because of her repeated detailed, insightful, and constructive reviewer comments. All remaining mistakes are my own.

In the course of my work, I have profited from generous financial support by the Wellcome Trust, the German Historical Institute in Washington, the Oxford Martin School, and the German National Academic Foundation. I would also like to thank John Swann (not a relative of Michael) and Guy Hall for their help in navigating the FDA and CDC archives. Numerous further librarians and academics aided my research at the National Archives and Record Administration in College Park and Atlanta, the Library of Congress, the Wellcome Collection, the Museum of English Rural Life, Cornell's Mann Library, Harvard's Countway Library, Yale's Beinecke Library, the Bodleian Libraries, the British Library, and the British National Archives.

At Oxford, it has been a great privilege to be a member of University College and Wolfson College. *Pyrrhic Progress*'s growth was aided by countless lunch and evening discussions with friends and scholars from every imaginable discipline. I would like to especially thank Ashley Maher for her help in correcting very early drafts of this book. Between 2013 and 2014, I also enjoyed the unique writing environment of the late Ian Moore's lab at the Department of Plant Sciences. Oxford's varied intellectual landscape has left indelible marks on both this book and its author.

Most importantly, I want to thank my family for their support, advice, and love. My parents, grandparents, siblings, parents-in-law, nieces, and nephews may have given up asking when the book will finally appear but always listened and encouraged me over the past seven years. It is, however, no exaggeration to say that the two people deserving my highest praise and thanks are my wife, Charlotte, and my daughter, Clara. Charlotte's advice, critique, and forbearance shaped this book from its first letter to the last. These pages would never have been filled without her infinite support and patience. Clara encountered this book a few hours after her birth when her father edited footnotes by her NHS crib. Over the past year, her lack of patience for academic distractions has helped keep me focused on the most important things in life.

Notes

1. The Sound of Coughing Pigs

1 Stacy Sneeringer et al., "Economics of Antibiotic Use in US Livestock Production," *USDA Economic Research Report* 200 (2015): 2.
2 Scott H. Podolsky, *The Antibiotic Era: Reform, Resistance and the Pursuit of a Rational Therapeutics* (Baltimore: Johns Hopkins University Press, 2015).
3 Thomas P. Van Boeckel et al., "Global Trends in Antimicrobial Use in Food Animals," *PNAS* 112, no. 18 (2015): 5649–5650.
4 Ellen K. Silbergeld, *Chickenizing Farms and Food: How Industrial Meat Production Endangers Workers, Animals, and Consumers* (Baltimore: Johns Hopkins University Press, 2016), 38–43.
5 William Boyd, "Making Meat. Technology and American Poultry Production," *Technology and Culture* 42, no. 4 (2001): 637.
6 Roger Horowitz, "Making the Chicken of Tomorrow: Reworking Poultry as Commodities and as Creatures, 1945–1990," in *Industrializing Organisms: Introducing Evolutionary History*, ed. Susan R. Schrepfer and Philip Scranton (New York and London: Routledge, 2004), 219–232.
7 Andrew Godley and Bridget Williams, "Democratizing Luxury and the Contentious 'Invention of the Technological Chicken' in Britain," *Business History Review* 83 (Summer 2009), 267–290; Andrew Godley, "The Emergence of Agribusiness in Europe and the Development of the Western European Broiler Chicken Industry, 1945–1973," *Agricultural History Review* 62, no. 2 (2014): 315–336.
8 William Boyd, *Putting Meat on the American Table: Taste, Technology, Transformation* (Baltimore: Johns Hopkins Press, 2005), 131–137.
9 Silbergeld, *Chickenizing Farms and Food*, 38–43.
10 Mark R. Finlay, "Hogs, Antibiotics, and the Industrial Environments of Postwar Agriculture," in *Industrializing Organisms: Introducing Evolutionary History*, ed. Susan R. Schrepfer and Philip Scranton (New York and London: Routledge, 2004), 237–260.
11 Thomas Jukes, *Antibiotics in Nutrition* (New York: Medical Encyclopedia, 1955), 41.

12 Tony Lawrence et al., *Growth of Farm Animals* (Wallingford, UK, and Cambridge, MA: CABI, 2012), 327.

13 Silbergeld, *Chickenizing Farms and Food*, 99–103; Jay P. Graham et al., "Growth Promoting Antibiotics in Food Animal Production: An Economic Analysis," *Public Health Reports* 122, no. 1 (2007): 79–87.

14 Abigail Woods, "Decentring Antibiotics: UK Responses to the Diseases of Intensive Pig Production (ca. 1925–65), *Palgrave Communications* 5, no. 41 (2019).

15 "US Broiler Performance," *National Chicken Council* (22.03.2019), https:// www.nationalchickencouncil.org/about-the-industry/statistics/u-s-broiler -performance/.

16 The EU restricted antibiotic sprays to use only in emergencies between 2002 and 2007.

17 Gilbert Wayne Gillespie Jr., "Antibiotic Animal Feed Additives and Public Policy: Farm Operators' Beliefs about the Importance of These Additives and Their Attitudes toward Government Regulation of Agricultural Chemicals and Pharmaceuticals," (doctoral thesis, Cornell University, 1987), 40.

18 Clare Chandler et al., *Addressing Antimicrobial Resistance through Social Theory: An Anthropologically Oriented Report* (London: LSHTM, 2016), 16–17.

19 Robert Bud, *Penicillin: Triumph and Tragedy* (Oxford: Oxford University Press, 2007), 105.

20 *No Time To Wait: Securing The Future From Drug-Resistant Infections. Report To The Secretary-General of the United Nations, April 2019* (New York: IACG, 2019).

21 Karl Drlica and David S. Perlin, *Antibiotic Resistance: Understanding and Responding to an Emerging Crisis* (Upper Saddle River, NJ: Pearson Education, 2011), 6–12, 31–37, 74–83.

22 Ibid., 91–98.

23 Ibid., 99–102.

24 Linus Sandegren, "Review: Selection of Antibiotic Resistance at Very Low Antibiotic Concentrations," *Uppsala Journal of Medical Sciences* 119, no. 2 (2014): 103–107.

25 Alan L. Olmstead and Paul W. Rhode, *Arresting Contagion: Science, Policy, and Conflicts over Animal Disease Control* (Cambridge, MA: Harvard University Press, 2015), 3.

26 Lance B. Price et al., "Staphylococcus aureus CC398: Host Adaptation and Emergence of Methicillin Resistance in Livestock," *mBio* 3, no. 1 (2012): 1–6.

27 Laura Kahn, *One Health and the Politics of Antimicrobial Resistance* (Baltimore: Johns Hopkins University Press, 2016), 7; Michael R. Gillings, "Lateral Gene Transfer, Bacterial Genome Evolution, and the Anthropocene," *Annals of the New York Academy of Sciences* 1389, no. 1 (2016): 20–36; Hannah Landecker, "Antibiotic Resistance and the Biology of History," *Body and Society* (2015).

28 Yi-Yun Liu et al., "Emergence of Plasmid-Mediated Colistin Resistance Mechanism MRC-1 in Animals and Human Beings in China: A Microbiological and Molecular Biological Study," *Lancet Infectious Diseases* 16, no. 2 (2016): 161–168.

29 Jingjing Quan, Xi Li, and Yan Chen, "Prevalence of mcr-1 in Escherichia coli and Klebsiella pneumoniae Recovered from Bloodstream Infections in China: A Multicentre Longitudinal Study," *Lancet Infectious Diseases* 17, no. 4 (2017): 400–410.

30 For reasons of space, aquaculture is not discussed in this book; for a short history

of the industry, see Claas Kirchhelle, "Pharming Animals: A Global History of Antibiotics in Food Production," *Palgrave Communications* 96, no. 4 (2018).

31 Ulrich Beck, *Weltrisikogesellschaft: Auf der Suche Nach der Verlorenen Sicherheit* (Frankfurt: Suhrkamp, 2007), 22–36.

32 Ulrich Beck, "The Reinvention of Politics: Towards a Theory of Reflexive Modernization," in *Reflexive Modernization: Politics, Tradition and Aesthetics in the Modern Social Order*, ed. Beck (Cambridge and Oxford: Polity Press, 1995), 9.

33 Beck, *Weltrisikogesellschaft*, 36; Beck, *Risikogesellschaft: Auf Dem Weg in Eine Andere Moderne* (Frankfurt: Suhrkamp, 1986), 29–31, 35.

34 Niklas Luhmann, *Die Gesellschaft Der Gesellschaft* (Frankfurt: Suhrkamp, 1997), 78, 94; see also the communication theories of Mancur Olson, *The Logic of Collective Action: Public Goods and the Theory of Groups* (Cambridge, MA: Harvard University Press, 1971), 165–166.

35 Carol Morris, Richard Helliwell, and Sujatha Raman, "Framing the Agricultural Use of Antimicrobial Resistance in UK National Newspapers and the Farming Press," *Journal of Rural Studies* 45 (2016): 43–55; Stephanie Begemann et al., "How Political Cultures Produce Different Antibiotic Policies in Agriculture: A Historical Comparative Case Study between the UK and Sweden," *Sociologica Ruralis* (2018); Paul Slovic et al., "Risk as Analysis and Risk as Feelings: Some Thoughts about Affect, Reason, Risk, and Rationality," *Risk Analysis* 24, no. 2 (2004): 311–322.

36 Sheila Jasanoff, *Designs on Nature: Science and Democracy in Europe & the United States*, 2nd ed. (Princeton, NJ: Princeton University Press, 2007), 9.

37 Alexander von Schwerin, "Prekäre Stoffe. Radiumökonomie, Risikoepisteme und die Etablierung der Radioindikatortechnik in der Zeit des Nationalsozialismus," *NTM* 17 (2009): 6.

38 Thomas Le Roux, "Governing the Toxics and the Pollutants: France, Great Britain, 1750–1850," *Endeavour* 40, no. 2 (2016): 70–81.

39 Nathalie Jas, "Adapting to "Reality: The Emergence of an International Expertise on Food Additives and Contaminants in the 1950s and Early 1960s," in *Toxicants, Health and Regulation Since 1945*, ed. Nathalie Jas and Soraya Boudia (London: Pickering & Chatto, 2013), 47–69; Nathalie Jas, "Public Health and Pesticide Regulation in France Before and After Silent Spring," *History and Technology* 23, no. 4 (2007): 369–388; Soraya Boudia, "Managing Scientific and Political Uncertainty. Environmental Risk Assessment in a Historical Perspective," in *Powerless Science? Science and Politics in a Toxic World*, ed. Soraya Boudia and Nathalie Jas (New York and Oxford: Berghahn, 2014), 95–112; Carsten Reinhardt, "Boundary Values," in Viola Balz et al., *Precarious Matters: The History of Dangerous and Endangered Substances in the 19th and 20th Centuries* (Berlin: Max Planck Institute for the History of Science, 2008), 39–50; Heiko Stoff and Alexander von Schwerin, "Einleitung—Lebensmittelzusatstoffe. Eine Geschichte gefährlicher Dinge und ihrer Regulierung 1950–1970," *Technikgeschichte* 81, no. 3 (2014): 215–228; Sheila Jasanoff, *Science at the Bar: Law, Science, and Technology in America* (Cambridge, MA, and London: Harvard University Press), 204–211; Heiko Stoff, *Gift in der Nahrung: Zur Genese der Verbraucherpolitike Mitte des 20. Jahrhunderts* (Stuttgart: Franz Steiner Verlag, 2015), 23–25.

40 Susan Rankin Bohme, *Toxic Injustice: A Transnational History of Exposure and Struggle* (Oakland: University of California Press, 2018).

41 Kendra Smith-Howard, *Pure and Modern Milk: An Environmental History since 1900* (Oxford: Oxford University Press, 2013), 3–12, 121–146.

42 Jasanoff, *Designs on Nature*, 131.

43 David Vogel, *The Politics of Precaution: Regulating Health, Safety and Environmental Risks in Europe and the United States* (Princeton, NJ, and Oxford: Princeton University Press, 2012); Nancy Langston, "Precaution and the History of Endocrine Disruptors," in *Powerless Science? Science and Politics in a Toxic World*, ed. Soraya Boudia and Nathalie Jas (New York and Oxford: Berghahn, 2014), 29–45.

44 Arthur Neslen, "European Parliament Approves Curbs on Use of Antibiotics on Farm Animals", *Guardian*, October 25, 2018.

45 "Sales of Veterinary Antimicrobial Agents in 29 European Countries in 2014: Trends from 2011 to 2014" (European Medicines Agency, 2016), 29; "Sales of Veterinary Antimicrobial Agents in 30 European Countries in 2016: Trends from 2010 to 2016" (European Medicines Agency, 2018), 26.

46 "2015 Summary Report on Antimicrobials Sold or Distributed for Use in Food-Producing Animals" (Washington, DC: DHSS/FDA, 2016), 26; "2017 Summary Report on Antimicrobials Sold or Distributed for Use in Food-Producing Animals" (Washington, DC: DHSS/FDA, 2018), 12.

47 Michelle Mart, *Pesticides, a Love Story: America's Enduring Embrace of Dangerous Chemicals* (Lawrence: University Press of Kansas, 2016), 7–9, 15–16.

2. Picking One's Poisons: Antibiotics and the Public

1 John E. Lesch, *The First Miracle Drugs: How the Sulfa Drugs Transformed Medicine* (Oxford: Oxford University Press, 2007), 1–11; Axel C. Hüntelmann, "Seriality and Standardization in the Production Of '606,'" *History of Science* xlvlii (2010), 435–460.

2 "History of the Word 'Antibiotic'/Discussion between S. A. Waksman and J. E. Flynn on 19 January 1962," *Journal of the History of Medicine and Allied Sciences* XXVIII, no. 3 (1973).

3 Bud, *Penicillin: Triumph and Tragedy* (Oxford: Oxford University Press, 2009), 13–17, 28–44; Peter Neushul, "Science, Government, and the Mass Production of Penicillin," *Journal of the History of Medicine and Allied Sciences* 48 (1993): 371–395.

4 Bud, *Penicillin*, 107–108, 120–128.

5 Robert A. Thom, "The Era of Antibiotics," *Great Moments in Pharmacy*, Painting No. 39, Parke, Davis and Company Advertising Series (1957).

6 Scott Podolsky, *The Antibiotic Era: Reform, Resistance and the Pursuit of a Rational Therapeutics* (Baltimore: Johns Hopkins University Press, 2015), 37–40.

7 "Advertisement—Super Anahist Antibiotic Nasal Spray," *San Bernardino Sun* [hereafter SBS], March 3, 1957, 11; "Advertisement—Yodora," *Life Magazine* [hereafter *Life*], March 28, 1955, 4; "Advertisement—Rexall," *Life*, September 6, 1954, 84, "Advertisement—Squibb," *Good Housekeeping*, February 1946, 74; "Advertisement—Parke Davis," *Good Housekeeping*, October 1950, 138.

8 "Advertisement—Hi and Dri," *Vogue*, September 1962, 29.

9 Podolsky, *Antibiotic Era*, 19–29.

10 Bud, *Penicillin*, 165–166; George A. Woods, "Potions for Pets," *New York Times* [hereafter *NYT*], July 25, 1954, SM20.

11 Hannah Landecker, "The Food of Our Food: Medicated Feed and the Industriali-
    sation of Metabolism," presented at Oxford Microbiome Seminar, February 15,
    2017, 10–14.

12 Mark R. Finlay, "Hogs, Antibiotics, and the Industrial Environments of
    Postwar Agriculture," in *Industrializing Organisms: Introducing Evolutionary
    History*, ed. Susan R. Schrepfer and Philip Scranton (New York: Routledge,
    2004), 243.

13 Thomas H. Jukes, *Antibiotics in Nutrition* (New York: Medical Encyclopedia,
    1955), 17–18; E. L. R. Stokstad et al., "The Multiple Nature of the Animal Protein
    Factor," *Journal of Biological Chemistry* 180, no. 2 (1949). For doubts about AGP
    efficacy claims, see Ellen K. Silbergeld, *Chickenizing Farms and Food: How
    Industrial Meat Production Endangers Workers, Animals, and Consumers*
    (Baltimore: Johns Hopkins University Press, 2016), 97–103; Jay P. Graham et al.,
    "Growth Promoting Antibiotics in Food Animal Production: An Economic
    Analysis," *Public Health Reports* 122, no. 1 (2007), 79–87.

14 Finlay, "Hogs, Antibiotics," 244; Waldemar Kaempffert, "Science in Review,"
    NYT, April 16, 1950, E9; "Farm Animals May Be Made to Grow Faster by
    'Wonder Drugs,'" *Madera Tribune* [hereafter *MT*], September 7, 1950, 12.

15 John W. Ball, "New Chicken Procedures Like Factory," *WP*, May 22, 1951, B2.

16 "Science: Pigs Without Moms," *Time*, December 3, 1951.

17 Ibid.

18 Ibid.

19 "Antibiotics Used on Livestock," *NYT*, December 13, 1951, 53.

20 Ellis M. Haller, "Farmyard Therapy: Drug Makers Stem the Tide of Animal
    Diseases," *Barron's National Business and Financial Weekly* [hereafter *Barron's*],
    April 12, 1954, 13.

21 "Mice Don't Scare Girls Nowadays," *SBS*, April 12, 1956, 18.

22 Foster Hailey, "More Care Urged in Antibiotics Use," *NYT*, March 25, 1954, 59.

23 N. S. Haseltine, "Drug Found to Seal in Flavor of Food as It Is Being Canned,"
    WP, May 18, 1950, B8; "Food News: Preservation Process," *NYT*, November 29,
    1955, 26.

24 "Antibiotics Now Fighting Plant Diseases," *WP*, September 11, 1953, 19.

25 "Have You heard the Latest Garden News?" *Better Homes and Gardens*,
    June 1952, 32.

26 "Antibiotics Promise Housewives Food that Will Keep for Weeks," *Desert Sun*
    [hereafter *DS*], October 19, 1956, 1.

27 "Now It's Toothsome Whaleburgers that May Come from Antibiotics," *DS*,
    October 20, 1956, 1.

28 Maryn McKenna, *Big Chicken: The Incredible Story of How Antibiotics Created
    Modern Agriculture and Changed the Way the World Eats* (Washington, DC:
    National Geographic, 2017), 75–82.

29 Kenneth B. Raper, "The Progress of Antibiotics," *Scientific American* [hereafter
    SciAm], April 1952, 49.

30 Ibid., 54.

31 Francis Joseph Weiss, "Chemical Agriculture," *SciAm*, August 1952, 18.

32 Ibid.

33 Irmi Seidl and Clem A. Tisdell, "Carrying Capacity Reconsidered: From Malthus'
    Population Theory to Cultural Carrying Capacity," *Ecological Economics*, 30, no. 3
    (1999): 395–408.

34 William Furlong, "Chemical Revolution on the Farm," *NYT*, October 4, 1959, 30, 37.

35 Harry S. Truman, "Truman's Inaugural Address," January 20, 1949, *Harry S. Truman Presidential Library*.

36 Nick Cullather, *The Hungry World: America's Cold War Battle against Poverty in Asia* (Cambridge, MA: Harvard University Press, 2010), 2–10.

37 "Scientist Predicts Antibiotic Farms," *Daily Illini* [hereafter *DI*], March 11, 1953, 3.

38 John Stuart, "American Farmer Still Making 'Hay,'" *NYT*, December 13, 1953, F1; Jack Ryan, "Farmers Reaping Bumper Crop of Chemicals," *NYT*, March 20, 1955, F1.

39 Aubrey Graves, "Ever Try to Stuff a Heifer with King-Size Antibiotics?" *WP*, January 4, 1953, B2.

40 "Russian Farm Leader Who Toured US Advocates More Exchanges," *SBS*, October 23, 1955, 51.

41 "Farm Animals May Be Made to Grow Faster by 'Wonder Drugs,'" *MT*, September 7, 1950, 12.

42 Scott Podolksy, "Historical Perspective on the Rise and Fall and Rise of Antibiotics and Human Weight Gain," *Annals of Internal Medicine* 166 (2017): 133–135.

43 Milton Levenson, "Six Latin Nations Study Nutrition," *NYT*, March 11, 1951, 27; trials with streptomycin were also considered; see also: "Question and Answer," *MT*, January 16, 1953, 9.

44 Podolksy, "Historical Perspective," 135.

45 Christopher C. Sellers, *Hazards of the Job: From Industrial Disease to Environmental Health Science* (Chapel Hill: University of North Carolina Press, 1997), 221–224.

46 Wallace F. Janssen, "FDA Since 1938: The Major Trends and Developments," *Journal of Public Law* 13, no. 1 (1964): 205–221.

47 "Investigation of the Use of Chemicals in Food Products: Report," *Union Calendar*, No. 1139, Report 3254, January 3, 1951, 1–11; Nancy Langston, *Toxic Bodies: Hormone Disruptors and the Legacy of DES* (New Haven, CT: Yale University Press, 2010), 80–81.

48 David Vogel, *The Politics of Precaution: Regulating Health, Safety and Environmental Risks in Europe and the United States* (Princeton, NJ: Princeton University Press, 2012), 45–46; Langston, *Toxic Bodies*, 15–42, 81.

49 Nate Haseltine, "72 Deaths Laid to Penicillin Use," *WP*, October 4, 1957, B1.

50 The National Archives PIN 20/216, "Sensitisation of Nursing Staffs to Antibiotics," extract from *Lancet* (July 4, 1953): 1–3; "Medicine: Hold that Penicillin," *Time*, October 30, 1950.

51 A rare exception is Louly Baer, "Keeping Foods Pure," *NYT*, February 9, 1952, 12.

52 "Indiscriminate Use of Drugs Dangerous," *True Republican* [hereafter *TR*], April 15, 1955, 7; "One Complication that May Follow Use of Antibiotics," *MT*, April 30, 1952, 9.

53 Jane Nickerson, "News of Food: Milk Plant Doubles Output," *NYT*, April 25, 1951, 45.

54 Nate Haseltine, "Milk Samplings Yield Traces of Penicillin," *WP*, February 22, 1956, 3.

55 Kendra Smith-Howard, "Antibiotics and Agricultural Change: Purifying Milk and Protecting Health in the Postwar Era," *Agricultural History* 84, no. 3 (2010): 327–351.

56 Peter Atkins, *Liquid Materialities: A History of Milk, Science and the Law* (Farnham, UK: Ashgate, 2010), 225–245.

57 Smith-Howard, "Antibiotics and Agricultural Change," 329–330, 32–33.

58 On US consumer activism, see Matthew Hilton, "Consumer Movements," in *The Oxford Handbook of the History of Consumption*, ed. Frank Trentmann (Oxford: Oxford University Press, 2012), 505–520.

59 Arnaldo Cortesi, "Cancer Is Traced to Food Additives," *NYT*, August 21, 1956, 31.

60 James Rorty and N. Philip Norman, *Tomorrow's Food: The Coming Revolution in Nutrition*, 2nd ed. (New York: Devin-Adair, 1956), v.

61 Ibid., vii.

62 Janssen, "FDA Since 1938," 209.

63 "Trace of DDT Found in 1958 Tests of Milk in Washington, Other Cities," *WP*, December 22, 1959, A1.

64 "US Pushes Fight to Rid Milk of Penicillin Dregs," *WP*, December 3, 1959, B2; "Penicillin and Other Antibiotics in Milk," *Journal of Home Economics*, December 1959, 889.

65 McKenna, *Big Chicken*, 91.

66 William Longgood, *The Poisons in Your Food* (New York: Simon and Schuster, 1960), 2.

67 Ibid., 72–73.

68 Ibid., 152.

69 Ibid., 154, 56.

70 William J. Darby, "Review, *The Poisons in Your Food* by William Longgood," *Science*, 131, no. 3405 (1960).

71 Ibid.

72 Andrew N. Case, *The Organic Profit: Rodale and the Making of Marketplace Environmentalism* (Seattle: University of Washington Press, 2018), 37–80.

73 Robin O'Sullivan, *American Organic: A Cultural History of Farming, Gardening, Shopping, and Eating* (Lawrence: University Press of Kansas, 2015), 66, 88.

74 J. I. Rodale et al., *The Complete Book of Food and Nutrition* (Emmaus, PA: Rodale, 1961), 121, 124, 263.

75 Christian Simon, *DDT. Kulturgeschichte einer Chemischen Verbindung* (Basel: Christian Merian Verlag, 1999); Edmund Russell, *War and Nature: Fighting Humans and Insects with Chemicals from World War I to Silent Spring* (Cambridge, UK: Cambridge University Press, 2001).

76 Rachel Carson, *Silent Spring* (New York: Houghton Mifflin, 1962).

77 Bookchin published under the pseudonym Lewis Herber, *Our Synthetic Environment* (New York: Knopf, 1962).

78 Ralph H. Lutts, "Chemical Fallout: Rachel Carson's Silent Spring, Radioactive Fallout, and the Environmental Movement," *Environmental Review* 9, no. 3 (1985): 211–225; Garry Kroll, "The 'Silent Springs' of Rachel Carson: Mass Media and the Origins of Modern Environmentalism," *Public Understanding of Science* 10 (2001): 403–420.

79 "Dairymen Warned on Pest Poisons," *WP*, December 11, 1962, A4.

80 Tom Mahoney, *The Merchants of Life; An Account of the American Pharmaceutical Industry* (New York: Harper, 1959).

81 John W. Finney, "The Drug Industry," *NYT*, December 13, 1959, E8.

82 Scott Podolsky, "Antibiotics and the Social History of the Controlled Clinical

Trial, 1950–1970," *Journal of the History of Medicine and Allied Sciences* 65, no. 3 (2010): 360–365.

83 "FDA Aide's Talk Edited by Ad Man," *NYT*, June 2, 1960, 25.

84 "Drug Aide Quits," *NYT*, May 20, 1960, 12.

85 Podolsky, "Antibiotics and the Social History," 364.

86 Rock Brynner and Trent Stephens, *Dark Remedy: The Impact of Thalidomide and Its Revival as a Vital Medicine* (New York: Perseus, 2001), ix, 5–20, 32–35.

87 Daniel Carpenter, *Reputation and Power. Organizational Image and Pharmaceutical Regulation at the FDA* (Princeton, NJ: Princeton University Press, 2010), 238–256.

88 Bridget M. Kuehn, "Frances Kelsey Honored for FDA Legacy," *JAMA*, 304, no. 19 (2010).

89 Carpenter, *Reputation*, 229, 592.

90 Nevin Scrimshaw, "Food," *SciAm*, March 1963, 75.

91 Ibid., 79.

92 "Food Additives: What They Are, How They Are Used" (Washington, DC: Manufacturing Chemists' Association, 1961), foreword.

93 Ibid., 9.

94 Ibid., 13.

95 Ibid., 17.

96 Ibid., 23.

97 The analysis was conducted using keyword searches for the terms "additive[s]," "chemical[s]," "natural," "organic," "antibiotic[s]," and "hormone[s]" in Cornell University's Home Economics Archive (Hearth) and the Hathi Trust's collection of digitized 1950s and 1960s cookbooks.

98 "Science Put the Meat on that Holiday Bird," *DS*, November 18, 1966, 14; Eva D. Wilson et al., *Principles of Nutrition* (New York: Wiley, 1959), 274–277.

99 Christoph Gradmann, "Magic Bullets and Moving Targets: Antibiotic Resistance and Experimental Chemotherapy, 1900–1940," *Dynamis* 31, no. 2 (2001).

100 George Gray, "The Antibiotics," *SciAm*, August 1949, 33.

101 René J. Dubos, "Microbiology," *Annual Review of Biochemistry* 11 (1942): 672.

102 Alexander Fleming, "Nobel Lecture," Nobel Prize, December 11, 1945, http://www.nobelprize.org/nobel_prizes/medicine/laureates/1945/fleming-lecture.html.

103 Podolsky, *Antibiotic Era*, 146–155; Christoph Gradmann, "Sensitive Matters: The World Health Organization and Antibiotic Resistance Testing, 1945–1975," *Social History of Medicine* 26, no. 3 (2013): 556–560.

104 "Will Antibiotics Be Abandoned," *NYT*, July 26, 1953, E7.

105 "Let Your Doctor Decide What Medicines You Need," *MT*, March 27, 1957, 4.

106 "Your Neighbors' Wonder Drugs Can Make You Sick," *SBS*, May 8, 1953, 35. See also "War on Bacteria Seen Backfiring," *NYT*, May 11, 1955, 25; "Drug-Resistant Germs at Home in Hospitals," *WP*, January 12, 1958, E2.

107 Kathryn Hillier, "Babies and Bacteria: Phage Typing Bacteriologists, and the Birth of Infection Control," *Bulletin of the History of Medicine* 80, no. 4 (2006): 733–761.

108 Nate Haseltine, "Hospital-Bred Germs Target of Drive Here," *WP*, October 29, 1958, B1.

109 Podolsky, *Antibiotic Era*; Podolsky, "Antibiotics and the Social History," 327.

110 "Medicine: Mixed Blessing," *Time*, May 18, 1959.

111 Podolsky, *Antibiotic Era*, 141–142, 152–154.
112 H. C. Newman, "Using Antibiotics," *WP*, August 21, 1952, 8.
113 See, for example, the absence of mentions of nonhuman antibiotic use in articles by Theodore Van Dellen, "How to Keep Well," *WP*, March 18, 1960, B8 and John A. Osmundsen, "Resistant Germs Reported on Rise," *NYT*, March 12, 1961, 55.
114 "Mass Murders Have Changed Bacteria," *MT*, January 4, 1962, 6.
115 Ibid.
116 Longgood, *Poisons in Your Food*, 154; Herber, *Our Synthetic Environment*; Rodale, *Complete Book of Food*, 845.
117 "Transferable Drug Resistance," *SciAm*, February 1966, 53.
118 "Infectious Drug Resistance," *NEJM* 275, no. 5 (1966): 277.
119 Ibid. See also Scott Podolsky and Anne Kveim Lee, "Futures and Their Uses," in *Therapeutic Revolutions: Pharmaceuticals and Social Change in the Twentieth Century*, ed. Jeremy Greene et al. (Chicago: University of Chicago Press, 2016), 18–42.
120 "New 'Silent Spring'?" *NYT*, August 12, 1966, 30.
121 "Excerpts from Report on Antibiotics Prepared for the FDA," *NYT*, August 22, 1966, 28.
122 "Bacteria: How Germs Learn to Live," *Time*, August 26, 1966.
123 "The Antibiotic Jungle," *WP*, August 23, 1966, A12.
124 Jane Brody, "Medicine: Too Many Antibiotics?," *NYT*, August 28, 1966, 178.
125 James Lavelle, "Letter to the Editor," *WP*, September 19, 1966, A16.
126 "Life on the Farm," *NYT*, October 21, 1961, 20.
127 "Agriculture: Phrenological Pickers and Such," *Time*, October 2, 1964.
128 Sue Cronk, "How Safe Is the Nation's Meat Supply?" *WP*, February 10, 1964, B5.

3. Chemical Cornucopia: Antibiotics on the Farm

1 Deborah Fitzgerald, *Every Farm a Factory: The Industrial Ideal in American Agriculture* (New Haven, CT: Yale University Press, 2003), 2–8.
2 Hannah Landecker, "The Food of Our Food: Medicated Feed and the Industrialisation of Metabolism," Presented at Oxford Microbiome Seminar, February 15, 2017.
3 Douglas R. Hurt, *Problems of Plenty: The American Farmer in the Twentieth Century* (Chicago: Ivan R. Dee, 2002), 46, 63.
4 Ibid., 69–78, 81, 83, 94.
5 Fitzgerald, *Every Farm a Factory*, 184.
6 Hurt, *Problems of Plenty*, 99–100.
7 Mark R. Finlay, "Hogs, Antibiotics, and the Industrial Environments of Postwar Agriculture" in *Industrializing Organisms: Introducing Evolutionary History*, ed. Susan R. Schrepfer and Philip Scranton (New York: Routledge, 2004), 237–241, 247–249, 252–253.
8 Chris Mayda, "Pig Pens, Hog Houses, and Manure Pits: A Century of Change in Hog Production," *Material Culture* 36, no. 1 (2004): 18–24; Mark Friedberger, "Cattlemen, Consumers, and Beef," *Environmental History Review* 18, no. 3 (1994): 37.
9 William Boyd, "Making Meat: Science, Technology, and American Poultry Production," *Technology and Culture* 42, no. 4 (2001): 635; Abigail Woods,

"Rethinking the History of Modern Agriculture: British Pig Production, C. 1910–65," *Twentieth Century British History*, 23, no. 2 (2012): 177.

10   P. P. Levine, "The Effect of Sulphanilamide on the Course of Experimental Avian Coccidiosis," *Cornell Veterinarian* 29 (1939): 309–320; C. Horton-Smith and E. L. Taylor, "Sulfamethazine and Sulfadiazine Treatment in Cecal Coccidiosis in Chickens," *Veterinary Record* [hereafter *VR*] 54 (1942): 516; Philip A. Hawkins and E. E. Kline, "The Treatment of Cecal Coccidiosis with Sulfamethazine," *Poultry Science* XXIV, no. 3 (1945): 277, 280–281; W. Malcolm Reid, "History of Avian Medicine in the United States", *Avian Diseases* 32, no. 3 (1990): 509–525; for bees, see Claas Kirchhelle, "Between Bacteriology and Toxicology: Agricultural Antibiotics and US Risk Regulation," in *Risk on the Table*, ed. Angela Creager and Jean-Paul Gaudillière (Chicago: University of Chicago Press), forthcoming.

11   Susan D. Jones, *Valuing Animals: Veterinarians and Their Patients in Modern America* (Baltimore: Johns Hopkins University Press, 2003), 96–104; John E. Lesch, *The First Miracle Drugs: How the Sulfa Drugs Transformed Medicine* (Oxford: Oxford University Press, 2007), 197, 287; Finlay, "Hogs, Antibiotics," 243–251; William C. Campbell, "History of the Discovery of Sulfaquinoxaline as a Coccidiostat," *Journal of Parasitology* 94, no. 4 (2008): 934–945.

12   Landecker, "The Food of Our Food."

13   Finlay, "Hogs, Antibiotics," 243; Ulrike Thoms, "Aus Wertlosem Wertvolles Schaffen: Die Mobilisierung der Fütterungswissenschaft zur Steigerung der Nahrungsmittelproduktion im Dritten Reich" (unpublished essay); Maureen Ogle, *In Meat We Trust: An Unexpected History of Carnivore America* (Harcourt: Houghton Mifflin, 2013), 109–111.

14   Thomas H. Jukes, "Some Historical Notes on Chlortetracycline," *Reviews of Infectious Diseases* 7, no. 5 (1985): 703; Ogle, *In Meat We Trust*, 108–109.

15   P. R. Moore et al., "Use of Sulfasuxidine, Streptothricin, and Streptomycin in Nutritional Studies with the Chick," *Journal of Biological Chemistry* 165 (1946): 437–441.

16   T. D. Luckey, "Sir Samurai T. D. Luckey," *Dose-Response* (2008): 4–5; T. D. Luckey, "Antibiotics in Nutrition," in *Antibiotics: Their Chemistry and Non-medical Uses*, ed. Herbert S. Goldberg (Princeton, NJ: Van Nostrand, 1959), 176. ; G. W. Newell et al., "The Value of Dried Penicillin Mycelium as a Supplement in Practical Chick Rations," *Poultry Science* XXVI, no. 3 (1947): 284–288. It was later speculated that purified experimental diets were responsible for the "universal myopia" about the commercial implications of the 1946 Wisconsin experiment. H. R. Bird, "Biological Basis for the Use of Antibiotics in Poultry Feeds," in *The Use of Drugs in Animal Feeds: Proceedings of a Symposium* (Washington, DC: National Academy of Sciences, 1969), 31.

17   Hannah Landecker, "It Is What It Eats: Chemically Defined Media and the History of Surrounds," *Studies in History and Philosophy of Biological and Biomedical Sciences* 57 (2016): 154; S. W. Page, "Current Use of Antimicrobial Growth Promoters in Food Animals: The Benefits," D. Barug et al., eds., *Antimicrobial Growth Promoters: Where Do We Go From Here* (Wageningen: Wageningen Academic, 2006), 21; Maureen Ogle, *In Meat We Trust*, 108–110.

18   Watson M. Laetsch et al., "Thomas H. Jukes, Integrative Biology: Berkeley," *In Memoriam* (2000), 109–111; Barry Shane and Kenneth Carpenter, "E. L. Stokstad (1913–1995)," *Journal of Nutrition* 127 (1997): 199–201.

19   Thomas H. Jukes, "Some Historical Notes on Chlortetracycline." *Reviews of*

Infectious Diseases 7, no. 5 (1985): 703–706; Finlay, "Hogs, Antibiotics," 243–245; Maryn McKenna, *Big Chicken: The Incredible Story of How Antibiotics Created Modern Agriculture and Changed the Way the World Eats* (Washington, DC: National Geographic, 2017), 37–42.

20 Ellen K. Silbergeld, *Chickenizing Farms and Food: How Industrial Meat Production Endangers Workers, Animals, and Consumers* (Baltimore: Johns Hopkins University Press, 2016), 99–105.

21 "Antibiotics May Be Missing Link," *Wallaces Farmer* [hereafter *WF*], May 6, 1950, 23.

22 Hurt, *Problems of Plenty*, 100–101.

23 Paul K. Conkin, *A Revolution Down on the Farm: The Transformation of American Agriculture Since 1929* (Lexington: University Press of Kentucky, 2009), 123.

24 J. A. Wakelam, "Vitamin B12 and Antibiotics in Animal Nutrition," *Manufacturing Chemist* 9 (1952): 375.

25 "Advertisement—Nutrena," *WF*, April 16, 1949, 499.15; Jewel Shasteen French, "Furfural—Wonder Products of Farm Waste," *Progressive Farmer* [hereafter *PF*] (April 1950): 65; "Advertisement—Sargent," *WF*, April 2, 1949, 463.55; "Advertisement—Staley Milling," *WF*, May 7, 1949, 599.13.

26 Rima Apple, *Vitamania: Vitamins in American Culture* (New Brunswick, NJ: Rutgers University Press, 1996), 19–26.

27 Jukes, "Historical Notes," 705.

28 "Advertisement—Gooch Feeds," *WF*, June 3, 1950, 45.

29 Ibid.

30 "What's the Lowdown on Aureomycin," *WF*, June 17, 1950, 16.

31 "Advertisement—Ames Reliable Products," *WF*, June 17, 1950, 20.

32 Jukes, "Historical Notes," 705–706.

33 Ibid.

34 "Advertisement—Ful-O-Pep," *WF*, February 17, 1951, 25; "Advertisement—Kraft," *WF*, October 20, 1951, 36.

35 "Advertisement—Allied Mills," *WF*, June 16, 1951, 20.

36 "Advertisement—Occident," *WF*, February 17, 1951, 56.

37 "Advertisement—Lederle," *WF*, September 1, 1951, 14.

38 Bird, "Biological Basis," 31.

39 Committee to Study the Human Effects of Subtherapeutic Antibiotic Use in Animals, "The Effects on Human Health of Subtherapeutic Use of Antimicrobials in Animal Feeds," (Washington, DC: National Academies Press, 1980), 8.

40 Bird, "Biological Basis," 33; "Antibiotic Supplements in Rations for Pigs," *VR*, February 23, 1957, 234; "Feed Additives Get Lead Role," *Farmers Weekly Review* [hereafter *FWR*], April 22, 1959, 3.

41 "Control Mastitis," *WF*, June 2, 1951, 50; "Mastitis Therapy Not a Substitute for Prevention," *FWR*, December 4, 1946, 5.

42 "Livestock Need Plenty of Water," *FWR*, February 23, 1955, 5.

43 John B. Herrick, "Drugs Can't Whip Old Lots," *WF*, August 18, 1951, 32.

44 B. F. Bullock, *Practical Farming for the South* (Chapel Hill: University of North Carolina Press, 1944), 345–367; R. A. Power, *100 Common Mistakes in Farming . . . And How to Correct Them* (Viroqua: National Farm, 1947), 9–29.

45 *Eastern States Farmers Handbook* (Springfield, IL: Eastern States Farmers' Exchange, 1947), 419: see also *Midwest Farm Handbook* (Ames: Iowa State College Press, 1949), 113, 419.

46 Keith Thomas, *Livestock Feeding Manual* (Danville, IL: Interstate, 1951), 9.

47 *Midwest Farm Handbook*, 2nd ed. (Ames: Iowa State College Press, 1951), 8, 53–54.

48 George P. Deyoe and J. L. Krider, *Raising Swine* (New York: McGraw-Hill, 1952), 141, 158–159, 172, 182; Rudolph Seiden, *Poultry Handbook: An Encyclopedia for Good Management of All Poultry Breeds* (New York: Van Nostrand, 1952), 145, 421–425, 428–430; *Colorado Agricultural Handbook* (Fort Collins: Colorado State University, 1952), 364.

49 Warren McMillen, *Hog Profits for Farmers* (Chicago: Windsor, 1952), 40.

50 Ibid., 61.

51 "Better Feeding of Livestock," *Farmers' Bulletin* 2052 (1952): 9, 28, 31–33, 39–41; *Agriculture Handbook* (Washington, DC: USDA, 1952), 3–4, 12–13; "Hogs Flourish on One-19," *News for Farmer Cooperatives*, April 1953, 2; "Feeding Cottonseed Products to Livestock," *Farmers' Bulletin* 1179 (1955): 9–11.

52 John David Anthony, *Diseases of the Pig and Its Husbandry* (Baltimore: Williams and Wilkins, 1955), 142, 158, 178.

53 W. B. Haubrich, "Bovine Sterility," *Cornell Veterinarian* XLV (1955): 304–315.

54 Kendra Smith-Howard, *Perfecting Nature's Food: A Cultural and Environmental History of Milk in the United States, 1900–1970*, dissertation, University of Wisconsin–Madison, 2007, 218.

55 "Bovine Mastitis Means Loss of Annual Income," *FWR*, February 22, 1955, 6.

56 "Drugs Plus Sense Lick Mastitis," *WF*, October 1, 1949, 1143.

57 "Advertisement—Lederle," *WF*, November 19, 1949, 1243.15.

58 "Udder Diseases of Dairy Cows," *Farmers' Bulletin* 1422 (1953): 12; C. M. Stowe et al., "A Survey of the Pharmacological Properties of Four Sulfonamides in Dairy Cattle," *Cornell Veterinarian* ILVII (1957): 469.

59 Janet M. Dewdney and R. G. Edwards, "Penicillin Hypersensitivity—Is Milk a Significant Hazard?" in *Antimicrobials and Agriculture. The Proceedings of the 4th International Symposium on Antibiotics in Agriculture: Benefits and Malefits*, ed. Malcolm Woodbine (London: Butterworths, 1984), 465.

60 "Mastitis Drug Labeling," *WF*, April 7, 1951, 45.

61 "Service Bureau—Throw Away Milk After 'Treating,'" *WF*, May 5, 1951, 40.

62 Smith-Howard, *Perfecting Nature's Food*, 220.

63 Robert Angelotti et al., "The Influence of Selected Antibiotics upon the Metabolic Activities of Streptococcus Lactis," *History of Randleigh Farm*, 8th ed. (Lockport, NY: W. R. Kenan, 1956), 176–181. I am indebted to Delphine Berdah for pointing out these experiments.

64 F. S. Thatcher and W. Simon, "The Resistance of Staphylococci and Streptococci Isolated from Cheese to Various Antibiotics," *Canadian Journal of Public Health* 46, no. 10 (1955): 407–409.

65 Jones, *Valuing Animals*, 104.

66 Jukes, "Historical Notes," 705.

67 Kendra Smith-Howard, "Healing Animals in an Antibiotic Age: Veterinary Drugs and the Professionalism Crisis, 1945–1970," *Technology and Culture* 58, no. 4 (2017): 722–748.

68 Jones, *Valuing Animals*, 109–14.

69 Homer Hush, "Makes Hogs of Runts," *WF*, May 5, 1951, 8.

70 W. R. Whitfield, "Ten Seconds per Bird per Day," *WF*, July 21, 1951, 36.

71 "Britisher Amazed at 'Miracle Drug' Pigs," *FWR*, February 27, 1952, 4.

72 "Antibiotic-Age Chicks," *FWR*, February 16, 1955, 3.

73 "Antibiotics Next for Crop Disease," *FWR*, October 28, 1959, 4.

74 "Two Vegetable Diseases Controlled with Streptomycin, USDA Reports," *Lancaster Farming* [hereafter *LF*], January 6, 1956, 9; "Bees Fight Fireblight by Carrying Antibiotic on Legs," *LF*, February 8, 1957, 3.

75 "Vitamin B12 and Antibiotics in Animal Nutrition: Annotated Bibliography" (Rahway, NJ: Merck, 1951) [reprinted with updates in 1952, 1954, and 1957]; Helen Maddock and Sterling Brackett, *The Continuous Feeding of Aureomycin to Swine* (New York: American Cyanamid, 1956); "Our Smallest Servants: The Story of Fermentation" (Brooklyn, NY: Pfizer, 1955); "Coccidiosis and Poultry Management' (Rahway, NJ: Merck, 1958).

76 "Use of Antibiotics, Other Drugs, and Vitamin B12 at Low Levels in Formula Feeds" (Washington, DC: USDA Agricultural Marketing Service, 1956), iii, 5.

77 "Effects on Human Health of Subtherapeutic Use," 8.

78 E. D. Griffin, "The US Feed Industry," in *Livestock Nutrition—Report of Four Conferences Held During the Feed Show at the US Trade Center, London* (London: Graham Cherry, 1962), 29–31.

79 E. H. Kampelmacher, "Some Aspects of the Non-medical Use of Antibiotics in Various Countries," in *Antibiotics in Agriculture: Proceedings of the University of Nottingham Ninth Easter School in Agricultural Science* (London: Butterworths, 1962), 317.

80 "Advertisement—Swift & Company: New ABC's of Animal Nutrition," *WF*, September 1, 1951, 25.

81 "Advertisement—Swift & Company: A Meaty Mouthful," *WF*, September 1, 1951, 25.

82 Hurt, *Plenty*, 111–112.

83 Ibid., 121.

84 Ibid., 123.

85 Conkin, *Revolution Down on the Farm*, 130–131.

86 "Question: How Big Are Family Farms?" *WF*, January 16, 1965, 9.

87 "1969 Census of Agriculture, Volume II" (Washington, DC: US Department of Commerce, Social and Economic Statistics, 1973), 11.

88 Boyd, "Making Meat," 635.

89 Mayda, "Pig Pens," 24–27.

90 Friedberger, "Cattlemen," 37–57; Alan Marcus, *Cancer from Beef: DES, Federal Food Regulation, and Consumer Confidence* (Baltimore: Johns Hopkins University Press, 1994), 23; L. S. Pope, "Animal Science in the Twentieth Century," *Agricultural History* 54, no. 1 (1980): 66.

91 "Can We Save the Family Farm?" *WF*, January 17, 1959, 13; "Should Your Family Move to Town?" *PF*, March 1960, 144.

92 "Khrushchev Sees Iowa Agriculture," *WF*, October 3, 1959, 8.

93 Henry Simons, "He Sells Hogs 24 Times a Year," *FJ*, April 1956, 56.

94 Ibid.

95 "Mighty New Germ Killer," *FJ*, February 1956, 160.

96 "Advertisement—Quaker Oats", *WF*, February 1965, 49; "Tips for Starting Pig Business—Number 8," *WF*, February 4, 1961, 16; "Antibiotics at Breeding Boosts Pig Numbers," *WF*, November 18, 1961, 36.

97 "Ready to Cook Poultry Gaining in Popularity," *LF*, February 15, 1957, 10.

98 "New Boost for Broilers," *FJ*, September 1956, 41.

99 "Is This the Only Way to Whip Mastitis?" *FJ*, August 1956, 35; "Farm Calendar," *LF*, February 21, 1958, 12.

100 "Finger Is Pointed at Penicillin in Milk," *FJ*, September 1956, 48; "Now Is the Time," *LF*, December 26, 1959, 4.

101 Smith-Howard, *Perfecting Nature's Food*, 222–224.

102 "To Clamp Down on All Farm Chemicals," *WF*, December 5, 1959, 8.

103 "Dairymen Are Warned—Keep Drugs Out of Milk," *LF*, December 12, 1959, 15; see also "Can You Guarantee Milk Has No Residue?" *PF*, August 1960, 8.

104 "Chemicals and Food," *WF*, November 21, 1959, 12.

105 "Cranberries an Example," *PF*, January 1960, 98.

106 "Food Cranks Can Hurt Us," *PF*, November 1960, 110.

107 "Scare Tactics—An Old Ruse," *LF*, July 30, 1960, 4.

108 "From Where We Stand," *LF*, August 1, 1964, 4.

109 "New Battle for Farmers," *WF*, July 21, 1962, 10.

110 Ibid.

111 "ISU Exhibit: The Safe and Profitable Use of Farm Chemicals," *WF*, September 4, 1965, 64.

112 John H. Harris, "Fight Yard Pests with Chemicals," *PF*, June 1960, 76.

113 "On Fruits and Vegetables—Avoid Harmful Residues," *PF*, June 1960, 78.

114 "Wallaces Farmers Poll—How Farmers Handle Chemicals," *WF*, March 20, 1965, 71.

115 Michelle Mart, *Pesticides, a Love Story: America's Enduring Embrace of Dangerous Chemicals* (Lawrence: University Press of Kansas, 2016), 2–5.

116 Andrew N. Case, *The Organic Profit: Rodale and the Making of Marketplace Environmentalism* (Seattle: University of Washington Press, 2018), 52–70.

117 Robin O'Sullivan, *American Organic: A Cultural History of Farming, Gardening, Shopping, and Eating* (Lawrence: University Press of Kansas, 2015), 17–51.

118 J. I. Rodale, "With the Editor—Is the Organic Method More Expensive," *Organic Farmer* [hereafter *OF*] 2, no. 4 (November 1950): 7; J. I. Rodale, "The Work of the Soil and Health Foundation Affects Everyone," *OF* 2, no. 12 (July 1951): 16; J. I. Rodale, "In Defense of the Organic Method," *OF* 3, no. 2 (September 1951): 17–19.

119 Ernest Colwell, "Your Invisible Friends," *OF* 4, no. 8 (March 1953): 38, 40; see also "Reader's Letter: Poultry Thrive on Natural Vitamin B-12," *OF* 3, no. 2 (September 1951): 6–7; J. I. Rodale, "How Difficult Is the Organic Method?" *OF* (September 1953): 11.

120 "Their Daily Bread Is Produced Naturally," *OF* 3, no. 10 (May 1952): 26.

121 Leonard Wickenden, *Gardening With Nature—How to Grow Your Own Vegetables, Fruits, and Flowers by Natural Methods* [British edition] (London: Faber and Faber, 1956), 266.

122 "Poultry Fed Antibiotics Can Poison Consumers," *OF* 3, no. 8 (March 1952): 43; see also "Turkeys Need Room to Grow," *OF* 2, no. 9 (April 1951): 59; "Antibiotics Do Not Stimulate Hens," *OF* 4, no. 5 (December 1952): 2.

123 Neil Womack Evans, "Poultry on Pasture," *OF* (August 1953): 30.

124 Ibid.

125 J. I. Rodale et al., *The Encyclopedia of Organic Gardening* (Emmaus, PA: Rodale, 1959), 39–41; "Why You Should Eat Organically Grown Foods," *Organic Gardening and Farming* [hereafter *OGF*], July 1958, 31.

126 "New Poisons Imperil Our Meat," *OGF*, March 1959, 102.

127 Ibid., 103; see also J. I. Rodale, "The Meat Steal," *OGF*, December 1959, 15–17.

128 Melvin Scholl, "Modern Herd Management: Is It Really Better?" *OGF*, January 1959, 82–87; "He Raises Better Turkeys," *OGF*, November 1962, 33–34; "Organic Beef—Naturally Better Meat," *OGF* October (1965): 42–45.

129 Mortimer P. Starr and Donald M. Reynolds, "Streptomycin Resistance of Coliform Bacteria from Turkeys Fed Streptomycin," *American Journal of Public Health*, November 1951, 1377.

130 Ibid., 1378.

131 Ibid., 1379.

132 K.R. Johansson et al., "Effects of Dietary Aureomycin upon the Intestinal Microflora and the Intestinal Synthesis of Vitamin B12 in the rat," *Journal of Nutrition* 49 (1953): 135–152; see also K. A. McKay, *A Study of the Intestinal Bacterial Flora of Swine Fed Terramycin Supplemented Rations*, graduate thesis, Toronto, 1954.

133 Podolsky, *Antibiotic Era*, 10.

134 C. A. Brandly, "Major Poultry Disease Problems," *Canadian Journal of Comparative* Medicine XVII, no. 8 (1953): 335; F. Vigue, "Relative Efficacy of Three Antibiotic Combinations in Bovine Mastitis," *Journal of the American Veterinary Medicine Association* 124 (1954): 377; L. Meyer Jones, "Suggestions for Antibiotic Therapy," *Cornell Veterinarian* XLV (1955): 323.

135 L. Meyer Jones, *Veterinary Pharmacology and Therapeutics*, 2nd ed. (Ames: Iowa State College Press, 1957), 837.

136 I. A. Merchant and R. A. Packer, *Veterinary Bacteriology and Virology*, 6th ed. (Ames: Iowa State University Press, 1961), 143.

137 See the notable exception of 1961 warnings by ARS deputy director T. C. Byerly, "The Role of Research in Solving the Poultry Condemnation Problem," *Disease, Environmental and Management Factors Related to Poultry Health* (Washington, DC: ARS, 1961), 7.

138 M. E. Ensminger, *The Stockman's Handbook* (Danville, IL: Interstate, 1955); Gustave F. Heuser, *Feeding Poultry*, 2nd ed. (New York: Wiley, 1955); Stanley Marsden and J. Holmes Martin, *Turkey Management*, 6th ed. (Danville, IL: Interstate, 1955); *Agricultural Yearbook* (Washington, DC: USDA, 1956); Earle W. Crampton, *Applied Animal Nutrition* (San Francisco: Freeman, 1956); Jules Haberman, *Poultry Farming for Profit* (Englewood Cliffs, NJ: Prentice Hall, 1956); C. R. Grau et al., *Principles of Nutrition for Chickens and Turkeys* (Berkeley: University of California, Division of Agricultural Sciences, 1956); Tony J. Cunha, *Swine Feeding and Nutrition* (New York: Interscience, 1957); *Farm Handbook*, 4th ed. (Ames: Iowa State College, 1957); Daniel Noorlander, *Milking Machines and Mastitis*, 2nd ed. (Madison, WI: Democratic Print, 1962); *Livestock Book, Revised and Edited by the Editors of Successful Farming* (Des Moines, IA: Meredith, 1957); Rudolph Seiden and W. H. Pfander, *The Handbook of Feedstuffs* (New York: Springer, 1957); B. L. Reid et al., *Antibiotics and Arsenicals in Poultry Nutrition* (College Station: Texas Agricultural Experiment Station, 1957); *Hoard's Dairyman Feed Guide: Dairy Cattle, Swine, Poultry* (Fort Atkinson, WI: Hoard's Dairyman, 1958); "Swine Production," *Farmer's Bulletin* 1437 (1958); "Broiler Feeding," *Agriculture Handbook* No. 151 (Washington, DC: USDA/ARS, 1959); M. E. Ensminger, *The Stockman's Handbook*, 2nd ed. (Danville, IL: Interstate, 1959); *History of Randleigh Farm*, 9th ed. (Lockport, NY: W. R. Kenan, 1959); Roscoe Snapp, A. L. Neumann, *Beef Cattle*, 5th ed. (New York: Wiley, 1960); M. E. Ensminger, *Swine Science* (Danville, IL: Interstate, 1961); *Georgia*

Agricultural Handbook (Athens: University of Georgia, 1961); Frank Morrison, *Feeds and Feeding* (Clinton, IA: Morrison, 1961); M. E. Ensminger, *The Stockman's Handbook*, 4th ed. (Danville, IL: Interstate, 1962); "Raising Dairy Calves and Heifers," *Farmers' Bulletin* 2176 (1962); *Midwest Farm Handbook*, 6th ed. (Ames: Iowa State University, 1964); "Finishing Beef Cattle," *Farmers' Bulletin* 2196 (1966).

139 L. Meyer Jones, "Antibiotics," *Yearbook of Agriculture* (Washington, DC: USDA, 1956), 96.

140 Noorlander, *Milking Machines and Mastitis*, 228.

141 Ibid., 230.

142 Maddock and Brackett, *Continuous Feeding*, 83.

143 "Production Performance of Swine at Experiment Stations Ten Years After the Introduction of Aureomycin" (Princeton, NJ: American Cyanamid, 1959), 15.

144 "Antibiotics May Lose Some Punch," *FWR*, March 21, 1962, 4.

145 Ibid.

146 "Antibiotics," *WF*, November 20, 1965, 24.

147 "Effects on Human Health of Subtherapeutic Use of Antimicrobials," 8.

148 "Golden Killer," *FWR*, April 15, 1959, 6; "News of New Penicillin," *WF*, March 4, 1961, 37; "Staph Pill," *WF*, November 3, 1962, 43.

149 "Salmonella Threaten Iowa Egg Industry," *WF*, August 21, 1965, 51.

150 "FDA Gets Calls, but Data Show No Need for Alarm," *Feedstuffs*, August 27, 1966, 1.

4. Toxic Priorities: Antibiotics and the FDA

1 Daniel Carpenter, *Reputation and Power: Organizational Image and Pharmaceutical Regulation at the FDA* (Princeton, NJ: Princeton University Press, 2010), 75, 80; Nancy Langston, *Toxic Bodies: Hormone Disruptors and the Legacy of DES* (New Haven, CT: Yale University Press, 2010), 19–21.

2 Christopher C. Sellers, *Hazards of the Job: From Industrial Disease to Environmental Health Science* (Chapel Hill: University of North Carolina Press, 1997), 194–195, 198–201, 211–220; Langston, *Toxic Bodies*, 21–27.

3 Langston, *Toxic Bodies*, 26–27; Carpenter, *Reputation and Power*, 73–75.

4 Langston, *Toxic Bodies*, 27; Alan Marcus, *Cancer from Beef: DES, Federal Food Regulation, and Consumer Confidence* (Baltimore: Johns Hopkins University Press, 1994), 71; Sellers, *Hazards of the Job*, 216–220. Carpenter, *Reputation and Power*, 118–123.

5 Langston, *Toxic Bodies*; Sarah A. Vogel, *Is It Safe? BPA and the Struggle to Define the Safety of Chemicals* (Berkeley: University of California Press, 2013).

6 William C. Campbell, "History of the Discovery of Sulfaquinoxaline as a Coccidiostat," *Journal of Parasitology* 94, no. 4 (2008): 938.

7 FDA State Cooperation Information Letter No. 16, July 27, 1949, Folder 432.73-11-432.97-.10, Box 1160, FDA General Subject Files [hereafter GS], Decimal Files [hereafter DF] A1/Entry 5, Record Group [hereafter RG] 88, National Archives and Records Administration, College Park [hereafter NARA]; J. P. Delaphane and J. H. Milliff, "The Gross and Micropathology of Sulphaquinoxaline Poisoning in Chickens," *American Journal of Veterinary Research* 92 (1948): 9.

8 "Proceedings FDA Conference on the Kefauver-Harris Amendments" (Washington, DC: HEW/FDA, 1963), 32.

9 Vogel, *Is It Safe?* 23.

10 Jukes, "Some Historical Notes," 705.

11 Ibid.

12 Mark R. Finlay and Alan I. Marcus, "Battles over Agricultural Antibiotics in the United States and Western Europe," *Agricultural History* 90, no. 2 (2016): 153.

13 The question of whether AGPs were drugs or additives, in which case regulators would have focused on residues, was only resolved by the 1968 Animal Drug Amendments; Gilbert Wayne Gillespie Jr., *Antibiotic Animal Feed Additives and Public Policy: Farm Operators' Beliefs About the Importance of These Additives and Their Attitudes Toward Government Regulation of Agricultural Chemicals and Pharmaceuticals*, Doctoral thesis, Cornell University, 1987, 244.

14 16 *Federal Register* [hereafter *FedReg*], 3647-3648 (April 28, 1951); see also Lisa Heinzerling, "Undue Process at the FDA," *Georgetown Public Law and Legal Theory Research Paper No. 13-016* (2013).

15 H. R. Bird, H.R., "The Biological Basis for the Use of Antibiotics in Poultry Feeds," in *The Use of Drugs in Animal Feeds: Proceedings of a Symposium* (Washington DC: National Academy of Sciences [NAS]), 1967, 31–42.

16 Jukes, "Some Historical Notes," 704.

17 "Investigation of the Use of Chemicals in Food Products," Select Committee to Investigate the Use of Chemicals in Food Products, US House of Representatives, 1951; "Chemicals in Food Products," House Select Committee to Investigate the Use of Chemicals in Food Products, US House of Representatives (Washington, DC: GPO, 1951), 129, 131, 496.

18 Moskey to Larrick, October 2, 1952, Box 1560, GS, DF A1/Entry 5, RG 88, NARA.

19 Larrick to W. P. Bomar, October 3, 1952, Box 1560, GS, DF A1/Entry 5, RG 88, NARA; see also Kendra Smith-Howard, "Healing Animals in an Antibiotic Age: Veterinary Drugs and the Professionalism Crisis, 1945–1970," *Technology and Culture* 58, no. 4 (2017): 722–748, 727–728.

20 Gillespie, *Antibiotic Animal Feed Additives*, 245–246; certification was discontinued in 1982.

21 Moskey to Lee, July 16, 1952, Folder 432.97.10-435, Box 1560, GS, DF A1/Entry 5, RG 88, NARA.

22 Ibid.

23 18 *FedReg*, 2335-2336 (April 22, 1953); Lisa Heinzerling, "Undue Process at the FDA," *Georgetown Public Law and Legal Theory Research Paper No. 13-016* (2013).

24 Gillespie, *Antibiotic Animal Feed Additives*, 249, 251; 20 *FedReg* 9696-9698 (December 20, 1955).

25 Memorandum of Interview, August 24, 1949, Folder 432.73-.11–432.97-.10, Box 1160, FDA GS, DF A1/Entry 5, RG 88, NARA; Rayfield to Atlanta District, July 8, 1949, Folder 432.10-432.4, Box 1160, GS, DF A1/Entry 5, RG 88, NARA; Elliott to St. Louis District Administration, October 11, 1949, Folder 432.10-432.4, Box 1160, GS, DF A1/Entry 5, RG 88, NARA.

26 Moskey to Lee, July 16, 1952, Folder 432.97.10-435, Box 1560, FDA GS, DF A1/Entry 5, RG 88, NARA.

27 On boundary values and tolerances, see Carsten Reinhardt, "Boundary Values," in *Precarious Matters: The History of Dangerous and Endangered Substances in the 19th and 20th centuries*, ed. Viola Balz, Alexander von Schwerin, Heiko Stoff, and Bettina Wahrig (Berlin: MPI for the History of Science, 2008), 39–41; Nathalie

Jas, "Adapting to 'Reality': The Emergence of an International Expertise on Food Additives and Contaminants in the 1950s and Early 1960s," in *Toxicants, Health and Regulation Since 1945*, ed. Nathalie Jas and Soraya Boudia (London: Pickering and Chatto, 2013), 47–69.

28   13 *FedReg*, 7403-7404 (December 4, 1948); Gillespie, *Antibiotic Animal Feed Additives*, 243–244.

29   18 *FedReg*, 1077 (February 25, 1953).

30   Ibid.

31   Langston, *Toxic Bodies*, 77–81; Vogel, *Is It Safe?* 15–42.

32   William Randall, "Antibiotic Residues," *Proceedings First International Conference on the Use of Antibiotics in Agriculture* (Washington, DC: National Academy of Sciences—National Research Council, 1956), 262–263.

33   Ella Barnes, "The Use of Antibiotics for the Preservation of Poultry and Meat," *Antibiotics in Agriculture—Proceedings of the Fifth Symposium of the Group of European Nutritionists, 1966* (Basel, CH: Karger, 1968), 64–65.

34   20 *FedReg*, 8776 (November 30, 1955); 21 *FedReg*, 8104 (October 23, 1956); E. H. Kampelmacher, "Some Aspects of the Non-Medical Use of Antibiotics in Various Countries," in *Antibiotics in Agriculture: Proceedings of the University of Nottingham Ninth Easter School in Agricultural Science* (London: Butterworths, 1962), 317.

35   "Excerpts from Report on Antibiotics," *New York Times* [hereafter *NYT*], August 22, 1966, 28.

36   F. E. Deatherage, "The Use of Antibiotics in the Preservation of Foods Other than Fish," *Proceedings First International Conference on the Use of Antibiotics in Agriculture* (Washington, DC: NAS-NRC, 1956), 221.

37   "Food Standards Committee Report on Preservatives in Food" (London: HMSO, 1959), 45.

38   Trials were conducted at the Norwegian whaling station in Steinchman; the author does not know whether whale steaks were actually sold. "Antibiotics Used to Preserve Food," *NYT*, October 20, 1956, 29; see also Johan Nicolay Tønnessen and Arne Odd Johnsen, *The History of Modern Whaling* (Berkeley: University of California Press, 1982), 694; Claas Kirchhelle, "Pharming Animals: A Global History of Antibiotics in Food Production, 1935–2017" *Palgrave Communications* 4, no. 96 (2018): 3.

39   *Poultry Inspectors' Handbook* (Washington, DC: GPO, 1957), 78.

40   "Food Standards Committee Report," 43–44.

41   Maryn McKenna, *Big Chicken: The Incredible Story of How Antibiotics Created Modern Agriculture and Changed the Way the World Eats* (Washington, DC: National Geographic, 2017), 82–92.

42   A. E. Murneek, "Thiolutin as a Possible Inhibitor of Fireblight," *Phytopathology* 42 (1952): 57; W. J. Zaumeyer, "Improving Plant Health with Antibiotics," *Proceedings First International Conference on the Use of Antibiotics in Agriculture* (Washington, DC: NAS-NRC, 1956), 172–174; M. H. Schroth et al., "Epidemiology and Control of Fire Blight," *Annual Review of Phytopathology* 12 (1974): 401–402.

43   Randall, "Antibiotic Residues," 260.

44   Ibid., 262.

45   "FDA's 1967 Look After 60 Years of Reorganization," *FDA Papers* 1, no. 1 (1967): 10.

46 "Proceedings First International Conference," 265–278.

47 J. H. Collins, "The Problem of Drugs for Food-Producing Animals and Poultry," *Food Drug Cosmetic Law Journal* 6 (1951): 876–877; Kendra Smith-Howard, *Perfecting Nature's Food: A Cultural and Environmental History of Milk in the United States, 1900–1970*, dissertation, University of Wisconsin–Madison, 2007, 223–224.

48 Randall, "Antibiotic Residues," 261.

49 Henry Welch, "Antibiotics in Food Preservation: Public Health and Regulatory Aspects," *Science* 126, no. 3284 (1957): 1160.

50 Randall, "Antibiotic Residues," 262.

51 Ibid., 261.

52 Smith-Howard, *Perfecting Nature's Food*, 222–226.

53 Kendra Smith-Howard, "Antibiotics and Agricultural Change: Purifying Milk and Protecting Health in the Postwar Era," *Agricultural History* 84, no. 3 (2010): 327–351, 339–340.

54 Eugene H. Holeman to Bill V. McFarland, December 12, 1959, enclosed in McFarland to Holeman, December 29, 1959, Folder 432.1-20-433.10, Box 2669, FDA GS, DF A1/Entry 5, RG 88, NARA; Robert Taft, *Residues in Dairy Products: Course Manual Milk and Food Training* (Cincinnati: US HEW/PHS, 1961), 1–8.

55 W. G. Huber, "The Impact of Antibiotic Drugs and Their Residues," *Advances in Veterinary Science and Comparative Medicine* 15 (1971), 107.

56 Claas Kirchhelle, "Between Bacteriology and Toxicology: Agricultural Antibiotics and US Risk Regulation," in *Risk on the Table*, ed. Angela Creager and Jean-Paul Gaudillière (Chicago: University of Chicago Press, forthcoming).

57 Maxwell Finland, "Emergence of Resistant Strains in Chronic Intake of Antibiotics: A Review," *First International Conference on Antibiotics in Agriculture, 19-21 October 1955* (NAS-NRC, 1956), 251.

58 Ibid., 265–278.

59 Countway Library of Medicine [hereafter CLM], Maxwell Finland Papers [hereafter FP], Series VIII, Box 13, Folder 51, Paul Weiss to Maxwell Finland (June 10, 1955).

60 CLM FP, Series VIII, Box 13, Folder 51, Jukes to Finland (July 1, 1955), Jukes to Finland (July 11, 1955).

61 CLM FP, Series VIII, Box 13, Folder 51, Finland to Jukes (July 14, 1955); see also CLM FP, Series VIII, Box 13, Folder 51, Finland to Damon Catron (August 22, 1955).

62 CLM FP, Series VIII, Box 13, Folder 51, Margarete [Framel] (Secretary to Dr. Jukes) to Finland (August 9, 1955).

63 CLM FP, Series VIII, Box 13, Folder 51, H. P. Broquest to Finland (August 31, 1955); Finland to Jukes (undated), enclosed in Margarete [Framel] (Secretary to. Jukes) to Finland (August 9, 1955).

64 CLM FP, Series VIII, Box 13, Folder 51, International Conference on the Use of Antibiotics in Agriculture. Information for Invited Participants; see also [booklet] "In Honor of the Participants in the International Conference on the Use of Antibiotics in Agriculture."

65 David Vogel, *The Politics of Precaution: Regulating Health, Safety and Environmental Risks in Europe and the United States* (Princeton, NJ: Princeton University Press, 2012), 45–46.

66 Langston, *Toxic Bodies*, 81; Wallace F. Janssen, "FDA Since 1938: The Major Trends and Developments," *Journal of Public Law* 13, no. 1 (1964): 208.

67 Walter Moses to Constance Winslade, June 21, 1961, Folder 432.1-10 January–December, Box 3041, GS, DF A1/Entry 5, RG 88, NARA, 1.

68 Langston, *Toxic Bodies*, 82.

69 Bill V. McFarland to Robert E. Rust, December 2, 1959, Folder PA 190#95, Box 3245, GS, DF A1/Entry 5, RG 88, NARA; required data was listed in Form FD-356. Homer R. Smith to Emil Lienert, [undated], Folder PA#95, Box 3245, GS, DF A1/Entry 5, RG 88, NARA, 1; J.F. Robens to Antonio Santos Ocampo, October 2, 1961, Folder 432.1 June–December, Box 3040, GS, DF A1/Entry 5, RG 88, NARA.

70 "Significant Dates in US Food and Drug Law History," *FDA History*; Vogel, *Is It Safe?*, 35.

71 Moses to Winslade, June 21, 1961, Folder 432.1-10 January–December, Box 3041, GS, DF A1/Entry 5, RG 88, NARA, 2.

72 Grove to Foster, October 2, 1961, Folder 432.1 June–December, Box 3040, GS, DF A1/Entry 5, RG 88, NARA. Detailed residue concentrations in different tissues are listed in section 121.1014 "Tolerances for Residues of Chlortetracycline, Subpart D—Food Additives Permitted in Animal Feed or Animal-Feed Supplements," reissued March 20, 1962, Folder PA 190#95, Box 3245, GS, DF A1/Entry 5, RG 88, NARA, 7.

73 Hubert S. Spungen to Brandenburg Brothers, April 12, 1966, Folder 88-75-1, Box 3846, GS, DF A1/Entry 5, RG 88, NARA.

74 Enclosure, "Drugs and Feed Additives," in LaVerne C. Harold, "Memo," enclosed in Smith to P. E. Poss, October 29, 1962, Folder 70A190#95, Box 3245, GS, DF A1/Entry 5, RG 88, NARA, 2, 43.

75 William E. Jester to Robert S. Roe, "Office Memorandum—Medicated Feeds," July 10, 1959, Folder 432.1-432.1-11, Box 2668, GS, DF A1/Entry 5, RG 88, NARA, 3.

76 Paul M. Sanders, "Summary of Some Differences and Sources of Confusion within the [FDA] and Their Jurisdiction over Medicated Feeds," April 8, 1959, enclosed in William E. Jester to Robert S. Roe, "Office Memorandum—Medicated Feeds," July 10, 1959, Folder 432.1-432.1-11, Box 2668, GS, DF A1/Entry 5, RG 88, NARA.

77 Charles G. Durbin to Office of the Commissioner, July 5, 1960, Folder 432.1 Dec.- 432.1 July, Box 2843, GS, DF A1/Entry 5, RG 88, NARA.

78 Paul M. Sanders, "Summary of Some Differences and Sources of Confusion within the [FDA] and Their Jurisdiction over Medicated Feeds," April 8, 1959, enclosed in William E. Jester to Robert S. Roe, "Office Memorandum—Medicated Feeds," July 10, 1959, Folder 432.1-432.1-11, Box 2668, GS, DF A1/Entry 5, RG 88, NARA.

79 Ibid.

80 Durbin to Quackenbush, July 13, 1959, Folder 432.1-432.1-11, Box 2668, GS, DF A1/Entry 5, RG 88, NARA.

81 Division of Pharmacology and Bureau of Medicine to Office of the Commissioner, "Veterinary Drugs Under the Food Additives Amendment," July 14, 1959, Folder 432.1-432.1-11, Box 2668, GS, DF A1/Entry 5, RG 88, NARA, 1.

82 Ibid.

83 Ibid., 2–3.

84 W. E. Glennon to Ralph F. Kneeland, October 19, 1959, Folder 432.1-432.1-11, Box 2668, GS, DF A1/Entry 5, RG 88, NARA.

85 "Successful Farming—December Issue (interview George P. Larrick)," enclosed in "Memorandum for File—Interview with G. P. Larrick," October 15, 1959, Folder 432.1-432.1-11, Box 2668, GS, DF A1/Entry 5, RG 88, NARA.

86 John W. Kuzmeski to W. E. Glennon, June 10, 1959, enclosed in Kuzmeski to Larrick, June 23, 1959, Folder 432.1-432.1-11, Box 2668, GS, DF A1/Entry 5, RG 88, NARA.

87 Ibid.

88 Ibid.

89 Quackenbush to Abbott Laboratories, February 07, 1959, enclosed in Bill V. McFarland to Eugene H. Holeman, December 29, 1959, Folder 432.1-20-433.10, Box 2669, GS, DF A1/Entry 5, RG 88, NARA.

90 Ibid.

91 Glennon to Larrick, December 6, 1960, 3, enclosed in Glennon to Morris Yakowitz, December 7, 1960, Folder 432.1 Dec.- 432.1 July, Box 2843, GS, DF A1/Entry 5, RG 88, NARA, 3; C. A. Armstrong to Dallas District, January 29, 1962, Folder 70A190#96, Box 3246, GS, DF A1/Entry 5, RG 88, NARA.

92 "Memorandum of Interview," December 4, 1962, Folder 432-432.80, Box 3245, GS, DF A1/Entry 5, RG 88, NARA.

93 Robert V. Marrs to A. Harris Kenyon, September 26, 1962, Folder 70A190#95, Box 3245, GS, DF A1/Entry 5, RG 88, NARA, 2.

94 Ibid.

95 Ibid.

96 K. L. Milstead to J. L. Harvey, May 28, 1962, Folder PA 190#95, Box 3245, GS, DF A1/Entry 5, RG 88, NARA.

97 Daniel DeCamp to Durbin, "Memorandum—Current Poultry Feeding Practices (Feed Supplies at the Farm)," September 18, 1962, Folder 70A190#96, Box 3246, FDA GS, DF Entry A1, RG 88, NARA, 3.

98 Efficacy claims were already being tested for veterinary drugs prior to the 1962 Amendment; CLM, FP, Series II, A. Professional Correspondence, 1929–1984, Box 2, Folder 25: FDA 1961–1965, Charles Durbin, "Veterinary Drugs," in *Proceedings. FDA Conference on the Kefauver-Harris Drug Amendments and Proposed Regulations*, February 15, 1963, 8–9.

99 John Harvey to Peter Dominick (House of Representatives), May 25, 1962, Folder PA 190#95, Box 3245, GS, DF A1/Entry 5, RG 88, NARA; Marcus, *Cancer from Beef*, 65–66.

100 Durbin to Bureau of Enforcement (Atten: C. Armstrong), April 27, 1962, Folder PA 190#95, Box 3245, GS, DF A1/Entry 5, RG 88, NARA; Levin to Larrick, Nov 29, 1962, Folder 70A190#95, Box 3245, GS DF A1/Entry 5, RG 88, NARA, 1; Marcus, *Cancer from Beef*, 68–71.

101 The history of veterinary drugs regulation in part undermines Daniel Carpenter's rehabilitation of Larrick; Carpenter, *Reputation and Power*, 9–11, 266–269.

102 Kirchhelle, "Between Bacteriology and Toxicology."

103 John. T. Logue, "The Public Health Significance of Antibiotics," in *Antibiotics—Their Chemistry and Non-Medical Uses*, ed. Herbert S. Goldberg (Princeton, NJ: Van Nostrand, 1959), 581–583; on the streptomycin trials, see also R. N. Goodman and H. S. Goldberg, "Antibiotics in Agriculture," *Science* 137, no. 3524 (1962): 135.

104 Henry Welch and Félix Marti-Ibanez, *The Antibiotic Saga* (New York: Medical

Encyclopedia, 1960), 49; see also Henry Welch, "Control of Antibiotics," *Food Drug Cosmetic Law Journal* 462 (1957): 462–468.

105 Antonio Santos Ocampo Jr. to G. V. Peacock, August 30, 1961, enclosed in J. F. Robens to Ocampo Jr., October 2, 1961, Folder 432.1 June–December, Box 3040, GS, DF A1/Entry 5, RG 88, NARA.

106 Ibid.

107 Robens to Ocampo Jr., October 2, 1961, Folder 432.1 June–December, Box 3040, GS, DF A1/Entry 5, RG 88, NARA.

108 The WHO report supported benign US assessments of AMR hazards: WHO, "The Public Health Aspects of the Use of Antibiotics in Food and Feedstuffs," *World Health Organization Technical Report Series* 260 (Geneva: WHO, 1963), 12–14.

109 Robens, "Memorandum of Conference," December 6, 1962, Folder 70A190#95, Box 3245, GS, DF A1/Entry 5, RG 88, NARA, 2.

110 John L. Harvey to Hubert H. Humphrey (US Senate), September 5, 1962, Folder 70A190#95, Box 3245, GS, DF A1/Entry 5, RG 88, NARA, 1.

111 "George P. Larrick—FDA Commissioners Page."

112 James S. Collins to Senator Humphrey, August 8, 1962, enclosed in Harvey to Humphrey (US Senate), September 5, 1962, Folder 70A190#95, Box 3245, GS, DF A1/Entry 5, RG 88, NARA, 1–2.

113 Ibid.

114 Ibid., 4.

115 Ibid., 1.

116 Ibid.

117 Ibid., 2–3.

118 Ibid., 2.

119 K. J. Davis, A. A. Nelson, B. J. Vos to Bureau of Enforcement, [undated], Folder 70A190#95, Box 3245, GS, DF A1/Entry 5, RG 88, NARA, 2.

120 Ibid.

121 "George P. Larrick—FDA Commissioners Page"; "Another Top Level Official to Leave FDA Next Week," *Feedstuffs*, December 11, 1965, 2, 75.

122 "James L. Goddard—FDA Commissioners Page."

123 31 *FedReg* 11141 (August 23, 1966).

124 Roger Berglund, "Industry Cautioned on Possible Salmonella, Chemical Residue Problems," *Feedstuffs*, February 13, 1965, 8, 73; "Residues in Swine Not Cause for Alarm," *Feedstuffs*, November 13, 1965, 6, 87.

125 *Manual of Meat Inspection Procedures of the USDA* (Washington, DC: USDA, 1964), 83.

126 CLM, FP, Series II, A. Alphabetical Correspondence, Box 2, Folder 25: FDA 1961–1965, Clem O. Miller to Finland (February 11, 1965); McKenna, *Big Chicken*, 91–92.

127 CLM, FP, Series II, A. Alphabetical Correspondence, Box 2, Folder 25: FDA 1961–1965, Finland to Miller (February 18, 1965).

128 CLM, FP, Series II, A. Alphabetical Correspondence, Box 2, Folder 25: FDA 1961–1965, William W. Wright to Finland (March 15, 1965), see also 31 *FedReg* 11141 (August 23, 1966).

129 Watanabe had already published on R-factor transfer in the 1950s and published a review in English in 1963: T. Watanabe, "Infective Heredity of Multiple Drug Resistance in Bacteria," *Bacteriological Reviews* 27, no. 1 (1963).

130 E. S. Anderson and Naomi Datta, "Resistance to Penicillins and Its Transfer in Enterobacteriaceae," *Lancet* 285, no. 7382 (1965): 407–409.

131 The list did not include Japanese publications on R-factors either; CLM FP, Series II, A. Alphabetical Correspondence, Box 2, Folder 25: FDA 1961–1965, William W. Wright to Finland (April 19, 1965). Finland's unchanged and outdated views become evident in his proposed reading contributions, see Finland to Miller (February 18, 1965); for Finland's very different influence on medical antibiotic regulation, see also Podolsky, *Antibiotic Era*, 17–18, 79, 127–29, 32–202.

132 CLM FP, Series II, A. Alphabetical Correspondence, Box 2, Folder 25: FDA 1961–1965, William W. Wright to Finland (July 26, 1965).

133 TNA MAF 386/44 (Draft Minutes. 4th Meeting Joint PHLS/AHD Advisory Committee, 24 Jan, 1967), 2.

134 "Excerpts from Report on Antibiotics," *NYT*, August 22, 1966, 28.

135 Ibid.

136 31 *FedReg* 11141 (August 23, 1966).

137 Ibid., 11141-11142.

138 W. N. Swain to Robert W. Kastenmeier (House of Representatives), December 12, 1966, Folder 88-75-1 Box 4 [sic], Box 3846, GS, DF A1/Entry 5, RG 88, NARA; Robert A. Baldwin and Laverne C. Harold, "Ecologic Effects of Antibiotics," *FDA Papers* 1, no. 1 (1967): 23–24; "Drug Residues Studied," *Science News* 90, no. 10 (1966): 148.

139 "The Use of Drugs in Animal Feeds" (Washington, DC: NAS, 1967), 60.

140 Ibid., 346.

141 Ibid., 349.

142 Ibid., 375.

143 Ibid., 359; Kirchhelle, "Between Bacteriology and Toxicology."

144 "The Use of Drugs in Animal Feeds," 304.

145 Ibid., 316.

146 Ibid., 324.

147 Ibid., 7.

148 Ibid.

149 Ibid., 8.

150 Fred Kingma, "Establishing and Monitoring Drug Residue Levels," *FDA Papers* 1, no. 6 (1967), 33.

151 "Use Medicated Feeds Carefully and Wisely," *FDA Papers* 1, no. 6 (1967).

152 C. D. Van Houweling, "Drugs in Animal Feed? A Question without an Answer," *FDA Papers* 1, no. 7 (1967): 15.

153 Ibid.

154 Ibid.

5. Fusing Concerns: Antibiotics and the British Public

1 Ina Zweiniger-Bargielowska, *Austerity in Britain: Rationing, Control and Consumption 1939–1955* (Oxford: Oxford University Press, 2002), 37, 73.

2 H. J. H. MacFie and H. L. Meiselman, *Food Choice, Acceptance and Consumption* (London: Blackie Academic and Professional, 1996), 377.

3 Stuart Anderson, *Making Medicines: A Brief History of Pharmacy and Pharmaceuticals* (London: Pharmaceutical Press, 2005), 248.

4   Robert Bud, *Penicillin: Triumph and Tragedy* (Oxford: Oxford University Press, 2009), 19, 59–72.

5   "A New Drug Speeds Pork Chops to Dining Table," *Daily Mirror* [hereafter *DM*], April 11, 1950, 6–7.

6   "Aureomycin Supplies," *VR*, May 6, 1950, 277.

7   G. R. H. Nugent, "The Twentieth-Century Hen," *Times*, July 30, 1951, 5; see also "Pigs Fattened by Antibiotics," *Times*, December 1, 1952, 3; "Feeding-Stuff Experiments," *Times*, July 8, 1952, 3.

8   "Second Reading—Therapeutic Substances (Prevention of Misuse) Bill (House of Lords)," May 13, 1953, *Hansard*, vol. 515, col. 1327–1343.

9   "Fatter Pigs on Penicillin," *Observer*, November 30, 1952, 3; "Mass Producing Antibiotics," *Financial Times* [hereafter *FT*], December 23, 1954.

10 "Antibiotic to Aid Egg Output," *FT*, April 1, 1959, 11; "Antibiotics Preserve Forage," *FT*, February 12, 1960, 13.

11 "New Method of Food Preservation," *Times*, April 11, 1956, 13.

12 "Use of Antibiotics in Keeping Fish Fresh," *FT*, June 11, 1956, 11.

13 Matthew Godwin et al., "The Anatomy of the Brain Drain Debate, 1950–1970s: Witness Seminar," *Contemporary British History* 23, no. 1 (2009): 35–60.

14 Bud, *Penicillin*, 67–72.

15 "Increasing Scope for Antibiotics," *FT*, October 13, 1952, xv.

16 "Over-production of Antibiotics," *Manufacturing Chemist* 12 (1952): 489; "Dietary Antibiotics," *Manufacturing Chemist* 10 (1952): 401.

17 Initially, companies like Lederle subcontracted parts of aureomycin production to Boots Pure Drug Company; "Mass Producing Antibiotics," *FT*, December 23, 1954.

18 "New Antibiotic Plant Opened," *Times*, October 1, 1955, 5.

19 "Advertisement—Cyanamid," *Times*, July 17, 1953, 5.

20 "Advertisement—Pfizer," *Times*, June 26, 1956, 2.

21 "Advertisement—Cyanamid," *Times*, June 29, 1956, 2.

22 "Advertisement—Cyanamid," *Times*, April 16, 1958, 7.

23 "Feeding the World," *Times*, September 19, 1962, ii; "Growing Role for the Chemist in Feeding 6,000m. by AD 2,000," *Guardian*, November 3, 1965, 5.

24 "Pigs Thrive on Antibiotics," *FT*, April 11, 1962, 11.

25 "Why the American Farmer Can Cope Single-Handed," *Times*, June 28, 1965, 14.

26 "Advertisement—Cyanamid," *Times*, April 14, 1961, 5.

27 "Advertisement—Cyanamid," *Times*, October 13, 1961, 5; "Advertisement—Cyanamid," *Times*, June 22, 1961, 5.

28 "Parliament, House of Lords, Wednesday, July 4," *Times*, July 5, 1951, 4.

29 Ibid.

30 "Therapeutic Substances (Prevention of Misuse) Bill," *VR*, February 21, 1953, 126.

31 "Antibiotics," *VR*, February 21, 1953, 127.

32 TNA MAF 287/299, Extract, House of Commons (P.Q. 3355), question put on 19th Feb, 1953.

33 Ibid.

34 "Fatter Pigs on Penicillin," *Observer*, November 30, 1952, 3.

35 Olive Whicher, "Penicillin for Pigs," *Observer*, December 28, 1952, 2; G. Pelham Reid, "Guidance Required," *Observer*, January 4, 1953, 3. Fears mirrored those of Germany's Gesellschaft für Vitalstofflehre; Claas Kirchhelle, "Toxic Confusion: The Dilemma of Antibiotic Regulation in West German Food Production," *Endeavor* 40, no. 2 (2016): 117.

36   Anne Hardy, "John Bull's Beef: Meat Hygiene and Veterinary Public Health in England in the Twentieth Century," *Review of Agricultural and Environmental Studies* 91, no. 4 (2010): 276–378.

37   "Farmers Given Assurance on Reactor Effects," *Times*, October 23, 1957, 6; on Windscale, see Soraya de Chadarevian, "Mice and the Reactor: The 'Genetics Experiment' in 1950s Britain," *Journal of the History of Biology* 39 (2006): 707–735.

38   James C. Thomson, *Constipation and Our Civilisation* (Edinburgh: Thorsons, 1943). Prominent books were G. T. Wrench, *Wheel of Health* (London: C. W. Daniel, 1938); Doris Grant, *Your Daily Bread* (London: Faber and Faber, 1944).

39   "What's in Our Food," *Daily Mail* [hereafter *DaMa*], May 28, 1959, 4.

40   "Farm Health Problems in New Methods," *Times*, September 11, 1961, 7.

41   "Drug Hazard in Dairy Milk," *Guardian*, May 30, 1963, 1.

42   "Keeping Milk Free of Antibiotics," *Times*, May 30, 1963, 18; "What Cures Cow Can Harm Milk," *Guardian*, February 12, 1965, 17; "Milk Warning," *DaMa*, May 30, 1963, 1.

43   Michael Winstanley, "Cow Punch," *Guardian*, June 25, 1963, 6.

44   "Who's to Blame for the Loss of Taste?" *DaMa*, August 16, 1962, 6; "Name Your Poison," *Spectator*, April 7, 1967, 393.

45   Franklin Bicknell, *Chemicals in Food and in Farm Produce: Their Harmful Effects* (London: Faber and Faber, 1960).

46   Franklin Bicknell, *The English Complaint or Your Fatigue and its Cure* (London: William Heinemann, 1952), 83–84.

47   Bicknell, *Chemicals in Food and in Farm Produce*, 71–75.

48   Ibid., 75–76.

49   Ibid., 78.

50   Doris Grant, *Housewives Beware* (London: Faber and Faber, 1958), 26–29.

51   Doris Grant, *Your Bread and Your Life* (London: Faber and Faber, 1961), 67–68, 79–80.

52   Ibid., 79.

53   "Farm Health Problems in New Methods," *Times*, September 11, 1961, 7; see also chapter 6.

54   "Working Out Policy for Disease Control," *Times*, September 14, 1959, 19.

55   Bud, *Penicillin*, 174–175; see also: "Farming Notes and Comments," *Times*, January 18, 1960, 21.

56   "Antibiotics for Farm Animals," *Lancet* 273, no. 7069 (1959): 402; "Drug-Resistant Staphylococci in the Farmyard," *Lancet* 275, no. 7138 (1960): 1338–1339.

57   Hilda Kean, *Animal Rights: Political and Social Change in Britain since 1800* (London: Reaktion, 1998), 35–136; Mieke Roscher, *Ein Königreich für Tiere. Die Geschichte der britischen Tierrechtsbewegung* (Marburg, DE: Tectum Verlag, 2009), 12, 206–257.

58   Roscher, *Ein Königreich für Tiere*, 252–262.

59   Clifford Selly, "Broilers Under Fire," *Observer*, March 8, 1959, 3.

60   Ibid.

61   G. B. Houston, "Letters to the Editor: Broiler Fowls," *Observer*, March 15, 1959, 4.

62   F. A. Dorris Smith, "Letters to the Editor: Broiler Fowls," *Observer*, March 15, 1959, 4.

63 John Archer, "Letters to the Editor: Broiler Fowls," *Observer*, March 22, 1959, 4.

64 Roscher, *Ein Königreich für Tiere*, 252–262, 294–295.

65 Ruth Harrison, *Animal Machines* (London: Vincent Stuart, 1964).

66 Richard D. Ryder, "Harrison, Ruth (1920–2000)," *Oxford Dictionary of National Biography* (Oxford: Oxford University Press, 2004); Claas Kirchhelle, *Ruth Harrison and the History of British Animal Welfare Science and Activism* (London: Palgrave Macmillan, forthcoming).

67 Harrison, *Animal Machines*, 116–120.

68 Ibid., 120.

69 YBL, Rachel Carson Papers, YCAL, MSS 46, Series I, Writings, Box 95, Folder 1669, Carson Preface for *Animal Machines* by Ruth Harrison, 1.

70 Ibid., 2.

71 YBL, Rachel Carson Papers, YCAL, MSS 46, Series II, General Correspondence, Box 103, Folder 1952, Ruth Harrison to Rachel Carson (July 10, 1963); Harrison to Carson (October 14, 1963).

72 Ruth Harrison, "Inside the Animal Factories," *Observer*, March 1, 1964, 21.

73 Ruth Harrison, "Fed to Death," *Observer*, March 8, 1964, 21.

74 Harrison, "Inside the Animal Factories," 21.

75 Ibid.

76 Ibid.

77 Harrison, "Fed to Death," 21.

78 Ibid.

79 Ibid.

80 Ibid.

81 "Views on Animal Factories," *Observer*, March 15, 1964, 30.

82 Helen M. Simpson, "Views on Animal Factories: Poles Apart," *Observer*, March 15, 1964, 30.

83 Sheila M. Mitchell, "Views on Animal Factories: Label them," *Observer*, March 15, 1964, 30.

84 Barbara Willard, "Views on Animal Factories: Try It on the Dog," *Observer*, March 15, 1964, 30.

85 John Hall, "Views on Animal Factories: Changing the Law," *Observer*, March 15, 1964, 30.

86 David Sainsbury, "Views on Animal Factories: Distorted," *Observer*, March 15, 1964, 30.

87 "Cruelty War by Church leader," *DM*, August 10, 1964, 3; "'Intensive Farming' Condemned," *Guardian*, August 10, 1964, 3.

88 "Get Rid of Farm Belsen," *Observer*, October 24, 1965, 9.

89 "Are Farmers Really as Cruel as this Housewife Says?" *DaMa*, March 9, 1964, 8; see also "Inquiry into Factory Farms," *DaMa* April 21, 1964, 11.

90 "Man, Food and Animals," *DaMa*, March 13, 1964, 10.

91 "MPs see author about factory farming," *DaMa*, May 12, 1964, 12; "Charter to Protect Farm Animals," *DaMa*, August 6, 1966, 9.

92 Karen Sayer, "Animal Machines: The Public Response to Intensification in Great Britain, C. 1960- C.1973," *Agricultural History* 87, no. 4 (2013): 482–483; Abigail Woods, "Rethinking the History of Modern Agriculture: British Pig Production, c. 1910–65," *Twentieth Century British History* 23, no. 2 (2012): 165–191.

93 "Hazard to Health in Food?" *Guardian*, March 28, 1964, 28.

94 "Starting New Charter for the Housewife," *DM*, September 10, 1964, 13.

95 "Miss Sandys and Her Book Stop the Prince," *DM*, October 29, 1965, 7.

96 Abigail Woods, "From Cruelty to Welfare: The Emergence of Farm Animal Welfare in Britain, 1964–71," *Endeavour* 36, no. 1 (2012): 18–20.

97 "Report of the Technical Committee to Enquire into the Welfare of Animals Kept Under Intensive Livestock Husbandry Systems" (London: HMSO, 1965), 13–14.

98 David F. Smith et al., *Food Poisoning, Policy and Politics: Corned Beef and Typhoid in Britain in the 1960s* (Woodbridge, UK: Boydell, 2005), 37–132; Anne Hardy, *Salmonella Infections, Networks of Knowledge, and Public Health in Britain 1880–1975* (Oxford: Oxford University Press, 2015), 217–218.

99 E. S. Anderson and H. R. Smith, "Chloramphenicol Resistance in the Typhoid Bacillus," *BMJ* 3, no. 5822 (1972): 329.

100 Smith et al., *Food Poisoning*, 85–87; 132.

101 Anderson and Datta, "Resistance to Pencillins."

102 Tsutomu Watanabe and Toshio Fukasawa, "Episome-Mediated Transfer of Drug Resistance in Enterobacteriaceae," *Journal of Bacteriology* 81, no. 5 (1961): 669–678; Tsutomu Watanabe, "Infective Heredity of Multiple Drug Resistance in Bacteria," *Bacteriological Review* 27 (1963): 87–115; Scott H. Podolsky, *The Antibiotic Era: Reform, Resistance and the Pursuit of a Rational Therapeutics* (Baltimore: Johns Hopkins University Press, 2015), 141–142, 154–156; Bud, *Penicillin*, 175–176.

103 Anderson and Datta, "Resistance to Pencillins," 409.

104 E. S. Anderson and M. J. Lewis, "Drug Resistance and Its Transfer in Salmonella Typhimurium," *Nature* 206, no. 4984 (1965): 579–583.

105 Ibid., 583.

106 E. S. Anderson, "Origin of Transferable Drug-Resistance Factors in the Enterobacteriaceae," *BMJ* 2, no. 5473 (1965): 1289.

107 "Transferable Antibiotic Resistance," *BMJ* 2, no. 5473 (1965): 1326. Restrictions of veterinary antibiotic use had already been proposed by Lawrence Garrod to curb residues in milk; L. P. Garrod, "Sources and Hazards to Man of Antibiotics in Foods," *Proceedings of the Royal Society of Medicine* 54 (1964): 1087–1088.

108 K. A. McKay, H. Louise Ruhnke, and D. A. Barnum, "The Results of Sensitivity Tests on Animal Pathogens Conducted over the Period 1956–1963," *Canadian Veterinary Journal* 6, no. 5 (1965): 103–111.

109 "Germ Survival in the Face of Antibiotics," *Times*, February 26, 1965, 15.

110 "Doctors Urge Ban on Animal Drugs," *DaMa*, April 15, 1965, 11.

111 "Animal Feeding-Stuffs (Antibiotics)," *VR*, June 19, 1965, 698.

112 John Davy, "New Health Fear on Super-farms," *Observer*, November 28, 1965, 5.

113 "Reconsidering Use of Antibiotics," *Times*, February 28, 1966, 13.

114 "Warning on Factory-Farm Bacteria," *Observer*, January 30, 1966, 4; see also Valerie Crofts and Margaret Cooper, "Letters to the Editor: Factory Farming," *Observer*, February 6, 1966, 30.

115 Lawrence Hills, *The Wholefood Finder* (Bocking, UK: Doubleday Research, 1968), 5.

116 Ibid., 9.

117 R. Braude, "Antibiotics in Animal Feeds in Great Britain," *Journal of Animal Science*, 46 (1978): 1427; Bud, *Penicillin*, 177–181.

118 TNA MAF 284/282 (P.Q. Mr. John Harr (Harborough), Oral, 26 Jul, 1967); TNA MAF 287/450 (House of Commons, Written Answer, Treatment of

Human Infections, Exclusive Use of Certain Antibiotics, No.84/1967/68, 13 Nov, 1967).

119 Herbert Williams Smith, "The Incidence of Infective Drug Resistance in Strains of *Escherichia coli* Isolated from Diseased Human Beings and Domestic Animals," *Journal of Hygiene* 64 (1966): 472.

120 Bud, *Penicillin*, 177–181.

121 Bernhard Dixon, "Antibiotics on the Farm—Major Threat to Human Health," *New Scientist* (October 5, 1967), 33.

122 Ibid., 34.

123 Bud, *Penicillin*, 178–181; for fatality numbers, House of Commons Debate, April 11, 1968, vol. 762, cols. 1619-1630, "Gastro-Enteritis (Tees-Side)."

124 N. S. Barron, "Letters to the Editor," *London Illustrated News* [hereafter *LIN*], December 6, 1967, 6; Robert Waller, "Letters to the Editor," *LIN*, December 30, 1967, 4; J. Bower, "Letters to the Editor," *LIN*, January 20, 1968, 31; F. Belsham, "Letters to the Editor," *LIN*, February 3, 1968, 6.

125 Tony Loftas, "How Do Germs Learn to Resist Drugs," *LIN*, January 27, 1968, 17.

126 E. S. Anderson, "Middlesbrough Outbreak of Infantile Enteritis and Transferable Drug Resistance," *BMJ* 1, no. 5887 (1968): 293.

127 Bud, *Penicillin*, 181.

128 Braude, "Antibiotics in Animal Feeds," 1427.

129 Leonard Amey, "Three Antibiotics Banned from Animal Food," *Times*, November 21, 1969, 2.

130 "Why Antibiotics Face Their Swann Song," *FT*, November 11, 1969, 4.

131 "Antibiotics Report Attacked," *FT*, November 26, 1969.

132 "What Are We Going to Feed 'Em?" *Times*, November 21, 1969, 11.

133 Leonard Amey, "Rapid Action on Farm Antibiotics," *Times*, November 10, 1969, 1; see also Leonard Amey, "Another Sort of Christmas", *Times*, December 22, 1969, 11; "Trouble on the Farm," *Economist*, November 15, 1969, 66.

134 Anthony Tucker, "Antibiotics to Be Banned from Animal Feeds," *Guardian*, November 21, 1969, 20.

6. Bigger, Better, Faster: Antibiotics and British Farming

1 John Martin, *The Development of Modern Agriculture: British Farming since 1931* (London: Macmillan and St. Martin's Press, 2000), 6–23; Michael Winter, *Rural Politics: Policies for Agriculture, Forestry and the Environment* (London: Routledge, 1996), 90-92.

2 Martin, *Development of Modern Agriculture*, 29, 33–35.

3 Graham Cox, Philip Lowe, and Michael Winter, "From State Direction to Self-Regulation: The Historical Development of Corporatism in British Agriculture," *Policy and Politics* 14, no. 4 (1986): 475–476.

4 Martin, *Development of Modern Agriculture*, 38–51.

5 Ibid., 54.

6 Cox et al., "State Direction," 475–476.

7 Martin, *Development of Modern Agriculture*, 61.

8 Holderness, *British Agriculture Since 1945* (Manchester: Manchester University Press, 1985), 12–16.

9 Martin, *Development of Modern Agriculture*, 91–92.

10 Cox et al., "State Direction," 480.

11 Ibid., 481–485.

12 Winter, *Rural Politics*, 115.

13 J. K. Bowers, "British Agricultural Policy Since the Second World War," *Agricultural History Review* 33 (1985): 68, 71; Francis Michael Longstreth Thompson, ed., *The Cambridge Social History of Britain 1750–1950, Volume 1: Regions and Communities* (Cambridge, UK: Cambridge University Press, 1990), 148–149; Holderness, *British Agriculture*, 21; Winter, *Rural Politics*, 109–111; Cox et al., "State Direction," 480–488.

14 "Heavy Borrowing from Banks," *FW*, October 6, 1950, 40; see also: "Millions More Spent on Buildings," *FW*, October 13, 1950, 36.

15 "Men and Machines," *FW*, November 10, 1950, 31; "Where Economy is False", *FW*, November 17, 1950, 27; "Cows Kept in All Year," *FW*, May 20, 1955, 93.

16 "Tribute to Britain," *FW*, November 17, 1950, 28; "British Tractors Work Hardest," *FW*, November 17, 1950, 36.

17 "Second Half", *FW*, December 29, 1950, 19, "Filling the Meat Gap," *FW*, December 1, 1950, 33; "World Output of Food Is Up by a Quarter," *FW*, September 16, 1955, 76.

18 Jack Hargreaves, "Never Farm Backwards," *FW*, July 1, 1955, 118–119, 121–122.

19 Alexander Tomey, "Not All Her Own Work," *FW*, October 27, 1950, 65.

20 A. Stewart, "Treat the Cow as Manufacturing Unit," *FW*, September 30, 1955, 48.

21 "The Urge to E-X-P-A-N-D," *FW*, September 2, 1960, 48.

22 "Europe: Meat Output Statistics," in *International Historical Statistics* (Basingstoke, UK: Palgrave Macmillan, 2013).

23 "Agriculture: Historical Statistics," *House of Commons Briefing Paper* 03339 (2016), 10–12.

24 Andrew Godley, "The Emergence of Agribusiness in Europe and the Development of the Western European Broiler Chicken Industry, 1945–1973," *Agricultural History Review* 62, no. 2 (2014): 315–336; Andrew Godley and Bridget Williams, "Democratizing Luxury and the Contentious 'Invention of the Technological Chicken' in Britain." *Business History Review* 83 (2009): 267–290.

25 Karen Sayer, "Animal Machines: The Public Response to Intensification in Great Britain, c. 1960- c. 1973," *Agricultural History* 87, no. 4 (2013): 482–483; Abigail Woods, "Rethinking the History of Modern Agriculture: British Pig Production, c. 1910–65," *Twentieth Century British History* 23, no. 2 (2012): 165–191; see also Paul Brassley, "Cutting Across Nature? The History of Artificial Insemination in Pigs in the United Kingdom," *Studies in History and Philosophy of Biological and Biomedical Sciences* 38 (2007): 446.

26 Anthony Phelps, "There's Still Money in Free Range," *FW*, August 5, 1955, 85. "Animal Crackers," *FW*, September 16, 1955, 46; "British Veterinary Congress— Has Man Put Animals' Health in the Balance?" *FW*, September 16, 1955, 76.

27 H. M. Rikard-Bell, *The Handbook of Modern Pig Farming* (London: Witherby, 1938).

28 T. A. B. Corley and Andrew Godley, "The Veterinary Medicine Industry in Britain in the Twentieth Century," *Economic History Review* 64, no. 3 (2011): 839–841; Claas Kirchhelle, "Pharming Animals: A Global History of Antibiotics in Food Production, 1935–2017," *Palgrave Communications* 4, no. 96 (2018).

29 James Herriot, *Vet in Harness* (London: Pan, 2012), ch. 33; reference taken from Corley and Godley, "Veterinary Medicine Industry." See similar descriptions in

Herriot, *Let Sleeping Vets Lie* (London: Pan, 2006), ch. 11; Herriot, *Every Living Thing* (London: Pan, 2013), ch. 2 and 29.

30 Stuart Anderson, "From 'Bespoke' to 'Off-the-Peg': Community Pharmacists and the Retailing of Medicines in Great Britain 1900 to 1970," *Pharmacy in History* 50, no. 2 (2008): 48.

31 Stuart Anderson, "Drug Regulation and the Welfare State: Government, the Pharmaceutical Industry and the Health Professions in Great Britain," in *Medicine, the Market and the Mass Media*, ed. Virginia Berridge and Kelly Loughlin (London: Routledge, 2005), 194–195; Viviane Quirke, "Thalidomide and Drug Safety Regulation," in *Ways of Regulating Drugs in the 19th and 20th Centuries*, ed. Jean-Paul Gaudillière and Volker Hess (Basingstoke, UK: Palgrave Macmillan, 2013), 153, 174.

32 I am indebted to Stuart Anderson for this information; see also "The Sale and Use of Penicillin," *VR* 60, no. 26 (1948): 316.

33 Anderson, "Drug Regulation," 195; Quirke, "Thalidomide," 153; "Aureomycin and Chloramphenicol Controlled," *VR* 63, no. 25 (1951): 432.

34 I am indebted to Stuart Anderson for this information.

35 Abigail Woods, "Science, Disease and Dairy Production in Britain," *Agricultural History Review* 62, no. II (2007): 301–302; A. W. Stableforth, "Bovine Mastitis with Particular Regard to Eradication of Streptococcus Agalactiae," *VR* 62, no. 15 (1950): 219–222; S. J. Edwards, "Antibiotics in the Treatment of Mastitis," *Antibiotics in Agriculture: Proceedings of the University of Nottingham Ninth Easter School in Agricultural Science* (London: Butterworths, 1962), 45.

36 Corley and Godley, "Veterinary Medicine Industry," 841; "The Use of Penicillin in Veterinary Practice," *VR* 60, no. 25 (1948): 309; Stableforth, "Bovine Mastitis," 220.

37 "Recent Advances in Chemotherapy as Applied to Practice," *VR* 60, no. 12 (1948): 132–133.

38 "Studies on Streptomycin in Relation to Its Possible Uses in Veterinary Medicine," *VR* 62, no. 18 (1950): 265; E. C. Hulse, "The Treatment of Acute Bovine Staphylococcal Mastitis," *VR* 63, no. 26 (1951): 439–442.

39 E. C. Hulse, "Aureomycin," *VR* 63, no. 22 (1951): 396.

40 Abigail Woods, "A Historical Synopsis of Farm Animal Disease and Public Policy in Twentieth Century Britain," *Philosophical Transactions of the Royal Society B: Biological Sciences* 366, no. 1573 (2011): 1943–1954.

41 "The Sale and Use of Penicillin," *VR* 60, no. 26 (1948): 316.

42 Ibid.

43 Ibid.

44 "The Use of Penicillin in Veterinary Practice," *VR* 60, no. 25 (1948): 309.

45 Corley and Godley, "Veterinary Medicine Industry": 838; see also, the growth of advice for routine antibiotic use in Primrose McConnell, *The Agricultural Notebook*, 12th ed. (London: Butterworths, 1953), 401–440.

46 "The Practitioner and His Drugs," *VR* 65, no. 21 (1953): 330.

47 "Unauthorised Sale of Penicillin," *VR*, May 12, 1956, 284.

48 "Developments in Animal Husbandry," *Country Life* [hereafter *CL*], September 26, 1952, 924–925, "Penicillin on the Farm," *CL*, March 28, 1952, 896.

49 Also see ongoing work by Delphine Berdah.

50 MERL, D 2NIRD ET 1/19, "B12 Experiment—Results up to and including Wednesday, 03 August 1949"; K. G. Mitchell to E. Lester Smith, May 12, 1949;

researchers later repeated the experiment by combining streptomycin residues with penicillin; Braude et al., "Antibiotics for Fattening Pigs," *British Journal of Nutrition* 5 (1951): viii.

51 Braude et al., "Antibiotics for Fattening Pigs," *British Journal of Nutrition* 5 (1951): viii; see also MERL D 2NIRD ET1/4 Antibiotics Fattening Pigs Experiment II (Vegetable Protein) 1951.

52 ARC, *Antibiotics in Pig Food* (London: HMSO, 1953), 3; Raphael Braude and K. G. Mitchell, "Antibiotics and Liver Extract for Suckling Pigs," *British Journal of Nutrition* 6, no. 1 (1952): 398–400; J. A. Wakelam, "Vitamin B12 and Antibiotics in Animal Nutrition," *Manufacturing Chemist* 9 (1952): 376; "Antibiotics in Animal Nutrition," *VR* 64, no. 39 (1952): 574; P. I. Shanks, "The Use of an Aureomycin Fermentation Product ('Aurofac') in the Treatment of Unthrifty Pigs," *VR* 64, no. 25 (1952): 365–366.

53 "Aureomycin in Poultry Feed," *Manufacturing Chemist* 6 (1952): 238.

54 "Growth Spurters," *Manufacturing Chemist* 6 (1952): 226.

55 "Antibiotics in Pig Food," 12; see also Barber et al., "Effect of Supplementing an All-Vegetable Pig-Fattening Meal with Antibiotics," *Chemistry and Industry* (1952): 713; Barber et al., "Antibiotics in the Diet of the Fattening Pig," *British Journal of Nutrition* 7, no. 4 (1953): 306–319; Braude et al., "The Value of Antibiotics in the Nutrition of Swine: A Review," *Antibiotics and Chemotherapy* 3 (1953): 271–291; MERL, D 2NIRD ETI/4, "Antibiotics Fattening Pigs Experiment II (Vegetable Protein)."

56 "Therapeutic Substances (Prevention of Misuse) Bill," *VR* 65, no. 21 (1953): 325; "Penicillin for Pigs and Poultry," *CL*, March 13, 1953, 742.

57 J. C. Leslie, "The World Supply Position of Animal Feedstuffs, and Its Influence on the Livestock Industry," *Conference on the Feeding of Farm Animals for Health and Production* [MERL] (London: BVA, 1954), 18; "The Use of Antibiotics in Animal Feeding-Stuffs," *VR* 65, no. 13 (1953): 199.

58 Abigail Woods, "Science, Disease and Dairy Production in Britain," *Agricultural History Review* 62, no. II (2007): 303; Leslie, "World Supply Position," 14.

59 "Vitamin B12, Animal Protein Factor and Antibiotics in Animal Nutrition," *VR* 64, no. 44 (1952): 654–656; J. A. Wakelam, "Vitamin B12 and Antibiotics in Animal Nutrition," *Manufacturing Chemist* 9 (1952): 376.

60 J. A. Wakelam, "Vitamin B12 and Antibiotics in Animal Nutrition," *Manufacturing Chemist* 9 (1952): 378.

61 "The Use of Antibiotics in Animal Feeding-Stuffs," *VR* 65, no. 13 (1953): 200.

62 Kenneth Blaxter and Noel Robertson, *From Dearth to Plenty—The Modern Revolution in Food Production* (Cambridge, UK: Cambridge University Press, 1995), 244.

63 TNA MAF 287/299 (Dugdale to Turner, 29 Jul, 1953), 1; (Draft Regulation Therapeutic Substances Bill, Meeting, 3rd Jul, 1953).

64 TNA MAF 287/299 (Antibiotics. Note of Meeting at Saughton to discuss TSA draft regulations, 4 Feb 1953), 4.

65 W. M. MacKay, "Discussion," *VR* 65, no. 47 (1953): 845.

66 Ibid.

67 I. A. M. Lucas, "Antibiotic Supplements in Rations for Pigs," *VR*, February 23, 1957, 245–246.

68 Abigail Woods, "Rethinking the History of Modern Agriculture: British Pig Production, c. 1910–65," *Twentieth Century British History* 23, no. 2 (2012):

179–181; Brassley, "Cutting Across Nature," 452, 459; see ongoing research by Alex Bowmer.

69   A. James, *Modern Pig-Keeping* (London: Cassell, 1952), 28–30; Connery Chappell, *Keeping Pigs* (London: Rupert Hart-Davis, 1953), 48–51, 109–110, 115.

70   V. C. Fishwick, *Pigs: Their Breeding, Feeding and Management*, 2nd ed. (London: Crosby Lockwood, 1945).

71   Ibid., 5th ed., 157–158.

72   Ibid., 6th ed., 160–161, 192–193.

73   W. R. Woodbridge, *Farm Animals in Health and Disease* (London: Crosby Lockwood, 1954), 339; *Conference on the Feeding of Farm Animals for Health and Production* [MERL] (London: BVA, 1954).

74   Lucas, "Antibiotic Supplements in Rations for Pigs," 243.

75   "Advertisement—ICI," *FW*, July 15, 1955, 96; "Advertisement—Glaxo," *FW*, July 20, 1962, 66–67; on inefficiencies in Britain's feed industry, see Godley and Williams, "Democratizing Luxury," 276.

76   "Advertisement—Cyanamid," *FW*, October 21, 1955, 76.

77   "Antibiotics Could Cut Pig Costs By £5m a Year," *FW*, October 7, 1955, 44.

78   "Artificial Rearing of Pigs," *FW*, May 13, 1955, 91.

79   "Antibiotics," *FW*, May 20, 1955, 45.

80   "Our Debt to the Chemist," *FW*, July 1, 1955, 101.

81   H. Williams Smith, "Drug-Resistant Bacteria in Domestic Animals, Presentation: Symposium on Epidemiological Risks of Antibiotics, February 21, 1958," *Proceedings of the Royal Society of Medicine* 51 (1958): 812.

82   Godley and Williams, "Democratizing Luxury," 269–277; see also A. Tessari and A. Godley, "Made in Italy, Made in Britain: Quality, Brands and Innovation in the European Poultry Market, 1950–80," *Business History* 56, no. 7 (2014): 1059–1060; Corley and Godley, "Veterinary Medicines Industry," 842–843.

83   John Martin, "The Commercialisation of British Turkey Production,' *Rural History* 20 (2009): 209–228, 209–210; John Martin, "The Transformation of Lowland Game Shooting in England and Wales Since the Second World War,' *Rural History* 22 (2011): 207–226.

84   S. F. M. Davies and S. B. Kendall, "Toxicity of Sulphaquinoxaline for Chickens," *VR* 65, no. 6 (1953): 86; C. Horton-Smith, "Additives for Disease Control in Poultry and Turkeys," *VR*, February 23, 1957, 167–168.

85   R. F. Gordon, "Problems Associated with (a) the Production of Table Poultry; (b) the Rearing of Turkey Poults," *Report of Proceedings—Conference on the Feeding of Farm Animals for health and Production* [MERL] (London: BVA, 1954), 139, 148–149.

86   *Practical Poultry Keeping*, 3rd ed. (London: Poultry World, 1946).

87   *Poultry Keeping for Profit* (London: Country Life, 1952), 109–115.

88   *Practical Poultry Keeping*, 5th ed. (London: Poultry World, 1955), 65.

89   Ibid.; see also "Antibiotics," *CL*, November 14, 1954, 111.

90   *Practical Poultry Keeping*, 5th ed., 65, 98–101, 107–109; *The Agricultural Notebook*, 13th ed. (London: Butterworths, 1958), 319, 435; "Modern Methods of Chick Rearing," *CL*, April 23, 1959, 894–895.

91   TNA FD1/8226 (Office Note observations on aspects of the use of antibiotics supplied by the CAFSMNA [ARC 574/60]), 1.

92   E. H. Kampelmacher, "Some Aspects of the Non-Medical Use of Antibiotics in Various Countries," in *Antibiotics in Agriculture: Proceedings of the University of*

Nottingham Ninth Easter School in Agricultural Science (London: Butterworths, 1962), 318; John T. Abrams, *Animal Nutrition and Veterinary Dietetics*, 4th ed. (Baltimore: Williams and Wilkins, 1962), 223.

93 Abrams, *Animal Nutrition and Veterinary Dietetics*, 222–223.

94 A. L. Bacharach, "Antibiotics and Hazards," *Food Drugs and Cosmetics Law Journal* 505 (1957): 508.

95 Corley and Godley, "Veterinary Medicines Industry," 838, 841–844.

96 Abigail Woods, "Decentring Antibiotics: UK Responses to the Diseases of Intensive Pig Production (ca. 1925–65)," *Palgrave Communications* 5, no. 1 (2019): 41.

97 P. Macey, "The Present Status of Antibiotics," *Antibiotics in Agriculture: Proceedings of the University of Nottingham Ninth Easter School in Agricultural Science* (London: Butterworths, 1962), 1; "Joint Committee on the Use of Antibiotics in Animal Husbandry and Veterinary Medicine" (London: HMSO, 1969), 65–66.

98 "Penicillin Spoils Milk for Cheese-Making," *FW* January 5, 1951, 32; "Toxicity of Sulphaquinoxaline for Chickens," *VR* 65, no. 6 (1953): 8.

99 A. W. Stableforth, "Bovine Mastitis with Particular Regard to Eradication of Streptococcus Agalactiae," *VR* 62, no. 15 (1950): 219–224.

100 "Can Stockmanship Replace Dairy Hygiene?" *FW*, March 9, 1951, 41.

101 L. H. Aynsley, "Sensitivity to Penicillin of Organisms Isolated from Cases of Bovine Mastitis," *VR* 65, no. 42 (1953): 663–665.

102 "Is There a New Mastitis Menace," *FW*, July 8, 1955, 47.

103 H. Williams Smith, "The Effects of the Use of Antibiotics on the Emergence of Antibiotic-Resistant Disease-Producing Organisms in Animals," *Antibiotics in Agriculture: Proceedings of the University of Nottingham Ninth Easter School in Agricultural Science* (London: Butterworths, 1962), 375.

104 T. Cornell Green, "More Milk—More Mastitis," *FW*, October 21, 1955, 99.

105 Proposed Amendments to the Penicillin Act," *VR* 65, no. 10 (1953): 166; see also "Proposed Amendments to the Penicillin Act," *VR* 65, no. 7 (1953): 116.

106 "The Use of Antibiotics in Animal Feeding-Stuffs," *VR* 65, no. 13 (1953): 200.

107 "Antibiotics as Dietary Supplements—Discussion," *VR*, February 23, 1957, 230–231; Blaxter Noel Robertson, *From Dearth to Plenty*, 244.

108 Robert White-Stevens, "Antibiotics as Dietary Supplements for Poultry," *VR*, February 23, 1957, 232.

109 Raphael Braude, "The Use of Hormones and Additives as Growth Promoters in Pigs," *VR*, February 23, 1957, 187.

110 "The General Discussion," *VR*, February 23, 1957, 248.

111 Ibid.

112 Ibid., 249.

113 J. M. Ross, "Antibiotics in Relation to Public Health," *VR*, February 23, 1957, 274.

114 Ibid., 277.

115 Herbert Williams Smith and W. E. Crabb, "The Effect of the Continuous Administration of Diets Containing Low Levels of Tetracyclines on the Incidence of Drug-Resistant Bacterium coli in the Faeces of Pigs and Chickens: The Sensitivity of the Bact. coli to Other Chemotherapeutic Agents,' *VR*, January 12, 1957, 30.

116 H. F. Marks and D. K. Britton, *A Hundred Years of British Food and Farming: A Statistical Survey* (London: Taylor and Francis, 1989), 12–13, 28.

117 Ibid., 62, 74–75, 81–83.

118 "Mass Production," *BF*, October 20, 1962, 3; Anthony Lisle, "Untouched by Hand," *FW*, July 6, 1962, 99; "A New 'Golden Age,'" *FW*, July 17, 1964, 31; W. G. R. Weeks, "Gear Up for the Supermarket Age," *FW*, September 7, 1962, 91.

119 Rupert Coles, "Points of Survival," *FW*, August 17, 1962, 101; Paul Atlee, "Nobody's Too Small," *FW*, October 19, 1962, 119, 121.

120 "End in Sight for the Family Farms?" *FW*, August 10, 1962, 41.

121 "'Factory' Farming," *BF*, February 15, 1964, 31.

122 Ibid.

123 "ARC," *FW*, July 8, 1960, 46.

124 Abigail Woods, "Is Prevention Better than Cure? The Rise and Fall of Veterinary Preventive Medicine, c. 1950–1980," *Social History of Medicine* 26, no. 1 (2012): 119–124. For a similar shift toward productivity-oriented animal health management in France and Germany, see Nicolas Fortané, "Naissance et déclin de l'ecopathologie. L'essor contrairié d'une médecine vétérinaire alternative (années 1970–1990)," *Regards Sociologiques* (2015); Ulrike Thoms, "Handlanger der Industrie oder berufener Schützer des Tieres? Der Tierarzt und seine Rolle in der Geflügelproduktion," in G. Hirschfelder et al., *Was der Mensch essen darf* (Heidelberg: Springer, 2015).

125 "Bugs and Drugs," *FW*, September 16, 1960, 133.

126 "Antibiotics in Milk," *VR*, February 13, 1960, 115.

127 Ibid.

128 "Drugs Without Vets' Move by Glos NFU," *FW*, October 19, 1962, 77.

129 W. R. Wooldridge, *Farm Animals in Health and Disease* (London: Crosby Lockwood, 1960), 114, 201, 203, 205–206; *Pictorial Poultry-Keeping* (London: C. Arthur Pearson, 1962), 144; *The Agricultural Notebook*, 14th ed. (London: Butterworths, 1962), 420–474, 499–500; M. Buckett, *Introduction to Livestock Husbandry* (Oxford: Pergamon, 1965), 21, 30, 105, 111–113; *Modern Poultry Keeping* (London: English Universities Press, 1965), 86–87; "Antibiotics in Livestock Feeding," *MAFF Advisory Leaflet* 418 (1960) [MERL PB 26081-2 JJ].

130 "Axe Will Fall on 'Antibiotic' Milk," *FW*, May 31, 1963, 41; "MMB Sends Out Warnings on Antibiotics," *FW*, June 7, 1963, 42; "Tube Trouble," *FW*, June 7, 1963, 40; "Antibiotics In Milk," *BF*, March 14, 1964, 45.

131 Frank Sykes, "Does Modern Farming Menace Public Health?" *CL*, November 8, 1962, 1140–1142.

132 "Killer Chemicals," *FW*, April 12, 1963, 82; John Sheail, "Pesticides and the British Environment: An Agricultural Perspective," *Environment and History* 19 (2013): 87–108.

133 Abigail Woods, "From Cruelty to Welfare: The Emergence of Farm Animal Welfare in Britain, 1964–71," *Endeavour* 36, no. 1 (2012): 21; see also Godley and Williams, "Democratizing Luxury," 284–285.

134 "Broiler Veal Not Cruel—says NFU," *FW*, July 22, 1960, 38; "Calves Don't Suffer," *FW*, July 29, 1960, 40.

135 A. G. Street, "Cruel to Their Kind?" *FW*, September 30, 1960, 83.

136 "Feather Heads," *BF*, March 28, 1964, 1.

137 "Techniques in Question," *FW*, March 13, 1964, 43.

138 Godley and Williams, "Democratizing Luxury," 286.

139 "Informed Climate Needed on Farm Poison Risks," *FW*, , March 20, 1964, 64.

140 "NFU Helped on Farming Film," *BF*, June 10, 1967, 5.

141 "The Union Makes a Film," *BF*, May 4, 1968, 22.

142 K. M. Petter Ropewind, "Battery Birds," *FW*, March 27, 1964, 41; A. H. Harris, "Obituary of a Calf," *FW*, July 17, 1964, supplement vii.

143 Hugh Sinclair, "Animal Machines," *Countryman* (Summer 1964): 292.

144 R. K. Cornwallis, "Animal Machines," *Countryman* (Summer 1964): 287, 289; see also a positive review of *Animal Machines* in *Country Life*, "Turning Farms into Factories," *CL*, March 19, 1964, 675.

145 "Misused Drugs Mask Disease," *FW*, March 6, 1964, 71.

146 TNA AJ 3/183 (An Enquiry into the Effect on Human Health on the Use of Antibiotics for Intensively Reared Animals with Special Reference to the Swann Committee's Report of December 1969, March 1970), 1.

147 *The Agricultural Notebook*, 15th ed. (London: Butterworths, 1968), 419, 462–474, 488–492, 501–510; "Getting Mastitis Under Control," *CL*, April 6, 1967, 791–792; "Farming for Profit or Posterity," *CL*, October 13, 1966, 945.

148 James Wentworth Day, "Misuse of Antibiotics," *BF*, July 9, 1966, 3.

149 TNA MAF 287/199 ("Diseases Master Drug Defences," *FW*, May 19, 1967; "Salmonellosis Will Be Target for the Next Stock Health Drive," *FW*, September 30, 1967; Williams Smith, "Salmonella," *FW*, January 13, 1967, 8).

150 "Infective Drug Resistance," *VR*, May 29, 1965, 611; H. Williams Smith, "Infective Drug Resistance," *VR*, March 20, 1965, 327–328.

151 H. Williams Smith and Sheila Halls, "Observations on Infective Drug Resistance in Britain," *VR*, March 19, 1966, 419.

152 Ibid., 419.

153 Ibid., 420.

154 "Antibiotics: Medical and Veterinary Control," *VR*, April 16, 1966, 570.

155 Ibid., 571.

156 J. R. Walton, "Antibiotics: Medical and Veterinary Control," *VR*, July 2, 1966, 27–28.

157 See for example Beecham brochures and the supposedly independent Farmer's Index: MERL TR BAP/P2/A5 Beecham Agricultural Products Leaflet "More Profit from Broilers (the Vitamealo Way)"; MERL Farmers Index.

158 "Advertisement—Cyanamid," *BF*, December 4, 1965, 46; "Advertisement—Cyanamid," *BF*, September 4, 1965, 39.

159 "Advertisement—Glaxo," *FW*, June 7, 1963, 54.

160 Christoph Gradmann, "Re-inventing Infectious Disease: Antibiotic Resistance and Drug Development at the Bayer Company 1945–1980," *Medical History* 60, no. 2 (2016): 155–180.

161 TNA FD1/8226 (ARC/MRC Joint Committee on Antibiotics, Scient. Subcommittee. Antibiotic for Animal Feeding Use Only, Suggestion by Bayer Products Ltd., [undated]).

162 "Growth Only from This Antibiotic," *BF*, April 8, 1967, 47.

163 MERL TR BAP/P2/A63-64, "In Controlling Mastitis—Orbenin"; TR BAP P2/A 71, "Orbenin in Mastitis Control," Beecham Technical Bulletin No. 3, 3; TR BAP/P2/AGi-62, Beecham booklet "In Controlling Mastitis."

164 "Antibiotic and Chemotherapeutic Agents," *VR*, August 20, 1966, 249.

165 R. More-Collyer, "Towards 'Mother Earth': Jorian Jenks, Organicism, the Right and the British Union of Fascists," *Journal of Contemporary History* (2004), 353–371.

166 Philip Conford and P. Holden, "The Soil Association," in William Lockeretz, *Organic Farming: An International History* (Wallingford, UK: CABI, 2001),

187–192; Philip Conford, *The Origins of the Organic Movement* (Edinburgh: Floris, 2001), 83–92, 146–151.

167 "Antibiotics in Animal Husbandry," *Mother Earth* [hereafter *ME*] 7, no. 2 (1953): 40–41.

168 Ibid., 41.

169 Hugh Corley, *Organic Farming* (London: Faber and Faber, 1957), 88.

170 Ibid.

171 Ibid., 169–170.

172 Elise Jerard, "American Food Additives," *SPAN—Soil Association News*, July 1967, 12–18; "Antibiotics Probe," *SPAN*, November 1967, 3; "Animal Welfare and Grazing," *ME*, January 1968, 15–16.

173 "Farm Antibiotics—Major Threat to Human Health," *ME*, January 1968, 47–48.

174 "Ecological Judgement," *ME*, April 1968, 68.

175 "Success Through Controlled Grazing," *ME*, January 1969, 274.

176 "Antibiotics on the Farm: The Swann Committee," *ME*, January 1969, 288; "Agriculture—the Suicidal Industry," *ME*, April 1969, 311.

177 TNA AJ 3/183 (Cecil Schwartz, "Vets Advise Swann," *New Scientist*, February 13, 1969, 348–349).

178 "Drugs: Good Servants, Bad Masters," *BF*, October 4, 1969, 45.

179 Ibid.

180 "Understanding Feed Additives," *VR*, August 9, 1969, 157–158.

181 "Antibiotics in Animal Husbandry" (London: Office of Health Economics, 1969), 29.

182 Ibid., 30.

183 "Putting on Weight," *FW* supplement, October 3, 1969, 27.

184 "Drugs and Bugs," *FW*, October 17, 1969, 110.

185 "Clamp on Antibiotics," *FW*, November 14, 1969, 30.

186 Ibid.

187 Ibid.; "Charter for Antibiotics Proposed," *FW*, November 14, 1969, 33.

188 "Likely Curb on Feed Drugs Worth £10 m," *BF*, November 1, 1969, 18.

189 "What Proof?" *FW*, November 14, 1969, 30.

190 "Little Evidence," *FW*, December 5, 1969, 77.

191 Ibid.

192 Peter Bell, "The Month," *BF*, December 6, 1969, 12.

193 "Blow to Antibiotics in Feed," *BF*, November 22, 1969, 3.

194 "Same Price for Additives," *FW*, November 28, 1969, 40; Brian Chester, "Drug Changes Will Be Made in Easy Stages," *FW*, November 28, 1969, 40.

195 "BVA Statement," *VR*, November 19, 1969, 620.

196 TNA AJ 3/183 (Joint Statement by the BVA [July 1970]), 1.

197 "Never a Dull Moment," *BF*, November 29, 1969, 1.

198 G. Armstrong, "'Closed Shop' Drugs," *FW*, December 5, 1969, 49.

199 "Drug Worry," *FW*, November 21, 1969, 33.

200 Collin Tudge, "Antibiotics—Farm Drugs with a Double Edge," *FW*, November 21, 1969, 41; see also "Opinion: Swann Song," *BF*, December 6, 1969, 11.

201 Brian Chester, "Drug Changes Will Be Made in Easy Stages," *FW*, November 28, 1969, 40; R. J. T. Holland, "'Safe Food' Promotion," *FW*, December 5, 1969, 49.

202 Wellcome Collections, WF/C/E/05/12/78 (Group Development Committee V.A., minutes of 94th meeting, December 8, 1969), 2.

7. Typing Resistance: Antibiotic Regulation in Britain

1 Parts of this chapter have informed Claas Kirchhelle, "Swann Song: Antibiotic Regulation in British Livestock Production (1953–2006)," *Bulletin of the History of Medicine* 92, no. 2 (2018): 317–350.

2 Penicillin Bill (House of Lords), House of Commons Debate June 9, 1947, *Hansard*, vol. 438, col. 822; I am indebted to Stuart Anderson for this reference.

3 "Dietary Antibiotics," *Manufacturing Chemist,* October 1952, 401.

4 Difficulties in defining what AGPs were in relation to existing drug and feed categories also complicated officials' task; TNA, MAF 119/23 (Minute Hill to Croxford, 19 April, 1952).

5 TNA, MAF 119/23 (Mr. Honnor, ARC, meeting 19 September, 1952), 3.

6 TNA MAF 119/23 (Dr. Magee; ARC, meeting, 25 Feb, 1952), 2.

7 J. A. Wakelam, "The General Discussion," *VR*, February 23, 1957, 231.

8 Ibid.

9 TNA MAF 119/23 (ARC Meeting, 19 September, 1952), 1; (W. G. Alexander; ARC, meeting, 25 Feb, 1952), 2; (Minute by P.J. Moss, 24 Sep. 1952), TNA MAF 287/299 (Meeting at Saughton to discuss TSA draft regulations, 4 Feb, 1953), 3.

10 TNA MAF 287/299 (Williams to Moss, 5 February 1953), 2.

11 TNA MAF 287/299 (Veterinary Interests, Meeting RCVS and BVA with MH and MAF, 12 Feb, 1953).

12 TNA, MAF 287/299 (Dugdale to Turner, 29 July, 1953), 2.

13 TNA MAF 119/23 (Sgd. A. Eden to O.A. Robertson, 2 Nov, 1953), 2.

14 TNA MAF 287/299 (Meeting at Saughton to discuss TSA draft regulations, 4 Feb, 1953), 1.

15 J. M. Ross, "Antibiotics in Relation to Public Health," *VR*, February 23, 1957, 271.

16 TNA MAF 119/23 (Draft: FG Raymond to GL Gray, 26 Nov, 1968, Attached: Therapeutic Substances Act 1956).

17 TNA MAF 284/281 (Minute 27, AB Bartlett, 10 Apr, 1956).

18 A. L. Bacharach, "UK Position on Use of Antibiotic Food Additives," *Chemical Age* 78 (1957).

19 TNA MAF 284/281 (Control of Antibiotics, Feb, 1969), 1.

20 A. L. Bacharach, "Antibiotics and Hazards," *Food Drug Cosmetic Law Journal* 505 (1957): 509.

21 Ibid., 510.

22 TNA MAF 284/281 (Note of Meeting held on 13.09.1956, to discuss the setting up of a Working Group on the use of Antibiotics in Agriculture and in Food Preservation).

23 TNA MAF 287/450 (Minute, Hensley to Bott, 9 Jan, 1967).

24 TNA MAF 287/450 (Minute, Macrae to Field, 18 Jan, 1967), 2.

25 TNA MAF 284/282 (Control of Antibiotics, Appendix III: List of relaxing regulations made under Part II of the therapeutic Substances Act 1956, Feb, 1959).

26 Sprays were not licensed for fire blight; trees infected with this notifiable disease had to be destroyed. *Antibiotics in Agriculture: Proceedings of the University of Nottingham Ninth Easter School in Agricultural Science* (London: Butterworths, 1962), 2.

27 TNA MAF 101/643 (Department of Scientific and Industrial Research. Preliminary report of the visit of Ella M. Barnes to the USA to investigate the use of

antibiotics for food preservation, 1956), 1; (Note: Antibiotics for Fish Preservation. Pilot Scale Sea Trials, undated); TNA MAF 260/82 (Reports of the Antibiotic Panel, 09 Apr, 1958), 1.

28 "Food Standards Committee Report on Preservatives in Food" (London: HMSO, 1959), 24–25, 34–35, 43–44, 64–69; TNA MAF 284/282 (Control of Antibiotics, Appendix III: List of relaxing regulations made under Part II of the therapeutic Substances Act 1956, Feb, 1959); Ella Barnes, "The Use of Antibiotics for the Preservation of Poultry and Meat," *Antibiotics in Agriculture—Proceedings of the Fifth Symposium of the Group of European Nutritionists, 1966* (Basel, CH: S. Karger, 1968), 62.

29 TNA MAF 284/282 (Control of Antibiotics, Appendix III: List of relaxing regulations made under Part II of the therapeutic Substances Act 1956, Feb, 1959); Ella M. Barnes, "The Use of Antibiotics for the Preservation of Poultry and Meat," *Antibiotics in Agriculture—Proceedings of the Fifth Symposium of the Group of European Nutritionists, 1966* (Basel: S. Karger, 1968), 62–65; J. M. Ross, "Antibiotics in Relation to Public Health," *VR*, February 23, 1957, 274; TNA MAF 260/82 Preservatives Sub-Committee papers.

30 TNA MAF 284/281 (Minute 30, GO Lace, 30 May, 1956).

31 TNA MAF 284/281 (Lush to Lace, 4 Jul, 1956).

32 TNA MAF 260/82 (Antibiotics Panel, Meeting, 20 Dec, 1956), 2.

33 TNA MAF 284/281 (Miller to Mills, 18 Mar, 1958).

34 TNA MAF 260/82 (Western European Union Sub-Committee on Health Control of Foodstuffs. Working Party on Poisonous Substances Used in Agriculture; Draft Paper by UK Delegation, 1956), 6–7.

35 TNA MAF 101/643 (Taylor to Berry, 9 Oct, 1957), 4; MAF 260/82 (Antibiotics Panel, Meeting, 20 Dec, 1956), 1.

36 A. L. Bacharach, "Antibiotics and Hazards," *Food Drug Cosmetic Law Journal* 505 (1957): 511.

37 Ibid.

38 Ross, "Antibiotics in Relation to Public Health," 272; MERL SR MMB A/1-4, "Antibiotics in Milk"—attached to Joint Committee of the MMB and the Central Milk Distributive Committee to Members of the Joint Committee, October 21, 1959, 1.

39 MERL SR MMB A/1-4, "Antibiotics in Milk"—attached to Joint Committee of the MMB and the Central Milk Distributive Committee to Members of the Joint Committee, October 21, 1959, 2.

40 Ibid.

41 MERL SR MMB A/1-4, Minutes of meeting of MMB and CMDC Joint Committee, September 16, 1959, 2; minutes of meeting on June 10, 1959, 2–3.

42 "Keeping Milk Free of Antibiotics," *Times*, May 30, 1963, 18; TNA MAF 251/369 (Minute 11, C. E. Coffin, 20th July, 1962).

43 Robert Bud, *Penicillin: Triumph and Tragedy* (Oxford: Oxford University Press, 2009), 171–173.

44 MERL SR MMB A/8 (Joint Committee MMB and CMDC—copy of exchange of letters between the Minister of Agriculture and W. R. Trehane, November 22, 1963, 1); minutes of meeting of Joint Committee, July 11, 1963, 2–3.

45 MERL SR MMB A/8 (Joint Committee of MMB and CMDC to all buyers of milk, December 31, 1963, 1).

46 Michele L. Williams and Jeffrey T. LeJeune, "Phages and Bacterial Epidemiology,"

in *Bacteriophages in Health and Disease*, ed. Paul Hyman and Stephen T. Abedon (Wallingford, UK: CABI, 2012), 76–85.

47 David F. Smith, H. Lesley Diack, and T. Hugh Pennington, *Food Poisoning, Policy and Politics: Corned Beef and Typhoid in Britain in the 1960s* (Woodbridge, UK: Boydell, 2005), 15–20; Kathryn Hillier, "Babies and Bacteria: Phage Typing Bacteriologists, and the Birth of Infection Control," *Bulletin of the History of Medicine* 80, no. 4 (2006): 733–761, 735; James Craigie, "Arthur Felix 1887–1956," *Biographical Memoirs of Fellows of the Royal Society* 3 (1957): 55–56.

48 R. E. O. Williams, *Microbiology for the Public Health* (London: PHLS, 1985), 21–22, 45; Smith et al., *Food Poisoning*, 18–20; Anne Hardy, *Salmonella Infections, Networks of Knowledge, and Public Health in Britain 1880–1975* (Oxford: Oxford University Press, 2015), 123–133.

49 Hillier, "Babies and Bacteria," 736; see also Flurin Condrau and Robert Kirk, "Negotiating Hospital Infections: The debate between Ecological Balance and Eradication Strategies in British Hospitals, 1947–1969," *Dynamis* 31, no. 2 (2011): 385–404.

50 The PHLS offered free phage-typing courses and sets; TNA MAF 189/390 (Spencer to Aldrige, August 31, 1954).

51 Naomi Datta, "Herbert Williams Smith: 3 May 1919–16 June 1987," *Biographical Memoirs of Fellows of the Royal Society* 34 (1988): 754–786, 754; Herbert Williams Smith and W. E. Crabb, "The Sensitivity of a Further Series of Strains of Bacterium coli from Cases of White Scours: The Relationship between Sensitivity Tests and Response to Treatment," *VR* 68 (1956): 274; Herbert Williams Smith and W. E. Crabb, "The Typing of Escherichia coli by Bacteriophage: Its Application in the Study of the E. coli Population of the Intestinal Tract of Healthy Calves and of Calves Suffering from White Scours," *Journal of General Microbiology* 15 (1956): 556.

52 H. Williams Smith, "The Effect of Diets Containing Tetracyclines and Penicillin on the Staphylococcus Aureus Flora of the Nose and Skin of Pigs and Chickens and Their Human Attendants," *Journal of Pathology and Bacteriology* 79 (1960): 243; "Drug-Resistant Staphylococci in the Farmyard," *Lancet* 275 (1960): 1338–1339; see also H. Williams Smith and W. E. Crabb, "The Effect of the Continuous Administration of Diets Containing Low Levels of Tetracyclines on the Incidence of Drug-Resistant Bacterium coli in the Faeces of Pigs and Chickens: The Sensitivity of the Bact. coli to Other Chemotherapeutic Agents," *VR* 69 (1959): 24.

53 K. A. McKay and H. D. Branion, "The Development of Resistance to Terramycin by Intestinal Bacteria of Swine," *Canadian Veterinary Journal* 1, no. 4 (1960): 144–150; Ella M. Barnes, "The Effects Of Antibiotic Supplements On The Faecal Streptococci (Lancefield Group D) Of Poultry," *BVJ* 114, no. 9 (1958), 333–344; S. D. Elliott and Ella M. Barnes, "Changes in Serological Type and Antibiotic Resistance of Lancefield Group D Streptococci in Chickens Receiving Dietary Chlortetracycline," *Journal of General Microbiology* 20 (1959): 426–433.

54 J. C. McDonald et al., "Nasal Carriers of Penicillin-Resistant Staphylococci in Recruits to the RAF," *Proceedings of the Royal Society of Medicine* 53 (1960): 255–258.

55 TNA FD 9/1458 (Porter to Clements, 17 Jul, 1959); Betty C. Hobbs et al., "Antibiotic Treatment of Poultry in Relation to Salmonella Typhi-Murium," *Monthly Bulletin of the Public Health Laboratory Service*, 19 (Oct 1960); Anderson appears as Principal Investigator.

56 TNA FD 9/1458 (Note on file, A.83/4, 9 Sept, 1959).

57 TNA FD 23/1936 (Report of the Joint Committee on Antibiotics in Animal Feeding, 1962).

58 TNA FD1/8226 (ARC/MRC Joint Committee on Antibiotics, The Antibiotic Sensitivity of Strains of Staphylococcus aureus Isolated from the Noses of Veterinary Surgeons and Farmers, Williams Smith & Crabb).

59 TNA FD1/8226 (ARC/MRC Joint Committee on Antibiotics, 2nd meeting Scient. Sub-Committee, 27 Jun, 1960), 2.

60 Ibid., 3.

61 TNA FD1/8226 (Information provided by the NFU, ARC 558/60), 1.

62 Ibid., 2.

63 TNA FD1/8226 (ARC/MRC Joint Committee on Antibiotics, 3rd meeting Scient. Sub-Committee, 18 Oct, 1960), 4.

64 Ibid., 5.

65 TNA FD1/8227 (ARC and MRC Joint Committee on Antibiotics in Animal Feeding. Report of the Scientific Sub-Committee).

66 TNA MAF 284/281 (Advisory Committee on Poisonous Substances, Meeting, 13 Nov, 1956; minutes Antibiotics Panel, comment Dr Barnes).

67 "The Public Health Aspects of the Use of Antibiotics in Food and Feedstuffs," World Health Organization Technical Report Series 260 (Geneva: WHO, 1963).

68 Bud, *Penicillin*, 176–183; Kirchhelle, *Swann Song*, 333–334.

69 Anthony Tucker, "E. S. Anderson," *Guardian*, March 22, 2006.

70 E. S. Anderson and Naomi Datta, "Resistance to Penicillins and Its Transfer in Enterobacteriaceae," *Lancet* 285, no. 7382 (1965): 407–409; E. S. Anderson and M. J. Lewis, "Drug Resistance and Its Transfer in Salmonella typhimurium," *Nature* 206, no. 4984 (1965): 583; E. S. Anderson, "Origin of Transferable Drug-resistance Factors in the Enterobacteriaceae," *BMJ* 2, no. 5473 (1965): 1289; Scott H. Podolsky, *The Antibiotic Era: Reform, Resistance and the Pursuit of a Rational Therapeutics* (Baltimore: Johns Hopkins University Press, 2015), 141–142, 154–156; Bud, *Penicillin*, 175.

71 TNA MAF 287/450 (Minute, Macrae to Field, January 18, 1967), 1.

72 TNA FD 1/8228 ARC 413/65—The Increase of drug resistance in S. typhimurium (June 1965).

73 TNA FD 1/8228 Note by Williams Smith (ARC443/65); based on Kirchhelle, "Swann Song."

74 TNA MAF 287/199 (Cattle Dealing and Salmonellosis), 1; Extract: "Diseases Master Drug Defences," *Farmers Weekly* (19.05.1967); Extract: Williams Smith, "Salmonella," *FW* (13.01.1967), 87; TNA FD 1/8228 (ARC/MRC Committee in Animal Feeding. Scient. Subcommittee December 13, 1965), 3; Kirchhelle, "Swann Song."

75 TNA MAF 287/199 (Agriculture Bill 17.01.1967, Report, 352); (Outbreak of salmonellosis Madam's Farm, Sussex); TNA MAF 386/45 (Minutes, 9th Meeting Joint PHLS/AHD Advisory Committee, 16 Oct, 1968), 4.

76 TNA MAF 287/199 (Geoghegan, "Milk-Borne Outbreak of Food Poisoning Due to S. Typhimurium," *Medical Officer*, 30.07.1965, 2). The 1959 Milk and Dairies (General) Regulations allowed MoHs to require heat treatment of milk; VICs were supposed to report salmonellosis threats to MoHs, but could take little direct action. TNA MAF 386/45 (Draft Reply to the Rural District Councils'

Association); MAF 386/46 (Minutes—24th Meeting PHLS/AHD Advisory Committee, 24 Jul, 1974), 1.

77 TNA FD 1/8228 (Anderson—Comments on ARC 561/65).

78 TNA FD 1/8228 (ARC/MRC Committee in Animal Feeding. Scient. Subcommittee June 22, 1965), 1.

79 Ibid., 2.

80 Ibid., 1–2.

81 Ibid., 2.

82 Ibid., 4–5; see also Kirchhelle, "Swann Song."

83 TNA FD 1/8228 (ARC/MRC Committee in Animal Feeding. Scient. Subcommittee Meeting on December 13, 1965. Brief), 1.

84 TNA FD 1/8228 (Further observations on infective drug resistance in S. typhimurium Type 29), 1.

85 Ibid., 3.

86 TNA MAF 287/450 (Annexe, ARC 22B/66, ARC & MRC. Joint Committee on Antibiotics in Animal Feeding. Second Report of the Scient. Subcommittee), 1–2.

87 Bud, *Penicillin*, 181.

88 TNA MAF 287/450 (Minute 4, Macrae to Field, 18 Jan, 1967), 2; TNA FD 7/899 (Note to Bunje, Note of a Meeting with the MAFF, February 13th, 1968).

89 TNA MAF 287/450 (Annexe, ARC 2546/66, ARC and MRC. Joint Committee on Antibiotics in Animal Feeding), 1.

90 TNA MAF 287/450 (Annexe, ARC 22B/66, ARC and MRC. Joint Committee on Antibiotics in Animal Feeding. Second Report of the Scientific Sub-Committee), 2.

91 TNA MAF 386/44 (Minutes—3rd Meeting of the Joint PHLS/AHD Advisory Committee, 2 Nov, 1966), 3.

92 TNA MAF 287/450 (Minute, Hensley to Bott, 9 Jan, 1967).

93 TNA MAF 287/450 (Minute, Hensley to Bott, 22 May, 1967).

94 TNA MAF 287/450 (Minute, Bott to Hensley, 23 May, 1967).

95 TNA MAF 287/450 (Minute, Hensley to Bott, 22 May, 1967).

96 D. W. Kent-Jones, "Obituary: Alastair Campbell Frazer," *Proceedings of the Society for Analytical Chemistry* 6 (1969): 209–210.

97 TNA MAF 284/282 (MAFF, Scientific Advisory Panel. The Use of Antibiotics in Agriculture and Food, Jan, 1967), 3–8, 12.

98 TNA MAF 284/282 (Minute, Bott to Parker, July 13, 1967).

99 TNA MAF 287/450 (Minute, BHB Dickinson to Hensley, 24 Jul, 1967).

100 TNA MAF 287/450 (Minute, Macrae to Field, 4 Oct, 1967).

101 TNA MAF 284/282 (Press Notice, 1 Sept, 1967).

102 TNA FD 7/899 (Note: Harold Himsworth, Antibiotics in Animal Foodstuffs, 20 Oct, 1967).

103 TNA MAF 287/450 (Minute, C.H.M. Wilcox to Hensley, 22 Dec, 1967; Minute, J. Hensley to TB Williamson, 29 Dec, 1967); (Minute, TB Williamson to Hensley, 25 Jan, 1968).

104 TNA FD 7/899 (TB Williamson to Hensley, 25 Jan, 1968).

105 TNA MAF 287/450 (Minute, FC Parker to Bott, Field, Macrae, 1 Feb, 1968).

106 TNA FD 7/899 (Note to Bunje, Note of Meeting with the MAFF, 21 Feb, 1968).

107 TNA MAF 287/450 (Note of Meeting "To Discuss the Second Report of the Joint ARC/MRC Committee on Antibiotics in Animal Feeding Stuffs," 13 Feb, 1968).

108 TNA MAF 287/450 (Minute, Macrae, Inter-Departmental Meeting on the Netherthorpe Committee Report, 19 Feb, 1968).

109 Ibid.

110 TNA MAF 287/450 (Minute, G.J.L. Avery, Joint Committee on Antibiotics, 25 April, 1968).

111 TNA MAF 287/450 (Minute, W.C. Tame to Secretary, 29 Apr, 1968).

112 TNA MAF 287/450 (G.J.L. Avery to Tame, 02 May, 1968); Anderson subsequently approached MP David Kerr to lobby for his nomination; TNA MAF 287/450 (David Kerr (MP) to Cledwyn Hughes (MAFF), 22 May, 1968).

113 TNA MAF 287/450 (Minute, J.G. Carnochan to Tame, 3 May, 1968).

114 TNA MAF 287/450 (Committee on the Use of Antibiotics in Animal Husbandry and Veterinary Medicine. Proposed Members).

115 "Action Sought on Antibiotics After Babies' Deaths," *Times*, April 14, 1969, 2.

116 See also MoH complaints about biased evidence collection: Wellcome Collections [hereafter WC] SA SMO L62 (James R. Preston to P. F. Cashman, August 19, 1968); (G. Ramage, "Memorandum," enclosed in Ramage to Cashman, August 23, 1968), 1–2; (The Society of MoH), 2.

117 WC PP EBC F 129 (F. P. Doyle, Memorandum, July 25, 1968), 1–2.

118 WC PP EBC F 129 (K. H. B. Wood to E. B. Chain, August 30, 1968).

119 WC PP EBC G 72 (The Problem of the Emergence of Bacterial Resistance to Antibiotics. Address to Marks & Spencer, April 10, 1967), 8–9.

120 Ibid., 10–11.

121 Ibid., 11. See also an earlier and milder attack on Anderson's findings in 1965, WC PP EBC G 62-63 (1st Draft—25 Years of Penicillin Research), 4.

122 WC PP EBC F 129 ("The Use of Antibiotics in Animal Husbandry—Summary of Conclusions," attached to K. H. B. Wood to Beecham pharmaceutical division management committee, members of the agricultural projects consultants committee, September 24, 1968), 1.

123 WC PP EBC F 129 (ABPI to study group on the Swann Committee—confidential notes on a meeting with the Swann Committee on May 29, 1969, June 5, 1969, enclosed in G. C. Brander to Agricultural Projects Consultants Committee), 1–3.

124 Ibid., 3.

125 Ibid.

126 Ibid., 5.

127 Ibid.

128 "Joint Committee on the Use of Antibiotics in Animal Husbandry and Veterinary Medicine" (London: HMSO, 1969), 40.

129 Ibid.

130 Ibid.

131 Abigail Woods, "Is Prevention Better than Cure? The Rise and Fall of Veterinary Preventive Medicine, c. 1950–1980," *Social History of Medicine* 26, no. 1 (2012): 113–131, 113.

132 "Joint Committee on the Use of Antibiotics in Animal Husbandry," 45.

133 Ibid., 46. Licensing of new drugs had been made mandatory by the 1968 Medicines Act.

134 Ibid., 34, 52.

135 Ibid., 49.

136 Ibid., 56.

137 Ibid.

138 TNA FD 7/900 (Note on file: BL to Bunje and Clements, 27 Sep, 1968); see also TNA FD 7/899 (Note: 10 Oct, 1969).

8. Marketplace Environmentalism: Antibiotics, Public Concerns, and Consumer Solutions

1 "Infectious Diseases: Trying Too Hard for the Fast Knockout," *Time*, January 6, 1967; "Germ Resistance to Drugs Studied," *New York Times* [hereafter *NYT*], March 26, 1967, 23; "UN Agency Warns Drugs Alone Can't Wipe Out VD," *NYT*, January 25, 1968, 13; Ian Maclean Smith, "Death from Staphylococci," *Scientific American* [hereafter *SciAm*], February 1968, 84–94; see also Scott H. Podolsky, *The Antibiotic Era: Reform, Resistance and the Pursuit of a Rational Therapeutics* (Baltimore: Johns Hopkins University Press, 2015), 112–119.

2 Harold M. Schmeck Jr., "Medicine: Now Bacteria Fight Back," *NYT*, May 28, 1967, E8.

3 "New Chemical Is Said to Breed Resistance to Potent Drugs," *Washington Post* [hereafter *WP*], December 10, 1967, A10.

4 Tsutomu Watanabe, "Infectious Drug Resistance," *SciAm*, December 1967, 26–27; "Animals Eat into Antibiotics," *WP*, October 6, 1968, F5.

5 Ovid Martin, "US Studies Additives in Animal Feeds," *WP*, July 9, 1967, A13; "Science Put the Meat on that Holiday Bird," *Desert Sun* [hereafter *DS*], November 1, 1966, 18; "Advances Fatten the Holiday Bird," *MT*, December 13, 1966, 4.

6 Harold Schmeck, "Scientists Study Feed Antibiotics," *NYT*, June 11, 1967, 54.

7 "To Save Spaceship Earth," *NYT*, June 2, 1968, E10; US Has One Restriction," *NYT*, November 21, 1969, 17; "Britain to Curb Antibiotic Feed," *NYT*, November 21, 1969, 17.

8 Sandra Blakeslee, "Food Safety a Worry in Era of Additives," *NYT*, November 9, 1969, 1.

9 Sarah A. Vogel, *Is It Safe? BPA and the Struggle to Define the Safety of Chemicals* (Berkeley: University of California Press, 2013), 43.

10 Ibid., 59.

11 Lizabeth Cohen, *A Consumers' Republic: The Politics of Mass Consumption in Postwar America* (New York: Vintage, 2004), 347, 298–299, 336–362.

12 Alan Marcus, *Cancer from Beef: DES, Federal Food Regulation, and Consumer Confidence* (Baltimore: Johns Hopkins University Press, 1994), 75.

13 Drew Pearson and Jack Anderson, "Lobby Battling US Meat Inspection," *WP*, July 18, 1967, B11.

14 Ibid.

15 Morton Mintz, "The Thundering Silence of Drug Consumers," *WP*, November 26, 1967, B2.

16 "Senate Witnesses Charge Antibiotic Killed Children," *NYT*, February 29, 1968, 39.

17 "When a Cure Is a Killer," *NYT*, March 3, 1968, E8.

18 Morton Mintz, "Antibiotic's Danger Seen as Underrated," *WP*, November 12, 1967, F7.

19 "FDA Bars Antibiotic from Market, Asks Recall," *WP*, January 20, 1968, E17.

20 Morton Mintz, "FDA Concedes Drug Curb May Fail," *WP*, March 1, 1968, A3; on chloramphenicol, see Thomas Maeder, *Adverse Reactions* (New York: Morrow, 1994).

21 Podolsky, *Antibiotic Era*, 105–111.
22 Vogel, *Is It Safe?* 51.
23 Barbara Resnick Troetel, *Three-Part Disharmony: The Transformation of the Food and Drug Administration in the 1970s*, dissertation, City University of New York, 1996, 31–52.
24 Richard D. Lyons, "Ousted FDA Chief Charges 'Pressure' from Drug Industry," *NYT*, December 31, 1969, 1; Vogel, *Is It Safe?* 53.
25 Morton Mintz, "New Commissioner Determined to Remake the Troubled FDA," *WP*, February 22, 1970, F1.
26 David Wallace, "Antibiotics Used in Meats Spur Study: Enough Regulation?" *WP*, June 14, 1970, K3.
27 Ibid.
28 David Wallace, "Monitoring of Meat Is Reduced Sharply," *WP*, November 22, 1970, A2.
29 James Turner, *The Chemical Feast: Ralph Nader's Study Group Report on the Food and Drug Administration* (New York: Grossman, 1970), 73–76.
30 Ibid.
31 Richard Lyons, "FDA: These Days It's Just Lurching from Crisis to Crisis," *NYT*, January 10, 1971, E2.
32 Warren Belasco, *Appetite for Change: How the Counterculture Took on the Food Industry* (Ithaca, NY: Cornell University Press 2007), 18.
33 Ibid., 18–28, 41, 76, 93–108.
34 Andrew G. Kirk, *Counterculture Green: The Whole Earth Catalog and American Environmentalism* (Lawrence: University Press of Kansas, 2007), 53–55.
35 Andrew N. Case, *The Organic Profit: Rodale and the Making of Marketplace Environmentalism* (Seattle: University of Washington Press, 2018), 150.
36 Ibid., 179; Robin O'Sullivan, *American Organic: A Cultural History of Farming, Gardening, Shopping, and Eating* (Lawrence: University Press of Kansas, 2015), 88.
37 *The Basic Book of Organically Grown Foods* (Emmaus, PA: Rodale, 1972), 318.
38 Ibid., 190.
39 Ibid., 191.
40 Jerome Goldstein, "Organic Force," *The New Food Chain: An Organic Link between Farm and City* (Emmaus, PA: Rodale, 1973), 3.
41 Sharon Cadwallader and Judi Ohr, *Whole Earth Cookbook* (San Francisco: San Francisco Book Co., 1973), xviii.
42 Ibid.
43 Frances Moore Lappé, *Diet for a Small Planet* (New York: Friends of the Earth/ Ballantine, 1971).
44 William Shurtleff and Akiko Aoyagi, *The Book of Tofu: Food for Mankind* (Brookline, MA: Autumn Press, 1975), 16–17, 28.
45 F. M. Lappé, "Fantasies of Famine," *Harper's*, February 1, 1975, 51–54, 87–90; Carl Alpert, "Eating What Comes Naturally," *Jewish Press*, December 22, 1978, 4; Nancy L. Ross, "It Tastes Like Meat Should," *WP*, December 14, 1972, M1.
46 O'Sullivan, *American Organic*, 85–87, 98.
47 Jean Hewitt, *The NYT Natural Foods Cookbook* (Chicago: Quadrangle Books, 1971).
48 Wade Greene, "Guru of the Organic Food Cult," *NYT*, June 6, 1971, SM30.
49 Daniel Yergin, "Supernutritionist," *NYT*, May 20, 1973, 32–33.
50 Ibid.

51 "Beauty: Food Beautiful Food," *Vogue*, June 1965, 108–109, 150–151, 160–161, 164.

52 "Beauty and Health: The Health Eaters," *Vogue*, May 1971, 168.

53 Lee Israel, "Health Foods—True or False?" *Cosmopolitan*, February 1969, 113–114; see also "The Neophyte Health-Nut's Guide to Health Foods," *Cosmopolitan*, April 1970, 160.

54 "The Move to Eat Natural," *Life*, December 1970, 45.

55 Ibid., 46.

56 Hedley Burrell, "Butz Hopes for Sales to China," *WP*, December 13, 1971, A3.

57 "Additives Aren't Always Villains," *DS*, May 13, 1977, A7; "Costly Protein," *LMC Experience*, March 14, 1974, 4.

58 George Will, "Embattled Farmers," *National Review* [hereafter *NR*], March 1973, 302.

59 M. Stanton Evans, "At Home," *NR*, February 1973.

60 Victor Cohn, "FDA Proposes Antibiotic Ban in Animal Feed," *WP*, February 1, 1972, A6.

61 Ibid.; see also "Use of Antibiotics on Farms Studied," *NYT*, June 4, 1970, 36.

62 Harold Schmeck, "Limitation on Antibiotics in Feed for Livestock Urged by FDA," *NYT*, February 1, 1972, 19.

63 Cohn, "FDA Proposes Antibiotic Ban," A6.

64 Schmeck, "Limitation on Antibiotics in Feed"; see also Thomas H. Jukes, "Antibiotics and Meat," *NYT*, October 2, 1972, 37.

65 Earl Ubell, "Are We Breeding an 'Andromeda Strain?'" *NYT*, February 6, 1972, E7.

66 Robert M. Bleiberg, "No Chicken Feed," *Barron's*, August 7, 1972, 7.

67 Morton Mintz, "Physicians Accused of Antibiotic Misuse," *WP*, December 8, 1972, A1; Royston C. Clowes, "The Molecule of Infectious Drug Resistance," *SciAm*, April 1973, 18–27.

68 Morton Mintz, "Safety Test Ordered on Animal Feed Drug," *WP*, April 19, 1973, A4.

69 This chronology roughly fits Podolsky's chronology for human medicine; Podolsky, *Antibiotic Era*, 155–160.

70 "World Biologists Tighten Rules on 'Genetic Engineering' Work," *NYT*, February 28, 1975, 1, 38.

71 Liebe F. Cavalieri, "New Strains of Life—or Death," *NYT*, August 22, 1976.

72 Stuart Auerbach, "Drug Resistant Bacteria Making Inroads," *WP*, March 2, 1975, 3.

73 Podolsky, *Antibiotic Era*, 162.

74 Stuart B. Levy, George B. Fitzgerald, and Ann B. Macone, "Changes in Intestinal Flora of Farm Personnel After Introduction of a Tetracycline-Supplemented Feed on a Farm," *NEJM* 295, no. 11 (1976): 583.

75 Stuart Auerbach, "Drug in Chicken Feed Is Traced in Humans," *WP*, June 1, 1975, 10.

76 Levy et al., "Changes in Intestinal Flora," 588; see also Stuart B. Levy, "Spread of Antibiotic Resistance Plasmids from Chicken to Chicken and from Chicken to Man," *Nature* 260 (1976): 40–42.

77 Harold Schmeck, "A Leapfrog War between Drugs and Their Targets," *NYT*, January 23, 1977, 145.

78 "FDA to Order Big Cuts in Penicillin for Animals," *NYT*, April 16, 1977, 12.

79 Vogel, *Is It Safe?* 57, 66–68.

80 "The Need for New Drugs," *NR*, August 1973, 859; see also M. Stanton Evans, "Taming the FDA," *NR*, February 17, 1978, 219.

81 William Tucker, "Of Mites and Men," *Harper's*, August 1, 1978, 45; "Medicine: The Drug Lag," *Time*, September 29, 1975; Vogel, *Is It Safe?* 64–68.

82 Robert Proctor, "Agnotology: A Missing Term to Describe the Cultural Production of Ignorance (and its study)," in *Agnotology: The Making and Unmaking of Ignorance*, ed. Robert Proctor and Londa Schiebinger (Stanford, CA: Stanford University Press, 2008), 1–33; Naomi Oreskes and Erik Conway, *Merchants of Doubt: How a Handful of Scientists Obscured the Truth on Issues from Tobacco Smoke to Global Warming* (New York: Bloomsbury, 2010).

83 Eliot Marshall, "Scientists Quit Antibiotics Panel at Cast," *Science* 203, no. 4382 (1979): 733; Mark R. Finlay and Alan I. Marcus, "'Consumerist Terrorists': Battles over Agricultural Antibiotics in the United States and Western Europe," *Agricultural History* 90, no. 2 (2016): 146–172. The group consisted of Roy Curtiss, Julian E. Davies, Richard Novick, Michael J. Haas, Raul Goldschmidt, and Vickers Hershfield.

84 Thomas H. Jukes, "Antibiotics in Feeds," *Science* 204, no. 4388 (1979): 8; Richard Novick, "Use in Animal Feed," *Science* 204, no. 4396, 908; Robert Bud, *Penicillin: Triumph and Tragedy* (Oxford: Oxford University Press, 2009), 186.

85 Maryn McKenna, *Big Chicken: The Incredible Story of How Antibiotics Created Modern Agriculture and Changed the Way the World Eats* (Washington, DC: National Geographic, 2017), 121.

86 Victor Cohn, "Antibiotics in Feeds Found Health Risk," *WP*, June 25, 1979, A2.

87 "Two Hands for Donald Kennedy," *NYT*, July 2, 1979, A16.

88 Ibid.

89 Ibid.

90 "Ban Urged on 2 Antibiotics in Animal Feed," *WP*, October 25, 1983, A17.

91 On the contemporary rise of US grass roots activism, see Richard Newman, *Love Canal: A Toxic History from Colonial Times to the Present* (Oxford: Oxford University Press, 2016).

92 "Medicine: Drugged Cows," *Time*, September 10, 1979; "New Worry over Drugs in Animals," *NYT*, June 17, 1980, C1; Lawrence Altman, "New Antibiotic Weapons in the Old Bacteria War," *NYT*, January 10, 1982, E9.

93 Bud, *Penicillin*, 189; Victor Cohn, "Worldwide Abuse of Antibiotics Poses Threat," *WP*, August 5, 1981, A2.

94 Cass Peterson, "Ban Urged on 2 Antibiotics in Animal Feed," *WP*, October 25, 1983, A17.

95 Scott D. Holmberg et al. "Animal-to-Man Transmission of Antimicrobial-Resistant Salmonella: Investigations of Us Outbreaks, 1971–1983," *Science* 225, no. 4664 (1984): 833.

96 Thomas O'Brien et al., "Molecular Epidemiology of Antibiotic Resistance in Salmonella from Animals and Human Beings in the United States," *NEJM* 307, no. 1 (1982): 1–6.

97 Scott D. Holmberg et al., "Drug-Resistant Salmonella from Animals Fed Antimicrobials," *NEJM* 311, no. 10 (1984): 617; "Poisoning Linked to Cattle Germs," *NYT*, September 6, 1984, A20; "Beware the Beef," *SciAm*, November 1984, 74–75.

98 Stuart Levy, "Playing Antibiotic Pool: Time to Tally the Score," *NEJM* 311, no. 10 (1984): 664.

99 "Feeding Beef Cattle Antibiotics Can Result in Illness, Study Says," *WSJ*, September 7, 1984, 1; "Antibiotic Burnout?" *Vogue*, September 1984, 456, 466; "Research Links Human Illness, Livestock Drugs," *WP*, September 6, 1984, A12.

100 Orville Schell, *Modern Meat: Antibiotics, Hormones and the Pharmaceutical Farm* (New York: Random House, 1985), 18–118; Finlay and Marcus, "Consumerist Terrorists," 157, 159.

101 "The Federal Report," *WP*, October 15, 1984, A21.

102 "Ties to Human Illness Revive Move to Ban Medicated Feed," *NYT*, September 16, 1984, 36.

103 Finlay and Marcus, "Consumerist Terrorists," 159–160.

104 "Decision on Feed Additives Angers Environmental Group," *WP*, November 22, 1985, A21.

105 Marrian Burros, "Shopping for Antibiotic-Free Meat," *NYT*, January 17, 2001, F2.

106 LuAnne Metzger, "Cattlemen Still Oppose Antibiotics Ban," *WF*, March 23, 1985, 2.

107 "New Meat-Packing Plant Tests Are Meant to Curb Harmful Drug Levels in Livestock," *Wall Street Journal* [hereafter *WSJ*], August 21, 1984, 1.

108 "Agency to Test Pigs for Excessive Level of Sulfamethazine," *WSJ*, February 5, 1988.

109 "FDA Chemist Asserts Agency Is Stalling on Tests for Milk Purity," *NYT*, February 7, 1990, A22; "Don't Drink Your Milk!" *WP*, February 17, 1990, A29.

110 "FDA's Regulation of Animal Drug Residues in Milk," Human Resources and Intergovernmental Relations Subcommittee of the Committee on Government Operations, US House of Representatives (Washington, DC: GPO, 1990), 117–127, 133–134; for FDA claims, see "US Calls Milk Free of Antibiotics," *NYT*, February 6, 1990, C13.

111 "FDA Chemist Asserts Agency Is Stalling on Tests for Milk Purity," *NYT*, February 7, 1990, A22; "House Probes Milk's Safety After Contamination Is Alleged," *WP*, February 7, 1990, A2.

112 "Cornell Institute for Social and Economic Research—Chemicals, 1990," *Polling the Nations* database, accessed August 1, 2017.

113 "Paradox over Produce Safety," *WP*, May 12, 1992, 20.

114 Mildred Howie, "No Bull Story: Raising Cattle on Natural Feed," *Healdsburg Tribune*, November 19, 1980, 4; Art Bock, "Letter–Trapped in Meat World," *Healdsburg Tribune*, June 4, 1986, A-4.

115 O'Sullivan, *American Organic*, 132.

116 Warren J. Belasco, *Appetite for Change: How the Counterculture Took on the Food Industry*, 2nd ed. (Ithaca, NY: Cornell University Press, 2007), 180-203.

117 O'Sullivan, *American Organic*, 116–132.

118 "Proposed Organic Certification Program," Joint Hearing of Subcommittee on Domestic Marketing, Consumer Relations and Nutrition and the Subcommittee on Department Operations, Research, and Foreign Agriculture of the Committee on Agriculture, US House of Representatives, June 19, 1990 (Washington, DC: GPO, 1991), 64–65.

119 "Organic? Industry Is Way Ahead of Government," *WP*, December 31, 1997, E1.

120 Catherine R. Greene, "US Organic Farming Emerges in the 1990s: Adoption of Certified Systems," *ERS Agriculture Information Bulletin* 770 (2001), 3.

121 Peter Hoffman, "Going Organic, Clumsily," *NYT*, March 24, 1998, A23; Marrian

Burros, "US to Subject Organic Foods, Long Ignored, to Federal Rules," *NYT,* December 15, 1997, A1, A14.

122 Rick Weiss, "'Organic' Label Ruled Out for Biotech, Irradiated Food," *WP,* May 1, 1998, A02.

123 "National Organic Program," https://www.ams.usda.gov/about-ams/programs -offices/national-organic-program.

124 Michelle Mart, *Pesticides, a Love Story: America's Enduring Embrace of Dangerous Chemicals* (Lawrence: University Press of Kansas, 2016), 7.

125 Stuart B. Levy, *The Antibiotic Paradox: How Miracle Drugs Are Destroying the Miracle* (New York: Plenum, 1992); "TB Takes a Deadly Turn," *Time,* December 2, 1991, 85; Sevgi O. Aral and King K. Holmes, "Sexually Transmitted Diseases in the AIDS Era," *SciAm,* February 1991, 62–68; John Nicholson, "Biotechnology and Agriculture: The Counterpoint," in *Poison in Your Food,* ed. Gary E. McCuen (Hudson, WI: G. E. McCuen, 1991), 150; "The Killer that Didn't Go Away," *WP,* June 6, 1993, 1, 14; "45 Infected with TB by Homeless Man," *WP,* August 15, 1995, 11; "The End of Antibiotics," *NW,* March 28, 1994, 46–52; "The Argument over Antibiotics," *WP,* June 23, 1993, E13; "Attack of the Superbugs," *Time,* August 31, 1992, p 62.

126 Daniel E. Koshland, "The Microbial Wars," *Science* 257, no. 5073 (1992): 1021; I am indebted to Christoph Gradmann for this reference.

127 Mitchell L. Cohen, "Epidemiology of Drug Resistance—Implications for a Post-Antimicrobial Era," *Science* 257, no. 5073 (1992): 1050–1055.

128 Carole Sugarman, "A Disease that's a Bite Away," *WP,* February 13, 1994, A1, A23; Susan D. Jones, *Valuing Animals: Veterinarians and Their Patients in Modern America* (Baltimore: Johns Hopkins University Press, 2003), 151–152.

129 "Agriculture Dept. Policy Blamed for Tainted Food," *NYT,* March 3, 1993, C1, C4; Maureen Ogle, *In Meat We Trust: An Unexpected History of Carnivore America* (New York: Houghton Mifflin Harcourt, 2013), 216–218.

130 For an overview of resistant 1990s outbreaks, see McKenna, *Big Chicken,* 139–147, 172–175.

131 Christoph Gradmann, "Re-inventing Infectious Disease: Antibiotic Resistance and Drug Development at the Bayer Company 1945–1980," *Medical History* 60, no. 2 (2016): 155–180, 155–180.

132 Laura H. Kahn, *One Health and the Politics of Antimicrobial Resistance* (Baltimore: Johns Hopkins University Press, 2016), 7–8.

133 Nicholas Wade, "Pax Antibiotica," *NYT,* October 15, 1995, SM30; Paul Epstein and Ross Gelbspan, "Should We Fear a Global Plague?" *WP,* March 19, 1995, C1.

134 "Scientists Say the Plague Is Coming," *Coronado Eagle,* March 8, 1995, 22.

135 Dick Thompson, "Drugged Chicks Hatch a Menace," *Time,* May 31, 1999, 81.

136 "The Bacterium and the Chicken," *NYT,* October 21, 1997, A26.

137 Sandra G. Boodman, "Poultry Peril," *WP,* December 9, 1997, 13.

138 Denise Grady, "A Move to Limit Antibiotic Use in Animal Feed," *NYT,* March 8, 1999, A13; see also "US Antibiotics Countered," *NYT,* May 20, 1999, A20; David Brown, "Drug Resistance in Food Chain," *WP,* May 20, 1999, A02; "Drugged Chicks Hatch a Menace," *Time,* May 31, 1999, 81.

139 Patricia Lieberman, "Letter: Control Antibiotic Use," *NYT,* November 7, 1999, WK14.

140 "FDA Announces Policy Designed to Curb Animal-Antibiotic Use," *WSJ,* October 24, 2003, A6.

141 Marc Kaufman, "Worries Rise over Effect of Antibiotics in Animal Feed," *WP*, March 17, 2000.

142 "Wonder Drugs at Risk," *WP*, April 19, 2001, A18.

143 Andrew Pollack, "Antibiotics Business Is Again Popular," *NYT*, November 13, 2001, B6.

144 Christopher Wanjek, "Cipromania," *WP*, October 23, 2001, F01.

145 Rick Weiss, "Demand Growing for Anthrax Vaccine," *WP*, September 29, 2001, A16; "Emphasis on Cipro Worries Officials," *WP*, October 19, 2001, A17; Shankar Vedantam, "Prescribing Cipro Is "Uncontrolled Experiment," *WP*, November 3, 2001, A15.

146 S. L. Gorbach, "Time to Stop," *NEJM* 345, no. 16 (2001).

147 "ABC News Polling Unit—Food," July 15, 2003, *Polling the Nations*.

148 Nicholas D. Kristof, "Pathogens in Our Pork," *NYT*, March 15, 2009, WK13.

149 "Healthy Growth for US Farms," *SciAm*, April 2009, 32.

150 Gardiner Harris, "Administration Seeks to Restrict Antibiotics in Livestock," *NYT*, July 14, 2009, A18.

151 Donald Kennedy, "Cows on Drugs," *NYT*, April 18, 2010, WK11.

152 "We Are What We Eat," *NYT*, September 22, 2010, A24; "Antibiotics and Agriculture," *NYT*, June 30, 2010, A30.

153 "We Are What We Eat," *NYT*, September 22, 2010, A24; see also Erik Eckholm, "Meat Farmers Brace for Limits on Antibiotics," *NYT*, September 15, 2010, A14.

154 Kahn, *One Health*, 65.

155 "FDA Moves to Cut Use of Antibiotics in Food-Producing Livestock," *WSJ*, December 11, 2013; "Meat Industry Won't Fight Antibiotics Rule," *WSJ*, December 12, 2013.

156 Kahn, *One Health*, 66–67.

157 Agricultural antibiotics were, however, problematized in the context of animal welfare. Matthew Scully, "Pro-Life, Pro-Animal," *NR*, November 2013, 35–38.

158 Mart, *Pesticides, a Love Story*, 227–229.

159 "Harris Poll—Superbugs, February 27, 2017," *Polling the Nations*.

160 Lizabeth Cohen, *A Consumers' Republic: The Politics of Mass Consumption in Postwar America* (New York: Vintage Books, 2004), 403–407, 409.

161 Andrew G. Kirk, *Counterculture Green: The Whole Earth Catalog and American Environmentalism* (Lawrence: University Press of Kansas, 2007), 189–190, 206–209.

162 "Harris Poll—Food, December 8, 2015," *Polling the Nations*.

163 Michael Pollan, *The Omnivore's Dilemma: A Natural History of Four Meals* (London: Penguin, 2006); O'Sullivan, *American Organic*, 149–152.

164 O'Sullivan, *American Organic*, 230, 259.

165 "US Organic Sales Set a New Record in 2016," *Organic Trade Association*, August 14, 2017, http://theorganicreport.com/us-organic-sales-set-new-record -2016.

166 O'Sullivan, *American Organic*, 167–179, 185, 194–195.

167 "Antibiotics in the Poultry Industry," *NYT*, February 13, 2002, A30; "McDonald's Takes Steps on Its Antibiotics Promise," *NYT*, January 12, 2005, F2; "A Chain that Pigs Would Die For," *NW*, May 12, 2008, 45–46.

168 "US Withdraws Approval for Tyson's Antibiotic-Free Label," *NYT*, November 20, 2007, C9; "Chain Reaction Report Urges Burger Restaurants to Beef Up Policies to Eliminate Routine Use of Antibiotics," *NRDC*, October 17, 2018.

169 Case, *Organic Profit*, 219–220.
170 Ulrich Beck, *Risikogesellschaft. Auf dem Weg in eine andere Moderne* (Frankfurt: Suhrkamp, 1986), 14, 17–19, 29–31, 35.
171 Mart, *Pesticides, a Love Story*, 7.
172 Rita Rubin, "Kosher Chickens Found to Have Lots of Antibiotics. Or Maybe Not," *Forward*, November 8, 2013, 1, 5; Jack M. Millman et al., "Prevalence of Antibiotic-Resistant E. coli in Retail Chicken: Comparing Conventional, Organic, Kosher, and Raised without Antibiotics, Version 2," *F1000Research* 2, no. 155, published online September 2, 2013.

9. Light-Green Reform: Antibiotic Change on American Farms

1 Paul K. Conkin, *A Revolution Down on the Farm: The Transformation of American Agriculture Since 1929* (Lexington: University Press of Kentucky, 2009), 131.
2 Cited according to Douglas R. Hurt, *Problems of Plenty: The American Farmer in the Twentieth Century* (Chicago: Ivan R. Dee, 2002), 132–133.
3 C. G. Struggs, "Will the 1970s Become the Era of Agricultural Capitalism," *Progressive Farmer* [hereafter *PF*], January 1970, 23.
4 Ibid.
5 "1992 US Census of Agriculture, vol. 1, part 51" (Washington, DC: USDA, 1994), 8.
6 Chris Mayda, "Pig Pens, Hog Houses, and Manure Pits: A Century of Change in Hog Production," *Material Culture* 36, no. 1. (2004): 28.
7 Mark Friedberger, "Cattlemen, Consumers, and Beef," *Environmental History Review* 18, no. 3 (1994): 37–57.
8 William Boyd, "Making Meat: Science, Technology, and American Poultry Production," *Technology and Culture* 42, no. 4 (2001): 635, 641–649.
9 Sarah A. Vogel, *Is It Safe? BPA and the Struggle to Define the Safety of Chemicals* (Berkeley: University of California Press, 2013), 59, 63.
10 Abraham L. Fairfax, "Favors DDT Ban," *PF*, December 1972, 1; Alex Bower, "The Mailbox—Kentuckians Differ on DDT," *PF*, October 1975, 14.
11 "Agriculture Has Big Stake in Crackdown on Pesticides," *PF*, February 1970, 152.
12 "Farm Chemicals," *Wallaces Farmer* [hereafter *WF*], April 8, 1972, 72; "The DDT Ban—What Does It Mean To You," *PF*, September 1972, 20, 42.
13 "Hood Says Fertilizers, Chemicals Are Essential to Food Production," *Farm Bureau News* [hereafter *FBNews*], March 13, 1972, 43.
14 "DDT, a victim of Ecological Fanatics," *PF*, September 1972, 90.
15 Kendra Smith-Howard, "Healing Animals in an Antibiotic Age: Veterinary Drugs and the Professionalism Crisis, 1945–1970," *Technology and Culture* 58, no. 4 (2017): 740.
16 43 *Federal Register*, 3034 (January 20, 1978).
17 "The Effects on Human Health of Subtherapeutic Use of Antimicrobials in Animal Feeds," *Committee to Study the Human Health Effects of Subtherapeutic Antibiotic Use in Animal Feeds* (Washington, DC: National Academies Press, 1980), 8.
18 *Yearbook of Agriculture* (Washington, DC: USDA, 1970), 60; "Swine Production," *Farmers' Bulletin* 2166 (1969): 10; "Finishing Beef Cattle," *Farmers' Bulletin* 2196 (1968): 21; "Antibiotics in Animal and Poultry Feeds—A Critical Review of Research" [ARS 44/237] (Washington, DC: USDA, 1972), 3–10.

19 "Antibiotics in Animal and Poultry Feeds—A Critical Review of Research"; John K. Matsushima, *Feeding Beef Cattle* (New York: Springer, 1979), 98–99, 117; Irwin Dyer and Clayton O'Mary, *The Feedlot*, 2nd ed. (Philadelphia: Lea and Febiger, 1977), 275, 277; Tilden Wayne Perry, *Beef Cattle Feeding and Nutrition* (New York: Academic Press, 1980), 235–238. Whereas the fourth and fifth and sixth editions of *Beef Cattle* had still been skeptical about antibiotic feed additives for cattle, this had changed in the seventh edition; Roscoe R. Snapp, *Beef Cattle*, 4th, 5th, 6th, 7th eds. (New York: Wiley, 1952, 1960, 1969, 1977).

20 Tilden Wayne Perry, *Beef Cattle Feeding and Nutrition* (New York: Academic Press, 1980), 322, 329; Frederic N. Owens, Jose Zorrilla-Rios and Paula Dubeski, "Effects of Ionophores on Metabolism, Growth Body Composition and Meat Quality," in *Growth Regulation in Farm Animals*, ed. A. M. Pearson and T. R. Dutson (London: Elsevier, 1991), 321, 323.

21 "Balanced Rations—Modern Concept," *Lancaster Farming* [hereafter *LF*], January 31, 1970, 10; "Dairymen Make Gains in Controlling Mastitis," *Madera Tribune*, March 8, 1968, 9; "Farm Poultry Management," *Farmers' Bulletin* 2197 (1969): 12; Glenn Geiger and Walter Russell, *Fact Sheet for Part Time Farmers and Gardeners* (Washington, DC: USDA, 1977), *Midwest Farm Handbook*, 7th ed. (Ames: Iowa State University Press, 1969), 349–351, 370–372, 381, 392–393; *Commercial Broiler Production* [Agriculture Handbook No. 320] (Washington, DC: USDA, 1967), 11, 32, 39–41; "Advertisement—Purina," *LF*, March 9, 1968, 6; "Advertisement—Red Rose Swine Feeds," *LF*, March 9, 1968, 13.

22 "From Where We Stand," *LF*, April 26, 1969, 4.

23 "Antibiotic Residues Hidden Risks" (Washington, DC: FDA, 1970); "Mastitis and Antibiotics Discussed at Dairy Day," *LF*, March 11, 1972, 21.

24 "FDA Proposes Ban on Antibiotics," *FBNews*, February 7, 1972, 24.

25 "Feeders Face Stricter Antibiotic Rules," *WF*, February 26, 1972, 12–13.

26 Ibid., 12.

27 Ibid., 13.

28 For FDA complaints, see: Van Houweling to American Cyanamid, December 17, 1970, Folder "Literature Notes," Box 3, Account No. 75A0106, RG 442, NARA Atlanta, 1.

29 "Across the Editor's Desk," *LF*, April 8, 1972, 10; a similar strategy was employed by the American Feed Manufacturers Association; "Guest Editorial," *LF*, April 29, 1972, 11, 13.

30 "Farm Bureau—For and Against," *PF*, February 1972, 63.

31 "Hogmen Could Lose Some Feed Additives," *WF*, September 23, 1972, 39.

32 "Deadline Near for Filing Thoughts on FDA Regulation," *Farmers Weekly Review* [hereafter *FWR*], March 23, 1972, 7; see also "Animal Health," *WF*, March 11, 1972, 44.

33 "Grassroots Opinions," *LF*, November 11, 1972, 10.

34 "Antibiotics in Animal and Poultry Feeds—A Critical Review of Research," 1–5.

35 "What's New In Washington," *PF*, July 1972, 6.

36 "Drug Certification Does Apply to You," *PF*, September 1972, 62–63.

37 "What's New in Washington," *PF*, August 1972, 8; Nancy Langston, *Toxic Bodies: Hormone Disruptors and the Legacy of DES* (New Haven, CT: Yale University Press, 2010), 97–100, 105–107.

38 "1992 US Census of Agriculture," 8.

39 J. L. Anderson, *Industrializing the Corn Belt: Agriculture, Technology and*

Environment, 1942–1972 (Dekalb: Northern Illinois University Press, 2009), 195; Hurt, *Problems of Plenty*, 134.

40 "Effects on Human Health of Subtherapeutic Use of Antimicrobials," 8.

41 "Facts for Dairymen," *LF*, July 29, 1978, 128.

42 Monte Sesker, "Disease Problems Plague Iowa's Hog Producers," *WF*, February 26, 1977, 10–11.

43 "Feed Additives Vital to Production," *LF*, June 30, 1973, 12–13.

44 Philip G. Connell, "Radicalism in Food Production," *Vital Speeches of the Day*, November 15, 1974, 70.

45 Ibid., 72; see also "Antibiotics Vital for Food Production," *WF*, February 26, 1977, 93.

46 "Butz Defends Modern Farming," *LF*, September 6, 1975, 21.

47 "Germans," *LF*, May 13, 1978, 38.

48 "Feedlot Additives Not Harmful to Environment," *LF*, September 6, 1975, 57; "War declared on antibiotic residues", *LF*, December 9, 1978, 103; Claas Kirchhelle, "Between Bacteriology and Toxicology: Agricultural Antibiotics and US Risk Regulation," in *Risk on the Table*, ed. Angela Creager and Jean-Paul Gaudillière (Chicago: University of Chicago Press, forthcoming).

49 "Farm Poultry Management," *Farmers' Bulletin* 2197 (1973): 27; "Farm Poultry Management", *Farmers' Bulletin* (revised 1977), 32; *Doane's Facts and Figures for Farmers*, 4th ed. (St Louis: Doane-Western, 1981), 130, 151,143; John K. Matsushima, *Feeding Beef Cattle* (New York: Springer, 1979), 108–117; Irwin Dyer and Clayton O'Mary. *The Feedlot*, 2nd ed. (Philadelphia: Lea and Febiger, 1977); Perry, *Beef Cattle*.

50 M. E. Ensminger, *The Stockman's Handbook*, 5th ed. (Danville, IL: Interstate, 1978), 225.

51 "Animal Health," *WF*, July 23, 1977, 29; "Hutchinson Speaks to Dairymen on Calf Care," *LF*, November 26, 1977, 35.

52 "Researchers Can't Agree . . . Penicillin Substitutes in Animal Feed," *WF*, October 8, 1977, 19; "PennAg Protests Proposed Antibiotic Ban," *LF*, June 18, 1977, 58; "Researchers Divided on Animal Feed Additives," *LF*, September 10, 1977, 36; "Pork Producers Attack Unfair Gov't Regulation," *FWR*, September 21, 1978, 2; "American Food Supply Safe, Convention Is Informed," *FWR*, February 2, 1978, 6; "Truth and Fact Wage War Against Fallacy," *LF*, September 23, 1978, 26.

53 Monte Sesker, "Another Threat to Livestock Producers," *WF*, June 25, 1977, 18; Susan D. Jones, *Valuing Animals: Veterinarians and Their Patients in Modern America* (Baltimore: Johns Hopkins University Press, 2003), 8, 104–106, 111–114.

54 "Antibiotics Treated Unfairly, Critics Charge," *LF*, April 1, 1978, 136.

55 "Asked to Write," *FWR*, March 30, 1978, 6; "NPPC Pushing for Awareness," *LF*, August 26, 1978, 107.

56 "Regulating Mother Nature," *Farm Journal* [hereafter *FJ*], January 1979, Hog32; "Nitrite Ban Would Boost Inflation," *FJ*, January 1979, Beef2; "What If They Ban Feed Drugs," *FJ*, January 1979, Beef2.

57 "CAST Considers Regulations," *FJ*, December 1979, Beef2.

58 "Regulators Talk Tough," *FJ*, November 1979, Hog28.

59 "Are Bacterial Infections the Culprit in MMA?" *FJ*, April 1979, Hog26; "A Veterinarian Looks at . . . Building Systems," *FJ*, April 1979, Hog30.

60 "Outlook/Washington," *FJ*, April 1980, 6.

61 "Antibiotic Feed Additives: The Prospect of Doing Without," *Farmline* 1, no. 9 (1980): 15.

62 Hurt, *Problems of Plenty*, 138.

63 Ibid., 150.

64 Gilbert Wayne Gillespie Jr., *Antibiotic Animal Feed Additives and Public Policy: Farm Operators' Beliefs About the Importance of These Additives and Their Attitudes Toward Government Regulation of Agricultural Chemicals and Pharmaceuticals*, doctoral thesis, Cornell University, 1987, 42.

65 *Nature's Ag School: The Thompson Farm* (Emmaus, PA: Regenerative Agriculture Association, 1986), 7.

66 Wilson G. Pond, "Modern Pork Production," *Scientific American* [hereafter *SciAm*], May 1983, 79.

67 Gillespie, *Antibiotic Animal Feed Additives*, 28–32.

68 Ellen K. Silbergeld, *Chickenizing Farms and Food: How Industrial Meat Production Endangers Workers, Animals, and Consumers* (Baltimore: Johns Hopkins University Press, 2016), 57–61.

69 Gillespie, *Antibiotic Animal Feed Additives*, 31.

70 Robert Rodale, "Organic Living," *LF*, May 31, 1975, 24; Robert Rodale, "Organic Living," *LF*, January 25, 1975, 41.

71 "Conventional Farmers Using Organic Farming Principles," *Indiana Prairie Farmer*, May 16, 1981, 47; see also "Pro's and Con's of Organic Farming," *LF*, April 12, 1980, C18, C20.

72 "Report and Recommendations on Organic Farming" (Washington, DC: USDA Study Team on Organic Farming, 1980), xi–xiv.

73 Ibid., 17.

74 Maureen Ogle, *In Meat We Trust: An Unexpected History of Carnivore America* (New York: Houghton Mifflin Harcourt, 2013), 231–242.

75 William D. Heffernan and Gary Green, "Appendix A—Opinions of Missouri Farmers Concerning Use of Antibiotics in Animal Feeds and the Consequences of Banning Their Use," *Antibiotics in Animal Feeds* [CAST Report No. 88] (Ames, IA: CAST, 1981), 72–75.

76 Gillespie, *Antibiotic Animal Feed Additives*, ii.

77 Ibid., 132–138, 147.

78 Ibid., 141–143.

79 "Antibiotics and Swine—Improving Efficiency, Reducing Costs," *LF*, October 22, 1983, D2; "Feed Antibiotics Have Safe Record for 30 Years," *FWR*, May 30, 1985, 2.

80 "Call to Ban Antibiotics in Feed Gets FDA Hearing Later in Month," *FBNews*, January 7, 1985, 3.

81 "FDA, Says FB, Needs More Evidence on Antibiotics in Feed Question," *FBNews*, January 28, 1985.

82 "Call to ban antibiotics in feed gets FDA hearing."

83 "FDA, Says FB, Needs More Evidence."

84 Gilespie, *Antibiotic Animal Feed Additives*, 291.

85 "Status Report," *FBNews*, April 29, 1985, 3.

86 "What's Behind the Salmonella Scare," *WF*, May 25, 1985, 14.

87 LuAnne Metzger, "Cattlemen Still Oppose Antibiotics Ban," *WF*, March 23, 1985, 2.

88 "Human Health Risks with the Subtherapeutic Use of Penicillin or Tetracyclines

in Animal Feed. 1988," Committee on Human Health Risk Assessment of Using Subtherapeutic Antibiotics in Animal Feeds, 1988, 69.

89 Conkin, *Revolution*, 134; Michael Bess, *The Light-Green Society: Ecology and Technological Modernity in France* (Chicago: University of Chicago Press, 2003), 3–4.

90 "2002 Census of Agriculture, vol. 1, part 51" (Washington, DC: USDA, 2004), 6.

91 William D. McBride and Nigel Key, "US Hog Production from 1992 to 2009: Technology, Restructuring, Productivity Growth," *Economic Research Report* 158 (2013): 1–13; Silbergeld, *Chickenizing*, 61–63; Mayda, "Pig Pens," 32–35; Stacy Sneeringer et al., "Economics of Antibiotic Use in US Livestock Production," *Economic Research Report* 200 (2015): 19.

92 "Farmers Willing to Reduce Pesticide, Fertilizer Use," *WF*, July 10, 1990, 28.

93 "Survival Strategies for the '90s," *WF*, March 13, 1990, The Hog Producer, H23; "National Agriculture Week TV Program," *WF*, March 13, 1990, 16–17.

94 "Farmers as Environmentalists," *FBNews*, May 7, 1990, 2.

95 Chester Peterson Jr., "Winning the War on Mastitis," *Successful Farming* [hereafter *SF*], October 1995, 34.

96 G. R. Carter, *Essentials of Veterinary Microbiology*, 5th ed. (Baltimore: Williams and Wilkins, 1995), 90-94.

97 "News and Notes," *WF*, March 13, 1990, The Hog Producer, H3.

98 *The Stockman's Handbook*, 7th ed. (Danville, IL: Interstate, 1992), 218.

99 Gail Damerow, *The Chicken Health Handbook* (North Adams, MA: Storey, 1994), 185.

100 Keith Propst, "Avoid Selective Information from Alternative Ag Study," *FBNews*, January 1, 1990, 8.

101 Vogel, *Is It Safe?* 121.

102 "Regs Hurting Agriculture Farm Bureau Tells Panel," *FBNews*, February 20, 1995, 1.

103 Vogel, *Is It Safe?* 122.

104 "Animal Drug Bill Hailed by Pork Producers," *SF*, January 1997, S16.

105 "H.R. 2508, Animal Drug Availability Act of 1996," https://www.congress.gov /bill/104th-congress/house-bill/2508/summary/36.

106 "Animal Drug Availability Act of 1996," *FDA Animal and Veterinary Acts, Rules and Regulations*, https://www.fda.gov/AnimalVeterinary /GuidanceComplianceEnforcement/ActsRulesRegulations/ucm105940.htm.

107 Alex Blanchette, "Herding Species: Biosecurity, Posthuman Labor, and the American Industrial Pig," *Cultural Anthropology* 30, no. 4 (2013): 657; see also Smith-Howard, "Healing Animals," 741–742.

108 Joann Alumbaugh, "Stay Committed—Come Out Stronger," *WF*, October 1998, H1.

109 "The Rebirth of the Blues," *SF*, November 1997, 62–64; "Maverick Ranch Association," *SF*, December 1997, 28; "Advertisement—Coleman Natural Beef," *SF*, December 1997, 27; "Successful Farm Family," *SF*, Mid-February 1999, 62–64; "Lowering Costs, Lowering Risks," *SF*, November 1999, 12–13.

110 "Premium Organic," *SF*, October 1997, 46A.

111 "Quite the Cropp," *SF*, September 1999, 22.

112 "Certified Organic and Total US Acreage, Selected Crops and Livestock, 1995 and 2011," *USDA Economic Research Service—Organic Production—Overview*;

Sneeringer et al., "Economics," 7; Bruce Ingersoll, "Rules on Organic Food Are Issued, Including a USDA Seal on Items," *WSJ*, December 21, 2000, B9.

113 "Infectious Diseases Mount a Comeback," *SF*, April 1999, 65.

114 "FDA May Tighten Antibiotic Rules," *SF*, December 1999, SB1; see also "Prudent Practices in the Cross-Hairs," *WF*, April 1999, D4; "New Animal Drug Guidelines Questioned," *WF*, April 1999, H7.

115 "Fire Blight Control, Nature's Way," *Agricultural Research* [hereafter *AR*], January 1998, 14; "Helping Pigs Resist Edema," *AR*, January 1998, 7; "Coupled Antibodies Fend Off Mastitis," *AR*, June 1998, 17; "Doctoring Fish—New Vaccines for Aquaculturists," *AR*, October 1999, 10–11; "Transgenic Cow to Resist Mastitis," *AR*, August 2001, 23; "A New Way to Control *E.coli* in Weaned Pigs," *AR*, March 2004, 9; "For You, the Consumer," *AR*, December 2002, 4; "Minimizing Mastitis," *SF*, January 2001, 49; W. E. Huff et al., "Prevention of *E. coli* Respiratory Infection in Broiler Chickens with Bacteriophage," *Poultry Science* 81 (2002): 437–441; "Turning the Phage on Produce Pathogens," *AR*, July 2001, 12.

116 Blanchette, "Herding Species," 657–663.

117 Lynne Finnerty, "Clucking About Agricultural Antibiotics Is Overblown," *FBNews*, April 28, 2002, 3.

118 William D. McBride, Nigel Key, and Kenneth H. Mathews, "Subtherapeutic Antibiotics and Productivity in US Hog Production," *Review of Agricultural Economics* 30, no. 2 (2008), 270–288.

119 "Farm Bureau: Antibiotics Are Needed to Keep Animals Healthy, Food Safe," *FBNews*, April 6, 2009, 3.

120 Ibid.

121 "Routine Animal Drugs May Threaten Human Health," *SF*, January 2001, 60.

122 "Sniffle or Sneeze? No Antibiotics Please," *FWR*, November 4, 2010, 8.

123 Alan Newport, "Let's Get Argument About Antibiotics Right," *WF* Beef Producer, April 2010, BP8.

124 Ibid.

125 "Raise Hogs without Antibiotics?" *WF*, December 2010, 61.

126 Ibid.; see also Alan Newport, "Pressure Is Mounting on Animal Antibiotic Use," *WF* Beef Producer, April 2011, BP8.

127 Sneeringer et al., "Economics of Antibiotic Use," 22–25, 59.

128 Ibid., 28–31.

129 Ibid., 32–36.

130 Ibid., iv–v.

131 "2014 Summary Report on Antimicrobials Sold or Distributed for Use in Food-Producing Animals" (Washington, DC: FDA/DHSS, December 2015); "2016 Summary Report."

132 Alan Newport, "Is Ceding Ground in Antibiotics Fray Smart?" *WF* Beef Producer, April 2013, BP3.

133 Robert Fears, "Make Wise Use of Antibiotics," *WF* Beef Producer, April 2014, BP6.

134 Melinda Moyer, "The Looming Threat of Factory-Superbugs," *SciAm*, December 2016, 78–79; on veterinarians' new control over antibiotic access, see Smith-Howard, "Healing Animals in an Antibiotic Age," 741.

135 Maryn McKenna, *Big Chicken: The Incredible Story of How Antibiotics Created Modern Agriculture and Changed the Way the World Eats* (Washington, DC: National Geographic, 2017), 264–284.

136 "Antibiotics in Agriculture: The Blurred Line between Growth Promotion and Disease Prevention," *Bureau of Investigative Journalism*, September 19, 2018.

137 Christina Dittmer, "The Reality of Antibiotic Resistance," *WF*, August 2014, 74.

138 "Preserving Antibiotic Access," *American Farm Bureau Federation—Priority Issues Antibiotics* (March 2015).

139 Moyer, "The Looming Threat of Factory-Superbugs."

140 "1982 Census of Agriculture, vol. 1, part 51" (Washington, DC: US Department of Commerce, 1984), 1; "2012 Census of Agriculture, vol. 1, part 51" (Washington, DC: USDA, 2014), 7.

10. Statutory Defeat: Voluntarism and the Limits of FDA Power

1 Barbara Resnick Troetel, *Three-Part Disharmony: The Transformation of the Food and Drug Administration in the 1970s*, dissertation, City University of New York, 1996, 14–15, 18.

2 Douglas Martin, "James L. Goddard, Crusading FDA Leader, Dies at 86," *NYT*, January 1, 2010.

3 Troetel, *Three-Part Disharmony*, 1–3, 31–52.

4 Alan Marcus, *Cancer from Beef: DES, Federal Food Regulation, and Consumer Confidence* (Baltimore: Johns Hopkins University Press, 1994), 70, 77–85.

5 Troetel, *Three-Part Disharmony*, 56, 63–65.

6 "Oral History Interview of C. D. Van Houweling by Ronald T. Ottes," History of the US Food and Drug Administration, FDA Oral History Transcripts, June 18, 1990, 1–3.

7 "Animal Drug Amendments of 1968," Report No. 1308 to accompany HR 3639 (Washington, DC: GPO, 1968), 2–3; "Animal Drug Amendments of 1968: Hearings Before the US Senate Committee on Labor and Public Welfare, May 24, 1968" (Washington, DC: GPO, 1968), 68, 85–86, 89; Daryl M. Freedman, "Reasonable Certainty of No Harm: Reviving the Safety Standard for Food Additives, Color Additives, and Animal Drugs," *Ecology Law Quarterly* 7, no. 2 (1978): 254–255.

8 "Animal Drug Amendments of 1968: Hearings Before the US Senate," 83–89, 101–102.

9 Gilbert Wayne Gillespie Jr., *Antibiotic Animal Feed Additives and Public Policy: Farm Operators' Beliefs About the Importance of These Additives and Their Attitudes Toward Government Regulation of Agricultural Chemicals and Pharmaceuticals*, doctoral thesis, Cornell University, 1987, 235, 261.

10 Freedman, "Reasonable Certainty," 247.

11 Ibid., 257.

12 Marcus, *Cancer from Beef*, 81.

13 Freedman, "Reasonable Certainty," 260–262.

14 Sheila Jasanoff, *The Fifth Branch: Science Advisers as Policymakers* (Cambridge, MA: Harvard University Press, 1994), 17, 37; see also Jean-Paul Gaudillière, "Food, Drug and Consumer Regulation—The Meat, DES and Cancer Debates in the US," in *Meat, Medicine and Human Health in the Twentieth Century*, ed. David Cantor et al. (London: Pickering and Chatto, 2010), 179–203.

15 Marcus, *Cancer from Beef*, 80–82.

16 C. D. Van Houweling, "Regulation in the United States," in *Factory Farming: A Symposium*, ed. J. R. Bellerby (London: Education Services, 1970), 32; 34 *Federal*

Register [hereafter *FedReg*], 7849–7850 (May 17, 1969); "Impacts of Antibiotic-Resistant Bacteria" (Washington, DC: Office of Technology Assessment, 1995), 165.

17 Marcus, *Cancer from Beef,* 79; "Cornelius D. Van Houweling," *JAVMAnews,* April 1, 2012.

18 Claas Kirchhelle, "Between Bacteriology and Toxicology: Agricultural Antibiotics and US Risk Regulation," in *Risk on the Table,* ed. Angela Creager and Jean-Paul Gaudillière (Chicago: University of Chicago Press, forthcoming).

19 Paul Pumpian to Birch Bayh, November 21, 1967, Folder A24B 88-75-3, Box 3966, GS, DF A1/Entry 5, RG 88, NARA, 1–2.

20 Robert A. Baldwin to Van Houweling, February 13, 1968, Folder 88-76 5V, Box 4103, GS, DF A1/Entry 5, RG 88, NARA.

21 Note, "Budget, Bureau of [432.1], Filed 251," February 20, 1968, Folder 88-76 5V, Box 4103, GS, DF A1/Entry 5, RG 88, NARA.

22 W. L. Margard, A. C. Peters, and J. H. Litchfield, "Summary Report on the Development and Transfer of Antibiotic Resistance in Enteric Microorganisms in Chickens and Swine—Battelle Memorial Institute," July 31, 1968, Folder FDA Interagency Task Force—Lectures, Box 3, Account No. 75A106, RG 442, NARA Atlanta, 1–4.

23 The FDA's Division of Microbiology criticized AGPs; Robert Angelotti to Herman Kraybill, March 27, 1970, Folder 128 1975 May–July 432.1, Box (FRC) 25, FDA GS, DF UD-WW/Entry 4, RG 88, NARA, 2.

24 David H. Smith, "Antibiotics in Agriculture and the Health of Man," *FDA Papers* (September 1968), 11.

25 Van Houweling to Dale R. Lindsay, November 26, 1969, Folder 88-76-80, Box 4214, GS, DF A1/Entry 5, RG 88, NARA, 1.

26 Ibid., 1–2.

27 Ibid., 3.

28 Troetel, *Three-Part Disharmony,* 56, 63–65; Lucas Richert, *Conservatism, Consumer Choice, and the Food and Drug Administration During the Reagan Era: A Prescription for Scandal* (Lanham, MD: Lexington, 2014), 26.

29 Troetel, *Three-Part Disharmony,* 76–79.

30 "Report to the Commissioner of the Food and Drug Administration by the FDA Task Force on the Use of Antibiotics in Animal Feeds," FDA, 1972, 2.

31 "FDA Task Force on the Use of Antibiotics in Animal Feeds," 12.

32 Ibid., 13-14.

33 "Note of a Meeting of Representatives of the MAFF and DHSS with the USA Task Force on Antibiotics in Feedingstuffs," February 1, 1971, Folder Literature Notes, Box 1, Account No. 75A0106, RG 442, NARA Atlanta.

34 "Notes on Meeting with Representatives of the MAFF and DHSS," February 1, 1971, Folder Task Force Antibiotics, Box 1, Account No. 75A0106, RG 442, NARA Atlanta, 1.

35 "MAFF—Note of a Meeting Held with US Task Force on Antibiotics," February 2, 1971, Folder Task Force Antibiotics, Box 1, Account No. 75A0106, RG 442, NARA Atlanta, 4.

36 Ibid.

37 "Notes on Meeting with Representatives of the MAFF and DHSS," February 1, 1971, Folder Task Force Antibiotics, Box 1, Account No. 75A0106, RG 442, NARA Atlanta, 2.

38 Ibid., 4.
39 "Meeting with Members of the CVL and MAFF," February 2, 1971, Folder Task Force Antibiotics, Box 1, Account No. 75A0106, RG 442, NARA Atlanta, 4.
40 "Meeting of February 3, 1971 at DSSH-PM," Folder Task Force Antibiotics, Box 1, Account No. 75A0106, RG 442, NARA Atlanta, 2.
41 Ibid.
42 Ibid.
43 Ibid., 1.
44 "Task Force Trip to London—Testimony of ES Anderson," Folder Task Force Antibiotics, Box 1, Account No. 75A0106, RG 442, NARA Atlanta, 1.
45 Ibid., 2.
46 "Task Force Trip to London—Testimony of ES Anderson," 3.
47 "Notes on Meeting with H Williams Smith," February 4, 1971, Folder Task Force Antibiotics, Box 1, Account No. 75A0106, RG 442, NARA Atlanta, 1–2.
48 Ibid., 2.
49 Ibid.
50 John V. Bennett to Van Houweling, February 23, 1971, Folder FDA Task Force 1971/ Task Force Reprints, Box 3, Account No. 75A0106, RG 442, NARA Atlanta, 1.
51 Ibid., 2.
52 "FDA Task Force on the Use of Antibiotics in Animal Feeds," 15.
53 "Task Force Will Recommend Against Antibiotic Use in Poultry," *Food Chemical News*, November 15, 1971, Folder FDA Task Force 1971, Box 3, Account No. 75A106, RG 442, NARA Atlanta, 1.
54 Ibid., 2. Final conflicts centered on industry-friendly evaluations of AGPs' economic benefits; medical and veterinary members opposed including quantitative evaluations in the final report; "Minority Report from Members of the FDA Task Force on Antibiotics in Animal Feeds," Wilcke to Van Houweling, April 21, 1971; Huber to Van Houweling, January 3, 1972; Saz to Van Houweling, December 21, 1971; Mercer to Van Houweling, December 28, 1971, Luhrs to Van Houweling, December 20, 1971, Folder FDA Task Force 1971, Box 3, Account No. 75A106, RG 442, NARA Atlanta.
55 "FDA Task Force on the Use of Antibiotics in Animal Feeds," 8.
56 Ibid., 9.
57 Ibid., 8.
58 Ibid., 9.
59 Ibid., 10.
60 37 *FedReg*, 2444–2445 (February 1, 1972).
61 "Abstract: Minority Report from Members of the FDA Task Force on Antibiotics in Animal Feeds," February 29, 1972, enclosed in "FDA FactSheet—Summary of the Report by the FDA Task Force," Folder 128 1975 May–July 432.1, Box (FRC) 25, GS, DF UD-WW/Entry 4, RG 88, NARA, 1–2.
62 Edward Press to Commissioner Edwards, March 10, 1972, Folder FDA Task Force 1971, Box 3, Account No. 75A106, RG 442, NARA Atlanta.
63 "Antibiotics in Animal and Poultry Feeds—A Critical Review of Research," 1–2.
64 "Economic Consequences of the Restricted Use of Antibiotics at Subtherapeutic Levels in Broiler and Turkey Production," *USDA—Farm Production Economics Division Working Paper* (November 1972), i–xii.

65 Joan Arehart-Treichel, "Antibiotics to Animal Feeds: Threat to Human Health?" *Science News* 101, no. 22 (1972): 348.

66 Ibid., 349.

67 Ibid.

68 CLM, FP, Series II, A. Alphabetical Correspondence, Box 2, Folder 26, Finland to Van Houweling (June 22, 1970).

69 CLM, FP Series VI, B. Veteran's Administration Committees and Projects Records, 1950–1983, Box 12, Folder 8, Duke Trexler to Finland (June 12, 1972); Finland to Trexler (June 22, 1972).

70 CLM, FP Series VI, B. Veteran's Administration Committees and Projects Records, 1950–1983, Box 12, Folder 8, "Recommendation Submitted for Approval of Ad Hoc Committee," enclosed in Finland to Trexler (October 19, 1972).

71 CLM, FP Series VI, B. Veteran's Administration Committees and Projects Records, 1950–1983, Box 12, Folder 8, H. George Mandel, "Report to the Drug Research Board" (October 18, 1972); W. Kalow to Trexler (November 22, 1972); Kalow to Finland (February 16, 1973).

72 Jukes to Edwards, January 17, 1973, enclosed in Edwards to Jukes, March 2, 1973, Folder 145 432.1 January–March, Box 4820, GS, DF A1/Entry 5, RG 88, NARA, 1–2.

73 CLM, FP Series VI, B. Veteran's Administration Committees and Projects Records, 1950–1983, Box 12, Folder 8, Finland to Jukes (December 3, 1973).

74 38 *Fed Reg*, 9811–9812 (April 20, 1973).

75 Philip Handler to Charles Edwards, August 3, 1972, Folder 128 1975 May–July 432.1, Box (FRC) 25, FDA GS, DF UD-WW/Entry 4, RG 88, NARA.

76 CLM, FP Series VI, B. Veteran's Administration Committees and Projects Records, 1950–1983, Box 12, Folder 8, Trexler to Finland (October 3, 1972); Edwards was acting on an earlier recommendation of the FDA's Commissioner for Planning and Evaluation; Gerald L. Barkdoll to the Deputy Commissioner, June 22, 1972, Folder 128 1975 May–July 432.1, Box (FRC) 25, GS, DF UD-WW/ Entry 4, RG 88, NARA.

77 "Report to the Commissioner of the Food and Drug Administration by the FDA Task Force on the Use of Antibiotics in Animal Feeds," 4–7.

78 John Jennings to the Commissioner, December 18, 1972, Folder 128 1975 May– July 432.1, Box (FRC) 25, GS, DF UD-WW/Entry 4, RG 88, NARA.

79 38 *FedReg*, 9811–9812 (April 20, 1973).

80 Ibid., 9812.

81 Ibid.

82 Ibid., 9813.

83 Ibid., 9812.

84 Ibid; see also Freedman, "Reasonable Certainty," 264–265.

85 Van Houweling to the Commissioner, "Salmonella Reservoir Data. Antibacterials in Animal Feeds—Action," August 28, 1974, Folder 432.1 July–September, Box 4983, GS, DF A1/Entry 5, RG 88, NARA, 2.

86 Gerald Guest, "Memorandum of Conference—Salmonella Reservoir Studies," August 23, 1974, Folder 432.1 July–September, Box 4983, GS, DF A1/Entry 5, RG 88, NARA.

87 Van Houweling to L. Paul Williams, January 30, 1974, Folder 432.1 January– March, Box 4985, GS, DF A1/Entry 5, RG 88, NARA; "Memorandum of Conference—Critique of Working Group Meeting on Antibacterials in Animal

Feeds," March 15, 1974, Folder 128 1975 May–July 432.1, Box (FRC) 25, GS, DF UD-WW/Entry 4, RG 88, NARA.

88 Van Houweling to the Commissioner, "Memorandum—Report of Joint US/ Canadian Fact Finding Visit to the United Kingdom—Action," July 17, 1974, enclosed in Van Houweling to Larry Stenswick, August 23, 1974, Folder 432.1 July–September, Box 4983, GS, DF A1/Entry 5, RG 88, NARA.

89 Van Houweling to the Commissioner, "Salmonella Reservoir Data. Antibacterials in Animal Feeds—Action," August 28, 1974, Folder 432.1 July–September, Box 4983, GS, DF A1/Entry 5, RG 88, NARA, 3.

90 "The Public Health Aspects of Antibiotics in Feedstuffs." Report of a Working Group convened by the Regional Office for Europe of the World Health Organization, Bremen, 1–5 October 1973. Geneva: WHO, 1974; "Statement of Gregory Ahart before the Subcommittee on Interstate and Foreign Commerce on FDA's Regulation of Antibiotics Used in Animal Feeds" (Washington, DC: GAO, 1977), 4.

91 D. Siegel et al., "Continuous Non-therapeutic Use of Antibacterial Drugs in Feed and Drug Resistance of the Gram-Negative Enteric Florae of Food-Producing Animals," *Antimicrobial Agents and Chemotherapy* 6, no. 6 (1974): 697–701.

92 D. Siegel et al., "Human Therapeutic and Agricultural Uses of Antibacterial Drugs and Resistance of the Enteric Flora of Humans," *Antimicrobial Agents and Chemotherapy* 8, no. 5 (1975): 538–543.

93 Special Assistant to the Director (BVM) to Assistant Commissioner for Professional & Consumer Programs, April 21, 1975, Folder 128 1975 May–July 432.1, Box (FRC) 25, GS, DF UD-WW/Entry 4, RG 88, NARA, 1.

94 Ibid., 2.

95 Ibid., 3.

96 Robert C. Wetherell, Jr. to Carl T. Curtis (US Senate), August 19, 1975, Folder 127 1975 432.1 August–December, Box (FRC) 25, GS, DF UD-WW/Entry 4, RG 88, NARA.

97 Gerald Guest to Caro Buckler, "Memorandum of Telephone Conversation— Antibiotics in Animal Feeds Subcommittee," December 6, 1976, Folder 103 1926 431.81–432.24, Box (FRC) 20, GS, DF UD-WW/Entry 15, RG 88, NARA.

98 "For Presentation by Donald Kennedy, Commissioner of FDA to the NAFDC, "Antibiotics Used in Animal Feeds," April 15, 1977, enclosed in David Martin to Dick C. Clark (US Senate), May 2, 1977, Folder 111 1977 432.1 January–May, Box (FRC) 22, GS, DF UD-WW/Entry 8, RG 88, NARA, 2.

99 Van Houweling to Acting Commissioner, March 7, 1977, Folder 111 1977 432.1 January–May, Box (FRC) 22, GS, DF UD-WW/Entry 8, RG 88, NARA, 4–5.

100 "For Presentation by Donald Kennedy, Commissioner of FDA to the NAFDC, "Antibiotics Used in Animal Feeds," April 15, 1977, enclosed in David Martin to Dick C. Clark (US Senate), May 2, 1977, Folder 111 1977 432.1 January–May, Box (FRC) 22, GS, DF UD-WW/Entry 8, RG 88, NARA, 2.

101 Richard Silver to Acting Commissioner, February 7, 1977, enclosed in Folder 111 1977 432.1 January–May, Box (FRC) 22, GS, DF UD-WW/Entry 8, RG 88, NARA, 4.

102 Rosa Gryder to Acting Commissioner, February 7, 1977, enclosed in Folder 111 1977 432.1 January–May, Box (FRC) 22, GS, DF UD-WW/Entry 8, RG 88, NARA, 3.

103 Falkow to Sherwin Gardner, February 14, 1977, Folder 111 1977 432.1 January–May, Box (FRC) 22, GS, DF UD-WW/Entry 8, RG 88, NARA.

104 Sarah A. Vogel, *Is It Safe? BPA and the Struggle to Define the Safety of Chemicals* (Berkeley: University of California Press, 2013), 68–69.

105 Troetel, *Three-Part Disharmony*, 245–246.

106 Richert, *Conservatism, Consumer Choice*, 27–28, 55–59.

107 Garrett Hardin, "The Tragedy of the Commons," *Science* 162, no. 3859 (1968).

108 "For Presentation by Donald Kennedy, Commissioner of FDA to NAFDC, 'Antibiotics Used in Animal Feeds,'" April 15, 1977, enclosed in David Martin to Dick C. Clark (US Senate), May 2, 1977, Folder 111 1977 432.1 January–May, Box (FRC) 22, GS, DF UD-WW/Entry 8, RG 88, NARA, 5.

109 Ibid.

110 "Statement of Gregory Ahard," 11.

111 Van Houweling to Acting Commissioner, March 7, 1977, Folder 111 1977 432.1 January–May, Box (FRC) 22, GS, DF UD-WW/Entry 8, RG 88, NARA, 4.

112 Ibid., 7.

113 Ibid.

114 Ibid.

115 Ibid.

116 Richard Geyer to Edward Allera, June 16, 1977, Folder 109 1977 432.1 June, Box (FRC) 22, GS, DF UD-WW/Entry 8, RG 88, NARA, 6.

117 Charles H. Stuart Harris and David M. Harris, eds., *The Control of Antibiotic Resistant Bacteria* (London: Academic Press, 1982), 196.

118 The FDA complained about misleading activism by R. C. Fish, Acting ARS Administrator for Livestock and Veterinary Sciences; Robert Wetherell to Bill Nichols (House of Representatives), July 29, 1977, Folder 108 432.1 1977 July, Box (FRC) 22, GS, DF UD-WW/Entry 8, RG 88, NARA.

119 See among many others Senator John Tower to Donald Kennedy, May 5, 1977, Folder 111 1977 432.1 January–May, Box (FRC) 22, GS, DF UD-WW/Entry 8, RG 88, NARA; Kennedy to Allan Grant (president AFBF), June 15, 1977, Folder 109 1977 432.1 June, Box (FRC) 22, GS, DF UD-WW/Entry 8, RG 88, NARA.

120 "News from Cyanamid (American Cyanamid Company)—For Immediate Release," April 22, 1977, Folder 109 1977 432.1 June, Box (FRC) 22, GS, DF UD-WW/Entry 8, RG 88, NARA, 3.

121 Ibid., 5.

122 Ibid., 7.

123 Kennedy to James T. Broyhill (House of Representatives), June 21, 1977, Folder 109 1977 432.1 June, Box (FRC) 22, GS, DF UD-WW/Entry 8, RG 88, NARA.

124 Van Houweling to the Commissioner, June 08, 1977, Folder 10 1977 4321.1 6-1-77/6-9-77, Box (FRC) 22, GS, DF UD-WW/Entry 8, RG 88, NARA, 1.

125 Ibid., 2.

126 Ibid.

127 42 *FedReg*, 43769–43793 (August 30, 1977); 42 *FedReg*, 56253–56289 (October 21, 1977).

128 Troetel, *Three-Part Disharmony*, 256–259.

129 Van Houweling to the Commissioner, April 23, 1977, Folder 111 1977 432.1 January–May, Box (FRC) 22, GS, DF UD-WW/Entry 8, RG 88, NARA.

130 Henry C. Gilliam Jr. and Rod Martin, "Economic Importance of Antibiotics in

Feeds to Producers and Consumers of Pork, Beef and Veal," *Journal of Animal Science* 40, no. 6 (1975).

131 Kennedy to Thomas S. Foley (House of Representatives) June 9, 1978 and Office of Planning and Evaluation estimates November 1976, both enclosed in Robert Wetherell to Associate Commissioner for Planning and Evaluation, August 4, 1978, Folder 91 432.1 1978 August–September, Box (FRC) 18, GS, DF UD-WW/ Entry II, RG 88, NARA.

132 "Briefing Paper: FDA Proposal to Restrict the Use of Selected Antibiotics at Subtherapeutic Levels in Animal Feeds" [ERS-662] (Washington, DC: USDA/ ERS, 1977), 3.

133 Ibid., 4–9.

134 J. B. Cordaro to Joseph A. Califano, August 8, 1977, Folder 106 1977 432.1 September, Box (FRC) 21, GS, DF UD-WW/Entry 8, RG 88, NARA, 1; Bob Bergland to Clark Burbee, Robert Brown, Roger Gerrits, Jon Spaulding, and Howard Teague, October 28, 1977, Folder 105 432.1 1977 October–December, Box (FRC) 21, GS, DF UD-WW/Entry 8, RG 88, NARA.

135 Robert Wetherell to John C. Culver, November 1, 1977, Folder 105 432.1 1977 October–December, Box (FRC) 21, GS, DF UD-WW/Entry 8, RG 88, NARA.

136 Van Houweling to the Commissioner, December 7, 1977, 4, Folder 105 432.1 1977 October–December, Box (FRC) 21, GS, DF UD-WW/Entry 8, RG 88, NARA.

137 43 *FedReg*, 3032–3045 (January 20, 1978).

138 G. Brayn Patrick Jr. to Hearing Clerk FDA, April 7, 1978, Folder 93 432.1 1978 April–May, Box (FRC) 18, GS, DF UD-WW/Entry II, RG 88, NARA.

139 US House of Representatives, Committee on Agriculture: Committee Resolution—Relative to the Use of Antibiotics in Animal Feeds, enclosed in Joseph Califano to Thomas Foley, September 7, 1978, Folder 91 432.1 1978 August–September, Box (FRC) 18, GS, DF UD-WW/Entry II, RG 88, NARA, 2.

140 Maryn McKenna, *Big Chicken: The Incredible Story of How Antibiotics Created Modern Agriculture and Changed the Way the World Eats* (Washington, DC: National Geographic, 2017), 121.

141 Van Houweling to the Commissioner, May 18, 1978, Folder 93 432.1 1978 April–May, Box (FRC) 18, GS, DF UD-WW/Entry II, RG 88, NARA, 2.

142 Susan E. Feinman to Robert Wetherell, enclosed in Wetherell to the Commissioner, July 27, 1978, Folder 92 432.1 1978 June–July, Box (FRC) 18, GS, DF UD-WW/Entry II, RG 88, NARA, 1–2.

143 US House of Representatives, Committee on Agriculture: Committee Resolution—Relative to the Use of Antibiotics in Animal Feeds, enclosed in Joseph A. Califano to Thomas S. Foley, September 7, 1978, Folder 91 432.1 1978 August–September, Box (FRC) 18, GS, DF UD-WW/Entry II, RG 88, NARA.

144 Feinman to Robert Wetherell, enclosed in Wetherell to the Commissioner, July 27, 1978, Folder 92 432.1 1978 June–July, Box (FRC) 18, GS, DF UD-WW/ Entry II, RG 88, NARA, 1–2; "Drugs in Livestock Feed," I—Technical Report (Washington, DC: Office of Technology Assessment, 1979), 20.

145 "Drugs in Livestock Feed"; "The Economic Effects of a Prohibition on the Use of Selected Animal Drugs," (Washington, DC: USDA Economics, Statistics and Cooperative Service, 1978).

146 "Effects on Human Health of Subtherapeutic Use of Antimicrobials," vii.

147 Eliot Marshall, "NAS Begins Study of Antibiotics in Feeds," *Science* 205, no. 4410 (1979): 982.
148 "Effects on Human Health of Subtherapeutic Use of Antimicrobials," xv.
149 Ibid.
150 "Antibiotic Resistance," Subcommittee on Investigations and Oversight of the Committee on Science and Technology, US House of Representatives, December 18–19, 1984, 81.
151 "FDA Stymied on Antibiotic Ban," *BioScience* 30, no. 5 (1980): 295–296.
152 "Antibiotics in Animal Feed: Hearings Before the Subcommittee on Health and the Environment on H.R. 7285," Subcommittee on Health and the Environment of the Committee on Interstate and Foreign Commerce, US House of Representatives (Washington, DC: GPO, 1980).
153 Eliot Marshall, "Health Committee Investigates Farm Drugs," *Science* 209, no. 4455 (1980): 481.
154 "Antibiotics in Animal Feed," 344.
155 Podolsky, *Antibiotic Era,* 132–136.
156 "Antibiotics in Animal Feed," 353.
157 Ibid., 427–428, 34. Dingel even sent a letter to *Science*; John D. Dingell, "Animal Feeds: Effect of Antibiotics," *Science* 209, no. 4461 (1980).
158 Gillespie, *Antibiotic Animal Feed Additives,* 280.
159 Richert, *Conservatism, Consumer Choice,* 97–98, 101.
160 Ibid., 83–91, 121.
161 Ibid., 95.
162 Gillespie, *Antibiotic Animal Feed Additives,* 281–287.
163 "Antibiotic Resistance: Hearings before the Subcommittee on Investigations and Oversight," 82; Gillespie, *Antibiotic Animal Feed Additives,* 288–290.
164 Richert, *Conservatism, Consumer Choice,* 116–121, 28–30.
165 "Antibiotic Resistance: Hearings before the Subcommittee on Investigations and Oversight," 83–96.
166 Ibid., 99.
167 Ibid., 107.
168 Ibid., 53; see also the presentation by Virgil Hays, ibid., 29.
169 Gillespie, *Antibiotic Animal Feed Additives,* 291–296.
170 Stuart Levy, *The Antibiotic Paradox: How the Misuse of Antibiotics Destroys Their Curative Powers* (Cambridge, MA: Perseus, 2002), 298–299; see also the description of the run-up to the decision in William A. Moats, ed., *Agricultural Uses of Antibiotics,* American Chemical Society Symposium Series 320 (Washington, DC: American Chemical Society, 1986), 104–109.
171 Committee on Government Operations, "Human Food Safety and the Regulation of Animal Drugs," US House of Representatives, 1985, 2.
172 Ibid., 5.
173 Ibid., 8; see also, Kirchhelle, "Between Bacteriology and Toxicology."
174 Richert, *Conservatism, Consumer Choice,* 167–82; Vogel, *Is It Safe?* 166-167.
175 "FDA's Regulation of Animal Drug Residues in Milk," *Hearing Before the Human Resources and Intergovernmental Relations Subcommittee of the Committee on Government Operations,* US House Of Representatives, February 6, 1990 (Washington, DC: GPO, 1990), 131.
176 Stanley E. Katz and Marietta Sue Brady, "Incidence of Residues in Foods of Animal Origin," in *Analysis of Antibiotic/Drug Residues in Food Products of*

Animal Origin, ed. Vipin K. Agarwal (New York: Plenum, 1992), 18; "Food Safety and Quality; FDA Strategy Needed to Address Animal Drug Residues in Milk" (Washington, DC: GAO, 1992).

177 "Problems with FDA Monitoring for Animal Drug Residues: Is Our Milk Safe?" Human Resources and Intergovernmental Relations Subcommittee of the Committee on Government Operations, US House of Representatives (Washington, DC: GPO, 1992), 69.

178 Ibid., 70.

179 "Food Safety: Risk-Based Inspections and Microbial Monitoring Needed for Meat and Poultry" (Washington, DC: GAO, 1994), 2-13; Susan D. Jones, *Valuing Animals: Veterinarians and Their Patients in Modern America* (Baltimore: Johns Hopkins University Press, 2003), 151.

180 "Food Safety: USDA's Role under the National Residue Program Should Be Reevaluated" (Washington, DC: GAO, 1994), 3.

181 Ibid., 4, 6, 32.

182 "Animal Medicinal Drug Use Clarification Act of 1994 (AMDUCA)," FDA Acts, Rules, and Regulations; see also Title 21, *Code of Federal Regulations*, Part 530 (21 CFR 530).

183 "Animal Drug Availability Act of 1996," FDA Acts, Rules, and Regulations.

184 "Human Health Risks with the Subtherapeutic Use of Penicillin or Tetracyclines in Animal Feed,1988." Institute of Medicine Committee on Human Health Risk Assessment of Using Subtherapeutic Antibiotics in Animal Feeds, 1989, iv-vi. A parallel CAST report pointed to antibiotics' benefits: Virgil W. Hays and Charles A. Black, *Antibiotics for Animals: The Antibiotic Resistance Issue: Comments from CAST* (Ames, IA: Council for Agricultural Science and Technology [CAST], 1989).

185 "Impacts of Antibiotic-Resistant Bacteria", 2.

186 Ibid., 156.

187 Ibid., 159–162, 164–165.

188 "Food Safety: The Agricultural Use of Antibiotics and Its Implications for Human Health," (Washington, DC: GAO, 1999), 11–12.

189 Donna U. Vogt and Brian A. Jackson, "Antimicrobial Resistance: An Emerging Public Health Issue," in *CRS Report for Congress*, Congressional Research Service, Library of Congress, January 24, 2001, CRS 24-25.

190 "Food Safety: The Agricultural Use of Antibiotics."

191 Ibid., 7; Geoffrey S. Becker, "Antibiotic Use in Agriculture: Background and Legislation," in *CRS Report for Congress*, Congressional Research Service, July 30, 2009, 7.

192 Stacy Sneeringer et al., "Economics of Antibiotic Use in US Livestock Production," *USDA Economic Research Report* 200 (2015): 11; Laura H. Kahn, *One Health and the Politics of Antimicrobial Resistance* (Baltimore: Johns Hopkins University Press, 2016), 56–57.

193 Matthew B. Brooks et al., "Survey of Antimicrobial Susceptibility Testing Practices of Veterinary Diagnostic Laboratories in the US," *JAVMA* 222, no. 2 (2003): 169; Kahn, *One Health*, 57.

194 "Food Safety: The Agricultural Use of Antibiotics," 12; McKenna, *Big Chicken*, 146–147.

195 Geoffrey S. Becker, "Animal Agriculture: Selected Issues in the 108th Congress," in *CRS Report for Congress*, Library of Congress, October 15, 2003, CRS 25-26; "H.R. 1549, Preservation of Antibiotics for Medical Treatment Act (PAMTA),"

Committee on Rules, US House of Representatives, July 13, 2009 (Washington, DC: GPO, 2009), 12; Vogt and Jackson, "Antimicrobial Resistance," CRS 1-2.

196 "Guidance for Industry #152: Evaluating the Safety of Antimicrobial New Animal Drugs with Regard to Their Microbiological Effects on Bacteria of Human Health Concern," FDA, 2003.

197 "Antibiotic Resistance: Federal Agencies Need to Better Focus Efforts to Address Risk to Humans from Antibiotic Use in Animals" (Washington, DC: GAO, 2004), 6–8.

198 PAMTA had originally been introduced in 2003/2004.

199 "Putting Meat on the Table: Industrial Farm Animal Production in America," PEW Commission on Industrial Farm Animal Production, 2008, 61–67.

200 Becker, "Antibiotic Use," 5.

201 Ibid., 1–2.

202 "Hearing to Review the Advances of Animal Health within the Livestock Industry," Subcommittee on Livestock, Dairy, and Poultry of the Committee on Agriculture, US House of Representatives, September 25, 2008 (Washington, DC: GPO, 2009), 24, 17–32.

203 "H.R. 1549, Preservation of Antibiotics for Medical Treatment Act (PAMTA)," 9.

204 Ibid.

205 Ibid.

206 Ibid., 10.

207 "Antibiotic Resistance and the Use of Antibiotics in Animal Agriculture," Subcommittee on Health of the Committee on Energy and Commerce, US House of Representatives, July 14, 2010 (Washington, DC: GPO, 2013), 28–42.

208 Sneeringer et al., "Economics of Antibiotic Use," 13.

209 "Guidance for Industry #209: The Judicious Use of Medically Important Antimicrobial Drugs in Food-Producing Animals," FDA, April 2012.

210 Ibid., 21–22.

211 "Guidance for Industry #213: New Animal Drugs and New Animal Drug Combination Products Administered in or on Medicated Feed or Drinking Water of Food-Producing Animals: Recommendations for Drug Sponsors for Voluntarily Aligning Product Use Conditions with GFI #209," FDA, December 2013, 5.

212 Ibid., 9.

213 Sneeringer et al., "Economics of Antibiotic Use," 13; Melinda Wenner Moyer, "The Looming Threat of Factory-Farm Superbugs," *SciAm*, December 2016, 76.

214 Margaret Hamburg, "FDA Strategies for Combating Antimicrobial Resistance," *Speeches by FDA Officials*.

215 Brian Grow and P. J. Huffstutter, "US Lawmakers Want to Curb Antibiotic Use on Farms," *Reuters*, September 16, 2014.

216 "Critical Antibiotics Still Used on US Farm Animals Despite Superbug Crisis," *Bureau of Investigative Journalism*, September 19, 2018.

217 "Three Senators Have Questions for Interagency Antibiotics Task Force," *Food Safety News*, December 17, 2014.

218 Kahn, *One Health*, 61.

219 Moyer, "Factory-Farm Superbugs," *SciAm*, December 2016, 77–78.

220 2017 Summary Report on Antimicrobials Sold or Distributed for Use in Food-Producing Animals" (Washington, DC: FDA/DHSS, 2018), 12, 19, 28, 29.

221 "Warning of Pig Zero", *NYT*, July 6, 2019.

11. Between Swann Patriotism and BSE: Antibiotics in the Public Sphere

1 Adam Lent, *British Social Movements Since 1945: Sex, Colour, Peace and Power* (Basingstoke, UK: Palgrave, 2001), 97; Meredith Veldman, *Fantasy, the Bomb and the Greening of Britain: Romantic Protest, 1945–1980* (Cambridge, UK: Cambridge University Press, 1994), 205.

2 Lent, *British Social Movements*, 100, 103.

3 "Critics of Farm Antibiotic Curbs Are 'Alarmist,'" *Financial Times* [hereafter *FT*], January 22, 1970; "Curbs on Farm Feed Antibiotics," *FT*, August 21, 1970, 4; Wayland Young, "Pollution: A 'Guardian' Special Report," *Guardian*, October 6, 1970, 14; Alan Long, "Food," *Observer*, July 19, 1970, 8; John Winter, "Scientists," *Daily Mail* [hereafter *DaMa*], January 26, 1971, 6; Bernard Dixon, "Drug Resistance," *Spectator*, October 7, 1972, 563; "Danes Act on Farm Antibiotics," *FT*, November 25, 1971, 4.

4 Frank Uekötter, *The Greenest Nation: A New History of German Environmentalism* (Cambridge, MA: MIT Press, 2014), 2–3, 6.

5 "Antibiotics: Farm Control Needed," *Times*, March 3, 1972, 14; "The Drugs Doctors Are Reluctant to Prescribe," *Times*, February 2, 1972, 8; "Bacteriology: Antibiotic Resistance," *Times*, May 30, 1974, 18; "Public Health: Resistant Bacteria," *Times*, September 2, 1974, 14.

6 "Antibiotics: Resistant Typhoid Strain," *Times*, August 4, 1972, 14; E. S. Anderson and H. R. Smith, "Chloramphenicol Resistance in the Typhoid Bacillus," *BMJ* 3, no. 5822 (1972): 329–331.

7 "Curb on Use of an Antibiotic Is Questioned," *Times*, August 1, 1974, 2; "Advertisement—Hindustan Antibiotics LTD," *Times*, November 30, 1970, XVIII; "Advertisement—Lederle," September 22, 1972, IV.

8 H. Williams Smith, "Persistence of Tetracycline Resistance in Pig E. Coli," *Nature* 258, no. 5536 (1975): 628–630.

9 C. L. Hartley and M. H. Richmond, "Antibiotic Resistance and Survival of E Coli in the Alimentary Tract," *BMJ* 4, no. 5988 (1975): 71.

10 R. Braude, "Antibiotics in Animal Feeds in Great Britain," *Journal of Animal Science* 46 (1978): 1434.

11 Ibid., 1430.

12 J. Bower, "Letter—Antibiotics in Poultry Feed," *FT*, March 14, 1973, 2; "Farmers Join Drug Racket," *DaMa*, October 15, 1971, 2.

13 Josée Doyère, "Sound Guarantees Needed on Additives and Hygiene," *Times*, November 7, 1974, VI.

14 Hugh Clayton, "Report Says Drugs Given to Cows May Be in Milk," *Times*, February 13, 1975, 3; see also residue allusions in Michael Denny, "The Milk of Kindness," *Observer (Sunday Plus)*, May 27, 1979, 40.

15 Mieke Roscher, *Ein Königreich für Tiere: Die Geschichte der britischen Tierrechtsbewegung* (Marburg: Tectum Verlag, 2009), 267, 294–297, 364–365.

16 "Animal Lovers Keep Their Eyes on Farmers," *Times*, November 17, 1975, 18; "This Is the Cruel Price of Breakfast," *DaMa*, July 26, 1979, 6.

17 "'Who Needs a Menu with a Handbag Like This?' *DaMa*, April 16, 1970, 4; "Harriet Crawley in America," *DaMa*, December 9, 1970, 6.

18 *Good Housekeeping Wholefoods Cook Book* (London: Ebury Press, 1971).

19 Vivien Quick and Clifford Quick, *Everywoman's Wholefood Cook Book* (Wellingborough, UK: Thorsons, 1974), 20; Peter Deadman and Karren Betteridge,

Nature's Foods (London: Rider, 1977), 66; Brenda O'Casey, *Natural Baby Food* (Duckworth: London, 1977), 28; Ursula M. Cavanagh, *Cooking and Catering the Wholefood Way* (London: Faber and Faber, 1970), 7; Ursula M. Cavanagh, *The Wholefood Cookery Book* (London: Faber and Faber, 1971).

20 Doris Grant, *Your Daily Food: Recipe for Survival* (London: Faber and Faber, 1973), 14, 27.

21 *The Eco-Cookbook*, 2nd ed. (Southampton, UK: Friends of the Earth, 1975), 5.

22 "Should This Little Piggy Go to Market?" *Radio Times*, May 5–11, 1979, cover page.

23 "Brass Tacks: It Shouldn't Happen to a Pig," British Film Institute National Archive.

24 Ibid.

25 "Furious Farmers Ready for Drugs Phone-In," *Guardian*, May 8, 1979, 2.

26 Anthony Tucker, "Illicit Drug Sales to Farmers Pose Threat to Public Health," *Guardian*, August 9, 1979, 2.

27 Patrick Marnham, "Postscript—Poisoned Eggs," *Spectator*, July 5, 1980, 27.

28 Denise Winn, "Scandal of Illegal Farm Drugs," *Daily Mirror* [hereafter *DM*], January 11, 1983, 8.

29 Rosemary Collins, "Dairy Farmers Overdo Quotas," *Guardian*, November 30, 1984, 6; see also "Pressure to Curb Irish Farm Drugs," *Observer*, May 2, 1982, 2.

30 Andrew Veitch, "Cattle Drug Ring Broken in Raids," *Guardian*, July 26, 1985, 28.

31 "Animal Drugs Ring Exposed," *Times*, October 29, 1985, 3; "Hunt for Animal Drugs Widens," *Guardian*, July 27, 1985, 2.

32 R. E. O. Williams, "Controlling Antibiotic Resistance without Eschewing Antibiotics," in *The Control of Antibiotic-Resistant Bacteria*, ed. Charles H. Stuart-Harris and David M. Harris (London: Academic Press, 1982), 232–234.

33 E. J. Threlfall et al., "Plasmid-Encoded Trimethoprim Resistance in Multiresistant Epidemic Salmonella Typhimurium Phage Types 204 and 193 in Britain," *BMJ* 280, no. 6225 (1980): 1211.

34 "Why Has Swann Failed?" *BMJ* 280, no. 6225 (1980): 1195.

35 H. Williams-Smith, "Why Has Swann Failed?" *BMJ* 280, no. 6230 (1980): 1537.

36 Hugh Clayton, "Britain Is 'Threatened by Food Super Germ,'" *Times*, October 28, 1981, 3.

37 Ibid.

38 "Salmonella Blamed on Antibiotics," *Times*, September 13, 1984, 3.

39 "'Death Risk' from Food in Hospitals," *Telegraph*, May 6, 1981, 16.

40 Simon Kinnersley, "Scandal of the Phoney Meat Merchants!" *DaMa*, January 28, 1981, 6.

41 James Erlichman, "UK Makes Mincemeat of Ban on Hormones," *Guardian*, November 15, 1985, 21.

42 Jan Walsh, *The Meat Machine* (London: Columbus, 1986), 10.

43 Michael Dineen, "The Answer Lies in Preserving the Soil," *Observer*, November 1, 1981, 21.

44 Natalie Graham, "The Food Revolution," *DaMa*, October 24, 1984, 23. Later *DaMa* reporting on organic food was mixed: "Health-Hype Fears of 'Organic Britain,'" *DaMa*, June 1, 1990, 19.

45 Sheila Dillon, "Real Meat, Real Money," *Guardian*, November 7, 1986, 19; "Bringing Home the Bacon: Good Food Guide," *Guardian*, November 15, 1985, 19.

46 David Canter, Kay Canter, and Daphne Swann, *The Cranks Recipe Book* (London:

J. M. Dent, 1982); Margaret Hanford, *The WI Book of Wholefood Cookery* (London: WI Books, 1985).

47 Gail Duff, *Good Housekeeping Wholefood Cookery* (London: Ebury Press, 1989), 6–7.

48 Nigel Dempster, "£200-a-Head Dinner," *DaMa*, September 23, 1987, 17; "Prince Charles's Ideal Home," *DaMa*, June 1, 1989, 23; "Greens for PM," *DaMa*, April 4, 1990, 22.

49 "The Green Guarantee for Organic Produce," *DaMa*, May 3, 1989, 18; "We'll Pay Price to Green, Say Shoppers," *DaMa*, June 13, 1989, 17.

50 John Young, "Malthus No: Malnutrition Yes," *Times*, November 20, 1985, 14.

51 Ibid.; see also John Ruxin, "The United Nations Protein Advisory Group," in *Food, Science, Policy and Regulation in the Twentieth Century: International and Comparative Perspectives*, ed. David F. Smith and Jim Philips (London: Routledge, 2000).

52 Libby Purves, "The Case for an Alternative Cure," *Times*, May 16, 1986, 9; John Young, "Bacteria Is Claimed to Aid Growth in Animals," *Times*, June 14, 1986, 15.

53 Denise Winn, "One Man's Meat May Be Everyone's Poison," *Times*, July 28, 1986, 11.

54 Peter Cox, *Why You Don't Need Meat* (Wellingborough, UK: Thorsons, 1986), 110.

55 "From Wonder Drug to Bitter Pill," *Times*, March 2, 1987, 10.

56 George Hill et al., "The Bitter Harvest," *Times*, March 4, 1987, 12.

57 James Erlichman, "Superbug Risk from Chemical Use in Milk," *Guardian*, April 3, 1987, 1, back page.

58 James Erlichman, "Cover-Up that May Help the Superbugs," *Guardian*, April 7, 1987, 29.

59 James Erlichman, "Drug Traces Found in Abattoir Carcasses," *Guardian*, January 18, 1988, 4.

60 James Erlichman, "Public 'Kept in Dark on Additives,'" *Guardian*, January 28, 1988, 2. By the 1990s, both supermarkets had reverted to conventionally produced meat; "Supermarkets Move to Ban Growth-Drug Meat from Shelves," *Guardian*, April 24, 1998, 4.

61 Jan Walsh, "Recipe for Danger," *DM*, February 4, 1988, 6.

62 "Hen Cull Could Halt Salmonella," *Guardian*, December 5, 1988, 24.

63 Ibid.

64 "Salmonella Eggs 'Kill One a Week,'" *Guardian*, December 19, 1988, 1.

65 "Poultry Purge Fails to Halt Tide of Salmonella," *Telegraph*, January 31, 1990, 2.

66 "When Salmonella Struck My Family," *Telegraph*, December 20, 1988, 11.

67 Colin Welch, "Scrambled Ideas," *DaMa*, December 20, 1988, 9.

68 Smith et al., *Food Poisoning, Policy and Politics: Corned Beef and Typhoid in Britain in the 1960s* (Woodbridge, UK: Boydell, 2005), 303–304; "How We Reported the Salmonella Crisis 25 Years Ago," *Farmers Weekly*, December 3, 2013.

69 "The Microbiological Safety of Food, Part II," (London: Committee on the Microbiological Safety of Food [Richmond Committee], 1990), 6.

70 "Pig Sick," *DM*, February 13, 1989, 1.

71 Peter Pallot, "Food Poisoning 'Out of Control,'" *Telegraph*, April 13, 1989, 1.

72 Frank Thorne and Anna Treacher, "Not Even Fit for Our Pigs," *DM*, February 13, 1989, 5.

73 "Cages of Cruelty," *DM*, February 13, 1989, 2.

74 Ibid.

75 "Poultry Purge Fails," 2; "32pc Increase in Food Poisoning from Poultry," *Telegraph*, January 30, 1991, 6.

76 "17pc Fall in Salmonella Cases Fails to End Worry," *Telegraph*, July 16, 1991, 7.

77 S. Poser, I. Zerr, and K. Felgenhauer, "Die Neue Variante Der Creutzfeldt-Jakob-Krankheit," *Deutsche Medizinische Wochenschrift* 127 (2002): 331–334.

78 Mark Harrison, *Contagion: How Commerce Has Spread Disease* (New Haven, CT: Yale University Press, 2012), 249.

79 Keir Waddington, "Mad and Coughing Cows: Bovine Tuberculosis, BSE and Health in Twentieth-Century Britain," in *Meat, Medicine and Human Health in the Twentieth Century*, ed. David Cantor, Christian Bonah, and Matthias Dörries (London: Pickering and Chatto, 2010), 160–161; "The Report of the Expert Group on Animal Feedingstuffs to the Minister of Agriculture, Fisheries and Food, the Secretary of State for Health and the Secretaries of State for Wales, Scotland and Northern Ireland (Lamming Report)," 1992, 5–15.

80 For my use of moral panic, see Nicolas Rasmussen, "Goofball Panic: Barbiturates, 'Dangerous' and Addictive Drugs, and the Regulation of Medicine in Postwar America," in *Writing, Filing, Using, and Abusing the Prescription in Modern America*, ed. Jeremy A. Greene and Elizabeth Siegel Watkins (Baltimore: Johns Hopkins University Press, 2012), 25.

81 Lucy Ellmann, "Off We All Traipse Like Those Mad Cows to the Slaughter," *Guardian*, February 5, 1990, 20.

82 "Man Is the Real Beast of the Field," *Observer*, July 10, 1994, D2; Tim Radford, "Old Enemies—Bacteria," *Guardian*, May 25, 1994, 2; Nigel Hawkes, "Bacteria that Eat the Flesh," *Times*, May 24, 1994, 15; "Small Epidemic, Not Many Dead," *Telegraph*, May 29, 1994, 24; Bernard Dixon, "We Say Tomato, They Say Flavr Savr," *Guardian*, May 21, 1994, 24; Polly Ghazi, "Fried Gene Tomatoes," *Observer*, September 25, 1994, D68; Javier Lezaun, "Genetically Modified Foods and Consumer Mobilization in the UK," *Technikfolgenabschätzung—Theorie und Praxis* 13, no. 3 (2004).

83 Alison Johnson, "Kind Food," *Times*, January 18, 1992, 44.

84 Thomas Stuttaford, "It's the Same Old Story with the 'Killer Bug,'" *Times*, May 31, 1994, 15.

85 Robert Bud, *Penicillin: Triumph and Tragedy* (Oxford: Oxford University Press, 2009), 199–201.

86 Javier Lezaun and Martijn Groenleer, "Food Control Emergencies and the Territorialization of the European Union," *European Integration* 28, no. 5 (2006): 439–440.

87 Patrick Holden, "Sacrificed on the Hi-Tech Altar," *Guardian*, March 27, 1996, 4.

88 Ibid.

89 Judy Jones and Anthony Bevins, "And What Food Is Safe?" *Observer*, March 31, 1996, 17.

90 Andrew Penman, "Cow that Proves We've Got Mad Farming Disease," *DM*, August 21, 1996, 6.

91 "You Can't Put a Price on Pets," *DaMa*, July 7, 1995, 22–23.

92 "Straining the Limits of Nature," *DaMa*, March 22, 1996, 8; see also "The Disturbing Questions You Won't Answer," *DaMa*, March 26, 1996, 8.

93 Clive Aslet, "How to Rescue British Beef," *Times*, May 3, 1996, 16.

94 Nigel Hawkes, "Scientists Fear 'Ominous' Spread of Mutant Bacteria," *Times*,

November 29, 1996, 4; Helen Nowicka, "World Warning over Antibiotics," *Guardian*, October 14, 1996, 3; "Beware: Mother's Little Helper Is Defecting," *Observer*, December 8, 1996, 21.

95 "EU to Ban Antibiotic in Feeds," *FT*, December 20, 1996, 20.

96 Laura H. Kahn, *One Health and the Politics of Antimicrobial Resistance* (Baltimore: Johns Hopkins University Press, 2016), 26–32.

97 Bud, *Penicillin*, 205.

98 Annika Ahnberg and Karl Erik Olsson, "Letter: EU Must Act to Allay Consumers' Fear over Food," *FT*, June 8, 1996, 6; see also "Letter: Farming Lessons," *DaMa*, June 28, 1996, 53.

99 "New Labour—Because Britain Deserves Better," 1997 Labour Party Manifesto, http://www.labour-party.org.uk/manifestos/1997/1997-labour-manifesto.shtml.

100 "Antibiotic Resistance: The Risk to Human Health and Safety from the Use of Antibiotics in Animal Production (CEG 98/2)," Consumers in Europe Group, 1998, 11.

101 "One in Ten Patients Infected in Hospital," *Telegraph*, May 16, 1997, 4; "Resisting Resistance," *Economist*, May 31, 1997, 99; "Treat All Raw Meat as a Danger, Say Doctors," *DaMa*, January 12, 1998, 25; Geoffrey Cannon, "Could Going into Hospital Be the Greatest Risk You Will Ever Face?" *DaMa*, April 24, 1998, 9.

102 Some Commentators Also Criticized the Reports: "Why the Superbug Isn't the End of the World," *Telegraph*, April 26, 1998, 37; Daniel Green, "Killer Bugs: Round Two," *FT*, April 25, 1998, 9.

103 Michael Hornsby, "Pig Feed Rules May Ban Use of Animal Remains," *Times*, November 13, 1998, 7; Michael Smith, "EU to Act on Antibiotics," *FT*, October 20, 1998, 7; "EU Bans Antibiotics in Fear of Superbugs," *DaMa*, December 15, 1998, 18; "Brussels Bans Antibiotics in Animal Feed," *FT*, December 15, 1998, 3.

104 "'Yoghurt' Medicine Tackles the Scourge of Calves," *FT*, November 24, 1998, 18.

105 "Europe 'in Science Dark Age,'" *FT*, November 8, 1999, 8.

106 "Planet Organic," *Observer*, June 27, 1999, G14; see also Frances Bissell, *The Organic Meat Cookbook* (London: Ebury, 1999), 23–25.

107 "World Markets for Organic Fruit and Vegetables," http://www.fao.org/docrep /004/y1669e/y1669e0f.htm, accessed June 17, 2019.

108 Libby Purves, "Called to Ordure," *Times*, October 26, 1999, 24.

109 Ibid.

110 "Supermarkets Move to Ban Growth-Drug Meat from Shelves," *Guardian*, April 24, 1998, 4.

111 "Chicken Firm Axes Growth Promoter," *Times*, September 2, 1999, 10. Industry's 2000 Assured Chicken Production scheme (Little Red Tractor Logo) loosened AGP bans in 2002 to allow preventive doses of growth promoters under veterinary supervision; members accounted for 85 percent of British poultry production in 2003. Andrew Purvis, "If Max Eats Up All His Chicken, He'll Grow to Be a Big, Strong Boy," *Observer*, August 10, 2003, F25.

112 James Meikle, "Shock at Food Drugs Ban," *Guardian*, September 2, 1999, 1.

113 James Meikle, "M&S Phases Out Antibiotics in Chicken," *Guardian*, November 19, 1999, 6; "Organic Sector Moves into Mainstream," *Marketing*, April 29, 1999, 14.

114 "Antibiotic Ban Upheld," *FT*, July 2, 1999, 2.

115 "Police Probe Irish Farm Drugs Racket," *Sunday Times*, July 4, 1999.

116 Ibid.; see also James Meikle, "Crackdown on Animal Drugs Scam," *Guardian*, September 7, 1999, 6.

117 "Food Safety and the Price of Human Folly," *DaMa*, August 19, 1999, 10; see also "Antibiotics Under Fire in Farming," *FT*, February 19, 1999, 28.

118 "Poultry Alert over Super-Salmonella," *DaMa*, January 3, 2000, 20.

119 David Pilling, "WHO Warns on Dangers in the Misuse of Antibiotics," *FT*, June 13, 2000.

120 Graham Harvey, "Good Food Needs Green Farms," *Times*, January 1, 2000, 16.

121 "Food for Life: Campaigning for Local and Organic School Meals and Food Education" (Bristol, UK: Soil Association); Tracey Harrison, "So Did You Enjoy Your Antibiotic Additive with Chemical Trimming?" *DM*, December 26, 2000, 6; "Organic Bites Back," *DaMa*, August 7, 2001, 33; "Jamie's Dinners Prime Cuts," *DaMa*, October 23, 2004, 90.

122 Stuart Jeffries, "True-Blue Green," *Guardian*, December 3, 2005, 31.

123 "Plan for Total Ban on Antibiotics in Animal Feed," *FT*, March 26, 2002, 7. The move was formally approved in 2003; (EC) No. 1831/2003.

124 "Danger Warning After Increase in Drug Residues Found in Eggs," *Guardian*, April 14, 2004, 2.

125 "Now Let's All Buy British," *DaMa*, October 25, 1999, 1; wider BSE boycotts had officially been lifted earlier that year.

126 George Parker, "Threat to Block Food Imports," *FT*, April 28, 1999, 8; "Will Britain Be the Loser in the Safe Food Race?" *DaMa*, April 13, 1999, 10.

127 Sue Bradford, "Free to Be Poisoned," *New Statesman*, May 12, 2008, 19; Liz Else, "The True Cost of Meat," *New Scientist* 183, no. 2460 (2004): 42–45; "Imported Foods Could Be Responsible for Growth of Superbugs," *Guardian*, September 12, 2005, 5; "A Bitter Taste of Honey," *Guardian*, July 21, 2004, 16.

128 Malcolm Dean, "Staring into the Abyss," *Guardian*, November 22, 2000, 7.

129 Sean Poulter, "Chickens Back on the Rooster Boosters," *DaMa*, May 28, 2003, 35.

130 Purvis, "If Max Eats Up All His Chicken."

131 Ibid.

132 Ibid.

133 "Riding Piggybank," *Economist*, December 1, 2007, 92–93.

134 "Chemical World," *Guardian*, May 15, 2004, Special Supplement, 32.

135 "European Parliament Approves Curbs on Use of Antibiotics on Farm Animals," *Guardian*, October 25, 2018.

12. Persistent Infrastructures: Antibiotic Reform and British Farming

1 Michael Winter, *Rural Politics: Policies for Agriculture, Forestry and the Environment* (London: Routledge, 1996), 114–128.

2 "Europe: Meat Output," in *International Historical Statistics*, vol. 3 (Basingstoke, UK: Palgrave Macmillan, 2013).

3 Berkeley Hill, *Farm Incomes, Wealth and Agricultural Policy: Filling the Cap's Core Information Gap* (Wallingford, UK: CABI, 2012), 119.

4 John Martin, *The Development of Modern Agriculture: British Farming Since 1931* (London: Macmillan, 2000), 103, 109; 139–145; B. A. Holderness, *British Agriculture Since 1945* (Manchester, UK: Manchester University Press, 1985), 37–40.

5 "Notes of Meeting with HC Mason (NFU), February 2, 1971," Folder Task Force Antibiotics, Box 3, Account No. 75A0106, RG 442, NARA Atlanta, 2.

6 T. A. B. Corley and Andrew Godley, "The Veterinary Medicines Industry in Britain, 1900–2000," *Economic History Review* 64 (2011): 838, 847.

7 R. Braude, "Antibiotics in Animal Feeds in Great Britain," *Journal of Animal Science* 46 (1978): 1429–1431.

8 P. St. John Howe, "The Illegal Sale of Veterinary Antibiotics," *Veterinary Record* [hereafter *VR*], November 27, 1971, 587.

9 Ibid.

10 Ibid., 588.

11 Ibid.

12 *The TV Vet Book for Pig Farmers*, 2nd ed. (Ipswich, UK: Farming Press, 1971), 30–34, 38–39; *Modern Poultry Keeping*, 2nd ed. (London: Teach Yourself, 1972), 89, 124, 135–147; *The Agricultural Notebook*, 16th ed. (London: Newnes-Butterworths, 1976), 601, 608–609, 613–617, 643–645, 672–673, 695, 698–704; M. Buckett, *Introduction to Livestock Husbandry*, 2nd ed. (Oxford: Pergamon, 1977), 31–32, 69, 90–91; Elisabeth Downing, *Keeping Pigs* (London: Pelham, 1978), 76; Stuart Banks, *The Complete Handbook of Poultrykeeping* (London: Ward Lock, 1979), 137–138, 148–151, 179–184.

13 Abigail Woods, "Science, Disease and Dairy Production in Britain," *Agricultural History Review* 62, no. II (2007): 307.

14 G. C. Brander, "Clinical Problems in Preventive Medicine: The Control of Mastitis," *British Veterinary Journal* [hereafter *BVJ*] 128 (1972): 58; see also J. K. L. Pearson, "Intramammary Therapy: Its Achievements And Limitations," in *Antibiotics and Antibiosis in Agriculture*, ed. Malcolm Woodbine (London: Butterworths, 1977), 217–228.

15 "Mastitis," *FW*, January 17, 1975, 100.

16 Museum of English Rural Life, T 72/54, May & Baker, brochures: Emtryl, Saquadil, Rovamcin (1972); TR BAP/P2/A23-24 Beecham Agriculture Products, brochures: Vitamealo (1970).

17 "Advertisement—Cyanamid," *Farmers Weekly* [hereafter *FW*], February 19, 1971, 52.

18 Joan Smith, "Health Ban on Chicken Sales," *Daily Mirror* [hereafter *DM*], May 8, 1973, 3.

19 Dennis Barker, "Producers Predict £1 a Dozen Eggs," *Guardian*, August 5, 1974, 1.

20 "New Rules Wanted for Egg Imports," *FW*, January 31, 1975, 35.

21 J. W. Murray, "A Crack at French Eggs," *Observer*, February 2, 1975, 1.

22 Woods, "Science, Disease and Dairy Production," 307–308; "Mastitis. We can do much better than this," *British Farmer and Stock Breeder* [hereafter *BFS*], January 4, 1975, 19; "Milk Hygiene. Super Dairymen Are Needed," *BFS*, March 29, 1975, 25.

23 "The Little Extra in Farm Feeds: Additives and Antibiotics," *Country Life* [hereafter *CL*], January 19, 1978, 158.

24 Ibid., 159.

25 G. C. Brander, "The Use of Antibiotics in the Veterinary Field in the 1970s," in *Antibiotics and Antibiosis in Agriculture*, ed. Malcolm Woodbine (London: Butterworths, 1977), 199–209.

26 Abigail Woods, "Is Prevention Better than Cure? The Rise and Fall of Veterinary Preventive Medicine, c. 1950–1980," *Social History of Medicine* 26, no. 1 (2012): 126–129.

27 B. Martin, "Farm Animal Disease: Veterinarian and Antibiotics," in *Antibiotics and Antibiosis in Agriculture*, ed. Malcolm Woodbine (London: Butterworths, 1977), 211.

28 See, for example, ibid., 214–216.

29 A. H. Linton, "Antibiotics and Man—An Appraisal of a Contentious Subject," in *Antibiotics and Antibiosis in Agriculture*, ed. Malcolm Woodbine (London: Butterworths, 1977), 325–334.

30 Ibid., 337.

31 Ibid., 339.

32 H. Williams Smith, "Antibiotic Resistance in Bacteria and Associated Problems in Farm Animals Before and After the 1969 Swann Report," in *Antibiotics and Antibiosis in Agriculture*, ed. Malcolm Woodbine (London: Butterworths, 1977), 346–349.

33 Ibid., 350–354.

34 H. F. Marks and D. K. Britton, *A Hundred Years of British Food and Farming: A Statistical Survey* (London: Taylor and Francis, 1989), 81–83; Martin, *Development of Modern Agriculture*, 127, 131.

35 Woods, "Science, Disease and Dairy Production," 298.

36 Marks and Britton, *Hundred Years*, 62–66.

37 Frank Sykes, "Pig Housing: A Fresh Approach," *CL*, September 2, 1976, 628; Abigail Woods, "Rethinking the History of Modern Agriculture: British Pig Production, c. 1910–65," *Twentieth Century British History* 23, no. 2 (2012): 191; Paul Brassley, "Cutting Across Nature? The History of Artificial Insemination in Pigs in the United Kingdom," *Studies in History and Philosophy of Biological and Biomedical Sciences* 38 (2007): 459; *The Agricultural Notebook*, 17th ed. (London: Butterworth Scientific, 1982), 392–394.

38 "Look Ahead in Search of a Road Out of Farming's Crisis," *FW*, January 17, 1975, 72; "Goodbye, Chemicals—Hello, Good Husbandry," *FW*, January 31, 1975, 77.

39 Martin, *Development of Modern Agriculture*, 149.

40 "Anger as Irish Exploit Their EEC Advantage," *BFS*, October 26, 1974, 7; "Hundreds Join Port Pickets," *BFS*, November 9, 1974, 8–9; "Caricature—In Germany the Small Farmer Is Still King," *BFS*, September 6, 1980, 6.

41 "A Caning from Their Lordships," *BFS*, April 26, 1980, 10.

42 Winter, *Rural Politics*, 127, 138–139, 239.

43 "School Meal Abolition Angers Farm Workers," *BFS*, August 16, 1980, 27.

44 Martin, *Development of Modern Agriculture*, 138–144; Marks and Britton, *Hundred Years*, 28.

45 G. H. Yeoman, "Review of Veterinary Practice in Relation to the Use of Antibiotics," in *The Control of Antibiotic-Resistant Bacteria*, ed. Charles H. Stuart-Harris and David M. Harris (London: Academic Press, 1982), 60.

46 Ibid.

47 Ibid., 61.

48 Ibid.

49 Ibid., 62.

50 Ibid., 64–65; for similar trends in dairy production, see Woods, "Science, Disease and Dairy Production," 308–309.

51 "NFU Annual Meeting: Issues of Concern," *BFS*, March 1, 1980, 25; Monty Keen, "Bitching About the Countryside," *BFS*, November, 22, 1980, 41.

52 Richard Norton-Taylor, "Furious Farmers Ready for Drugs Phone-In," *Guardian*, May 8, 1979, 2.

53 "Not Pure Science Fiction," *VR*, June 9, 1979.

54 Ibid.; see also "Salmonellosis—An Unhappy Turn of Events," *Lancet* 313, no. 8124 (1979): 1009–1010.

55 "Conflict Over Antibiotics," *BFS*, October 4, 1980, 14.

56 Ibid.; J. R. Walton, "Antimicrobials and Agriculture," *BVJ* 142 (1986): 198–199.

57 *The UK Animal Health Products Market: A Status Report* (Richmond, UK: V&O, 1984), 38.

58 Ibid., 32.

59 Veterinarians had consistently opposed PML sales: "Government Changes Plans for Animal Medicines," *VR*, March 27, 1976, 262; "Animal Medicines in the Market Place," *VR*, November 6, 1976; "Stricter Controls over Medicine," *BFS*, February 2, 1980, 25.

60 H. S. Harrison, *The Law on Medicines—A Comprehensive Guide*, vol. 1 (Lancaster, UK: MTP Press, 1986), 163–166.

61 *The UK Animal Health Products Market*, 30–32.

62 Yeoman, "Review of Veterinary Practice"—Discussion, 70.

63 *The UK Animal Health Products Market*, 21–33.

64 Hugh Clayton, "Milk Penalties Increased," *Times*, November 14, 1979, 3.

65 "Letter—Obsesses with Penalties," *BFS*, August 16, 1980, 6.

66 "Tougher Antibiotic Tests for Milk Are Coming," *BFS*, July 19, 1980, 14.

67 Ibid.

68 Clayton, "Milk Penalties Increased."

69 Rosemary Collins, "British Milk Has Highest Antibiotic Level in Europe," *Guardian*, January 27, 1982, 4; J. R. D. Allison, "Antibiotic Residues in Milk," *BVJ* 141, no. 1 (1985): 10–11.

70 J. M. Booth, "Intramammary Antibiotic Preparations and Their Withholding Times," *VR*, January 11, 1986, 34-35; "Antibiotics Scheme a Success," *VR*, March 29, 1986, 348.

71 Winter, *Rural Politics*, 146–159.

72 Quoted according to Martin, *Development of Modern Agriculture*, 178.

73 "The Soil Association in 1974," *The Soil Association* 3, no. 1 (1975): 1; Philip Conford and P. Holden. "The Soil Association," in *Organic Farming: An International History*, ed. William Lockeretz, 187–200 (Wallingford, UK: CABI, 2001), 187, 192.

74 "Towards the Stationary State—An Ecological Critique of the Religion of Productivity," *Journal of the Soil Association* 16, no. 2 (1970): 73–81; Jim Worthington, *Natural Poultry-Keeping* (London: Crosby Lockwood, 1971); *Alternative Agriculture: Organic Farming* (Haughley, UK: Soil Association, 1974); "Agrochemicals—Our Point of View," *The Soil Association* 2, no. 1 (1974): 5–6; "Thoughts on Chicken Breeding," *The Soil Association* 2, no. 3 (1974): 13; "Pigs in Organic Farming," *The Soil Association* 2, no. 7 (1974): 13, *Self-Sufficient Small Holding* (Haughley, UK: Soil Association, 1976); Michael Allaby with Colin Tudge, *Home Farm Complete Food Self-Sufficiency* (London: Macmillan, 1977).

75 Sam Mayall, *Farming Organically* (Haughley, UK: Soil Association, 1976); B. Stonehouse, *Biological Husbandry: A Scientific Approach to Organic Farming* (London: Butterworths, 1981); Francis Blake, *The Handbook of Organic Husbandry* (Marlborough, UK: Crowood Press, 1987).

76 Conford and Holden, "The Soil Association," 196.

77 Simon Wright and Diane McCrea, *Handbook of Organic Food Processing and Production* (Oxford: Blackwell Science, 2000), 11–12.

78 See for example the "Night Soil Association" in Rintoul Booth, *The Farming Handbook to End All Farming Handbooks* (London: Wolfe, 1970), 119.

79 Starting with Sir Henry Plumb, NFU presidents had attended Soil Association conferences from the mid-1970s onward; "Commentary," *The Soil Association* 2, no. 9 (1974): 5.

80 "Register for Hormone-Free Beef," *FW,* February 21, 1986, 14; "Cutting Out Hormones Wins Premium Prices," *FW,* February 21, 1986, 16; "Organic Opportunity," *FW,* March 14, 1986, 13.

81 Conford and Holden, "The Soil Association," 197. Soil Association organic standards dated back to 1973; EEC Regulation 2092/91 harmonized European organic standards in 1991.

82 A. T. Chamberlain et al., eds., *Organic Meat Production in the '90s* (Maidenhead, UK: Chalcombe, 1989), 10.

83 "Hormones Procedure Challenged," *FW,* March 14, 1986, 32.

84 "An Utter Waste of Merrie England," *FW,* March 14, 1986, 41.

85 Audrey Curran, "Don't Blame Buyers for Your Methods," *FW,* February 28, 1986, 52.

86 Robert Gair, "Hopes Rise for Not So Silent Spring," *FW,* October 30, 1987, 37.

87 Ibid.

88 *The Agricultural Notebook,* 17th ed. (London: Butterworth Scientific, 1982), 418, 447–448, 456, 472–475; J. Batty, *Pictorial Poultry-Keeping,* 3rd ed. (Hindhead, UK: Spur, 1980), 144; Eddie Straiton, *Pig Ailments: Recognition and Treatment—TV Book for Pig Farmers,* 6th ed. (Ipswich, UK: Farming Press, 1988), 24, 26, 29–30, 37–38, 51–52, 60–61.

89 Bill Weeks, "Letter—Public Must Be Given the Full Farming Facts," *FW,* May 23, 1986, 47.

90 "Drug Residues in Meat Are at Absolute Rock-Bottom," *FW,* November 20, 1987, 3.

91 "No Evidence for Food Scaremongers," *FW,* December 25, 1987, 4.

92 John R. Walton, "Personal Viewpoint—Antibiotic Residues in Meat," *BVJ* 143 (1987): 486.

93 "Common Sense About Antibiotics," *VR,* July 4, 1987, 1; see also: "Antibiotics 'Myth,'" *VR,* July 4, 1987, 2.

94 "NOAH: A New Voice in Animal Medicines," *VR,* March 15, 1986.

95 Peter Mitchelmore, *The Pigkeeper's Guide* (Newton Abbot, UK: David and Charles, 1981), 109.

96 *The Agricultural Notebook,* 18th ed. (London: Butterworths, 1988), 471–472.

97 Ibid., 473; see also Alice Stern, *Poultry and Poultry-Keeping* (London: Merehurst, 1988), 124–128.

98 Stephen R. Wharfe, "Satisfy the Public—Not Scientists," *FW,* May 23, 1987, 48.

99 "No-Additive Feed," *FW,* May 23, 1986, 55; see also "Probiotics Soothe Stressful Calves," *FW,* May 2, 1986, 26.

100 "Tainted Milk," *FW,* February 28, 1986, 21.

101 H. U. Bertschinger, "Antibiotic Resistance," *VR,* January 21, 1989, 79; see also R. A. LL Brown, "Control of Antibiotics," *VR,* January 17, 1987, 71.

102 "Don't Panic over BSE," *FW, October* 30, 1987, 24.

103 "BSE Thrives on Rumours," *FW*, November 20, 1987, 7.

104 Winter, *Rural Politics*, 160, 225–229.

105 Martin, *Development of Modern Agriculture*, 161–166.

106 Alois Seidl, *Deutsche Agrargeschichte* (Frankfurt: DLG Verlag, 2006), 311–314.

107 Council Directive 92/43/EEC.

108 Winter, *Rural Politics*, 161–165, 255, 308; Martin, *Development of Modern Agriculture*, 165, "Agriculture: Historical Statistics," *House of Commons Library Briefing Paper* 03330 (2016), 12.

109 "Agriculture: Historical Statistics," 8–10; *The Insiders Guide to Agriculture—Outdoor Pigs* (Agricola Training, 1992); *The Insiders Guide to Agriculture—Pigs* (Agricola Training, 1993), 2; Martin, *Development of Modern Agriculture*, 124–125.

110 "Agriculture: Historical Statistics," 7, 9, 11; Martin, *Development of Modern Agriculture*, 122, 127; *The Insiders Guide to Agriculture—Broilers* (Agricola Training, 1993), 2; *The Agricultural Notebook*, 19th ed. (Oxford: Blackwell Science, 1995), 489.

111 "On-Farm Mixers to Be Listed," *BF*, August 1992, 20; "Residues in Meat," *BF*, March 1992, 20. On-farm mixing and PML rules were reformed with Veterinary Written Directions (VWD) in 1995: "New Vet Rules Affect Feed Rations," *BF*, July/August 1995, 5.

112 Caroline Waldegrave, "Love Me, Love My Supermarket," *BF*, April 1991, 31; Michael Hornsby, "A Two-Tier Future," *BF*, May 1991, 31.

113 David Naish, "Farming for the Environment," *BF*, April 1992, 5.

114 R. Barrow, "Letter—Forget the Greenies," *BF*, June 1992, 6.

115 Tessa Gates, "Carving a Niche in Landscape," *FW*, July 24, 1992, 18; "Farming and Nature Living Side by Side," *FW*, August 7, 1992, 66–67.

116 Simon Wright, *Handbook of Organic Food Processing and Production* (Glasgow: Chapman and Hall, 1994), 14–15; Wright and McCrea, *Handbook of Organic Food Processing*, 12.

117 David Fleming, "The Kindest Cut?" *CL*, March 3, 1994, 32–35.

118 "A Growing Niche for 'Green Pigs,'" *BF*, June 1991, 13.

119 Ibid.

120 "The Headstart Challenge," *BF*, June 1991, 12.

121 Richard Young, "Healing Art of Homeopathy," *CL*, June 27, 1991, 110–111; Jonathon Porritt, "Sustain the Future," *CL*, May 23, 1991, 92–94; Richard Body, "Keeping Pigs Happy," *CL*, March 28, 1991, 70.

122 *The Insiders Guide to Agriculture—Pigs, Outdoor Pigs, Beef, Dairy, Eggs, Broilers* (Agricola Training, 1992–1993).

123 *The Agricultural Notebook*, 19th ed. (Oxford: Blackwell Science, 1995), 496.

124 "Sales of Antimicrobial Products Used as Veterinary Medicines, Growth Promoters and Coccidiostats in the UK from 1993–1998" (London: VMD, 1999).

125 Andrew Gordon, "It Must Have Been Staggers," *BF*, August 1992, 13.

126 Ibid.

127 "Consumer Confidence," *BF*, April 1996, 12–13.

128 Philip Clarke, "Supermarket Price Cuts Tempt Back Beef Buyers," *FW*, April 5–11, 1996, 21.

129 David Naish, "Letter to Readers," *BF*, April 1996, 3.

130 Ibid.

131 "UK Incinerators Cannot Cope," *FW*, April 5–11, 1996, 8; "Suicide Fear Grows," *FW*, March 29–April 4, 1996, 10.

132 John Pidsley, "Total Slaughter Is Unnecessary," *FW*, April 5–11, 1996, 85.

133 Worried farmer from Gloucestershire, "Feed-Makers the Real Villains," *FW*, April 5–11, 1996, 85.

134 Anthony Carter, "Don't Tamper with Nature," *FW*, April 5–11, 1996, 87.

135 Ibid.

136 "Setting Standards," *BF*, April 1996, 15.

137 "Patrick Holden—Country Life Interview," *CL*, May 2, 1996, 80–81.

138 "Avoparcin Feed Ban," *FW*, January 10–16, 1997, 34.

139 Jessica Buss, "Balanced Pig Diets Overcome Effects of Swedes' GP ban," *FW*, January 10–16, 1997, 34; "Effective Stockmanship Halves Vet Costs," *FW*, March 28–April 3, 1997, 44.

140 T. L. J. Lawrence and V. R. Fowler, *The Growth of Farm Animals* (Wallingford, UK: CABI, 1997), 315.

141 Jonathan Riley, "Leaders Reject Consumer Attack on Intensive Area," *FW*, March 13, 1998, 7.

142 Jonathan Riley, "NOAH Rounds on Sweden," *FW*, June 19, 1998, 14.

143 Ibid.

144 Ibid.

145 Ibid.

146 Shelley Wright, "MAFF's Antibiotic 'Smokescreen,'" *FW*, July 31, 1998, 13.

147 Philip Clarke, "FEDESA Refutes Antibiotic Claim," *FW*, September 11, 1998, 10; "Antibiotics for Growth Attack," *FW*, September 18, 1998, 14.

148 Jonathan Riley, "Leaders Reject Consumer Attack On Intensive Area," *FW*, March 13, 1998, 7; "Industry Says Yes to Use of Antibiotics for Growth," *FW*, March 22, 1998, 42.

149 "Ban Antibiotics as GPS, Urges MPs' Committee," *FW*, May 1, 1998, 6.

150 Sandy Mitchell, "Superbugs (and Supermarkets)," *CL*, April 30, 1998, 69.

151 Ibid.

152 Ibid.

153 Ben Gill, "Antibiotic Rules Need a Closer Look," *CL*, May 14, 1998, 152.

154 Ibid.

155 Jonathan Riley, "Antibiotic Restrictions Are Urged," *FW*, December 11, 1998, 6.

156 Simon Wragg, "Conversion Is Now Name of the Game," *FW*, September 4, 1998, 47; Emma Penny, "Pre-empting Antibiotic Cut," *FW*, September 18, 1998, 14; "Waitrose in Discussion with Suppliers," *FW*, May 8, 1998, 42.

157 Philip Clarke, "Antibiotic Use as Growth Promoters Set to Be Banned," *FW*, November 20, 1998, 8.

158 Sue Rider, "Don't Look for One Alternative. AGPs," *FW*, May 21, 1999, 52.

159 "Questioning the Need for AGPs Warning About Weaners," *FW*, February 5, 1999, 38.

160 Ibid.

161 "Sales of Antimicrobial Products Used as Veterinary Medicines, Growth Promoters and Coccidiostats in the UK from 1993–1998"; "UK Veterinary Antibiotic Resistance and Sales Surveillance Report (UK-VARSS 2015)" (New Haw, UK: VMD, 2016), 68.

162 "Euro Briefs," *FW*, September 13, 2002, 8.

163 "EU Scientists Push for End of AGP Use," *FW*, June 4, 1999, 8.

164 "Just Learn To Live Without AGPs," *FW*, May 21, 1999, 46.

165 Ibid.

166 "Plan for Total Ban on Antibiotics in Animal Feed," *FT*, March 26, 2002, 7; "GP Ban Could Cost Dear," *FW*, April 5, 2002, 43.

167 "Ban on Antibiotics as Growth Promoters in Animal Feed Enters into Effect," EU Press releases database.

168 T. J. Lawrence et al., *Growth of Farm Animals*, 2nd ed. (Wallingford, UK: CABI, 2002), 328.

169 "Antimicrobial Sales Rise Seen as a Warning," *FW*, March 15, 2002, 39; "Loophole in Law Makes Mockery of Ban on AGPs," *FW*, September 5, 2003, 3; "Growing Threat of Antibiotic Resistance," *FW*, February 11–17, 2005, 31.

170 Laura H. Kahn, *One Health and the Politics of Antimicrobial Resistance* (Baltimore: Johns Hopkins University Press, 2016), 23.

171 David Burch, "Problems of Antibiotic Resistance in Pigs in the UK," *VR—In Practice* 27 (2005): 37–39.

172 *The Agricultural Notebook*, 20th ed. (Oxford: Blackwell, 2003), 557–559, 566–580.

173 Sam Leadley, "Understanding Why Antibiotics Fail," *FW*, December 24–30, 2004, Livestock, 27.

174 Ibid.

175 "Your Pneumonia Know-How Could Win Weighing Kit," *FW*, November 16, 2001, 44.

176 Ibid.

177 Ibid.

178 "Your Pneumonia Know-How Could Win Weighing Kit," *FW*, November 23, 2001, 38.

179 "Your Pneumonia Know-How Could Win Weighing Kit," *FW*, November 30, 2001, 43.

180 "Farmers Weekly Academy," *FW*, September 16, 2005, 52–53.

181 Y. E. Jones et al., "Antimicrobial Resistance in Salmonella Isolated from Animals and Their Environment in England and Wales from 1988 to 1999," *VR* 150 (2002): 649.

182 Burch, "Problems of Antibiotic Resistance in Pigs in the UK," 39–42; "Control of Antibiotic Resistant *Escherichia coli*," *VR* 161 (2007): 576; Richard Young, "Use of Antimicrobials," *VR* 164 (2009): 788.

183 Ibid.

184 L. A. Brunton et al., "A Survey of Antimicrobial Usage on Dairy Farms and Waste Milk Feeding Practices in England and Wales," *VR* 171 (2012): 296.

185 C. Wray et al., "Feeding Antibiotic-Contaminated Waste Milk to Calves," *BVJ* 146 (1990): 80–87; "Waste Milk Feeding, Animal By-Products Regulations and Antibiotic Resistance," *VR* 172 (2013): 166.

186 Brunton et al., "Survey of Antimicrobial Usage," 296.

187 M. W. H. Wulf et al., "Prevalence of Methicillin-Resistant *Staphylococcus aureus* Among Veterinarians: An International Study," *Clinical Microbiology and Infection* 14 (2008): 29–34.

188 "What Is the Superbug LA-MRSA CC398 and Why Is It Spreading on Farms?" *Guardian*, June 18, 2015; "MRSA Superbug Found in Supermarket Pork Raises Alarm over Farming Risks," *Guardian*, June 18, 2015.

189 See for example: "Antimicrobial Resistance—Why the Irresponsible Use of Antibiotics in Agriculture Must Stop: A Briefing from the Alliance to Save Our Antibiotics," Alliance to Save Our Antibiotics, 2015.

190 "Antibiotic Use in Livestock: Time to Act," *BEUC Position Paper* (Brussels: BEUC, August 2014), 2–3.

191 RUMA website, https://www.ruma.org.uk, accessed July 25, 2019.

192 Brian Jennings (NFU), "Path of Most Resistance," *CL*, September 23, 1999, 144; Roger Cook (NOAH/RUMA), "The Food-Chain Is Safe," *CL*, November 4, 1999, 164; "Green Lobby Critical of Antibiotics Guidelines," *FW*, July 2, 1999, 12.

193 Quoted in Charles Arthur, "Farmers 'Failing' to Limit Drug Use," *The Independent*, June 28, 1999.

194 "Raising Awareness of Resistance," *VR*, November 8, 2008, 551; "BVA Poster Promotes Responsible Use of Antimicrobials," *VR*, November 21, 2009, 609; "FVE Welcomes Decision on Prescription of Antibiotics," *VR* 171 (2012): 4; Patrick Butaye, "Measuring Antibiotic Use: A Way Forward," *VR* 171 (2012): 322–323; "Sharing Responsibility for Tackling Antibiotic Resistance," *VR* 171 (2012): 391–392; "New Client Leaflets on Antibiotics," *VR* 173 (2017) 562.

195 Robert Russell, "Use of Antibiotics," *VR* 166 (2010): 182.

196 "Antibiotic Use in Animal Health—'As Little as Possible, but as Much as Necessary,'" *VR* 164 (2009): 444.

197 "Calls to Restrict Veterinary Use of Antimicrobials," *VR* 171 (2012): 490.

198 Ibid. See also "Soil Association Calls for Antibiotic Use on Farms to Be Halved in Five Years," *VR* 170 (2012), 348.

199 "Ability to Use Antibiotics Preventively Must Be Maintained, says BVA," *VR* 175 (2014): 56; "Ensuring Access to Working Antimicrobials," House of Commons Science and Technology Committee, First Report of Session 2014-15/HC 509 (2014), 24–39; "RUMA Clarifies Position on Antibiotic Use for Livestock," *VR* 175 (2014), 265.

200 "EU Ban on Prophylactic Use of Antibiotics 'Overly Restrictive,' *Farmers Guardian*, October 30, 2018.

201 "BVA Highlights Vets' Concerns About Antimicrobial Resistance," *VR* 175 (2014): 496.

202 L. A. Coyne et al., "Understanding Antimicrobial Use and Prescribing Behaviours by Pig Veterinary Surgeons and Farmers: A Qualitative Study," *VR* 175 (2014): 593.

203 UK-VARSS 2015, Executive Summary, 19–26; "UK Veterinary Antibiotic Resistance and Sales Surveillance Report (UK-VARSS 2017)" (New Haw, UK: VMD, 2018).

204 Broom, Alex, et al., "Improvisation, Therapeutic Brokerage and Antibiotic (Mis) use in India: A Qualitative Interview Study of Hyderabadi Physicians and Pharmacists," *Critical Public Health* (2018): 1–12.

205 "Overuse of Antibiotics in Farming," *Sustain—Save Our Antibiotics*, http://www.sustainweb.org/antibiotics/overuse_of_antibiotics_in_farming/, accessed June 17, 2019; Corley and Godley, "Veterinary Medicines Industry," 844; "Joint Committee on the Use of Antibiotics in Animal Husbandry and Veterinary Medicine" (London: HSMO, 1969), 6.

206 "UK Could Use Brexit to Avoid EU Ban on Antibiotics Overuse in Farming," *Guardian*, September 27, 2018.

13. Swann Song: British Antibiotic Policy After 1969

1   TNA MAF 416/67 (Minute, E Doling to Cruickshank, 28 May, 1970); Laura H. Kahn, *One Health and the Politics of Antimicrobial Resistance* (Baltimore: Johns Hopkins University Press, 2016), 20.

2   W. M. McKay, "The Use of Antibiotics in Animal Feeds in the UK: The Impact and Importance of Legislative Controls," *World's Poultry Science Journal* 31, no. 2 (1975): 119.

3   TNA MAF 416/67 (Submission to Minister, Swann Report—Current Position and Further Action, Appendix III: Consultations About Withdrawal Date for Penicillin and the Tetracyclines, 13 Jul, 1970), 2.

4   TNA MAF 416/67 (Ernest Hesse to John Mackie, 20 May, 1970), 1.

5   Ibid.

6   Garry Kroll, "The 'Silent Springs' of Rachel Carson: Mass Media and the Origins of Modern Environmentalism," *Public Understanding of Science* 10 (2001): 414–415; Thomas Jukes, "The Right to Be Heard," *BioScience* 18, no. 4 (1968), Thomas Jukes, "DDT: Affluent Enemy or Beneficial Friend," *BioScience* 19, no. 7 (1969).

7   TNA MAF 284/283 (Press Information. Cyanamid of Great Britain Limited, "Scientists Deplore 'Instant Decision' by Governments," 20 Jan, 1970), 1.

8   Ibid., 1–2.

9   TNA MAF 284/283 (Press Information, Cyanamid of Great Britain Limited, "New Evidence Casts Doubt on Link Between Farm Antibiotics and Human Disease," 20 Jan, 1970), 1.

10 Ibid.

11 WC PP EBC F186 (Enclosed Transcript in: AT Mennie to Ernst Boris Chain, 27.01.1970).

12 Ibid.

13 WC PP EBC F186 (WE McAlister to Chain, 04.03.1970).

14 WC PP EBC F186 (Chain to WE McAlister, 06.03.1970).

15 TNA MAF 284/283 (Minute, I Armstrong to Mr Dawes, 21 Jan, 1970).

16 Ibid.

17 Ibid. The *Financial Times* interviewed Herbert Williams Smith to refute Cyanamid claims; David Fishlock, "Critics of Farm Antibiotic Curbs Are 'Alarmist,'" *FT*, January 22, 1970.

18 TNA MAF 416/67 (Submission to Minister, Swann Report—Current Position and Further Action, Appendix: Background (Animal Health Division II, 8 Jul, 1970), 3.

19 WC PP EBC F186 (WE McAlister to Chain, November 6, 1970).

20 TNA MAF 416/67 (Submission to Minister, Swann Report—Current Position and Further Action, Appendix: Background (Animal Health Division II, 8 Jul, 1970), 13.

21 Ibid., 7.

22 Ibid., 12.

23 TNA MAF 284/283 (Minute, Doling to Evans, 26 May, 1970); MAF 416/67 (Submission to the Minister, Swann Report—Current Position and Further Action, 13 Jul, 1970), 2.

24 TNA MAF 416/67 (Minute, PW Murphy to Doling, 14 Aug, 1970).

25 Ibid.

26 TNA MAF 416/67 (Press Notice, Antibiotics, Further Implementation of Recommendations of the Swann Committee, 20 Aug, 1970), 1.

27 TNA MAF 416/67 (K.P. Grainger to Jim M.L. Prior [sic], 10 Aug, 1970), 1.

28 Ibid.

29 Ibid., 2.

30 Ibid.

31 Ibid.

32 Ibid., attachment, 1.

33 TNA MAF 416/67 (Minute, E. Doling to Mr. Carnochan, 15 Jun, 1970).

34 Thomas H. Jukes, "The Present Status and Background of Antibiotics in the Feeding of Domestic Animals," *Annals of the New York Academy of Sciences* 182 (1971): 362–364, 76.

35 Howard Jarolmen, "Experimental Transfer of Antibiotic Resistance in Swine," *Annals of the New York Academy of Sciences* 182 (1971): 79.

36 C. D. Van Houweling, "The Food, Drug, and Cosmetic Act," *Annals of the New York Academy of Sciences* 182 (1971): 412.

37 TNA MAF 416/67 (AB Paterson to Van Houweling, 22 Oct, 1970).

38 Ibid.

39 Ibid.

40 McKay, "Use of Antibiotics," 119–120.

41 TNA MAF 260/678 (NW Taylor to Departments—MAFF, 24 Aug, 1971), 1.

42 TNA MH 149/2484 (DHSS Medicines Division, Paper A: Medicines Act—Proposed Orders As To Antibiotics, First Draft, 16 May, 1972), 3–4.

43 TNA MH 149/2484 (A Note of a meeting Held to Discuss the Establishment of a Committee on Antibiotics, 20 Mar, 1972; Minute Hogg to Williamson, 22 Feb, 1970).

44 TNA MAF 260/678 (NW Taylor to Departments—MAFF, 24 Aug, 1971), 2; TNA MAF 461/34 (Note of Meeting on the Future of the JSC on Antimicrobial Substances, 28 Sep, 1979), 1–2; TNA MH 149/2484 (CSG Russell (VPC) and EF Scowen (CSM) to Sir James Howie, 22 Dec, 1972); (Medicines Commission: Antibiotics, Second Draft, 9 Jun, 1972), 4.

45 TNA MH 149/2484 (AMR Nelson to CSG Grunsell, 14 Mar, 1973).

46 TNA MH 149/2484 (RJ Blake to JB Brown, 27 Mar, 1973).

47 TNA BN 116/71 (JSC on Antimicrobial Substances, 1st meeting, 2 Jul, 1973); TNA MAF 461/34 (Minute, Pamela Green to members of meeting between MAFF and MH, Sep, 1979), 1.

48 TNA MH 149/2484 (Terms of Reference of the JSC on Antimicrobial Substances).

49 TNA BN 116/71 (Appendix A, ML 11, Extract from the report of the Joint Committee on the Use of Antibiotics in Animal Husbandry and Veterinary Medicine).

50 TNA MAF 461/34 (James Howie to Chairmen of Committee on Safety of Medicines (CSM) and Veterinary Products Committee (VPC), 7 Aug, 1979, 2.

51 Ibid.

52 TNA MAF 461/34 (Minute, Paul Ditchfield to Lawson (14 Apr, 1980).

53 "Helping Aunt Sally," *Veterinary Record* [hereafter *VR*], July 28, 1979, 67.

54 TNA MAF 461/34 (Note of Meeting on the Future of the JSC on Antimicrobial Substances, 28 Sep, 1979), 2.

55 TNA BN 116/119 (JSC on Antimicrobial Substances, Minutes of Meeting on

December 3rd, 1980), 3, 7; see also John Harvey and Liz Mason, *The Use and Misuse of Antibiotics in UK Agriculture, Part 1: Current Usage* (Bristol, UK: Soil Association, 1998), 10.

56 C. Wray, Y. E. Beedell, and I. M. McLaren, "A Survey of Antimicrobial Resistance in Salmonellae from Animals in England and Wales During 1984–1987," *British Veterinary Journal* 147 (1991): 356. Previous testing had been hampered by unstandardized sensitivity tests; TNA MAF 386/44 (Draft Minutes—5th Meeting of Joint PHLS/AHD Advisory Committee, 21 Apr, 1967), 2–3.

57 MAFF—Note of a Meeting Held with US Task Force on Antibiotics, February 2, 1971, Folder Task Force Antibiotics, Box 1, Account No. 75A0106, RG 442, NARA Atlanta, 1–2.

58 TNA MAF 386/45 (Minutes, 14th Meeting PHLS/AHD Advisory Committee, 21 Jan 1970), 3.

59 TNA MAF 416/85 (Minute, E. Doling, 24 Apr, 1970), 1.

60 TNA MAF 416/85 (Minute, F.C. Parker to Doling, 26 Feb, 1970).

61 TNA MAF 416/85 (Minute, Doling, 24 Apr, 1970), 3.

62 TNA MAF 416/85 (Minute, Doling, 24 Apr, 1970), 4.

63 TNA MAF 416/85 (Minute, J.G. Carnochan, 30 Apr, 1970); see also CVL skepticism; TNA MAF 386/45 (Minutes—14th Meeting PHLS/AHD Advisory Committee, 21 Jan, 1970), 1–2.

64 TNA MAF 416/85 (Minute, R.J. Blake, 27 Aug, 1970).

65 TNA MAF 416/85 (Minute, Doling, 1 Sep, 1970).

66 TNA MAF 416/85 (Minute, Doling, 18 Nov, 1970), 1–2.

67 TNA MAF 416/85 (Minute, D. Stoker, 16 Jun, 1971).

68 TNA MAF 416/86 (Minute, JN Jotchan, 20 Sep, 1972).

69 Claas Kirchhelle, "Toxic Confusion: The Dilemma of Antibiotic Regulation in West German Food Production," *Endeavor* 40, no. 2 (2016): 121.

70 TNA MAF 416/86 (Draft Letter For Signature by the Parliamentary Secretary to P. Blaker, attached to Minute JN Jotcham, 20 Sep, 1972).

71 TNA MAF 416/86 (E.S. Anderson to J. Jotcham, 17 Apr, 1972; attached, Interdepartmental Working Party on the implementation of the Swann Report. Examination of imported meat for contamination with drug-resistant Escherichia coli), 2–3.

72 Ibid., 8; attached, Table 1: E. coli from Irish beef and pork carcasses.

73 TNA MAF 416/86 (E.S. Anderson to J. Jotcham, 17 Apr, 1972; attached, Interdepartmental Working Party on the implementation of the Swann Report. Examination of imported meat for contamination with drug-resistant Escherichia coli), 3.

74 Ibid., 9.

75 Ibid.

76 Ibid.

77 TNA MAF 416/86 (Minute, ARM Kidd to WT Barker, 1 Jun, 1972; Barker to Anderson, 20 Jun, 1974).

78 TNA MAF 416/86 (RW McQuiston to Barker, 8 May, 1972).

79 TNA MAF 416/86 (Working Group on the Monitoring of Imported Meat for Antibiotic Resistant Enterobacteria, Meeting, 13 Jun, 1972), 2.

80 TNA MAF 416/86 (ARM Midd, Antibiotic Resistance in the UK and Belgium, 13 Jan, 1973), 1.

81 TNA MAF 416/86 (JN Jotcham to AB Paterson, 9 Apr, 1973).

82 Ibid.

83 TNA MAF 282/186 (Steering Group on Food Surveillance, Sub-Group on Antibiotic Residues in Food, Secretariat, Jun 1975).

84 McKay, "Use of Antibiotics," 124–125.

85 TNA MAF 386/46 (Minutes 25th Meeting PHLS/AHD Advisory Committee, 23 Oct, 1974), 5; (Minutes 24th Meeting PHLS/AHD Advisory Committee, 24 Jul, 1974), 4.

86 TNA MAF 416/70 (Submission to Minister, Swann Report—Recommendation on Advertising, January 1972).

87 Ibid.

88 Ibid.

89 TNA MAF 416/70 (Draft for the Minister's Signature to Lord Aberdare); see also (Lord Aberdare to James Prior, 21 Mar, 1972).

90 TNA MAF 416/71 (Minute John H Drury to Nelson, 23 Jan, 1974); TNA MAF 416/71 (Minute RJ Blake to WT Barker, 17 Apr, 1974).

91 TNA MAF 416/71 (Submission to the Minister, Swann Report—Recommendation On Advertising, 20 Jun, 1974), 5.

92 TNA MAF 416/71 (Minute, Shillito to Evans, 9 Jul, 1974).

93 TNA MAF 416/71 (Minute, Nelson to RJ Blake, 9 Jan, 1975).

94 A. H. Linton, "The Swann Report and Its Impact," in *The Control of Antibiotic Resistant Bacteria*, ed. Charles H. Stuart Harris and David M. Harris (London: Academic Press, 1982), 183–194, 197–198.

95 E. J. Threlfall and B. Rowe, "Antimicrobial Drug Resistant in Salmonellae in Britain—A Real Threat to Public Health," in *Antimicrobials and Agriculture*: *The Proceedings of the 4th International Symposium on Antibiotics in Agriculture* (London: Butterworths, 1984), 515, 519, 522.

96 Linton, "Swann Report and Its Impact," 191–192.

97 Ibid.

98 Ibid., 195.

99 TNA MAF 461/67 (Steering Group on Food Surveillance, Sub-Group Antibiotic Residues in Food, Antibiotics Price Deduction Scheme, Jan, 1976).

100 Anne Hardy, "John Bull's Beef: Meat Hygiene and Veterinary Public Health in England in the Twentieth Century," *Review of Agricultural and Environmental Studies* 91, no. 4 (2010): 386–387.

101 TNA MAF 284/282 (Committee on Medical Aspects of Food Policy, meeting, 4 Mar, 1968), 6.

102 The practice of preserving food with antibiotics other than nisin was no longer endorsed in 1975; TNA MAF 282/186 (SGFS, Sub-Group on Antibiotic Residues in Food. Minutes of the 1st meeting of the Sub-Group held on 30 Apr, 1976), 3.

103 TNA MAF 284/282 (LC Gaskell to JG Kelsey, 19 Sep, 1968).

104 TNA MAF 260/678 (EJ Mehen to NW Taylor, 2 Sep, 1971).

105 TNA MAF 260/678 (S Simmons to WJD Williams, 9 Aug, 1971); (LR Maddock to Williams, 4 Feb, 1971; TNA MH 148/934 (Medicines Act 1968: Antibiotic Residues in Milk), 4.

106 Peter A. Koolmes, "Veterinary Inspection and Food Hygiene in the Twentieth Century," in *Food, Science, Policy and Regulation in the Twentieth Century: International and Comparative Perspectives*, ed. David F. Smith and Jim Philips (London: Routledge, 2000), 60, 62; Kirchhelle, "Toxic Confusion," 122.

107 Directive 72/462/EEC; the Third Country directive banned residue-tainted non-EEC meat consignments in 1972.

108 TNA MAF 461/67 (SGFS, Sub-Group on Antibiotic Residues in Food, 1st Meeting, 30 Apr, 1976), 5 and 7; TNA MAF 282/186 (Norman D Baird to RV Blamaire, "Residues in Meat—Some Pertinent Facts," 4 Aug, 1976), 2.

109 TNA MAF 282/186 (Baird to Blamaire, "Residues in Meat—Some Pertinent Facts," 4 Aug, 1976), 4; see also TNA MAF 461/67 (Davies to Tugwell, 16 Nov, 1977), 1.

110 TNA MAF 461/67 (SGFS, Sub-Group on Antibiotic Residues in Food. Draft Terms of Reference, Mar, 1976).

111 TNA MAF 461/67 (Notes of meeting, UK Drug Residue Monitoring Programme, 1977), 1–2; see also MAF 461/69 (Minute, AW Hubbard to W Barker, 3 Aug, 1978), 2–3; MAF 461/68 (SGFS, Working Party on Veterinary Residues in Meat and Meat Products, 1st Meeting, 29 Jul, 1977), 3.

112 TNA MAF 461/68 (Working Party on Veterinary Residues in Meat and Meat Products. Antibiotic Residues in Meat Taken from Export-Licensed Abattoirs in UK). Annexe: Antibiotic Residues in Imported Meat), 1; see also: (Working Party on Veterinary Residues, Draft Report to the SGFS, Note of 2nd meeting, 17 Jul, 1978).

113 TNA MAF 461/67 (Minute, J. Morey to Giles, 30 Jan, 1978), 2.

114 TNA MAF 282/198/1 (Submission to the Parliamentary Secretary, MAFF National Meat Residue Monitoring Programme, Food Science Division, Appendix C, February 1980); draft directives 4850/VI/77 (1977) and 728/VI/78 (1978).

115 TNA MAF 461/67 (Notes of meeting, UK Drug Residue Monitoring Programme, 1977), 1; (Minute, E Owen to J Morey, 16 Dec, 1977), 1; (Minute (undated), Monitoring of Drug Residues, FCN 228); TNA MAF 461/68 (Meat Residue Monitoring Programme), 3.

116 TNA MAF 461/68 (EEC, Summary Report of Meeting with Representatives of Community Institutions or Of Member Governments, "Working Party 'Veterinary Legislation' Sub-Group 'Residues'" (25-26.05.1978), "Draft Directives on Undesirable Residues in Fresh Meat," 30 May, 1978), 2.

117 TNA MAF 461/68 (EEC: Summary Report Of Meeting: "Working Party 'Veterinary Legislation' Sub-Group 'Residues'" (25.-26.05.1978)), 2.

118 TNA MAF 282/198/1 (Submission to the Parliamentary Secretary, MAFF National Meat Residue Monitoring Programme, Food Science Division, Appendix C 64/433/EEC, February 1980; Sampling of Meat for Residue Investigations).

119 TNA MAF 461/70 (Working Party on Veterinary Residues in Meat and Meat Products, National Meat Monitoring Programme Year 1 Results, 13 Oct, 1982), 7.

120 TNA MAF 282/198/1 (Submission to the Parliamentary Secretary, MAFF National Meat Residue Monitoring Programme, Food Science Division, February 1980), 2; TNA MAF 282/199 (Memorandum: From DVO MAFF to JA Grisedale, DRVO (Reading) and JM Threlkeld, RVO (Reading), 9 Feb, 1983); (RS Beynon to Jenkinson, 18 Feb, 1983), 1; DG Lindsay, "Monitoring and Testing for Residues of Therapeutics in Meat," *VR*, May 14, 1983, 469–471.

121 TNA MAF 461/70 (Working Party on Veterinary Residues in Meat and Meat Products, Antimicrobial Agents in Muscle and Kidney, results from Abattoirs exporting to West Germany, Feb 1982).

122 TNA MAF 282/199 (Minute, LG Mitchell to Fry, 10 Aug, 1984; attached, Meat Inspection Review, EEC proposals to Control Residues in Meat for

Intra-Community Trade), 1–2; attached, Meat Inspection Review, SVS National
Surveillance Scheme for Residues in Meat), 1–4; attached, Meat Inspection
Review, EEC proposals to Control Residues in Meat for Intra-Community Trade;
Directive 83/90/EEC; Directive 83/91/EEC; Directive 81/602/EEC), 1–3.

123 "Drug Traces Found in Abattoir Carcasses," *Guardian*, January 18, 1988, 4.

124 Hardy, "John Bull's," 387.

125 "Report on Microbial Antibiotic Resistance in Relation to Food Safety," Advisory
Committee on the Microbiological Safety of Food, 1999, 146.; Richard Young
et al., *The Use and Misuse of Antibiotics in UK Agriculture, Part 2: Antibiotic
Resistance and Human Health* (Bristol, UK: Soil Association, 1999), 42.

126 Hardy, "John Bull's," 387.

127 C. Wray et al., "Survey of Antimicrobial Resistance," 356–369; surveys were
published from 1984 onward.

128 "The Report of the Expert Group on Animal Feedingstuffs to the Minister of
Agriculture, Fisheries and Food, the Secretary of State for Health and the
Secretaries of State for Wales, Scotland and Northern Ireland (Lamming
Report)," 1992, 3.

129 Ibid., 70, 75.

130 Ibid., 47.

131 Ibid., 48.

132 Ibid., 45–47.

133 "The Path of Least Resistance," Standing Medical Advisory Committee, Sub-
group on Antimicrobial Resistance, 1998, 78.

134 "Use of Quinolones in Food Animals and Potential Impact on Human Health:
Report of a WHO Meeting, Geneva, Switzerland, June 2–5, 1998," 7.

135 Ibid., 6–7.

136 Uwe Petersen, "Entwicklungen Im Deutschen Futtermittelrecht," in *Meilensteine
Für Die Futtermittelsicherheit*, ed. Bundesforschungsanstalt Für Landwirtschaft
(Fal) (Braunschweig: Petersen et al., 2007), 6.

137 M. Dettenkofer, M. Ackermann, M. Eikenberg, and H. Merkel, *Auswirkungen
Des Einsatzes Von Antibiotika Und Substanzen Mit Antibiotischer Wirkung in Der
Landwirtschaft Und Im Lebensmittelsektor* (Freiburg: Institut für Umweltmedizin
und Krankenhaushygiene, 2004), 24.

138 Laura H. Kahn, *One Health and the Politics of Antimicrobial Resistance* (Balti-
more: Johns Hopkins University Press, 2016), 33–35, 58; M. Setbon and A. Castot,
"Interdiction de l'avoparcine face au risqué d'antibiorésistance," *Comité national
de sécurité sanitaire: Risques et sécurité sanitaire* (2001), 55–57; J. J. Dibner and J. D.
Richards, "Antibiotic Growth Promoters in Agriculture: History and Mode of
Action," *Poultry Science* 84 (2005), 634; Petersen, "Entwicklungen," 6; "Written
Answers to Questions, 10.06.1996," House of Commons Written Questions and
Answers, 1996, col. 10.

139 "The Path of Least Resistance," 78.

140 Robert Bud, *Penicillin: Triumph and Tragedy* (Oxford: Oxford University Press,
2009), 205.

141 David F. Smith et al., *Food Poisoning, Policy and Politics: Corned Beef and Typhoid
in Britain in the 1960s* (Woodbridge, UK: Boydell, 2005), 307–308.

142 "The Medical Impact of Antimicrobial Use in Food Animals: Report of a WHO
Meeting, Berlin, Germany, 13–17 October 1997"; Setbon and Castot, "Interdiction," 61.

143 "Antimicrobial Resistance Monitoring: Information Exchange and Opportunities for Collaboration: Report of the Second Joint WHO/IFPMA Meeting, Geneva 2–3 April 1998," 1.
144 "Use of Quinolones in Food Animals," 17.
145 "Resistance to Antibiotics and Other Antimicrobial Agents," House of Lords, Select Committee Appointed to Consider Science and Technology, 1998, 11.18.
146 Bud, *Penicillin*, 203.
147 "Resistance to Antibiotics and Other Antimicrobial Agents," 11.19.
148 Ibid., 11.23.
149 Ibid., 11.21.
150 Ibid.
151 Select Committee on Agriculture, "Fourth Report. Food Safety," House of Commons, 1998, IV.123.m.
152 "The Path of Least Resistance," 80.
153 Tore Midtvedt, "The Microbial Threat: The Copenhagen Recommendation," *Microbial Ecology in Health and Disease* 10 (1998): 66–67.
154 "Decision No 2119/98/Ec," *Official Journal L 298* (October 3, 1998).
155 The organization is responsible to the European Centre for Disease Control (ECDC) and is now known as EARS-Net.
156 "Report on Microbial Antibiotic Resistance in Relation to Food Safety," 147, 151–152, 172–173.
157 Dibner and Richards, "Antibiotic Growth Promoters in Agriculture," 635.
158 Kahn, *One Health*, 42–43.
159 "Defra Antimicrobial Resistance Coordination (DARC) Group," responsibility for animal antimicrobial resistance policy passed to the VMD in 2011.
160 I am indebted to the VMD for answering my questions about antibiotic legislation (correspondence September 2, 2013); see also Kahn, *One Health*, 43–45.
161 Correspondence with VMD.
162 "Sales of Antimicrobial Products Authorised for Use as Veterinary Medicines in the UK in 2011," Veterinary Medicines Directorate, 2012, 8; Kahn, *One Health*, 45.
163 Nicolas Fortané and Frédéric Keck, "How Biosecurity Reframes Animal Surveillance," *Revue d'anthropologie des connaissances* 9, no. 2 (2015): a–d, g–h.
164 "Veterinary Medicines Regulations," "Veterinary Medicines Regulations—Changes to Advertising Rules," October 11, 2012.
165 Fergus Walsh, "Antibiotics Resistance 'as Big as Terrorism'—Medical Chief," *BBC News*, March 11, 2013.
166 "Collection Antimicrobial Resistance (AMR)," https://www.gov.uk/government /collections/antimicrobial-resistance-amr-information-and-resources #parliamentary-inquiries, accessed July 20, 2019.
167 "Tackling Drug-Resistant Infections Globally: Final Report and Recommendations: The Review of Antimicrobial Resistance (O'Neill AMR Review)," May 2016, 4–7.
168 Ibid. 28–32.
169 "Antimicrobials in Agriculture and the Environment: Reducing Unnecessary Use and Waste: The Review on Antimicrobial Resistance (O'Neill AMR Review)," December 2015, 29–30.
170 "UK Secures Historic UN Declaration on Antimicrobial Resistance," UK Department of Health—News Story, September 21, 2016.

171 Arthur Neslen, "European Parliament Approves Curbs on Use of Antibiotics on Farm Animals," *Guardian*, October 25, 2018.

Conclusion: Antibiotics Unleashed

1 Ellen K. Silbergeld, *Chickenizing Farms and Food: How Industrial Meat Production Endangers Workers, Animals, and Consumers* (Baltimore: Johns Hopkins University Press, 2016), 74–75.

2 Ibid., 80–83.

3 Thomas Van Boeckel et al., "Global Trends in Antimicrobial Use in Food Animals," *PNAS* 112, no. 18 (2015): 5649–5650.

4 Laura H. Kahn, *One Health and the Politics of Antimicrobial Resistance* (Baltimore: Johns Hopkins University Press, 2016), 36–40, 45–49, 59.

5 Ibid., 7.

6 Kate S Baker, "The Extant World War 1 Dysentery Bacillus NCTC1: A Genomic Analysis," *Lancet* 384, no. 9955 (2014): 1691–1697; see also: Hannah Landecker "Antimicrobials before Antibiotics: War, Peace, and Disinfectants," *Palgrave Communications* 5, no. 45 (2019).

7 Catriona P. Harkins et al., "Methicillin-Resistant *Staphylococcus aureus* Emerged Long Before the Introduction of Methicillin into Clinical Practice," *Genome Biology* 18, no. 130 (2017); Alicia Tran-Dien et al., "Early Transmissible Ampicillin Resistance in Zoonotic *Salmonella* Serotype Typhimurium in the Late 1950s: A Retrospective, Whole-Genome Sequencing Study," *Lancet Infectious Diseases* 18, no. 2 (2018): 207–214; for a general overview, see Stephen Baker et al., "Genomic Insights into Emergence and Spread of Antimicrobial-Resistant Bacterial Pathogens," *Science* 360 (2018): 733–738.

8 Notes on Meeting with H. Williams Smith, February 4 1971, Folder Task Force Antibiotics, Box 1, Account No. 75A0106, RG 442, NARA Atlanta, 2.

9 Christian W. McMillen, *Discovering Tuberculosis: A Global History 1900 to the Present* (New Haven, CT: Yale University Press, 2015), 226.

10 Ibid., 225–227.

11 For an overview of recent historical research, see José Ramón Bertomeu-Sánchez and Ximo Guillem-Llobat, "Following Poisons in Society and Culture (1800–2000): A Review of Current literature," *Actes d'història de la ciència i de la tècnica* 9 (2017): 9–36; Claas Kirchhelle, "Toxic Tales—Recent Histories of Pollution, Poisoning, and Pesticides (ca. 1800–2010)," *NTM Zeitschrift für Geschichte der Wissenschaften, Technik und Medizin* 26, no. 2 (2018): 213–29.

12 Kahn, *One Health*, 5–6.

13 Clare Chandler et al., *Addressing Antimicrobial Resistance Through Social Theory: An Anthropologically Oriented Report*, London: LSHTM, 2016, 17.

14 Scott H. Podolsky, *The Antibiotic Era: Reform, Resistance and the Pursuit of a Rational Therapeutics* (Baltimore: Johns Hopkins University Press, 2015), 141–142.

15 "Interventions on Antibiotic Use Not without Consequence, Warns RUMA," *Veterinary Record*, 178, no. 17 (2016): 406; Frank Aarestrup, "Sustainable Farming: Get Pigs Off Antibiotics," *Nature* 486 (2012): 465–466.

16 Alex Broom et al., "Improvisation, Therapeutic Brokerage and Antibiotic (Mis) Use in India: A Qualitative Interview Study of Hyderabadi Physicians and Pharmacists," *Critical Public Health* (2018): 1–12.

17 Scott H. Podolsky et al., "History Teaches Us that Confronting Antibiotic Resistance Requires Stronger Global Collective Action," *Journal of Law, Medicine and Ethics* 42, no. 2 (2015): 27–32.

18 For an overview of policy suggestions, see Thomas P. Van Boeckel et al., "Reducing Antimicrobial Use in Food Animals," *Science* 357, no. 6358 (2017): 1350–1352.

Bibliography

Archives

British National Archives. TNA, Richmond, UK
Countway Library of Medicine. CLM, Boston, USA
Museum of English Rural Life. MERL, Reading, UK
National Archives and Records Administration. NARA, Atlanta, USA
National Archives and Records Administration. NARA, College Park, USA
Wellcome Collections. WC, London, UK
Yale Beinecke Library. YBL, New Haven, USA

Databases

Cornell University's Home Economics Archive. HEARTH
Hathi Trust
Polling the Nations

Published Serial Sources

Agricultural Research
Barron's National Business and Financial Weekly
BBC News
Better Homes and Gardens
British Farmer
British Farmer and Stock Breeder
British Veterinary Journal
Consumer Reports
Cornell Veterinarian
Coronado Eagle and Journal
Cosmopolitan
Country Life
Countryman
Daily Illini

Daily Mail
Daily Mirror
Desert Sun
Economist
Farm Bureau News
Farm Journal
Farm Journal and Country Gentleman
Farmline
Farmers' Bulletin
Farmers Weekly
Farmers Weekly Review
FDA Veterinarian
Federal Register
Feedstuffs
Food Chemical News
Food Drug Cosmetic Law Journal
Food Safety News
Forward
Good Housekeeping
Guardian
Harper's
Healdsburg Tribune
Indiana Prairie Farmer
Jewish Press
Journal of Home Economics
Journal of the Soil Association
Life
LMC Experience
London Illustrated News
Madera Tribune
Mother Earth
National Review
New England Business
New Scientist
New Statesman
News for Farmer Cooperatives
Newsweek
New York Times
Observer
Organic Farmer
Organic Gardening and Farming
Poultry Science
Prairie Farmer
Progressive Farmer
Reuters
Rodale Press Guide to Organic Living
San Bernardino Sun
Scientific American
Soil Association Quarterly Review

SPAN—Soil Association News
Successful Farming
Telegraph
The Soil Association
Times (London)
True Republican
USDA Agriculture Handbook
Veterinary Record
Vogue
Wallaces Farmer
Wall Street Journal
Washington Post
Yearbook of Agriculture

Governmental and Nongovernmental Organization Publications

"1969 Census of Agriculture Vol. 2." Washington, DC: US Department of Commerce, 1973.
"1982 Census of Agriculture. Vol. 1, Part 51." Washington, DC: US Department of Commerce, 1984.
"1992 Census of Agriculture Vol. 1, Part 51." Washington, DC: USDA, 1994.
"2002 Census of Agriculture Vol. 1, Part 51." Washington, DC: USDA, 2004.
"2012 Census of Agriculture Vol. 1, Part 51." Washington, DC: USDA, 2014.
"2013 Summary Report on Antimicrobials Sold or Distributed for Use in Food-Producing Animals." Washington, DC: FDA/DHSS, 2015.
"2014 Summary Report on Antimicrobials Sold or Distributed for Use in Food-Producing Animals." Washington, DC: FDA/DHSS, 2015.
"2015 Summary Report on Antimicrobials Sold or Distributed for Use in Food-Producing Animals." Washington, DC: FDA/DHSS, 2016.
"2016 Summary Report on Antimicrobials Sold or Distributed for Use in Food-Producing Animals." Washington, DC: FDA/DHSS, 2017.
"2017 Summary Report on Antimicrobials Sold or Distributed for Use in Food-Producing Animals." Washington, DC: FDA/DHSS, 2018.
"Agriculture: Historical Statistics." House of Commons Briefing Paper 03339, 2016.
Allaby, Michael, with Colin Tudge. *Home Farm Complete Food Self-Sufficiency.* London: Macmillan, 1977.
Allen, George, and Clark Burbee. "Economic Consequences of the Restricted Use of Antibiotics at Subtherapeutic Levels in Broiler and Turkey Production." USDA—Farm Production Economics Division Working Paper, November 1972.
"Animal Drug Amendments of 1968." Report No. 1308. Washington, DC: GPO, 1968.
"Animal Drug Amendments of 1968: Hearings Before the US Senate Committee on Labor and Public Welfare, May 24th, 1968." Washington, DC: GPO, 1968.
"Animal Drug Availability Act of 1996." https://www.fda.gov/AnimalVeterinary/GuidanceComplianceEnforcement/ActsRulesRegulations/ucm105940.htm.
"Animal Medicinal Drug Use Clarification Act of 1994 (AMDUCA)." https://www.fda.gov/AnimalVeterinary/GuidanceComplianceEnforcement/ActsRulesRegulations/ucm085377.htm.
"Antibiotic Residues Hidden Risks." Washington, DC: FDA, 1970.
"Antibiotic Resistance." Subcommittee on Investigations and Oversight of the

Committee on Science and Technology, US House of Representatives, December 18, 19, 1984. Washington, DC: GPO, 1985.

"Antibiotic Resistance and the Use of Antibiotics in Animal Agriculture." Subcommittee on Health of the Committee on Energy and Commerce, July 14, 2010, US House of Representatives. Washington, DC: GPO, 2013, 28–42.

"Antibiotic Resistance: Federal Agencies Need to Better Focus Efforts to Address Risk to Humans from Antibiotic Use in Animals." Washington, DC: GAO, 2004.

"Antibiotic Resistance: The Risk to Human Health and Safety from the Use of Antibiotics in Animal Production (CEG 98/2)." Consumers in Europe Group, 1998.

"Antibiotics in Animal and Poultry Feeds—A Critical Review of Research (ARS 44/237)." Washington, DC: USDA, 1972.

"Antibiotics in Animal Feed: Hearings Before the Subcommittee on Health and the Environment on H.R. 7285." Subcommittee on Health and the Environment of the Committee on Interstate and Foreign Commerce, US House of Representatives. Washington, DC: GPO, 1980.

"Antibiotics in Animal Husbandry." London: Office of Health Economics, 1969.

"Antibiotics in Livestock Feeding." MAFF Advisory Leaflet 418, 1960.

"Antibiotics in Pig Food." London: HMSO/ARC, 1953.

"Antibiotic Use in Food Animals." PEW Antibiotic Resistance Project. https://www.pewtrusts.org/en/projects/antibiotic-resistance-project.

"Antibiotic Use in Livestock: Time to Act." BEUC Position Paper. Brussels: BEUC, August 2014.

"Antimicrobial Resistance Monitoring: Information Exchange and Opportunities for Collaboration." Report of the Second Joint WHO/IFPMA Meeting, Geneva, April 2–3, 1998.

"Antimicrobial Resistance—Why the Irresponsible Use of Antibiotics in Agriculture Must Stop: A Briefing from the Alliance to Save Our Antibiotics." Alliance to Save Our Antibiotics, 2015.

"Antimicrobials in Agriculture and the Environment: Reducing Unnecessary Use and Waste." Review on Antimicrobial Resistance (O'Neill AMR Review), December 2015.

"Ban on Antibiotics as Growth Promoters in Animal Feed Enters into Effect." European Commission press release, December 22, 2005. http://europa.eu/rapid/press-release_IP-05-1687_en.htm.

"Certified Organic and Total US Acreage, Selected Crops and Livestock, 1995 and 2011." USDA Economic Research Service, Organic Production. http://www.ers.usda.gov/data-products/organic-production.aspx.

"Chemicals in Food Products." House Select Committee to Investigate the Use of Chemicals in Food Products, US House of Representatives. Washington, DC: GPO, 1951.

"Coccidiosis and Poultry Management." Rahway, NJ: Merck, 1958.

"Collection Antimicrobial Resistance (AMR)." https://www.gov.uk/government/collections/antimicrobial-resistance-amr-information-and-resources#parliamentary-inquiries, accessed July 20, 2019.

"Decision No 2119/98/EC." *Official Journal L 298*, October 3, 1998. http://eur-lex.europa.eu/LexUriServ/LexUriServ.do?uri=CELEX:31998D2119:EN:HTML, 0001–0007.

"DEFRA Antimicrobial Resistance Coordination. DARC Group." https://www.gov
.uk/government/groups/defra-antimicrobial-resistance-coordination-darc-group.

"Drugs in Livestock Feed." Washington, DC: Office of Technology Assessment, 1979.

"EARS-Net." https://ecdc.europa.eu/en/about-us/partnerships-and-networks/disease
-and-laboratory-networks/ears-net.

"Ensuring Access to Working Antimicrobials." House of Commons Science and
Technology Committee, First Report of Session 2014–15/HC 509, 2014.

"Extralabel Use and Antimicrobials." FDA Antimicrobial Resistance, August 24,
2018. https://www.fda.gov/animal-veterinary/antimicrobial-resistance/extralabel
-use-and-antimicrobials.

"FDA Proposal to Restrict the Use of Selected Antibiotics at Subtherapeutic Levels in
Animal Feeds (ERS-662). Washington, DC: USDA/ERS, 1977. "FDA's 1967 Look
after 60 Years of Reorganization." FDA Papers 1, no. 1. 1967: 9–12.

"FDA's Regulation of Animal Drug Residues in Milk." Human Resources and
Intergovernmental Relations Subcommittee of the Committee on Government
Operations, US House of Representatives. Washington, DC: GPO, 1990.

"Food Additives: What They Are, How They Are Used." Washington, DC: Manufacturing Chemists' Association, 1961.

"Food for Life: Campaigning for Local and Organic School Meals and Food Education." Bristol, UK: Soil Association.

"Food Safety and Quality: FDA Strategy Needed to Address Animal Drug Residues
in Milk." Washington, DC: GAO, 1992.

"Food Safety: Risk-Based Inspections and Microbial Monitoring Needed for Meat
and Poultry." Washington, DC: GAO, 1994.

"Food Safety: The Agricultural Use of Antibiotics and Its Implications for Human
Health." Washington, DC: GAO, 1999.

"Food Safety: USDA's Role Under the National Residue Program Should Be Reevaluated." Washington, DC: GAO, 1994.

"Food Standards Committee Report on Preservatives in Food." London: HMSO,
1959.

"Fourth Report: Food Safety." Select Committee on Agriculture, House of Commons, 1998.

"Gastro-Enteritis. Tees-Side." Parliamentary Debates vol. 762, April 11, 1968, col.
1619–1630. http://hansard.millbanksystems.com/commons/1968/apr/11/gastro
-enteritis-tees-side.

"George P. Larrick," FDA. https://www.fda.gov/about-fda/fdas-evolving-regulatory
-powers/george-p-larrick.

"Guidance for Industry #152: Evaluating the Safety of Antimicrobial New Animal
Drugs with Regard to Their Microbiological Effects on Bacteria of Human Health
Concern." FDA, October 2003.

"Guidance for Industry #209: The Judicious Use of Medically Important Antimicrobial Drugs in Food-Producing Animals." FDA, April 2012.

"Guidance for Industry #213: New Animal Drugs and New Animal Drug Combina-
tion Products Administered in or on Medicated Feed or Drinking Water of
Food-Producing Animals: Recommendations for Drug Sponsors for Voluntarily
Aligning Product Use Conditions with GFI #209." FDA, December 2013.

"Hearing to Review the Advances of Animal Health within the Livestock Industry."
Subcommittee on Livestock, Dairy, and Poultry of the Committee on Agriculture,
US House of Representatives, September 25, 2008. Washington, DC: GPO, 2009.

"H.R. 1549, Preservation of Antibiotics for Medical Treatment Act (PAMTA)." Committee on Rules, US House of Representatives, July 13, 2009. Washington, DC: GPO, 2009.

"H.R. 2508 Animal Drug Availability Act of 1996." https://www.congress.gov/bill /104th-congress/house-bill/2508/summary/36.

"Human Food Safety and the Regulation of Animal Drugs." Committee on Government Operations, US House of Representatives, 1985.

"Impacts of Antibiotic-Resistant Bacteria." Office of Technology Assessment, US Congress, September 1995.

"Interview of C. D. Van Houweling, by Ronald T. Ottes." *History of the US Food and Drug Administration, FDA Oral History Transcripts*, June 18, 1990. https://www .fda.gov/media/81081/download.

"Investigation of the Use of Chemicals in Food Products." Select Committee to Investigate the Use of Chemicals in Food Products, US House of Representatives, 1951.

"James L. Goddard." FDA. https://www.fda.gov/about-fda/fdas-evolving-regulatory -powers/james-l-goddard-md.

"Joint Committee on the Use of Antibiotics in Animal Husbandry and Veterinary Medicine." London: HMSO, 1969.

"Milestones in US Food and Drug Law History." FDA. https://www.fda.gov/about -fda/fdas-evolving-regulatory-powers/milestones-us-food-and-drug-law-history, accessed June 17, 2019.

"New Labour—Because Britain Deserves Better." 1997 Labour Party Manifesto. http://www.labour-party.org.uk/manifestos/1997/1997-labour-manifesto.shtml.

"Organic Standards." USDA. https://www.ams.usda.gov/grades-standards/organic -standards, accessed June 17, 2019.

"Our Smallest Servants: The Story of Fermentation." Brooklyn: Pfizer, 1955.

"Overuse of Antibiotics in Farming." Sustain—Save Our Antibiotics. http://www .sustainweb.org/antibiotics/overuse_of_antibiotics_in_farming/, accessed June 17, 2019.

"Penicillin Bill." Parliamentary Debates vol. 438, June 6, 1947, col. 822.

"Post Penicillin Antibiotics: From Acceptance to Resistance? A Witness Seminar Held at the Wellcome Institute for the History of Medicine, London, on 12 May 1998." London: Wellcome Trust, 2000.

"Preserving Antibiotic Access," *American Farm Bureau Federation—Priority Issues Antibiotics* (March 2015).

"Problems with FDA Monitoring for Animal Drug Residues: Is Our Milk Safe?" Human Resources and Intergovernmental Relations Subcommittee of the Committee on Government Operations, US House of Representatives. Washington, DC: GPO, 1992.

"Proceedings, FDA Conference on the Kefauver-Harris Amendments." Washington, DC: HEW/FDA, 1963.

"Proceedings of the First International Conference on the Use of Antibiotics in Agriculture." Washington, DC: NAS-NRC, 1955.

"Production Performance of Swine at Experiment Stations Ten Years after the Introduction of Aureomycin." Princeton, NJ: American Cyanamid Company, 1959.

"Proposed Organic Certification Program." Joint Hearing of Subcommittee on Domestic Marketing, Consumer Relations and Nutrition and the Subcommittee on Department Operations, Research, and Foreign Agriculture of the Committee on Agriculture, US House of Representatives, June 19, 1990. Washington, DC: GPO, 1991.

"Putting Meat on the Table: Industrial Farm Animal Production in America." PEW Commission on Industrial Farm Animal Production, 2008.

"Report and Recommendations on Organic Farming," USDA Study Team on Organic Farming, 1980. "Report of a Study: Human Health Risks with the Subtherapeutic Use of Penicillin or Tetracyclines in Animal Feed, 1988." Institute of Medicine Committee on Human Health Risk Assessment of Using Subtherapeutic Antibiotics in Animal Feeds, 1989.

"Report of the Joint Committee on the Use of Antibiotics in Animal Husbandry and Veterinary Medicine, 1969–1970." London: HSMO, 1969.

"Report of the Technical Committee to Enquire into the Welfare of Animals Kept Under Intensive Livestock Husbandry Systems." London: HMSO, 1965.

"Report on Microbial Antibiotic Resistance in Relation to Food Safety." London: Advisory Committee on the Microbiological Safety of Food, 1999.

"Report to the Commissioner of the Food and Drug Administration by the FDA Task Force on the Use of Antibiotics in Animal Feeds." FDA, 1972.

"Resistance to Antibiotics and Other Antimicrobial Agents." House of Lords, Select Committee Appointed to Consider Science and Technology, 1998.

"Revised Figures for Sales of Antimicrobial Products Used as Growth Promoters in the UK." VMD, 2005.

"Sales of Antimicrobial Products Authorised for Use as Veterinary Medicines in the UK in 2011." VMD, 2012.

"Sales of Antimicrobial Products Used as Veterinary Medicines, Growth Promoters and Coccidiostats in the UK from 1993–1998." VMD, 2000.

"Sales of Veterinary Antimicrobial Agents in 29 European Countries in 2014: Trends from 2011 to 2014." European Medicines Agency, 2016.

"Sales of Veterinary Antimicrobial Agents in 30 European Countries in 2016: Trends from 2010 to 2016." European Medicines Agency, 2018.

"Second Reading Therapeutic Substances. Prevention of Misuse Bill." Parliamentary Debates vol. 515, May 13, 1953, col. 1327–1343.

"Sir Alexander Fleming: Nobel Lecture." Nobel Prize, December 11, 1945. http://www.nobelprize.org/nobel_prizes/medicine/laureates/1945/fleming-lecture.html.

"Statement of Gregory Ahard before the Subcommittee on Interstate and Foreign Commerce on FDA's Regulation of Antibiotics Used in Animal Feeds." Washington, DC: GAO, 1977.

"Tackling Drug-Resistant Infections Globally: Final Report and Recommendations." Review of Antimicrobial Resistance (O'Neill AMR Review), May 2016.

"The Economic Effects of a Prohibition on the Use of Selected Animal Drugs." USDA Economics, Statistics and Cooperative Service, 1978.

"The Effects on Human Health of Subtherapeutic Use of Antimicrobials in Animal Feeds." Committee to Study the Human Health Effects of Subtherapeutic Antibiotic Use in Animal Feeds. Washington, DC: National Academies Press, 1980.

"The Medical Impact of Antimicrobial Use in Food Animals." Report of a WHO Meeting, Berlin, October 13–17, 1997.

"The Microbiological Safety of Food, Part II." Committee on the Microbiological Safety of Food. Richmond Committee, 1990.

"The Path of Least Resistance." Standing Medical Advisory Committee, Subgroup on Antimicrobial Resistance, 1998.

"The Public Health Aspects of Antibiotics in Feedstuffs". Report of a Working Group convened by the Regional Office for Europe of the World Health Organization, Bremen, 1-5 October 1973. Geneva: WHO, 1974.

"The Public Health Aspects of the Use of Antibiotics in Food and Feedstuffs." World Health Organization Technical Report Series 260. Geneva: WHO, 1963.

"The Report of the Expert Group on Animal Feedingstuffs to the Minister of Agriculture, Fisheries and Food, the Secretary of State for Health and the Secretaries of State for Wales, Scotland and Northern Ireland (Lamming Report)." 1992.

"The Use of Drugs in Animal Feeds." Washington, DC: NAS, 1967.

"UK Veterinary Antibiotic Resistance and Sales Surveillance (UK-VARSS 2013)." New Haw: VMD, 2014.

"UK Veterinary Antibiotic Resistance and Sales Surveillance Report (UK-VARSS) 2015." New Haw: VMD, 2016.

"UK Veterinary Antibiotic Resistance and Sales Surveillance Report (UK-VARSS) 2017." New Haw: VMD, 2018.

"US Broiler Performance." National Chicken Council (March 22, 2019). http://www.nationalchickencouncil.org/about-the-industry/statistics/u-s-broiler-performance/, accessed June 17, 2019.

"Use Medicated Feeds Carefully and Wisely." *FDA Papers* 1, no. 6. July–August 1967: 45.

"Use of Antibiotics, Other Drugs, and Vitamin B12 at Low Levels in Formula Feeds". Washington, DC: USDA Agricultural Marketing Service, 1956.

"Use of Quinolones in Food Animals and Potential Impact on Human Health." Report of a WHO Meeting, Geneva, June 2–5, 1998.

"US Organic Sales Set a New Record in 2016." Organic Trade Association, August 14, 2017. http://theorganicreport.com/us-organic-sales-set-new-record-2016.

"Veterinary Medicines Regulations." August 15, 2017. https://www.gov.uk/guidance/veterinary-medicines-regulations.

"Veterinary Medicines Regulations: Changes to Advertising Rules." October 11, 2012.

"Vitamin B12 and Antibiotics in Animal Nutrition: Annotated Bibliography." Rahway, NJ: Merck, 1951, reprints 1952, 1954, 1957.

"World Markets for Organic Fruit and Vegetables." http://www.fao.org/docrep/004/y1669e/y1669eof.htm, accessed June 17, 2019.

"Written Answers to Questions, 10.06.1996." House of Commons Written Questions and Answers, 1996, col. 9.

Published Primary Sources

Abrams, John T. *Animal Nutrition and Veterinary Dietetics*, 4th ed. Baltimore: Williams and Wilkins, 1962.

Agricola Training. *The Insiders Guide to Agriculture—Broilers* (1993).

———. *The Insiders Guide to Agriculture—Outdoor Pigs* (1992).

———. *The Insiders Guide to Agriculture—Pigs* (1993).

Alliance to Save Our Antibiotics. "Farm Antibiotic Use in the Netherlands." http://www.saveourantibiotics.org/media/1751/farm-antibiotic-use-in-the-netherlands.pdf.

Alternative Agriculture. Organic Farming. Haughley: Soil Association, 1974.

Anderson, E. S. "Middlesbrough Outbreak of Infantile Enteritis and Transferable Drug Resistance." *BMJ* 1, no. 5887 (1968): 293.

———. "Origin of Transferable Drug-resistance Factors in the Enterobacteriaceae." *BMJ* 2, no. 5473 (1965): 1289–1291.

Anderson, E. S., and Naomi Datta. "Resistance to Penicillins and Its Transfer in Enterobacteriaceae." *Lancet* 285, no. 7382 (1965): 407–409.

Anderson, E. S., and M. J. Lewis. "Drug Resistance and Its Transfer in Salmonella typhimurium." *Nature* 206, no. 4984 (1965): 579–583.

Anderson, E. S., and H. R. Smith. "Chloramphenicol Resistance in the Typhoid Bacillus." *BMJ* 3, no. 5822 (1972): 329.

Anthony, John David. *Diseases of the Pig and Its Husbandry.* Baltimore: Williams and Wilkins, 1955.

"Antibiotics for Farm Animals." *Lancet* 273, no. 7069 (1959): 402.

Arehart-Treichel, Joan. "Antibiotics to Animal Feeds: Threat to Human Health?" *Science News* 101, no. 22 (1972): 349–350.

Atkinson, Barbara, and Victor Lorian. "Antimicrobial Agent Susceptibility Patterns of Bacteria in Hospitals from 1971 to 1982." *Journal of Clinical Microbiology* 20, no. 4 (1984): 791–796.

Bacharach, A. L. "Antibiotics and Hazards." *Food Drug Cosmetic Law Journal* 505 (1957): 505–513.

———. "UK Position on Use of Antibiotic Food Additives." *Chemical Age* 78 (1957): 176.

Baldwin, Robert A., and LaVerne C. Harold. "Ecologic Effects of Antibiotics." *FDA Papers* 1, no. 1 (February 1967): 20–24.

Banks, Stuart. *The Complete Handbook of Poultrykeeping.* London: Ward Lock Limited, 1979.

Barber, R. S. "Antibiotics in the Diet of the Fattening Pig." *British Journal of Nutrition* 7, no. 4 (1953): 306–319.

———. "Effect of Supplementing an All-Vegetable Pig-Fattening Meal with Antibiotics." *Chemistry and Industry* (1952): 713.

Barnes, Ella M. "The Use of Antibiotics for the Preservation of Poultry and Meat." *Antibiotics in Agriculture—Proceedings of the Fifth Symposium of the Group of European Nutritionists, 1966.* Basel: S. Karger, 1968.

Basic Book of Organically Grown Foods, The. Emmaus, PA: Rodale, 1972.

Batty, J. *Pictorial Poultry-Keeping,* 3rd ed. Hindhead: Spur Publications, 1980.

Becker, Geoffrey S. "Animal Agriculture: Selected Issues in the 108th Congress." CRS Report for Congress. Washington, DC: Congressional Research Service, 2003.

———. "Antibiotic Use in Agriculture: Background and Legislation." CRS Report for Congress. Washington, DC: Congressional Research Service, 2009.

———. "Food Safety: Selected Issues and Bills in the 110th Congress." CRS Report for Congress. Washington, DC: Congressional Research Service, 2007.

Bicknell, Franklin. *Chemicals in Food and in Farm Produce: Their Harmful Effects.* London: Faber and Faber, 1960.

———. *The English Complaint or Your Fatigue and its Cure.* London: William Heinemann, 1952.

Bird, H. R. "Biological Basis for the Use of Antibiotics in Poultry Feeds." *The Use of Drugs in Animal Feeds: Proceedings of a Symposium.* Washington, DC: NAS, 1969, 31–41.

Bissell, Francis. *The Organic Meat Cookbook*. London: Ebury Press, 1999.

Blake, Francis. *The Handbook of Organic Husbandry*. Marlborough: Crowood Press, 1987.

Booth, Rintoul. *The Farming Handbook to End All Farming Handbooks*. London: Wolfe, 1970.

Brander, G. C. "The Use of Antibiotics in the Veterinary Field in the 1970s." In *Antibiotics and Antibiosis in Agriculture*, edited by Malcolm Woodbine, 199–209. London: Butterworths, 1977.

Brandly, C. A., "Major Poultry Disease Problems." *Canadian Journal of Comparative Medicine* XVII, no. 8 (1953): 332.

Braude, R. "Antibiotics in Animal Feeds in Great Britain." *Journal of Animal Science* 46 (1978): 1425–1436.

Braude, R., S. K. Kon, and K. G. Mitchell. "The Value of Antibiotics for Fattening Pigs." *British Journal of Nutrition* 5 (1951): viii.

Braude, R., and K. G. Mitchell. "Antibiotics and Liver Extract for Suckling Pigs." *British Journal of Nutrition* 6, no. 1 (1952): 398–400.

Braude, R., H. D. Wallace, and T. J. Cunha. "The Value of Antibiotics in the Nutrition of Swine: A Review." *Antibiotics and Chemotherapy* 3 (1953): 271–291.

Brooks, M. B., P. S. Morley, D. A. Dargatz, and D. R. Hyatt. "Survey of Antimicrobial Susceptibility Testing Practices of Veterinary Diagnostic Laboratories in the US." *JAVMA* 222, no. 2 (2003): 168–173.

Buckett, M. *Introduction to Livestock Husbandry*. Oxford: Pergamon Press, 1965, 1977.

Bullock, B. F. *Practical Farming for the South*. Chapel Hill: University of North Carolina Press, 1944.

Byerly, T. C. "The Role of Research in Solving the Poultry Condemnation Problem." In *Disease, Environmental and Management Factors Related to Poultry Health*, 6–9. Washington, DC: ARS, 1961.

Cadwallader, Sharon, and Judi Ohr. *Whole Earth Cookbook*. San Francisco: San Francisco Book Co., 1973.

Canter, David, Kay Canter, and Daphne Swann. *The Cranks Recipe Book*. London: J. M. Dent, 1982.

Carson, Rachel. *Silent Spring*. New York: Houghton Mifflin, 1962.

Carter, G. R. *Essentials of Veterinary Microbiology*, 5th ed. Baltimore: Williams and Wilkins, 1995.

CAST. "The Phenoxy Herbicides." *Weed Science* 23, no. 3 (May 1975): 259.

Cavanagh, Ursula M. *Cooking and Catering the Wholefood Way*. London: Faber and Faber, 1970.

———. *The Wholefood Cookery Book*. London: Faber and Faber, 1971.

Chamberlain, A. T., J. M. Walsingham, and B. A. Stark, eds. *Organic Meat Production in the '90s*. Maidenhead, UK: Chalcombe, 1989.

Chappell, Connery. *Keeping Pigs*. London: Rupert Hart-Davis, 1953.

Cohen, Mitchell L. "Epidemiology of Drug Resistance—Implications for a Post-Antimicrobial Era." *Science* 257, no. 5073 (1992): 1050–1055.

Collins, J. H. "The Problem of Drugs for Food-Producing Animals and Poultry." *Food Drug Cosmetic Law Journal* 6 (November 1951): 872–878.

Colorado Agricultural Handbook. Fort Collins: Colorado State University, 1952.

Commercial Broiler Production [Agriculture Handbook 320]. Washington, DC: USDA, 1967.

Connell, Philip G. "Radicalism in Food Production." *Vital Speeches of the Day* (November 11, 1974): 70–72.

Corley, Hugh. *Organic Farming.* London: Faber and Faber, 1957.

"Cornelius D. Van Houweling." *JAVMAnews*, April 1, 2012.

Cox, Peter. *Why You Don't Need Meat.* New York: Thorsons, 1986.

Craigie, J. "Arthur Felix 1887–1956." *Biographical Memoirs of Fellows of the Royal Society* 3 (1957): 52–79.

Crampton, Earle W. *Applied Animal Nutrition.* San Francisco: W. H. Freeman, 1956.

Cunha, Tony J. *Swine Feeding and Nutrition.* New York: Interscience, 1957.

Damerow, Gail. *The Chicken Health Handbook.* Pownall, VT: Storey Communications, 1994.

Darby, William J. "Review, *The Poisons in Your Food* by William Longgood." *Science* 131, no. 3405 (1960): 979.

Datta, Naomi. "Herbert Williams Smith: 3 May 1919–16 June 1987." *Biographical Memoirs of Fellows of the Royal Society* 34 (1988): 754–786.

Deadman, Peter, and Karren Betteridge. *Nature's Foods.* London: Rider, 1977.

Deatherage, F. E. "The Use of Antibiotics in the Preservation of Foods Other than Fish." In *Proceedings First International Conference on the Use of Antibiotics in Agriculture*, 211–222. Washington, DC: NAS-NRC, 1956.

Delaphane, J. P., and J. H. Milliff. "The Gross and Micropathology of Sulphaquinoxaline Poisoning in Chickens." *American Journal of Veterinary Research* 92 (1948): 9.

Dettenkofer, M., M. Ackermann, M. Eikenberg, and H. Merkel. *Auswirkungen des Einsatzes von Antibiotika und Substanzen mit antibiotischer Wirkung in der Landwirtschaft und im Lebensmittelsektor. Ein Literatur-Review.* Ernährungs Wende Produkte, Materialienband Nr. 4; Freiburg: Institut für Umweltmedizin und Krankenhaushygiene am Universitätsklinikum Freiburg, 2004.

Deyoe, George P., and J. L. Krider. *Raising Swine.* New York: McGraw-Hill, 1952.

Dingell, John D. "Animal Feeds: Effect of Antibiotics." *Science* 209, no. 4461 (1980): 1069.

Doane's Facts and Figures for Farmers, 4th ed. St. Louis: Doane-Western, 1981.

Downing, Elisabeth. *Keeping Pigs.* London: Pelham, 1978.

"Drug-Resistant Staphylococci in the Farmyard." *Lancet* 275, no. 7138 (1960): 1338–1339.

Dubos, René J. "Microbiology." *Annual Review of Biochemistry* 11 (1942): 659–678.

Duff, Gail. *Good Housekeeping Wholefood Cookery.* London: Ebury, 1989.

Dyer, Irwin, and Clayton O'Mary. *The Feedlot*, 2nd ed. Philadelphia: Lea and Febiger, 1977.

Eastern States Farmers' Exchange. *Eastern States Farmers Handbook*, 1947.

The Eco-Cookbook, 2nd ed. Southampton, UK: Friends of the Earth, 1975.

Edwards, S. J. "Antibiotics in the Treatment of Mastitis." In *Antibiotics in Agriculture: Proceedings of the University of Nottingham Ninth Easter School in Agricultural Science*, 43–57. London: Butterworths, 1962.

Ehrlich, Paul. *The Population Bomb.* New York: Ballantine, 1968.

Elliott, S. D., and Ella M. Barnes. "Changes in Serological Type and Antibiotic Resistance of Lancefield Group D Streptococci in Chickens Receiving Dietary Chlortetracycline." *Journal of General Microbiology* 20 (1959): 426–433.

Ensminger, M. E. *The Stockman's Handbook*, 1st, 2nd, 3rd, 5th, 7th eds. Danville, IL: Interstate, 1955, 1959, 1962, 1978, 1992.

———. *Swine Science.* Danville, IL: Interstate, 1961.

Farm Handbook, 4th ed. Ames: Iowa State University Press, 1957.

"FDA Stymied on Antibiotic Ban." *BioScience* 30, no. 5 (1980): 295–296.

Finland, Maxwell. "Emergence of Resistant Strains in Chronic Intake of Antibiotics: A Review." In *First International Conference on Antibiotics in Agriculture, 19–21 October 1955*, 233–258. NAS-NRC, 1956.

Fishwick, V. C. *Pigs: Their Breeding, Feeding and Management*, 2nd, 5th, 6th eds. London: Crosby Lockwood, 1945, 1953, 1956.

Flynn, J. E. "History of the Word 'Antibiotic'/Discussion between S. A. Waksman and J. E. Flynn on 19 January 1962." *Journal of the History of Medicine and Allied Sciences* XXVIII, no. 3 (1973): 284–286.

Garrod, Lawrence P. "Sources and Hazards to Man of Antibiotics in Foods." *Proceedings of the Royal Society of Medicine* 54 (1964): 1087–1088.

Geiger, Glenn, and Walter Russell. *Fact Sheet for Part Time Farmers and Gardeners*. Washington, DC: USDA, 1977.

Georgia Agricultural Handbook. Athens: University of Georgia, 1961.

Gillespie, Gilbert Wayne, Jr. *Antibiotic Animal Feed Additives and Public Policy: Farm Operators' Beliefs About the Importance of These Additives and Their Attitudes Toward Government Regulation of Agricultural Chemicals and Pharmaceuticals*. Doctoral thesis, Cornell University, 1987.

Gilliam, Henry C., Jr., and Rod Martin. "Economic Importance of Antibiotics in Feeds to Producers and Consumers of Pork, Beef and Veal." *Journal of Animal Science* 40, no. 6 (1975): 1241–1255.

Goldstein, Jerome. "Organic Force." In *The New Food Chain: An Organic Link Between Farm and City*, 1–19. Emmaus, PA: Rodale, 1973.

Good Housekeeping Wholefoods Cook Book. London: Ebury Press, 1971.

Goodman, R. N., and H. S. Goldberg. "Antibiotics in Agriculture." *Science* 137, no. 3524 (1962): 133–135.

Gorbach, S. L. "Time to Stop." *NEJM* 345, no. 16 (2001): 1202–1203.

Gordon, R. F. "Problems Associated with a. The Production of Table Poultry; b. The Rearing of Turkey Poults." In *Report of Proceedings: Conference on the Feeding of Farm Animals for Health And Production* (MERL). London: BVA, 1954.

Gould, J. C. "Origin of Antibiotic-Resistant Staphylococci." *Nature* 180, no. 4580 (1957): 282–283.

Grant, Doris. *Housewives Beware*. London: Faber and Faber, 1958.

———. *Your Bread and Your Life*. London: Faber and Faber, 1961.

———. *Your Daily Bread*. London: Faber and Faber, 1944.

———. *Your Daily Food: Recipe for Survival*. London: Faber and Faber, 1973.

Grau, C. R., F. H. Kratzer, and W. E. Newlon. *Principles of Nutrition for Chickens and Turkeys*. Berkeley: University of California, Division of Agricultural Sciences, 1956.

Greene, Catherine R. "US Organic Farming Emerges in the 1990s: Adoption of Certified Systems." *ERS Agriculture Information Bulletin* 770 (2001): 3.

Griffin, E. D. "The US Feed Industry." In *Livestock Nutrition—Report of Four Conferences Held During the Feed Show at the US Trade Center, London*. London: Graham Cherry, 1962.

Haberman, Jules. *Poultry Farming for Profit*. Englewood Cliffs, NJ: Prentice Hall, 1956.

Hamburg, Margaret. "FDA Strategies for Combating Antimicrobial Resistance." *Speeches by FDA Officials*, 2014.

Hanford, Margaret. *The WI Book of Wholefood Cookery*. London: WI Books, 1985.

Harris, Richard. *The Real Voice.* New York: Macmillan, 1964.

Harrison, H. S. *The Law on Medicines—A Comprehensive Guide,* vol. 1. Boston: MTP, 1986.

Harrison, Ruth. *Animal Machines.* London: Vincent Stuart, 1964.

Hartley, C. L., and M. H. Richmond. "Antibiotic Resistance and Survival of E Coli in the Alimentary Tract." *BMJ* 4, no. 5988 (1975): 71–74.

Harvey, John, and Liz Mason. *The Use and Misuse of Antibiotics in UK Agriculture. Part 1: Current Usage.* Bristol: Soil Association, 1998.

Haseman, Leonard. "Sulfathiazole Control of American Foulbrood." *University of Missouri Agricultural Experiment Station Circular* 341 (1949): 1–10.

Hays, Virgil W., and Charles A. Black. *Antibiotics for Animals: The Antibiotic Resistance Issue. Comments from CAST.* Ames, IA: CAST, 1989.

Heffernan, William D., and Gary Green. "Appendix A: Opinions of Missouri Farmers Concerning Use of Antibiotics in Animal Feeds and the Consequences of Banning Their Use." In *Antibiotics in Animal Feeds* (CAST Report No. 88), 72–75. Ames, IA: CAST, 1981.

Herber, Lewis [Pseudonym for Murray Bookchin]. *Our Synthetic Environment.* New York: Knopf, 1962.

Herriot, James. *Every Living Thing.* London: Pan, 1992.

———. *Let Sleeping Vets Lie.* London: Pan, 1973.

———. *Vet in Harness.* London: Pan, 1974.

Heuser, Gustave. *Feeding Poultry,* 2nd ed. New York: Wiley, 1955.

Hewitt, Jean. *The NYT Natural Foods Cookbook.* Chicago: Quadrangle, 1971.

Hills, Lawrence D. *The Wholefood Finder.* Bocking, UK: Henry Doubleday, 1968.

Hoard's Dairyman Feed Guide: Dairy Cattle, Swine, Poultry. Fort Atkinson, WI: Hoard's Dairyman, 1958.

Hobbs, Betty C., et al. "Antibiotic Treatment of Poultry in Relation to Salmonella typhimurium." *Monthly Bulletin of the Ministry of Health and the Public Health Laboratory* 19 (October 1960): 178–192.

Holmberg, Scott D., M. T. Osterholm, K. A. Senger, and M. L. Cohen. "Drug-Resistant Salmonella from Animals Fed Antimicrobials." *NEJM* 311, no. 10 (1984): 617–622.

Holmberg, Scott D., Joy G. Wells, and Mitchell L. Cohen. "Animal-to-Man Transmission of Antimicrobial-Resistant Salmonella: Investigations of US Outbreaks, 1971–1983." *Science* 225, no. 4664 (1984): 833–835.

Huber, W. G. "The Impact of Antibiotic Drugs and Their Residues." *Advances in Veterinary Science and Comparative Medicine* 15 (1971): 101–130.

"Infectious Drug Resistance." *NEJM* 275, no. 5 (1966): 277.

International Historical Statistics. Basingstoke, UK: Palgrave Macmillan, 2013.

James, A. *Modern Pig-Keeping.* London: Cassell, 1952.

Janssen, Wallace F. "FDA Since 1938: The Major Trends and Developments." *Journal of Public Law* 13, no. 1 (1964): 205–221.

Jarolmen, Howard. "Experimental Transfer of Antibiotic Resistance in Swine." *Annals of the New York Academy of Sciences* 182 (1971): 72–79.

Johansson, K. R. et al. "Effects of Dietary Aureomycin upon the Intestinal Microflora and the Intestinal Synthesis of Vitamin B12 in the Rat." *Journal of Nutrition* 49 (1953): 135–152.

Jukes, Thomas H. "Antibiotics in Feeds." *Science* 204, no. 4388 (1979): 8.

———. *Antibiotics in Nutrition.* New York: Medical Encyclopedia, 1955.

———. "DDT: Affluent Enemy or Beneficial Friend." *BioScience* 19, no. 7 (1969): 640–641.

———: "The Present Status and Background of Antibiotics in the Feeding of Domestic Animals." *Annals of the New York Academy of Sciences* 182 (1971): 362–379.

———. "Some Historical Notes on Chlortetracycline." *Reviews of Infectious Diseases* 7, no. 5 (1985): 702–707.

Jukes, Thomas H., and Richard S. Miller. "The Right to Be Heard." *BioScience* 18, no. 4 (1968): 272.

Kampelmacher, E. H. "Some Aspects of the Non-Medical Use of Antibiotics in Various Countries." In *Antibiotics in Agriculture: Proceedings of the University of Nottingham Ninth Easter School in Agricultural Science*, 315–332. London: Butterworths, 1962.

Katz, Stanley E., and Marietta Sue Brady. "Incidence of Residues in Foods of Animal Origin." In *Analysis of Antibiotic/ Drug Residues in Food Products of Animal Origin*, edited by Vipin K. Agarwal, 5–21. New York: Plenum, 1992.

Kenan, William Rand. *History of Randleigh Farm*, 8th, 9th ed. Lockport, NY: W. R. Kenan, 1956, 1959.

Kent-Jones, D. W. "Obituary: Alastair Campbell Frazer." *Proceedings of the Society for Analytical Chemistry* 6 (1969): 209–210.

Kingma, Fred J. "Establishing and Monitoring Drug Residue Levels." *FDA Papers* 1, no. 6 (July–August 1967): 8–9, 31–33.

Koshland, Daniel E. "The Microbial Wars." *Science* 257, no. 5073 (1992): 1021–1022.

Kuehn, Bridget M. "Frances Kelsey Honored for FDA Legacy." *JAMA* 304, no. 19 (2010): 2109–2112.

Lappé, Frances Moore, *Diet for a Small Planet*. New York: Friends of the Earth/ Ballantine, 1971.

Leslie, J. C. "The World Supply Position of Animal Feedstuffs, and Its Influence on the Livestock Industry." *Report of Proceedings—Conference on the Feeding of Farm Animals for Health and Production* (MERL). London, 1954, 13–18.

Levy, Stuart B. *The Antibiotic Paradox: How Miracle Drugs Are Destroying the miracle*. New York: Plenum, 1992.

———. *The Antibiotic Paradox: How the Misuse of Antibiotics Destroys Their Curative Powers*. Cambridge, MA: Perseus, 2002.

———. "Playing Antibiotic Pool: Time to Tally the Score." *NEJM* 311, no. 10 (1984): 663–665.

———. "Spread of Antibiotic Resistance Plasmids from Chicken to Chicken and from Chicken to Man." *Nature* 260 (1976): 40–42.

Levy, Stuart B., George B. Fitzgerald, and Ann B. Macone. "Changes in Intestinal Flora of Farm Personnel After Introduction of a Tetracycline-Supplemented Feed on a Farm." *NEJM* 295, no. 11 (1976): 583–588.

Linton, A. H. "Antibiotics and Man—An Appraisal of a Contentious Subject." In *Antibiotics and Antibiosis in Agriculture*, edited by Malcolm Woodbine, 315–343. London: Butterworths, 1977.

———. "The Swann Report and its Impact." In *The Control of Antibiotic Resistant Bacteria*, edited by Charles H. Stuart-Harris and David M. Harris, 183–194. London: Academic Press, 1982.

Livestock Book. Successful Farming, eds. Des Moines, IA: Meredith, 1957.

Logue, John T. "The Public Health Significance of Antibiotics." In *Antibiotics—Their*

Chemistry and Non-medical Uses, edited by Herbert S. Goldberg, 561–598. Princeton, NJ: Van Nostrand, 1959.

Longgood, William. *The Poisons in Your Food*. New York: Simon and Schuster, 1960.

Luckey, T. D. "Antibiotics in Nutrition." In *Antibiotics Their Chemistry and Non-medical Uses*, edited by Herbert S. Goldberg, 174–321. Princeton, NJ: Van Nostrand, 1959.

———. "Sir Samurai T. D. Luckey." *Dose-Response* (2008): 1–16.

Maddock, Helen, and Sterling Brackett. *The Continuous Feeding of Aureomycin to Swine*. New York: American Cyanamid, 1956.

Mahoney, Tom. *The Merchants of Life: An Account of the American Pharmaceutical Industry*. New York: Harper, 1959.

Manual of Meat Inspection Procedures of the USDA. Washington, DC: USDA, 1964.

Marsden, Stanley, and J. Holmes Martin. *Turkey Management*, 6th ed. Danville, IL: Interstate, 1955.

Marshall, Eliot. "Health Committee Investigates Farm Drugs." *Science* 209, no. 4455 (1980): 481.

———. "NAS Begins Study of Antibiotics in Feeds." *Science* 205, no. 4410 (1979): 982.

———. "Scientists Quit Antibiotics Panel at CAST." *Science* 203, no. 4382 (1979): 732–733.

Martin, B. "Farm Animal Disease: Veterinarian and Antibiotics." In *Antibiotics and Antibiosis in Agriculture*, edited by Malcolm Woodbine, 211–216. London: Butterworths, 1977.

Matsushima, John K. *Feeding Beef Cattle*. New York: Springer, 1979.

Mayall, Sam. *Farming Organically*. Haughley, UK: Soil Association, 1976.

McBride, William D., and Nigel Key. "US Hog Production from 1992 to 2009: Technology, Restructuring, Productivity Growth." *Economic Research Report* 158 (2013).

McBride, William D., Nigel Key, and Kenneth H. Matthews. "Subtherapeutic Antibiotics and Productivity in US Hog Production." *Review of Agricultural Economics* 30, no. 2 (2008): 270–288.

McConnell, Primrose. *The Agricultural Notebook*, 12th–20th eds. London: Butterworths, 1953, 1958, 1962, 1968, 1976, 1982, 1988, 1995, 2003.

McDonald, J. C., D. L. Miller, M. P. Jevons, and R. E. O. Williams. "Nasal Carriers of Penicillin-Resistant Staphylococci in Recruits to the RAF." *Proceedings of the Royal Society of Medicine* 53 (1960): 255–258.

McKay, K. A. *A Study of the Intestinal Bacterial Flora of Swine Fed Terramycin Supplemented Rations*, graduate thesis, Toronto, 1954.

McKay, K. A., and H. D. Branion. "The Development of Resistance to Terramycin by Intestinal Bacteria of Swine." *Canadian Veterinary Journal* 1, no. 4 (1960): 144–150.

McKay, K. A., Louise H. Ruhnke, and D. A. Barnum. "The Results of Sensitivity Tests on Animal Pathogens Conducted OVER THE Period 1956–1963." *Canadian Veterinary Journal* 6, no. 5 (1965): 103–111.

McKay, W. M. "The Use of Antibiotics in Animal Feeds in the UK: The Impact and Importance of Legislative Controls." *World's Poultry Science Journal* 31, no. 2 (1975): 116–128.

McMillen, Warren. *Hog Profits for Farmers*. Chicago: Windsor, 1952.

Merchant, I. A., and R. A. Packer. *Veterinary Bacteriology and Virology*, 6th ed. Ames: Iowa State University Press, 1956, 1961.

Meyer Jones, L. *Veterinary Pharmacology and Therapeutics.* Ames: Iowa State University Press, 1957.

Midtvedt, Tore. "The Microbial Threat: The Copenhagen Recommendation." *Microbial Ecology in Health and Disease* 10 (1998): 65–67.

Midwest Farm Handbook, 1st, 2nd, 6th, 7th eds. Ames: Iowa State University Press, 1949, 1951, 1964, 1969.

Millman, Jack M., et al., "Prevalence of Antibiotic-Resistant E. coli in Retail Chicken: Comparing Conventional, Organic, Kosher, and Raised without Antibiotics. Version 2." *F1000Research 2, no. 155.* Published online September 2, 2013.

Mitchelmore, Peter. *The Pigkeeper's Guide.* Newton Abbot, UK: David and Charles, 1981.

Moats, William A., ed. *Agricultural Uses of Antibiotics.* American Chemical Society Symposium Series 320. Washington, DC: American Chemical Society, 1986.

Modern Poultry Keeping, 1st, 2nd eds. London: English Universities Press, 1965, 1972.

Moore, P. R., A. Evenson, T. D. Luckey, E. McCoy, C. A. Elvehjem, and E. B. Hart. "Use of Sulfasuxidine, Streptothricin, and Streptomycin in Nutritional Studies with the Chick." *Journal of Biological Chemistry* 165 (1946): 437–441.

Morrison, Frank. *Feeds and Feeding.* Clinton, IA: Morrison, 1961.

Murneek, A. E. "Thiolutin as a Possible Inhibitor of Fireblight." *Phytopathology* 42 (1952): 57.

Napach, Bernice. "How Four Companies Control Nearly All the Meat You Eat." *Yahoo Finance*, February 19, 2014. http://finance.yahoo.com/blogs/daily-ticker /how-four-companies-control-the-supply-and-price-of-beef--pork-and-chicken-in -the-u-s-eat-prices-224406080.html.

National Resources Defense Council. "Chain Reaction Report Urges Burger Restaurants to Beef Up Policies to Eliminate Routine Use of Antibiotics." Press release, October 17, 2018.

Nature's Ag School: The Thompson Farm. Emmaus, PA: Regenerative Agriculture Association, 1986.

Nicholson, John. "Biotechnology and Agriculture: The Counterpoint." In *Poison in Your Food*, edited by Gary McCuen, 148–153. Hudson, WI: G. E. McCuen, 1991.

Noorlander, Daniel. *Milking Machines and Mastitis*, 2nd ed. Madison, WI: Democratic Print, 1962.

Novick, Richard. "Use in Animal Feed." *Science* 204, no. 4396 (1979): 908.

O'Brien, Thomas, et al. "Molecular Epidemiology of Antibiotic Resistance in Salmonella from Animals and Human Beings in the United States." *NEJM* 307, no. 1 (1982): 1–6.

O'Casey, Brenda. *Natural Baby Food.* London: Duckworth, 1977.

Owens, F. N., J. Zorrilla-Rios, and P. Dubeski. "Effects of Ionophores on Metabolism, Growth Body Composition and Meat Quality." In *Growth Regulation in Farm Animals*, edited by A. M. Pearson and T. R. Dutson, 321–342. London: Elsevier, 1991.

Page, S. W. "Current Use of Antimicrobial Growth Promoters in Food Animals: The Benefits." In *Antimicrobial Growth Promoters: Where Do We Go from Here*, edited by D. Barug, J. Dejong, A. K. Kies, and M. W. A. Verstegen. Wageningen, NL: Wageningen Academic, 2006.

Pearson, J. K. L. "Intramammary Therapy: Its Achievements and Limitations." In *Antibiotics and Antibiosis in Agriculture*, edited by Malcolm Woodbine, 217–228. London: Butterworths, 1977.

Perry, Tilden Wayne. *Beef Cattle Feeding and Nutrition.* New York: Academic Press, 1980.

Pictorial Poultry-Keeping. London: Pearson, 1962.

Poultry Inspectors' Handbook. Washington, DC: GPO, 1957.

Poultry Keeping for Profit. London: Country Life, 1952.

Power, R. A. *100 Common Mistakes in Farming and How to Correct Them.* Viroqua, WI: National Farm, 1947.

Practical Poultry Keeping, 3rd, 5th eds. London: Poultry World, 1946, 1955.

"Pranks and Greens." *The Simpsons*, season 21, episode 6, 2009.

Quick, Vivien, and Clifford Quick. *Everywoman's Wholefood Cook Book.* Wellingborough, UK: Thorsons, 1974.

Randall, William A. "Antibiotic Residues." In *Proceedings First International Conference on the Use of Antibiotics in Agriculture*, 259–263. Washington, DC: NAS-NRC, 1956.

Reid, B. L. *Antibiotics and Arsenicals in Poultry Nutrition.* College Station: Texas Agricultural Experiment Station, 1957.

Rifkin, Jeremy. *Beyond Beef: The Rise and Fall of the Cattle Culture.* London: Penguin Books, 1992.

Rikard-Bell, H. M. *The Handbook of Modern Pig Farming.* London: H. F. and G. Witherby, 1938.

Rodale, J. I., et al., *The Complete Book of Food and Nutrition.* Emmaus, PA: Rodale, 1961.

———. *The Encyclopedia of Organic Gardening.* Emmaus, PA: Rodale, 1959.

Rorty, James, and N. Philip Norman. *Tomorrow's Food: The Coming Revolution in Nutrition*, 2nd ed. New York: Devin-Adair, 1956.

"Salmonellosis—An Unhappy Turn of Events." *Lancet* 313, no. 8124 (1979): 1009–1010.

Schell, Orville. *Modern Meat: Antibiotics, Hormones and the Pharmaceutical Farm.* New York: Random House, 1985.

Schroth, M. H., S. V. Thomson, D. C. Hildebrand, and W. J. Moller. "Epidemiology and Control of Fire Blight." *Annual Review of Phytopathology* 12 (1974): 389–412.

Seiden, Rudolph. *Poultry Handbook: An Encyclopedia for Good Management of All Poultry Breeds.* New York: Van Nostrand, 1952.

Seiden, Rudolph, and W. H. Pfander. *The Handbook of Feedstuffs.* New York: Springer, 1957.

Self-Sufficient Small Holding. Haughley, UK: Soil Association, 1976.

Shane, Barry, and Kenneth Carpenter. "E. L. Stokstad: 1913–1995." *Journal of Nutrition* 127 (1997): 199–201.

Siegel, D., W. G. Huber, and S. Drysdale. "Human Therapeutic and Agricultural Uses of Antibacterial Drugs and Resistance of the Enteric Flora of Humans." *Antimicrobial Agents and Chemotherapy* 8, no. 5 (1975): 538–543.

Siegel, D., W. G. Huber, and F. Enloe. "Continuous Non-Therapeutic Use of Antibacterial Drugs in Feed and Drug Resistance of the Gram-Negative Enteric Florae of Food-Producing Animals." *Antimicrobial Agents and Chemotherapy* 6, no. 6 (1974): 697–701.

Snapp, Roscoe, and A. L. Neumann. *Beef Cattle*, 4th, 5th, 6th eds. New York: Wiley, 1952, 1960, 1969.

Sneeringer, Stacy, James MacDonald, Nigel Key, William McBride, and Ken

Mathews. "Economics of Antibiotic Use in US Livestock Production." *USDA Economic Research Report* 200 (2015).

Starr, Mortimer P., and Donald M. Reynolds. "Streptomycin Resistance of Coliform Bacteria from Turkeys Fed Streptomycin." *American Journal of Public Health* (November 1951): 1377–1379.

Stern, Alice. *Poultry and Poultry-Keeping.* London: Merehurst, 1988.

Stokstad, E. L. R., T. H. Jukes, J. Pierce, A. C. Page Jr., and A. L. Franklin. "The Multiple Nature of the Animal Protein Factor." *Journal of Biological Chemistry* 180, no. 2 (1949): 640–654.

Stonehouse, B. *Biological Husbandry: A Scientific Approach to Organic Farming.* London: Butterworths, 1981.

Straiton, Eddie. *Pig Ailments: Recognition and Treatment—TV Book for Pig Farmers,* 6th ed. Ipswich, UK: Farming Press, 1988.

Stuart-Harris, Charles H., and David M. Harris, eds. *The Control of Antibiotic Resistant Bacteria.* London: Academic Press, 1982.

Taft, Robert. *Residues in Dairy Products: Course Manual Milk and Food Training.* Cincinnati: HEW/PHS, 1961.

Thatcher, F. S., and W. Simon. "The Resistance of Staphylococci and Streptococci Isolated from Cheese to Various Antibiotics." *Canadian Journal of Public Health* 46, no. 10 (1955): 407–409.

Thom, Robert A. "The Era of Antibiotics." *Great Moments in Pharmacy,* Painting No. 39. Parke, Davis and Company Advertising Series, 1950s.

Thomas, Keith. *Livestock Feeding Manual.* Danville, IL: Interstate, 1951.

Thomson, James C. *Constipation and Our Civilisation.* Edinburgh: Thorsons, 1943.

Threlfall, E. J., and B. Rowe. "Antimicrobial Drug Resistant in Salmonellae in Britain—A Real Threat to Public Health." In *Antimicrobials and Agriculture: The Proceedings of the 4th International Symposium on Antibiotics in Agriculture: Benefits and Malefits,* edited by Malcolm Woodbine, 513–524. London: Butterworths, 1984.

Threlfall, E. J., L. R. Ward, A. S. Ashley, and B. Rowe. "Plasmid-Encoded Trimethoprim Resistance in Multiresistant Epidemic Salmonella Typhimurium Phage Types 204 and 193 in Britain." *BMJ* 280, no. 6225 (1980): 1210–1211.

"Transferable Antibiotic Resistance." *BMJ* 2, no. 5473 (1965): 1326.

Truman, Harry S. "Truman's Inaugural Address," January 20, 1949. Harry S. Truman Presidential Library and Museum. https://www.trumanlibrary.org/whistlestop /50yr_archive/inagural20jan1949.htm.

Turner, James. *The Chemical Feast: Ralph Nader's Study Group Report on the Food and Drug Administration.* New York: Grossman, 1970.

The TV Vet Book for Pig Farmers, 2nd ed. Ipswich, UK: Farming Press, 1971.

The UK Animal Health Products Market: A Status Report. Richmond, UK: V. and O., 1984.

UK Department of Health. "UK Secures Historic UN Declaration on Antimicrobial Resistance." News story, September 21, 2016. https://www.gov.uk/government /news/uk-secures-historic-un-declaration-on-antimicrobial-resistance.

Van Houweling, C. D. "Drugs in Animal Feed? A Question without an Answer." *FDA Papers* 1, no. 7 (September 1967): 11–15.

———. "The Food, Drug, and Cosmetic Act, Animal Drugs, and the Consumer." *Annals of the New York Academy of Sciences* 182 (1971): 411–415.

———. "Regulation in the United States." In *Factory Farming: A Symposium,* edited by J. R. Bellerby. London: Education Services, 1970, 29–36.

Vigue, F. "Relative Efficacy of Three Antibiotic Combinations in Bovine Mastitis." *JAVMA* 124 (1954): 377–378.

Vogt, Donna U., and Brian A. Jackson. "Antimicrobial Resistance: An Emerging Public Health Issue." In *CRS Report for Congress*. Washington, DC: Congressional Research Service, 2001.

Wakelam, J. A. "Vitamin B12 and Antibiotics in Animal Nutrition." *Manufacturing Chemist* (September 1952): 375.

Walsh, Fergus. "Antibiotics Resistance 'as Big a Risk as Terrorism'—Medical Chief." *BBC News*, March 11, 2013. http://www.bbc.co.uk/news/health-21737844.

Walsh, Jan. *The Meat Machine*. London: Columbus, 1986.

Wasley, Andrew, Ben Stockton, Natalie Jones, and Alexandra Heal. "Critical Antibiotics Still Used on US Farm Animals Despite Superbug Crisis." Bureau of Investigative Journalism, September 19, 2018.

Watanabe, Tsutomo. "Infective Heredity of Multiple Drug Resistance in Bacteria." *Bacteriological Reviews* 27, no. 1 (1963): 87–116.

Watanabe, Tsutomu, and Toshio Fukasawa. "Episome-Mediated Transfer of Drug Resistance in Enterobacteriaceae." *Journal of Bacteriology* 81, no. 5 (1961): 669–678.

Welch, Henry. "Antibiotics in Food Preservation: Public Health and Regulatory Aspects." *Science* 126, no. 3284 (1957): 1159–1161.

Welch, Henry, and Félix Marti-Ibanez. *The Antibiotic Saga*. New York: Medical Encyclopedia, 1960.

"Why Has Swann Failed?" *BMJ* 280, no. 6225 (1980): 1195–1196.

Wickenden, Leonard. *Gardening with Nature—How to Grow Your Own Vegetables, Fruits, and Flowers by Natural Methods*. London: Faber and Faber, 1956.

Williams, R. E. O. "Controlling Antibiotic Resistance without Eschewing Antibiotics." In *The Control of Antibiotic-Resistant Bacteria*, edited by Charles H. Stuart-Harris and David M. Harris, 227–235. London: Academic Press, 1982.

Williams Smith, Herbert. "Antibiotic Resistance in Bacteria and Associated Problems in Farm Animals Before and After the 1969 Swann Report." In *Antibiotics and Antibiosis in Agriculture*, edited by Malcolm Woodbine, 344–357. London: Butterworths, 1977.

———. "Drug-Resistant Bacteria in Domestic Animals." *Proceedings of the Royal Society of Medicine* 51 (1958): 812–813.

———. "The Effect of the Continuous Administration of Diets Containing Tetracyclines and Penicillin on the Number of Drug-Resistant and Drug-Sensitive Clostridium Welchii in the Faeces of Pigs and Chickens. *Journal of Pathology and Bacteriology* 77 (1959): 79–93.

———. "The Effects of the Use of Antibiotics on the Emergence of Antibiotic-Resistant Disease-Producing Organisms in Animals." In *Antibiotics in Agriculture: Proceedings of the University of Nottingham Ninth Easter School in Agricultural Science*, 374–388. London: Butterworths, 1962.

———. "The Incidence of Infective Drug Resistance in Strains of Escherichia coli Isolated from Diseased Human Beings and Domestic Animals." *Journal of Hygiene* 64 (1966): 465–474.

———. "Persistence of Tetracycline Resistance in Pig E. coli." *Nature* 258, no. 5536 (1975): 628–630.

———. "Why Has Swann Failed?" *BMJ* 280, no. 6230 (1980): 1537.

Williams Smith, Herbert, and W. E. Crabb. "The Effect of the Continuous Administration of Diets Containing Low Levels of Tetracyclines on the Incidence of

Drug-Resistant Bacterium coli in the Faeces of Pigs and Chickens: The Sensitivity of the Bact. coli to Other Chemotherapeutic Agents." *Veterinary Record* 69 (1957): 24–30.

———. "The Effect of Diets Containing Tetracyclines and Penicillin on the Staphylococcus Aureus Flora of the Nose and Skin of Pigs and Chickens and Their Human Attendants." *Journal of Pathology and Bacteriology* 79 (1960): 243–249.

———. "The Sensitivity of a Further Series of Strains of Bacterium coli from Cases of White Scours: The Relationship between Sensitivity Tests and Response to Treatment." *Veterinary Record* 68 (1956): 274–277.

———. "The Typing of Escherichia coli by Bacteriophage: Its Application in the Study of the E. coli Population of the Intestinal Tract of Healthy Calves and of Calves Suffering from White Scours." *Journal of General Microbiology* 15 (1956): 556–574.

Williams Smith, Herbert, and Sheila Halls. "Observations on Infective Drug Resistance in Britain." *Veterinary Record* (March 19, 1966): 419–420.

Wilson, Eva D., Katherine H. Fisher, and Mary E. Fuqua. *Principles of Nutrition*. New York: Wiley, 1959.

Woodbine, Malcolm, ed. *Antimicrobials and Agriculture: The Proceedings of the 4th International Symposium on Antibiotics in Agriculture: Benefits and Malefits*. London: Butterworths, 1984.

Wooldbridge, W. R. *Farm Animals in Health and Disease*, 1st, 2nd eds. London: Crosby Lockwood, 1954, 1960.

Worthington, Jim. *Natural Poultry-Keeping*. London: Crosby Lockwood, 1971.

Wrench, G. T. *Wheel of Health*. London: C. W. Daniel, 1938.

Wright, Simon. *Handbook of Organic Food Processing and Production*. Glasgow: Chapman and Hall, 1994.

Wright, Simon, and Diane McCrea. *Handbook of Organic Food Processing and Production*. Oxford: Blackwell Science, 2000.

Wulf, M. W. H., et al., "Prevalence of Methicillin-Resistant Staphylococcus aureus among Veterinarians: An International Study." *Clinical Microbiology and Infection* 14 (2008): 29–34.

Yeoman, G. H. "Review of Veterinary Practice in Relation to the Use of Antibiotics." In *The Control of Antibiotic-Resistant Bacteria*, edited by Charles H. Stuart-Harris and David M. Harris, 57–69. London: Academic Press, 1982.

Young, Richard, Alison Cowe, Cóilín Nunan, John Harvey, and Liz Mason. *The Use and Misuse of Antibiotics in UK Agriculture, Part 2: Antibiotic Resistance and Human Health*. Bristol, UK: Soil Association, 1999.

Zaumeyer, W. J. "Improving Plant Health with Antibiotics." In *Proceedings First International Conference on the Use of Antibiotics in Agriculture*, 171–187. Washington, DC: NAS-NRC, 1956.

Secondary Literature

Aarestrup, Frank. "Sustainable Farming: Get Pigs Off Antibiotics." *Nature* 486 (2012): 465–466.

Anderson, J. L. *Industrializing the Corn Belt: Agriculture, Technology and Environment, 1942–1972*. Dekalb: Northern Illinois University Press, 2009.

Anderson, Stuart. "Drug Regulation and the Welfare State: Government, the Pharmaceutical Industry and the Health Professions in Great Britain." In

Medicine, the Market and the Mass Media, edited by Virginia Berridge and Kelly Loughlin, 192–217. London: Routledge, 2005.

———. "From 'Bespoke' to 'Off-the-Peg': Community Pharmacists and the Retailing of Medicines in Great Britain 1900 to 1970." *Pharmacy in History* 50, no. 2 (2008): 43–69.

———. *Making Medicines: A Brief History of Pharmacy and Pharmaceuticals*. London: Pharmaceutical Press, 2005.

Apple, Rima. *Vitamania: Vitamins in American Culture*. New Brunswick, NJ: Rutgers University Press, 1996.

Atkins, Peter. *Liquid Materialities: A History of Milk, Science and the Law*. Farnham, UK: Ashgate, 2010.

Baker, Kate S., et al. "The Extant World War 1 Dysentery Bacillus NCTC1: A Genomic Analysis." *Lancet* 384, no. 9955 (2014): 1691–1697.

Baker, Stephen, Nicholas Thomson, François-Xavier Weill, and Kathryn E. Holt. "Genomic Insights into the Emergence and Spread of Antimicrobial-Resistant Bacterial Pathogens." *Science* 360 (2018): 733–738.

Beck, Ulrich. "The Reinvention of Politics: Towards a Theory of Reflexive Modernization." In Beck, Anthony Giddens, and Scott Lash, *Reflexive Modernization: Politics, Tradition and Aesthetics in the Modern Social Order*, 1–55. Cambridge: Polity Press, 1994.

———. *Risikogesellschaft. Auf dem Weg in eine andere Moderne*. Frankfurt: Suhrkamp, 1986.

———. *Weltrisikogesellschaft. Auf der Suche nach der verlorenen Sicherheit*. Frankfurt: Suhrkamp, 2007.

Begemann, Stephanie, Elizabeth Perkins, Ine Van Hoyweghen, Robert Christley, and Francine Watkins. "How Political Cultures Produce Different Antibiotic Policies in Agriculture: A Historical Comparative Case Study between the UK and Sweden." *Sociologica Ruralis* (2018).

Belasco, Warren J. *Appetite for Change: How the Counterculture Took on the Food Industry*, 2nd ed. Ithaca, NY: Cornell University Press, 2007.

Bertomeu-Sánchez, José Ramón, and Ximo Guillem-Llobat. "Following Poisons in Society and Culture (1800–2000): A Review of Current Literature." *Actes d'història de la ciència i de la tècnica* 9 (2017): 9–36.

Bess, Michael. *The Light-Green Society: Ecology and Technological Modernity in France*. Chicago: University of Chicago Press, 2003.

Blanchette, Alex. "Herding Species: Biosecurity, Posthuman Labor, and the American Industrial Pig." *Cultural Anthropology* 30, no. 4 (2013): 640–669.

Blaxter, Kenneth, and Noel Robertson. *From Dearth to Plenty: The Modern Revolution in Food Production*. Cambridge, UK: Cambridge University Press, 1995.

Boudia, Soraya, and Nathalie Jas, eds. *Powerless Science? Science and Politics in a Toxic World*. New York: Berghahn, 2014.

Bowers, J. K. "British Agricultural Policy Since the Second World War." *Agricultural History Review* 33 (1985): 71.

Boyd, William. "Making Meat: Science, Technology, and American Poultry Production." *Technology and Culture* 42, no. 4 (2001): 631–664.

———. *Putting Meat on the American Table. Taste, Technology, Transformation*. Baltimore: Johns Hopkins University Press, 2005.

Brassley, Paul. "Cutting Across Nature? The History of Artificial Insemination in Pigs in the United Kingdom." *Studies in History and Philosophy of Biological and Biomedical Sciences* 38 (2007): 442–461.

Broom, Alex, et al., "Improvisation, Therapeutic Brokerage and Antibiotic (Mis) Use in India: A Qualitative Interview Study of Hyderabadi Physicians and Pharmacists." *Critical Public Health* (2018): 1–12.

Brynner, Rock, and Trent Stephens. *Dark Remedy: The Impact of Thalidomide and Its Revival as a Vital Medicine.* New York: Perseus, 2001.

Bud, Robert. *Penicillin: Triumph and Tragedy.* Oxford: Oxford University Press, 2009.

Campbell, William C. "History of the Discovery of Sulfaquinoxaline as a Coccidiostat." *Journal of Parasitology* 94, no. 4 (2008): 934–945.

Carpenter, Daniel. *Reputation and Power: Organizational Image and Pharmaceutical Regulation at the FDA.* Princeton, NJ: Princeton University Press, 2010.

Case, Andrew N. *The Organic Profit: Rodale and the Making of Marketplace Environmentalism.* Seattle: University of Washington Press, 2018.

Chadarevian, Soraya de. "Mice and the Reactor: The 'Genetics Experiment' in 1950s Britain." *Journal of the History of Biology* 39 (2006): 707–735.

Chandler, Clare, Eleanor Hutchinson, and Coll Hutchison. *Addressing Antimicrobial Resistance Through Social Theory: An Anthropologically Oriented Report.* London: LSHTM, 2016.

Chen, L., R. Todd, J. Kiehlbauch, M. Walters, and A. Kallen. "Notes from the Field: Pan-resistant New Delhi Metallo-Beta-Lactamase-Producing Klebsiella pneumoniae—Washoe County, Nevada, 2016." *Morbidity and Mortality Weekly Report.* CDC 66, no. 1 (2017): 33.

Cohen, Lizabeth. *A Consumers' Republic: The Politics of Mass Consumption in Postwar America.* New York: Vintage Books, 2004.

Condrau, Florin, and Robert Kirk. "Negotiating Hospital Infections: The Debate between Ecological Balance and Eradication Strategies in British Hospitals, 1947–1969." *Dynamis* 31, no. 2 (2011): 385–404.

Conford, Philip. *The Origins of the Organic Movement.* Edinburgh: Floris, 2001.

Conford, Philip, and P. Holden. "The Soil Association." In *Organic Farming: An International History*, edited by William Lockeretz, 187–200. Wallingford, UK: CABI, 2001.

Conkin, Paul K. *A Revolution Down on the Farm: The Transformation of American Agriculture Since 1929.* Lexington: University Press of Kentucky, 2009.

Corley, T. A. B., and Andrew Godley. "The Veterinary Medicines Industry in Britain, 1900–2000." *Economic History Review* 64 (2011): 832–854.

Cox, Graham, Philip Lowe, and Michael Winter. "From State Direction to Self-Regulation: The Historical Development of Corporatism in British Agriculture." *Policy and Politics* 14, no. 4 (1986): 475–476.

Creager, Angela N. H. "Adaptation or Selection? Old Issues and New Stakes in the Postwar Debates over Bacterial Drug Resistance." *Studies in History and Philosophy of Biology and Biomedical Sciences* 38 (2007): 159–190.

Cullather, Nick. *The Hungry World: America's Cold War Battle Against Poverty in Asia.* Cambridge, MA: Harvard University Press, 2010.

Dibner, J. J., and J. D. Richard. "Antibiotic Growth Promoters in Agriculture: History and Mode of Action." *Poultry Science* 84 (2005): 634.

Drlica, Karl, and David S. Perlin. *Antibiotic Resistance: Understanding and Responding to an Emerging Crisis.* Upper Saddle River, NJ: Pearson Education, 2011.

Finlay, Mark R. "Hogs, Antibiotics, and the Industrial Environments of Postwar Agriculture." In *Industrializing Organisms: Introducing Evolutionary History*, edited by Susan R. Schrepfer and Philip Scranton, 237–260. New York: Routledge, 2004.

Finlay, Mark R., and Alan I. Marcus. "'Consumerist Terrorists': Battles over Agricultural Antibiotics in the United States and Western Europe." *Agricultural History* 90, no. 2 (2016): 146–172.

Fitzgerald, Deborah. *Every Farm a Factory: The Industrial Ideal in American Agriculture*. New Haven, CT: Yale University Press, 2003.

Fortané, Nicolas. "Naissance et déclin de l'ecopathologie: L'essor contrairié d'une médecine vétérinaire alternative, années 1970–1990." *Regards Sociologiques* (2015).

Fortané, Nicolas, and Frédéric Keck. "How Biosecurity Reframes Animal Surveillance." *Revue d'anthropologie des connaissances* 9, no. 2 (2015): a–k.

Freedman, Daryl M. "Reasonable Certainty of No Harm: Reviving the Safety Standard for Food Additives, Color Additives, and Animal Drugs." *Ecology Law Quarterly* 7, no. 2 (1978): 245–284.

Friedberger, Mark. "Cattlemen, Consumers, and Beef." *Environmental History Review* 18, no. 3 (1994): 37–57.

Gallagher, Jason C., and Conan MacDougall. *Antibiotics Simplified*. Boston: Jones and Bartlett, 2009.

Gaudillière, Jean-Paul. "Food, Drug and Consumer Regulation—The Meat, DES and Cancer Debates in the US." In *Meat, Medicine and Human Health in the Twentieth Century*, edited by David Cantor, Christian Bonah, and Mathias Dörries, 179–203. London: Pickering and Chatto, 2010.

Gillings, Michael R. "Lateral Gene Transfer, Bacterial Genome Evolution, and the Anthropocene." *Annals of the New York Academy of Sciences*, 1389, no. 1 (2016): 20–36.

Godley, Andrew. "The Emergence of Agribusiness in Europe and the Development of the Western European Broiler Chicken Industry, 1945–1973." *Agricultural History Review* 62, no. 2 (2014): 315–336.

Godley, Andrew, and Bridget Williams. "Democratizing Luxury and the Contentious 'Invention of the Technological Chicken' in Britain." *Business History Review* 83 (2009): 267–290.

Godwin, Matthew, Jane Gregory, and Brian Balmer. "The Anatomy of the Brain Drain Debate, 1950–1970s: Witness Seminar." *Contemporary British History* 23, no. 1 (2009): 35–60.

Gradmann, Christoph. "Magic Bullets and Moving Targets: Antibiotic Resistance and Experimental Chemotherapy, 1900–1940." *Dynamis* 31, no. 2 (2001): 305–321.

———. "Re-inventing Infectious Disease: Antibiotic Resistance and Drug Development at the Bayer Company 1945–1980." *Medical History* 60, no. 2 (2016): 155–180.

———. "Sensitive Matters: The World Health Organization and Antibiotic Resistance Testing, 1945–1975." *Social History of Medicine* 26, no. 3 (2013): 555–574.

Graham, Jay P., John J. Boland, and Ellen Silbergeld. "Growth Promoting Antibiotics in Food Animal Production: An Economic Analysis." *Public Health Reports* 122, no. 1 (2007): 79–87.

Hardin, Garrett. "The Tragedy of the Commons." *Science* 162, no. 3859 (1968): 1243–1248.

Hardy, Anne. "John Bull's Beef: Meat Hygiene and Veterinary Public Health in

England in the Twentieth Century." *Review of Agricultural and Environmental Studies* 91, no. 4 (2010): 276–378.

———. *Salmonella Infections, Networks of Knowledge, and Public Health in Britain 1880–1975*. Oxford: Oxford University Press, 2015.

Harkins, Catriona P., et al. "Methicillin-Resistant Staphylococcus aureus Emerged Long Before the Introduction of Methicillin into Clinical Practice." *Genome Biology* 18, no. 130 (2017).

Harrison, Mark. *Contagion: How Commerce Has Spread Disease*. New Haven, CT: Yale University Press, 2012.

Heinzerling, Lisa. "Undue Process at the FDA." *Georgetown Public Law and Legal Theory Research Paper No. 13-016* (2013).

Hill, Berkeley. *Farm Incomes, Wealth and Agricultural Policy: Filling the CAP's Core Information Gap*, 4th ed. Wallingford, UK: CABI, 2012.

Hillier, Kathryn. "Babies and Bacteria: Phage Typing Bacteriologists, and the Birth of Infection Control." *Bulletin of the History of Medicine* 80, no. 4 (2006): 733–761.

Hilton, Matthew. "Consumer Movements." In *The Oxford Handbook of the History of Consumption*, edited by Frank Trentmann, 505–520. Oxford: Oxford University Press, 2012.

Holderness, B. A. *British Agriculture Since 1945*. Manchester, UK: Manchester University Press, 1985.

Horowitz, Roger. "Making the Chicken of Tomorrow. Reworking Poultry as Commodities and as Creatures, 1945–1990." In *Industrializing Organisms: Introducing Evolutionary History*, edited by Susan R. Schrepfer and Philip Scranton, 215–235. New York: Routledge, 2004.

———. *Putting Meat on the American Table: Taste, Technology, Transformation*. Baltimore: Johns Hopkins University Press 2006.

Hüntelmann, Axel C. "Seriality and Standardization in the Production Of '606.'" *History of Science* xlvlii (2010): 435–460.

Hurt, Douglas R. *Problems of Plenty: The American Farmer in the Twentieth Century*. Chicago: Ivan R. Dee, 2002.

Jas, Nathalie. "Adapting to 'Reality': The Emergence of an International Expertise on Food Additives and Contaminants in the 1950s and Early 1960s." In *Toxicants, Health and Regulation Since 1945*, edited by Nathalie Jas and Soraya Boudia, 47–69. London: Pickering and Chatto, 2013.

———. "Public Health and Pesticide Regulation in France Before and After Silent Spring." *History and Technology* 23, no. 4 (2007): 369–388.

Jasanoff, Sheila. *Designs on Nature: Science and Democracy in Europe and the United States*, 2nd ed. Princeton, NJ: Princeton University Press, 2007.

———. *The Fifth Branch. Science Advisers as Policymakers*. Cambridge, MA: Harvard University Press, 1994.

———. *Science at the Bar: Law, Science, and Technology in America*. Cambridge, MA: Harvard University Press, 1997.

Jones, Susan D. *Valuing Animals: Veterinarians and Their Patients in Modern America*. Baltimore: Johns Hopkins University Press, 2003.

Kahn, Laura H. *One Health and the Politics of Antimicrobial Resistance*. Baltimore: Johns Hopkins University Press, 2016.

Kean, Hilda. *Animal Rights: Political and Social Change in Britain Since 1800*. London: Reaktion, 1998.

Kirchhelle, Claas. "Between Bacteriology and Toxicology: Agricultural Antibiotics and US Risk Regulation." In *Risk on the Table*, edited by Angela Creager and Jean-Paul Gaudillière. Chicago: University of Chicago Press, forthcoming.

———. "Pharming Animals: A Global History of Antibiotics in Food Production, 1935–2017." *Palgrave Communications* 4, no. 96 (2018).

———. *Ruth Harrison and the History of British Animal Welfare Science and Activism*. London: Palgrave Macmillan, forthcoming.

———. "Swann Song: Antibiotic Regulation in British Livestock Production, 1953–2006." *Bulletin of the History of Medicine* 92, no. 2 (2018): 317–350.

———. "Toxic Confusion: The Dilemma of Antibiotic Regulation in West German Food Production." *Endeavor* 40, no. 2 (2016): 114–127.

———. "Toxic Tales—Recent Histories of Pollution, Poisoning, and Pesticides (ca. 1800–2010)." *NTM Zeitschrift für Geschichte der Wissenschaften, Technik und Medizin* 26, no. 2 (2018): 213–29.

Kirk, Andrew G. *Counterculture Green. The Whole Earth Catalog and American Environmentalism*. Lawrence: University Press of Kansas, 2007.

Koolmes, Peter A. "Veterinary Inspection and Food Hygiene in the Twentieth Century." In *Food, Science, Policy and Regulation in the Twentieth Century: International and Comparative Perspectives*, edited by David F. Smith and Jim Philips, 53–68. Oxford: Routledge, 2000.

Kroll, Garry. "The 'Silent Springs' of Rachel Carson: Mass Media and the Origins of Modern Environmentalism." *Public Understanding of Science* 10 (2001): 403–420.

Krücken, Georg. *Risikotransformation. Die politische Regulierung technisch-ökologischer Gefahren in der Risikogesellschaft*. Opladen/Wiesbaden: Westdeutscher Verlag, 1997.

Kuehn, Bridget M. "Frances Kelsey Honored for FDA Legacy." *JAMA* 304, no. 19 (2010): 2109–2112.

Laetsch, Waston M. "Thomas H. Jukes, Integrative Biology: Berkeley." *In Memoriam* (2000): 109–111.

Landecker, Hannah. "Antibiotic Resistance and the Biology of History." *Body and Society* (2015): 1–34.

———. "Antimicrobials before Antibiotics: War, Peace, and Disinfectants." *Palgrave Communications* 5, no. 45 (2019).

———. "The Food of Our Food: Medicated Feed and the Industrialisation of Metabolism." Presented at Oxford Microbiome Seminar, February 15, 2017.

———. "It Is What It Eats: Chemically Defined Media and the History of Surrounds." *Studies in History and Philosophy of Biological and Biomedical Sciences* 57 (2016): 148–160.

Langston, Nancy. "Precaution and the History of Endocrine Disruptors." In *Powerless Science? Science and Politics in a Toxic World*, edited by Soraya Boudia and Nathalie Jas, 29–45. New York: Berghahn, 2014.

———. *Toxic Bodies: Hormone Disruptors and the Legacy of DES*. New Haven, CT: Yale University Press, 2010.

Lawrence, Tony, and Vernon R. Fowler. *Growth of Farm Animals*. Wallingford, UK: CABI, 1997.

Lawrence, Tony, Vernon R. Fowler, and Jan E. Novakofski. *Growth of Farm Animals*, 2nd ed. Wallingford, UK: CABI, 2012.

Lent, Adam. *British Social Movements Since 1945: Sex, Colour, Peace and Power*. Basingstoke, UK: Palgrave, 2001.

Le Roux, Thomas. "Governing the Toxics and the Pollutants: France, Great Britain, 1750–1850." *Endeavour* 40, no. 2 (2016): 70–81.

Lesch, John E. *The First Miracle Drugs: How the Sulfa Drugs Transformed Medicine.* Oxford: Oxford University Press, 2007.

Lezaun, Javier. "Genetically Modified Foods and Consumer Mobilization in the UK." *Technikfolgenabschätzung—Theorie und Praxis* 13, no. 3 (2004): 49–56.

Lezaun, Javier, and Martijn Groenleer. "Food Control Emergencies and the Territorialization of the European Union." *European Integration* 28, no. 5 (2006): 437–455.

Lezaun, Javier, and Natalie Porter. "Containment and Competition: Transgenic Animals in the One Health Agenda." *Social Science and Medicine* 129 (2015): 96–105.

Lobao, Linda M., and Pamela Thomas. "Political Beliefs in an Era of Economic Decline: Farmers' Attitudes Toward State Economic Intervention, Trade, and Food Security." *Rural Sociology* 57, no. 4 (1992): 453–475.

Luhmann, Niklas. *Die Gesellschaft der Gesellschaft.* Frankfurt: Suhrkamp, 1997.

Lutts, Ralph H. "Chemical Fallout: Rachel Carson's Silent Spring, Radioactive Fallout, and the Environmental Movement." *Environmental Review* 9, no. 3 (1985): 211–225.

Macfie, H. J. H., and Herbert L. Meiselman. *Food Choice, Acceptance and Consumption.* London: Blackie Academic and Professional, 1996.

Maeder, Thomas. *Adverse Reactions.* New York: William Morrow, 1994.

Marcus, Alan. *Cancer from Beef: DES, Federal Food Regulation, and Consumer Confidence.* Baltimore: Johns Hopkins University Press, 1994.

Marks, H. F., and D. K. Britton. *A Hundred Years of British Food and Farming: A Statistical Survey.* London: Taylor and Francis, 1989.

Mart, Michelle. *Pesticides, a Love Story: America's Enduring Embrace of Dangerous Chemicals.* Lawrence: University Press of Kansas, 2016.

Martin, John. "The Commercialisation of British Turkey Production." *Rural History* 20 (2009): 209–228.

———. *The Development of Modern Agriculture: British Farming Since 1931.* London: Macmillan, 2000.

———. "The Transformation of Lowland Game Shooting in England and Wales Since the Second World War." *Rural History* 22 (2011): 207–226.

Mayda, Chris. "Pig Pens, Hog Houses, and Manure Pits: A Century of Change in Hog Production." *Material Culture* 36, no. 1. (2004): 18–42.

McKenna, Maryn. *Big Chicken: The Incredible Story of How Antibiotics Created Modern Agriculture and Changed the Way the World Eats.* Washington, DC: National Geographic, 2017.

McMillen, Christian W. *Discovering Tuberculosis: A Global History 1900 to the Present.* New Haven, CT: Yale University Press, 2015.

More-Collyer, R. "Towards 'Mother Earth': Jorian Jenks, Organicism, the Right and the British Union of Fascists." *Journal of Contemporary History* 39, no. 3 (2004): 353–371.

Morris, Carol, Richard Helliwell, and Sujatha Raman. "Framing the Agricultural Use of Antibiotics and Antimicrobial Resistance in UK National Newspapers and the Farming Press." *Journal of Rural Studies* 45 (2016): 43–55.

Neushul, Peter. "Science, Government, and the Mass Production of Penicillin." *Journal of the History of Medicine and Allied Sciences* 48 (1993): 371–395.

Newman, Richard. *Love Canal: A Toxic History from Colonial Times to the Present.* Oxford: Oxford University Press, 2016.

Ogle, Maureen. *In Meat We Trust: An Unexpected History of Carnivore America.* New York: Houghton Mifflin Harcourt, 2013.

Olmstead, Alan L., and Paul W. Rhode. *Arresting Contagion. Science, Policy, and Conflicts over Animal Disease Control.* Cambridge, MA: Harvard University Press, 2015.

Olson, Mancur. *The Logic of Collective Action: Public Goods and the Theory of Groups.* Cambridge, MA: Harvard University Press, 1971.

Oreskes, Naomi, and Erik M. Conway. *Merchants of Doubt: How a Handful of Scientists Obscured the Truth on Issues from Tobacco Smoke to Global Warming.* New York: Bloomsbury, 2010.

O'Sullivan, Robin. *American Organic: A Cultural History of Farming, Gardening, Shopping, and Eating.* Lawrence: University Press of Kansas, 2015.

Petersen, Uwe. "Entwicklungen im deutschen Futtermittelrecht." In *Meilensteine für die Futtermittelsicherheit: Vortragsveranstaltung im Forum der FAL am 16./17,* edited by Uwe Petersen, Sabine Kruse, Sven Dänicke, and Gerhard Flachowsky, 2–8. Braunschweig: Bundesforschungsanstalt für Landwirtschaft, 2007.

Podolsky, Scott H. *The Antibiotic Era: Reform, Resistance and the Pursuit of a Rational Therapeutics.* Baltimore: Johns Hopkins University Press, 2015.

———. "Antibiotics and the Social History of the Controlled Clinical Trial, 1950–1970." *Journal of the History of Medicine and Allied Sciences* 65, no. 3 (2010): 327–367.

Podolsky, Scott H., et al. "History Teaches Us that Confronting Antibiotic Resistance Requires Stronger Global Collective Action." *Journal of Law, Medicine and Ethics* 43, no. 2, Special Supplement (2015): 27–32.

Podolsky, Scott H., and Anne Kveim Lie. "Futures and Their Uses: Antibiotics and Therapeutic Revolutions." In *Therapeutic Revolutions: Pharmaceuticals and Social Change in the Twentieth Century,* edited by Jeremy A. Greene, Flurin Condrau, and Elizabeth Siegel Watkins, 18–42. Chicago: University of Chicago Press, 2016.

Pollan, Michael. *The Omnivore's Dilemma: A Natural History of Four Meals.* London: Penguin, 2006.

Pope, L. S. "Animal Science in the Twentieth Century." *Agricultural History* 54, no. 1 (1980): 64–70.

Porter, Roy. "Doing Medical History from Below." *Theory and Society* 14, no. 2 (1985): 175–198.

Poser, S., I. Zerr, and K. Felgenhauer. "Die neue Variante der Creutzfeldt-Jakob-Krankheit." *Deutsche Medizinische Wochenschrift* 127 (2002): 331–334.

Price, Lance B., et al. "Staphylococcus aureus CC398: Host Adaptation and Emergence of Methicillin Resistance in Livestock." *mBio* 3, no. 1 (2012): 1–6.

Proctor, Robert N. "Agnotology: A Missing Term to Describe the Cultural Production of Ignorance and Its Study." In *Agnotology: The Making and Unmaking of Ignorance,* edited by Robert N. Proctor and Londa Schiebinger, 1–33. Stanford, CA: Stanford University Press, 2008.

Quan, Jingjing, Xi Li, and Yan Chen. "Prevalence of mcr-1 in Escherichia coli and Klebsiella Pneumonia Recovered from Bloodstream Infections in China: A Multicentre Longitudinal Study." *Lancet Infectious Diseases* 17, no. 4 (2017): 400–410.

Quirke, Viviane. "Thalidomide and Drug Safety Regulation." In *Ways of Regulating*

Drugs in the 19th and 20th Centuries, edited by Jean-Paul Gaudillière and Volker Hess, 151–180. Basingstoke, UK: Palgrave Macmillan, 2013.

Rasmussen, Nicolas. "Goofball Panic: Barbiturates, 'Dangerous' and Addictive Drugs, and the Regulation of Medicine in Postwar America." In *Writing, Filing, Using, and Abusing the Prescription in Modern America*, edited by Jeremy A. Greene and Elizabeth Siegel Watkins, 23–45. Baltimore: Johns Hopkins University Press, 2012.

Reid, W. Malcolm. "History of Avian Medicine in the United States." *Avian Diseases* 32, no. 3 (1990): 509–525.

Reinhardt, Carsten. "Boundary Values." In *Precarious Matters: The History of Dangerous and Endangered Substances in the 19th and 20th centuries*, edited by Viola Balz, Alexander von Schwerin, Heiko Stoff, and Bettina Wahrig, 39–50. Berlin: MPI for the History of Science, 2008.

Richert, Lucas. *Conservatism, Consumer Choice, and the Food and Drug Administration during the Reagan Era: A Prescription for Scandal*. Lanham, MD: Lexington, 2014.

Roscher, Mieke. *Ein Königreich für Tiere: Die Geschichte der britischen Tierrechtsbewegung*. Marburg: Tectum Verlag, 2009.

Russell, Edmund. *War and Nature: Fighting Humans and Insects with Chemicals from World War I to Silent Spring*. Cambridge, UK: Cambridge University Press, 2001.

Ruxin, John. "The United Nations Protein Advisory Group." In *Food, Science, Policy and Regulation in the Twentieth Century: International and Comparative Perspectives*, edited David F. Smith and Jim Philips, 151–168. Oxford: Routledge, 2000.

Ryder, Richard D. "Harrison, Ruth, 1920–2000." *Oxford Dictionary of National Biography* (2004) [online edition May 2005].

Sandegren, Linus. "Review: Selection of Antibiotic Resistance at Very Low Antibiotic Concentrations." *Upsala Journal of Medical Sciences* 119, no. 2 (2014): 103–107.

Sayer, Karen. "Animal Machines: The Public Response to Intensification in Great Britain, c. 1960- c. 1973." *Agricultural History* 87, no. 4 (2013): 473–501.

Schwerin, Alexander von. "Prekäre Stoffe. Radiumökonomie, Risikoepisteme und die Etablierung der Radioindikatortechnik in der Zeit des Nationalsozialismus." *NTM* 17 (2009): 5–33.

Seidl, Alois. *Deutsche Agrargeschichte*. Frankfurt: DLG Verlag, 2006.

Seidl, Irmi, and Clem A. Tisdell. "Carrying Capacity Reconsidered: From Malthus' Population Theory to Cultural Carrying Capacity." *Ecological Economics* 30, no. 3 (1999): 395–408.

Sellers, Christopher C. *Hazards of the Job: From Industrial Disease to Environmental Health Science*. Chapel Hill: North Carolina Press, 1997.

Setbon, M., and A. Castot. "Interdiction de l'avoparcine face au risqué d'antibiorésistance." *Comité national de sécurité sanitaire: Risques et sécurité sanitaire* (2001): 55–69.

Sheail, John. "Pesticides and the British Environment: An Agricultural Perspective." *Environment and History* 19 (2013): 87–108.

Silbergeld, Ellen K. *Chickenizing Farms and Food: How Industrial Meat Production Endangers Workers, Animals, and Consumers*. Baltimore: Johns Hopkins University Press, 2016.

Simon, Christian. *DDT. Kulturgeschichte einer Chemischen Verbindung*. Basel: Christian Merian Verlag, 1999.

Slovic, Paul, Melissa L. Finucane, Ellen Peters, and Donald G. MacGregor. "Risk as Analysis and Risk as Feelings: Some Thoughts about Affect, Reason, Risk, and Rationality." *Risk Analysis* 24, no. 2 (2004): 311–322.

Smith, David F., H. Lesley Diack, and T. Hugh Pennington. *Food Poisoning, Policy and Politics: Corned Beef and Typhoid in Britain in the 1960s.* Woodbridge, UK: Boydell, 2005.

Smith-Howard, Kendra. "Antibiotics and Agricultural Change: Purifying Milk and Protecting Health in the Postwar Era." *Agricultural History* 84, no. 3 (2010): 327–351.

———. "Healing Animals in an Antibiotic Age: Veterinary Drugs and the Professionalism Crisis, 1945–1970." *Technology and Culture* 58, no. 4 (2017): 722–748.

———. *Perfecting Nature's Food: A Cultural and Environmental History of Milk in the United States, 1900–1970.* Dissertation, University of Wisconsin–Madison, 2007.

———. *Pure and Modern Milk: An Environmental History Since 1900.* Oxford: Oxford University Press, 2013.

Stoff, Heiko. *Gift in der Nahrung. Zur Genese der Verbraucherpolitike Mitte des 20. Jahrhunderts.* Stuttgart: Franz Steiner Verlag, 2015.

Stoff, Heiko, and Alexander von Schwerin. "Einleitung—Lebensmittelzusatstoffe. Eine Geschichte gefährlicher Dinge und ihrer Regulierung 1950–1970." *Technikgeschichte* 81, no. 3 (2014): 215–228.

Tessari, A., and A. Godley. "Made in Italy, Made in Britain: Quality, Brands and Innovation in the European Poultry Market, 1950–80." *Business History* 56, no. 7 (2014): 1057–1083.

Thompson, Francis Michael Longstreth, ed. *The Cambridge Social History of Britain 1750–1950. Volume 1: Regions and Communities.* Cambridge, UK: Cambridge University Press, 1990.

Thoms, Ulrike. "Aus Wertlosem Wertvolles schaffen: Die Mobilisierung der Fütterungswissenschaft zur Steigerung der Nahrungsmittelproduktion im Dritten Reich," unpublished essay.

———. "Handlanger der Industrie oder berufener Schützer des Tieres? Der Tierarzt und seine Rolle in der Geflügelproduktion." In *Was der Mensch essen darf,* edited by Günther Hirschfelder, Angelika Ploeger, Jana Rückert-John, and Gesa Schönberger, 173–192. Heidelberg: Springer, 2015.

Tønnessen, Johan Nicolay, and Arne Odd Johnsen. *The History of Modern Whaling.* Translated by R. I. Christophersen. Berkeley: University of California Press, 1982.

Tran-Dien, Alicia, S. Le Hello, C. Bouchier, and F. X. Weill. "Early Transmissible Ampicillin Resistance in Zoonotic Salmonella Serotype Typhimurium in the Late 1950s: A Retrospective, Whole-Genome Sequencing Study." *Lancet Infectious Diseases* 18, no. 2 (2018): 207–214.

Troetel, Barbara Resnick. *Three-Part Disharmony: The Transformation of the Food and Drug Administration in the 1970s.* Dissertation, City University of New York, 1996.

Ueköter, Frank. *Die Wahrheit ist auf dem Feld. Eine Wissensgeschichte der deutschen Landwirtschaft.* Göttingen: Vandenhoeck and Ruprecht, 2010.

———. *The Greenest Nation: A New History of German Environmentalism.* Cambridge, MA: MIT Press, 2014.

Van Boeckel, Thomas P., et al. "Global Trends in Antimicrobial Use in Food Animals." *PNAS* 112, no. 18 (2015): 5649–5650.

Van Boeckel, Thomas P., et al. "Reducing Antimicrobial Use in Food Animals." *Science* 357, no. 6358 (2017): 1350–1352.

Veldman, Meredith. *Fantasy, the Bomb and the Greening of Britain: Romantic Protest, 1945–1980.* Cambridge, UK: Cambridge University Press, 1994.

Vogel, David. *The Politics of Precaution: Regulating Health, Safety and Environmental*

Risks in Europe and the United States. Princeton, NJ: Princeton University Press, 2012.

Vogel, Sarah A. *Is It Safe? BPA and the Struggle to Define the Safety of Chemicals*. Berkeley: University of California Press, 2013.

Waddington, Keir. "Mad and Coughing Cows: Bovine Tuberculosis, BSE and Health in Twentieth-Century Britain." In *Meat, Medicine and Human Health in the Twentieth Century*, edited by David Cantor, Christian Bonah, and Matthias Dörries, 159–177. London: Pickering and Chatto, 2010.

Williams, Michele L., and Jeffrey T. Lejeuen. "Phages and Bacterial Epidemiology." In *Bacteriophages in Health and Disease*, edited by Paul Hyman and Stephen T. Abedon, 76–85. Wallingford, UK: CABI, 2012.

Williams, R. E. O. *Microbiology for the Public Health*. London: PHLS, 1985.

Winter, Michael. *Rural Politics: Policies for Agriculture, Forestry and the Environment*. London: Routledge, 1996.

Woods, Abigail. "From Cruelty to Welfare: The Emergence of Farm Animal Welfare in Britain, 1964–71." *Endeavour* 36, no. 1 (2012): 14–22.

———.Decentring Antibiotics: UK Responses to the Diseases of Intensive Pig Production (ca. 1925–65). *Palgrave Communications* 5, no. 41 (2019).

———. "A Historical Synopsis of Farm Animal Disease and Public Policy in Twentieth Century Britain." *Philosophical Transactions of the Royal Society Series B: Biological Sciences* 366, no. 1573 (2011): 1943–1954.

"Is Prevention Better than Cure? The Rise and Fall of Veterinary Preventive Medicine, c. 1950–1980." *Social History of Medicine* 26, no. 1 (2012): 113–131.

———. "Rethinking the History of Modern Agriculture: British Pig Production, c. 1910–65." *Twentieth Century British History* 23, no. 2 (2012): 165–191.

———. "Science, Disease and Dairy Production in Britain." *Agricultural History Review* 62, no. II (2007): 294–314.

Yi-Yun, Liu, et al. "Emergence of Plasmid-Mediated Colistin Resistance Mechanism MCR-1 in Animals and Human Beings in China: A Microbiological and Molecular Biological Study." *Lancet Infectious Diseases* 16, no. 2 (2016): 161–168.

Zweiniger-Bargielowska, Ina. *Austerity in Britain: Rationing, Control and Consumption 1939–1955*. Oxford: Oxford University Press, 2002.

Index

Animal Medicinal Drug Use Clarification
Act (AMDUCA), 178, 207
animal protein factor (APF): ad campaigns
for, 38; vitamin B$_{12}$ and, 20, 36–38, 99–100.
See also antibiotic growth promoter
animal welfare, 4, 240–241, 250, 256,
284–285; *Animal Machines* and, 84–86,
112; factory farms and, 77–91; public
and, 75, 83–84
antibiotic-free, 179, 215, 217, 226, 228–230,
232, 248–249, 258, 285, 289–290
antibiotic growth promoter (AGP), 3–4;
American Cyanamid, 170–171, 194,
200–201, 260; antimicrobial resistance
(AMR) and health fears and, 31, 50–51,
63, 68–70, 72, 81–83, 88, 107–111,
113–114, 128, 133, 151, 207–208, 239;
avoparcin and, 227–229, 250, 273, 280,
283; bans and 90, 117–118, 120, 215–216,
233, 251–253, 259–263, 274–275 (UK),
141, 149–155, 158–169, 165, 168–171, 178,
181–183, 187–205, 209–213, 267 (US);
BSE and, 249–252; licensing and
labelling of, 56–57, 65–66, 75, 78–79,
121–124, 130; Netherthorpe Committee
and 129–131, 134; nontherapeutic
antibiotics and 115, 175, 180, 234, 237,
249, 264; public announcement and
efficacy of, 20, 22, 37–41, 52, 92, 99–101,
246; sales boom of, 56–57, 102, 105; Soil
Association and, 250–251, 255–256;
human growth promotion, 22–23;
Swann committee and, 118–120, 138–139,
farmer surveys and 174–175, 181–182.
antibiotic infrastructure, 4, 8, 15, 38–45, 47,
53, 75, 84, 106, 116, 119, 139–140, 163,
171–172, 176–177, 182, 212–213, 221, 230,
232, 240, 249, 253, 257, 275, 281–282,
286–287
Antibiotic Paradox, The (Levy, S.), 157
antibiotic preservatives, 20–21, 26, 29, 47,
54, 59–60, 68, 78, 82, 109, 125–126, 128;
ban on, 71, 125, 133, 154; in fish and
poultry, 70–72, 78
Antibiotics in Animal Feeds Subcommittee
(AAFC), US, 197–199
Antibiotics in Nutrition (Jukes, T.), 64
Antibiotics Panel of Advisory Committee on
Poisonous Substances, UK, 124–125, 130

antibiotic sprays (plant protection), 59–62,
294n16
Antibiotics Task Force (1970–1972), FDA,
US, 149–150; 168–169, 183, 190–196,
200, 213.
antibiotic stewardship, 223, 234, 257,
280–281, 283–285, 290; AMR
debates and, 12, 131, 151, 179, 262;
reform and, 160, 231, 233–234,
277–278; short-termism and, 7,
282–283; voluntarism and, 161–162,
179, 185, 208–209.
antibiotic use: agricultural origins of, 1–4,
33–38; black market, 219–221, 224,
228–229, 241–243, 265, 269; commercial
pressure in, 45–53; corporatism and,
129–136; in food production, 2–3, 59;
illegal, 71, 113 129, 133, 156, 206, 219, 234,
243; MAFF and, 260–273; in plant
protection, 59–62; public concerns of,
80–91; regulation of, 121–126; residue in
meat, 63, 72–74; risk normalization and,
58–64; Scandinavian reform of, 180,
226–227, 255, 270, 274–276, 280; Swann
report and, 136–140; therapeutic and
veterinary, 4, 35–36, 40–44, 54, 56, 58,
83, 90, 95–97, 101–102, 104–105, 107–109,
113, 115, 117–119, 124, 130–134, 139, 161,
165–166, 170, 177, 181–182, 187, 190–192,
203, 210–213, 215–217, 221, 228–230,
233–234, 237–239, 241, 245–246, 249, 253,
255, 260, 264, 266, 268–269, 275, 277,
280, 282, 285–289; UK expansion and
sales of, 77–80, 101–107, 233–234,
242–243, 249, 250, 252–253, 257–258; UK
licensing of, 121–123; UK reform of, 232,
240–246, 253–258; US sales and, 15,
52–53, 205; US reform of 57–58, 64–66,
68, 70–74, 186–189, 206–211. See *also*
antibiotic growth promoter
antimicrobial resistance (AMR): apocalyptic warnings of, 30–31, 157–158, 225,
229–230, 285, 290; bacteriophage-typing
and, 108, 127–129, 131–132; FDA bans
and, 151–155; FDA complacency and,
149–151; hospitals and 6, 31, 82–83, 89,
human medicine and, 6, 27, 30, 51, 53, 64,
82, 87–89, 109, 128–129, 139, 144, 158,
191–192, 196–198; 218, 241–242, 246, 251,

About the Author

CLAAS KIRCHHELLE (DPhil, Oxon) is a historian at the University of Oxford. His award-winning research explores the history of antibiotics and the development of modern risk perceptions, microbial surveillance, and international drug regulation.

Available titles in the Critical Issues in Health and Medicine series:

Karen Seccombe and Kim A. Hoffman, *Just Don't Get Sick: Access to Health Care in the Aftermath of Welfare Reform*

Leo B. Slater, *War and Disease: Biomedical Research on Malaria in the Twentieth Century*

Dena T. Smith, *Medicine over Mind: Mental Health Practice in the Biomedical Era*

Matthew Smith, *An Alternative History of Hyperactivity: Food Additives and the Feingold Diet*

Paige Hall Smith, Bernice L. Hausman, and Miriam Labbok, *Beyond Health, Beyond Choice: Breastfeeding Constraints and Realities*

Susan L. Smith, *Toxic Exposures: Mustard Gas and the Health Consequences of World War II in the United States*

Rosemary A. Stevens, Charles E. Rosenberg, and Lawton R. Burns, eds., *History and Health Policy in the United States: Putting the Past Back In*

Barbra Mann Wall, *American Catholic Hospitals: A Century of Changing Markets and Missions*

Frances Ward, *The Door of Last Resort: Memoirs of a Nurse Practitioner*

Shannon Withycombe, *Lost: Miscarriage in Nineteenth-Century America*

Printed and bound by CPI Group (UK) Ltd, Croydon, CR0 4YY

27/10/2024

14580230-0003